THE ASTRONOMICAL REVOLUTION

The Astronomical Revolution

COPERNICUS — KEPLER — BORELLI

ALEXANDRE KOYRÉ

Translated by

Dr R. E. W. Maddison, F.S.A.

Librarian of the Royal Astronomical Society

DOVER PUBLICATIONS, INC.

New York

Copyright © 1973 by Hermann, Paris.
All rights reserved under Pan American and International Copyright Conventions.

This Dover edition, first published in 1992, is an unabridged and unaltered republication of the English translation originally published in 1973 by Methuen & Co. Ltd., London, and Cornell University Press, Ithaca, N.Y. The work was originally published by Hermann, Paris, in 1961 under the title *La révolution astronomique* as part of the series, *Histoire de la pensée*. This edition is published by special arrangement with Hermann SA, 293 rue Lecourbe, Paris 75015.

Manufactured in the United States of America
Dover Publications, Inc., 31 East 2nd Street, Mineola, N.Y. 11501

Library of Congress Cataloging-in-Publication Data

Koyré, Alexandre, 1892–1964.
[Révolution astronomique. English]
The astronomical revolution / Alexandre Koyré
p. cm.
Originally published: Ithaca, N.Y. : Cornell University Press, 1973.
Translations of: La révolution astronomique.
Includes bibliographical references and index.
ISBN 0-486-27095-5 (pbk.)
1. Astronomy—History. I. Title.
QB29.K6913 1992
520′.9′03—dc20 92-10477
CIP

Library
University of Texas
at San Antonio

Contents

Foreword

In these three studies—Copernicus, Kepler, Borelli—which are brought together in this volume, it is not my intention to review the history of astronomy in the sixteenth and seventeenth centuries from Copernicus to Newton, but only the history of the 'astronomical revolution', that is to say, the history of the evolution and transformation of the key-concepts by means of which endeavours to reduce to order, or 'to save', phenomena—*σώζειν τὰ φαινόμενα*, *salvare phenomena*—by clearly setting forth the facts that underlie and explain the chaos of sentient appearances.

The history of astronomy should include that of observational astronomy, which was revived in the West by Regiomontanus and Bernhard Walter: some attention should be given to the foundation and work of the earliest observatories, such as those of the Landgraf Wilhelm IV of Hessen-Kassel, and especially of Tycho Brahe; the upheaval caused by the invention of the astronomical refracting telescope, which opened vistas never before seen by human eye, should be described; and the impact of the great discoveries made by Galileo and his rivals, requires analysis. . . . It is rather strange that the astronomical revolution, not only as regards its origin—the observational data of Copernicus are almost the same as those of Ptolemy—but also as regards its development, was almost entirely independent of the development of observational astronomy. Kepler, tells us that the eight minutes difference between theory and Tycho Brahe's observations are at the root of this *Astronomia Nova* . . . but, in fact, this difference was of importance to him only because he was trying to explain it within the framework of an astronomy which he describes as *αἰτιολογητός*, that is to say, within the framework of a celestial physics conceived long before he had access to Tycho Brahe's observations. Interpreted within the framework of astral kinematics—the framework of Tycho Brahe—the difference in question would have had no serious consequence.

The same applies to Borelli. He is assured of a place beside Copernicus and Kepler in the history of this revolution, not on account of his telescopic observations of the motion of the Medicean planets, but because of his decision to extend the principles of the new dynamics to astronomy—something that Galileo had not even attempted, and in which Descartes had not succeeded.† This decision was not motivated by his own observations, nor by those of others. At most, one can say that it resulted from the cumulative effect of telescopic observations which sounded the knell of the closed universe, and substituted the boundless universe.

In fact, the astronomical revolution was accomplished in three stages, each one of which was linked with the work of one man. With Copernicus, who brought the Sun to a standstill and launched the Earth into the skies—heliocentrism replaced geocentrism. With Kepler, whose celestial dynamics—unfortunately Aristotelian—replaced the kinematics of circles and spheres used by Copernicus and the Ancients. As a result, obsession with circularity was partially overcome (though it could never be completely so in the case of a closed universe), and 'astronomy of the ellipse' made its triumphal entry into the universe. Finally, with Borelli, in a universe henceforth unbounded and ruled by dynamics, the unification of celestial and terrestrial physics was completed and made manifest by abandonment of the circle in favour of the straight line to infinity. These ideas of Kepler and Borelli were fructified by the genius of Newton, and brought forth his immortal work: but that is another story, which cannot be told here.

One further remark! As far as possible, I have allowed authors, especially Kepler and Borelli, to speak for themselves in these studies. The writings of Copernicus are relatively accessible to readers, but those of Kepler and Borelli are not. For a history of scientific thought, provided, of course, that it is not treated as a catalogue of errors or of achievements, but as the entrancing, instructive history of the efforts of the human mind in its search for truth, nothing can take the place of original sources and texts. They alone enable us to catch the spiritual and intellectual atmosphere of the period under study; they alone enable us to appreciate at their true value the motives and incentives which guided and impelled the authors of them;

† Cf. my *Études Galiléennes*, Paris, 1939. Cartesian ideas, which fostered continental opposition to Newton for a hundred years, were important for cosmology and physics, but not in the least for astronomy.

they alone enable us to understand the powerful nature of the obstacles that were erected on the difficult, tortuous, uncertain path which had led them to abandon the ancient truths in order to discover and proclaim fresh truths.

Itinerarium mentis in veritatem is not a straight line. The road must be traversed, no matter how circuitous or mazelike; blind alleys must be negotiated; wrong paths must be retraced in order to discover the facts of the quest and hence the truth. Then, with Kepler, we can acknowledge that the ways by which the mind attains the truth are even more wonderful than the achievement itself.

Paris–Princeton.

I

COPERNICUS
AND THE COSMIC OVERTHROW

Introduction

The year 1543 is an important one in the history of human thought,[2] for that year witnessed the publication of *De Revolutionibus Orbium Coelestium*, and the death of its author, Nicolas Copernicus.[1]

There is a temptation to regard this date as marking 'the end of the Middle Ages and the beginning of modern times', because it symbolizes the end of one world and the beginning of another to a far greater extent than does the conquest of Constantinople by the Turks, or the discovery of America by Christopher Columbus.

I sometimes wonder if we ought not to go further and say that the break caused by Copernicus signifies far more than the end of the Middle Ages. Indeed, it determines the end of a period which includes not only the Middle Ages but classical antiquity too; for only since the time of Copernicus has man ceased to be the centre of the Universe, and the Cosmos ceased to be regulated around him.[3]

It is very difficult at the present time to comprehend and appreciate the magnitude of intellectual effort, boldness and moral courage involved in this work of Copernicus.[4] To do so, we must forget the intellectual development of succeeding centuries, and reorientate ourselves in such a way that common sense accepts with naive, confident certitude the direct evidence of the Earth's immobility and the heavens' mobility.[5]

However, to step back in time is not enough: to the force of visual evidence must be added that of the threefold teaching and tradition of science, philosophy and theology, as well as the threefold authority of calculation, reason and revelation.[6] Only then shall we be capable of fully appreciating the incomparable daring of Copernican thought, which tore the Earth from its foundations and launched it into the heavens.[7]

If it be well nigh impossible to realize the magnitude of the achievement of Copernicus without making the effort of imagination outlined above, it is just as difficult to grasp the extraordinary and profound impression that his contemporaries must have experienced on reading the work: a work which involved the destruction of a world that everything—science, philosophy, religion—represented as being centred on man, and created for him[8]; the collapse of the hierarchical order, which, placing this sublunary world in opposition to the heavens, nevertheless united them by this very opposition.

The impact was too great. This fresh conception of the Universe seemed too hare-brained to be taken seriously. Furthermore, the greater part, if not all, of the book was too difficult for readers lacking adequate mathematical and astronomical background.[9] Copernicus said to himself: *mathemata mathematicis scribuntur.* Let us leave the mathematics to the mathematicians. It is only another hypothesis, new, yet old. A calculating device having no more claim to truth, and therefore of no greater importance, than those that had been previously devised by astronomers. That was the general (and comforting) interpretation for half a century,[11] though there were some few exceptions,[10] amongst whom we may mention Tycho Brahe and Giordano Bruno, notwithstanding the *Narratio Prima* of Georg Joachim Rheticus (published in 1540) which resolutely affirmed the Copernican concept. This interpretation, moreover, was put forward as that of Copernicus himself in the very clever introduction—*Ad Lectorem de Hypothesibvs hvivs Operis*—which the publisher, Andreas Osiander, wrote and inserted anonymously at the beginning of *De Revolutionibus*, and which allowed 'mathematicians', *i.e.*, professional astronomers, to give high praise to the genius of Copernicus, and even to use his data and methods, whilst rejecting the cosmological truth of his system. As for the non-mathematicians, unable to appreciate the greatness and value of the work of Copernicus, and who read the *Narratio Prima* of Rheticus (or read nothing at all), these laymen mocked the madman who doubted the immobility of the Earth; philosophers opposed it with all the old Aristotelian arguments, which, in their opinion, Copernicus had not refuted; and theologians cited the undeniable authority of Holy Writ.

In 1539, *before* publication of *De Revolutionibus* by Copernicus, and even of the *Narratio Prima* by Rheticus, Luther, no doubt informed by rumour, harshly condemned the new doctrine; in 1541 Philip Melanchthon, having read the *Narratio Prima*, was even more severe in his condemnation.[12] Strange to say, the Roman Catholic Church made no move.

Only at a much later date, when it became evident that this work of Copernicus was not intended for mathematicians alone; when it became clear that the blow to the geocentric and anthropocentric Universe was deadly; when certain of its metaphysical and religious implications were developed in the writings of Giordano Bruno, only then, did the old world react. There was a twofold reaction: the condemnation of Copernicus in 1616 and of Galileo in 1632,[13] which tried to suppress the new concept of the Universe: and there was the *Pensées* of Pascal, which essayed a reply.

I

First Outline: the *Commentariolus*

It would be of inestimable value for the history and phenomenology of human thought were it possible to reconstruct step by step the development of the Copernican mind. Unfortunately, it is impossible to do so. Copernicus has left no autobiography describing his mental development; even the biography written by his pupil, Georg Joachim Rheticus, is lost; and the scant indications given in his introduction to *De Revolutionibus*, as well as in his fine, noble letter to Pope Paul III, which forms the preface thereto, are most brief, and consequently difficult to interpret. As for the work itself, it is presented, exactly like its first draft, the *Commentariolus*,[1] in a state of maturity which is the despair of historians.

However, if we must abandon any hope of writing a history of Copernican thought, we ought, nevertheless, to try and grasp its historical significance and nature; we ought to expose the hidden and acknowledged motives and incentives; and yet avoid any modernization.[2] To succeed in this, we must try not to regard Copernicus as the forerunner of Galileo and Kepler, and so explain him in the light of later developments,[3] even though it is precisely these developments that assure the work of Copernicus its recognized position in the history of ideas.

This highly important task has been rather neglected so far. In fact, only occasionally has there been any serious consideration of the physics of Copernicus,[6] although there are excellent reviews of his astronomy[4]; and a wealth of learning and ingenuity has been spent on secondary biographical problems,[5] or, more precisely, much effort has been wasted in trying to prove that Copernicus was Polish or German (the question, strictly speaking is of no importance, and even meaningless). However, as Schiaparelli[7] with his usual insight has said already, the Copernican problem is essentially a problem of physics and cosmology.

Nicolaus Copernicus[8] was born 19 February 1473 at Toruń (Thorn) in West Prussia, then a Polish province. His father, also called Nicolaus, was a rich citizen of Cracow, who had settled at Toruń some years before this town had been annexed to the Kingdom of Poland; he acquired citizenship and even became a magistrate of Toruń. His mother, Barbara Watzelrode, came from an old aristocratic family of the same town, and was apparently of Silesian origin. Having lost his parents at the age of ten, Nicolaus, as well as his brother Andreas, was adopted by his maternal uncle, Lucas Watzelrode,[9] who was at that time a Canon of Frauenburg Cathedral, and later (in 1489) became Bishop of Warmia (Ermland).

Was Copernicus Polish or German? I must admit that this question, or rather the importance attached to it in some quarters, appears to me to be somewhat irrelevant. Copernicus lived at a period which, with few exceptions and despite the increased use of the vernacular, was not so poisoned by nationalism and nationalistic passions as at the present time; the period in which he lived still enjoyed the blessings of a common culture and a common literary language.[10] Moreover, he lived in an Empire (medieval Poland was not a National State) whose subjects of various nationalities speaking different languages were united by and in their common adherence to the State (*Res Publica*); they knew nothing of the problem of minorities, nor of their rights (which they enjoyed to the full); they frequently intermarried and, generally speaking, got on tolerably well with each other.

Were it possible to ask Copernicus himself, 'Are you Polish or German?', it is certain that he would not understand the question,[11] and would probably reply that he was *Pruthenus*, meaning thereby that he was a citizen of Toruń, a good Catholic, Canon of Frauenburg in Warmia, and therefore a subject of the King of Poland. He would probably add, that he used the language of learning and culture, namely Latin, for his literary correspondence[12]; but that in dealing with the common people or princes[13] he naturally used the vernacular—Polish in Cracow, German in Frauenburg or in Heilsberg,[14] and, no doubt, Italian in Padua or in Bologna.

Were we to persist in trying to make him understand that we are looking for something else and that we are trying to explain his work and genius by his 'nationality' and his 'race', he would probably become annoyed and would declare that the idea of characterizing any individual by the specific or the general is quite absurd; that such a procedure is valid only when dealing with general (or specific) peculiarities of individual beings; that it can be successfully applied in those realms of nature which lack genuine individuality (for example, minerals, plants, animals), but not in the case of man, who, considered as spiritual substance, cannot be characterized by features and factors belonging to his material (or social) being, I fear that were we to allow him to become a trifle anachronistic (as we are being at the moment), he would say that all historical, Marxist or sociological conceptions by which a man, or a work, of genius may be 'explained' by heredity, race, class, position, background and a particular moment in history, could only have been invented and maintained by barbarians completely devoid of philosophical upbringing. No doubt, he would add that if one wanted at all costs to 'explain' him, Copernicus, in these ridiculous terms, then the enquirer should look towards Italy, and not Poland or Germany.

However, as we cannot entirely avoid the question, I must say that there is in fact no reason to suppose that Copernicus was not Polish. Furthermore, until the middle of the nineteenth century hardly anyone doubted it.[15] Only with the increasing growth of nationalism in European thought and historical writing have certain German historians, ready to serve the aims of German Imperialism, made claims to Copernicus; a claim which has no more basis than those concerning Leonardo da Vinci, Dante, Michelangelo and many others.[16]

In 1491, we find Copernicus at the University of Cracow, where he remained for three or four years. There is no need to seek the reasons which guided his choice: the University of Cracow enjoyed a very high reputation at that time. In fact, since the decline of the University of Prague, it had become the most important university in the east, being justly famous as a centre of scientific and classical culture. Moreover, Cracow was not far from Toruń, and it may be assumed that Copernicus still had some relations in the town where his father was born.

We have no information about the course of studies followed by Copernicus at Cracow. Though it is certain that he made a thorough study of astronomy, we have no reason to think that he did not follow the usual curriculum of the Faculty of Arts, based on dialectics and philosophy. Indeed, Copernicus, even less so than Peurbach or Regiomontanus, is not a narrow specialist, an 'astronomico'-technician, but a man deeply imbued with the entire, rich culture of his period, an artist, *savant*, scholar, man of action: a *humanist* in the best sense of the word.

One of the most famous professors at the University of Cracow was Albert de Brudzewo (Brudzewski), a distinguished astronomer and mathematician, and author (in 1482) of a commentary on Peurbach's *Theoricae Novae Planetarum*.[17] Consequently, historians interested in Copernicus have not failed to establish a connection between the two of them, and to explain the inclination Copernicus had for astronomy as the result of Brudzewski's influence. This is possible, of course, even though Copernicus does not seem to have attended the lectures of Brudzewo, who, moreover, after 1490 no longer taught astronomy as such, but 'read' Aristotle's *De Coelo*: therefore, it was not with Brudzewo, but with one of the numerous other 'readers' who taught the rudiments of astronomy to would-be practisers of the science, that Copernicus must have served his apprenticeship.[18] Albert de Brudzewo left Cracow in 1494; and most biographies of Copernicus state that he left at about the same time also, or even a little later in 1495; it is very likely that he did, but it is by no means certain. Be that as it may, in 1496 he was once more with his uncle, Lucas Watzelrode, Bishop of Warmia,[19] who tried in vain to get him nominated Canon of Frauenburg Cathedral. In this same year, 1496, he departed for Italy in order to study law, as his uncle Lucas had done before him.[20] On 6 January 1497 his name was inscribed on the *Natio Germanorum* register of the University of Bologna. Although great importance has frequently been ascribed to this fact, it does not by any means imply that Copernicus ever considered himself to be a German.[21] The '*nationes*' of a medieval university had nothing in common with nations in the modern sense of the word.[22] Students who were natives of Prussia and Silesia were automatically described as belonging to the *Natio Germanorum*. Furthermore, at Bologna, this was the privileged 'nation'; consequently, Copernicus had very good reason for inscribing himself on its register. Copernicus spent about three years at Bologna, where he obtained the degree of *magister artium* after one year's study. Whilst there, he no doubt continued his studies in astronomy, though he seems to have been sufficiently advanced in this subject for the well-known astronomer Domenico Maria Novara to describe him as *non tam discipulus quam adjutor et testis observationum*.[23] He studied many other subjects too; law, of course, but medicine and philosophy as well; he learned Greek[24]; he read Plato; he absorbed the 'renaissant' mentality and was greatly influenced by the Pythagorean and Neo-Platonic inspiration which then held sway in Italy, being centred at Florence, and of which his master, Novara, was an earnest exponent.[25] Once again, it should be stressed that Copernicus was not an 'astronomer', but a humanist and one of the most cultivated minds of his time.

In 1500 he went to Rome (1500 was the Jubilee Year which attracted thousands of pilgrims to the capital of Christendom), where, as Rheticus tells us, he gave a course of lectures on mathematics which undoubtedly means astronomy. It seems that Copernicus was not anxious to return home. However, he was obliged to do so in order to assume the office of Canon of Frauenburg Cathedral, to which office Lucas Watzelrode had finally succeeded in having him elected (in 1497, apparently).

In 1501 he returned to Poland, but not for long. After being officially installed in his canonry, he asked for leave of absence to study medicine; it was granted, 27 July 1501, and he returned to Italy. This time he went to Padua. However, it was not at Padua, but at Ferrara, that he received his doctorate in canon law (31 May 1503). After receiving the degree, he returned to Padua.[26]

Copernicus seems to have been very happy in Italy. Unfortunately, all good things come to an end, especially leave of absence. Ecclesiastical chapters do not appreciate non-resident members, and Copernicus, even though he was the bishop's nephew, had to abide by the rules. In 1503 he returned to his diocese, where he remained for the rest of his life.

The life that awaited Copernicus on his return home bore no resemblance to the popular conception of a canon's life, peaceful and serene, devoted to prayer, study and meditation. It was, on the contrary, one of activity. A medieval bishopric, particularly one like that of Warmia,[27] was as much a political as an ecclesiastical institution, and was obliged to concern itself with the affairs of this world just as much as with those of the world to come. Moreover, Copernicus was for many years secretary and physician to Lucas Watzelrode, until the death of the latter in 1512; then he was administrator of the chapter's assets (it owned one-third of the country), and resident delegate at Allenstein (Olsztyn) during which period of his career he produced a treatise on currency.[28] Furthermore, he had never given up the practice of medicine[29]; so, he would not have had much time to meditate, study or make calculations. Consequently, his work progressed and took form extremely slowly.

I have used the words, *meditate* and *make calculations*—not *make observations*. It must not be thought that Copernicus spent his nights sweeping the skies in order to discover new facts to account for the position of the celestial bodies. Copernicus was not a Tycho Brahe. He did make observations,[30] of course, and they are more numerous than was formerly believed; 63 are known to date,[31] of which only a small number was used in *De Revolutionibus*.[32] Instead of this work, he had planned at one time to limit himself to publishing a set of astronomical tables on the lines of the *Alphonsine Tables*. In 1535, he had even prepared an *Almanach* calculated in accordance with the heliocentric system; the data it contained were far more accurate than those of its predecessors, and could be regarded to some extent as proof of the reality of the system.[33]

Notwithstanding this work of a practising astronomer, and his continual care to have agreement with observable 'phenomena'—(Rheticus gives persuasive evidence[34] of this in his *Narratio Prima*: 'My master always has in front of him a properly compiled catalogue of observations made at all periods of time, and keeps it with his own observations.')—it is nevertheless true, and has long been recognized, that the greatness of Copernicus does not depend on his contribution of new facts, but on the conception and development of a new theory. It happens, that his theory, or rather his system, is based on old data, mainly of Ptolemy, much more than on fresh data. In fact, his system—a fresh interpretation of observational data—at least as far as the calculation of observable phenomena is concerned, does not agree with them much better than does Ptolemy's system which he was striving to replace.[35]

This statement may seem surprising. However, it is a fact, that Ptolemy's astronomy is relatively satisfactory, whilst that of Copernicus is hardly much better[36] in practice for calculating the positions of planets, when one takes into account the very large margin of error inherent in the star catalogues then available.

It must not be forgotten that observations were made not only with the naked eye, but also with instruments whose accuracy left much to be desired[37]—at least up to the time of Tycho Brahe; nor must it be forgotten that the Ptolemaic system from the mathematical point of view is one of the finest and most outstanding works of the human mind. In fact, the combination of circular motions he conceived or perfected in conjunction with eccentrics and epicycles enabled him to reproduce almost any closed curve, though necessarily in a rather complicated manner, and to establish a mathematical relationship between any observational data almost as well as can be done by modern mathematical methods.[38] It was only necessary to assemble a sufficient number of circular motions, without any restrictions as to the speed allotted to them, until they fitted the data. This was not the case in the astronomical system of Hipparchus, who remained faithful to the scheme imposed on astronomy by Plato, and used uniform circular motions exclusively. The brilliant modification by Ptolemy of the principle of uniform circular motion,[39] by his invention of equants, enabled him to reproduce planetary motion with surprising accuracy without unduly increasing the number of 'circles', and on the whole with sufficient accuracy for practical purposes.[40] From the point of view of mathematical methods, Copernicus discovered practically nothing. By shifting the centre of all motion from the Earth to the Sun, or more exactly, to the centre of the Earth's orbit he inverted the system of the Universe, but not the mathematical structure of astronomical knowledge; and from this point of view he was the most important follower of Ptolemy, or even Hipparchus.[41] However, there is a slight difference; Copernicus followed the Arabic method in trigonometry, and used *sines* (though he did not use the term) instead of *chords*, which were used by Greek astronomers. In this respect, his mathematical work was superior to that of his teachers.

Nevertheless, astronomy is much more than a purely mathematical doctrine: the Sun, Moon and planets are real objects; and even though Ptolemaic astronomy gives complete satisfaction from the purely geometric and practical points of view, it is not so when we come to cosmology. The astronomer is then obliged to treat his circles and orbits as real objects in real space. Certain difficulties then arise; the circles and orbits revolve on themselves without having anything as their centre. This is a distinct contradiction of Aristotle's cosmology according to which such motion is quite impossible; and, by placing the Earth at the centre of the Universe and planetary motion, allocates to each planet its particular homocentric sphere. For this reason, particularly since the revival of interest in Ptolemaic astronomy in the second half of the fifteenth century,[42] the contrary views of 'philosophers' and 'mathematicians' gave rise to (a) attempts to harmonize the Aristotelian and Ptolemaic systems,[43] and (b) attempts to suppress the problem entirely by representing mathematical astronomy purely as a calculating device devoid of any cosmic reality, yet enabling advance calculation, or 'prediction', of planetary positions to be made. This motion, moreover, favoured complete mathematical equivalence of the system of simple or movable eccentrics and the system of epicycles: and this equivalence seemed to confirm the entirely artificial nature of both these systems.[44]

Copernicus would seem to have formed the main idea of his system at a quite early date, possibly even before he left Italy. In his letter to Pope Paul III, which serves as preface to *De Revolutionibus Orbium Coelestium*, he says that he had kept his work secret not only for nine years (as recommended by the Pythagoreans[45]), but for four times nine years.[46] This statement cannot be taken literally, certainly not as regards the book itself; it can only relate to his basic idea, namely heliocentrism. Consequently, it must be understood that developing and perfecting the theory occupied 'four times nine years'. Copernicus realized that it was not enough to try and formulate fresh ideas, or in his case to try and resuscitate ancient Pythagorean concepts—or those that he believed to be such.[47] He was well aware, that to succeed, he had to do something quite different—something that no-one had ever done before, namely, provide a *theory* of planetary motion—a theory as complete and serviceable as that elaborated by Ptolemy and his medieval successors and commentators,[48] that is to say, a theory by which observable celestial motions could be predicted and calculated.

We do not know exactly when Copernicus finished *De Revolutionibus Orbium Coelestium*. It was definitely not before 1530, seeing that he uses observations made in 1529; nor was it later, at least not much later, than 1531, seeing that he does not use observations of 1532.[49]

The *De Revolutionibus Orbium Coelestium* represents the final stage in the development of its author's ideas, and offers us an astronomical system in its final, fully developed form. However, this was not the only exposition of heliocentric cosmology to come from the pen of Copernicus. Long before *De Revolutionibus Orbium Coelestium* had been perfected, he had written, and no doubt had circulated amongst his friends,[50] a short outline or first draft of his conception of the Universe; it was the *Commentariolus*, which I have already mentioned.[51]

From the astronomical point of view, *i.e.*, definite arrangement of orbits and motion of celestial bodies, the *Commentariolus* is not identical with *De Revolutionibus Orbium Coelestium*[52]; but it is extremely interesting in that it reveals far better than the latter work some of the considerations that guided the mind of Copernicus. He starts by reminding us that the main purpose of astronomy, as understood by our ancestors, was to reduce the apparent motion of heavenly bodies to regular motion, for it seemed quite absurd that heavenly bodies endowed with perfect sphericity should not move always in a uniform manner. He reminds us that Callippos and Eudoxos used concentric spheres only, and failed,[53] because it was necessary to account for changes in distance between the planets and Earth, as well as changes in apparent position (in direction). He points out also that the Ptolemaic system, although it has been highly successful from the point of view of computation, nevertheless has one grave defect, namely, that it is unable to represent planetary motion as the resultant of uniform circular motion. Consequently, in order to retain some semblance of uniformity. Ptolemy was obliged to introduce into his calculations the very dubious concept of 'equant'; as a result, the motion of planets in his system is not uniform either with respect to their deferents, or their own centres—a state of affairs which is not consistent with reason.[54] Therefore, astronomy must try to find something else, another arrangement of circles which will safeguard the uniformity of motion. Copernicus considers that he has satisfactorily solved this difficult problem by using much simpler, and much more suitable means than his predecessors. His solution requires that seven 'axioms' or 'petitions' (postulates) be assumed or conceded to allow of the construction of a system of celestial motion 'in which everything shall move *uniformly about its own proper centre* as the law of absolute motion requires'.[55]

The first axiom states that there is not one single common centre for all the circles and celestial spheres.[56] The second states that 'the centre of the Earth is not the centre of the Universe, but only the centre of gravity and of the lunar sphere'.

The third says that 'all the spheres revolve about the Sun taken as their centre-point, and therefore the Sun is the centre of the Universe'.[57]

The fourth and fifth axioms state that the distance separating the Earth from the Sun is insignificant compared with that separating the Sun from the sphere of the fixed stars; and that the common motion of celestial phenomena arises not from the motion of the firmament but from that of the Earth.

The sixth axiom proclaims that the Sun is stationary, and that its apparent motion is no more than the projection of the Earth's motion on to the sky. The seventh axiom states that the stations, retrograde motions and certain other particularities of planetary motion are no more than apparent, and not true, motions, which likewise result from the projection of the Earth's annual motion on to the firmament.

In seven short chapters Copernicus sets forth the sequence of the celestial spheres; deals with the Earth's triple motion[58]; explains the advantage of referring all motions to the fixed stars, and not, as is usual, to the equinoctial points (precisely because these are not *fixed*, but change their position, though extremely slowly); describes the mechanism of planetary motion (the Moon, the superior planets, Venus, Mercury, Earth), and gives data for the dimensions of the epicycles and circles. No proofs of any kind are given, as they were reserved for a larger work, *majori volumini*.

The *Commentariolus* closes with the triumphant declaration: 'Thus, Mercury moves on seven circles, Venus on five, Earth on three and around it the Moon on four; and finally, Mars, Jupiter and Saturn each on five. Therefore, thirty-four circles in all suffice to explain the entire structure of the universe as well as the graceful motion of the planets.'[59]

Did the *Commentariolus* have a wide circulation? Personally, I think not. Copernicus undoubtedly communicated it to his uncle and some trustworthy friends. It is very unlikely that he wanted—or authorized—a wide circulation,[60] when we remember his refusal to publish or even propagate *De Revolutionibus Orbium Coelestium*, and his fear of attack from theologians and philosophers (to say nothing of the common people) which is so eloquently expressed in his dedication to Pope Paul III (there is a suggestion of it, too, in Osiander's letters which I quote *infra*). Moreover, no-one speaks of the writings or ideas of Copernicus, with the single exception of Matthias of Miechow who was probably 'in the know' and clearly had not publicized it amongst his circle of acquaintances; he had certainly not shown it to the professors at Cracow. Had he done so, the fact would be known.[61]

The secret seems to have been well kept, for it was not till some twenty years after the drafting of the *Commentariolus* that the ideas of Copernicus started to become known. In 1533, the Austrian chancellor, Johann Albrecht von Widmanstadt (or Widmanstetter[62]), who had probably secured a copy of this small work, gave an account of the new astronomy at Rome, and explained its principles to Pope Clement VII and his entourage. The Pope rewarded him with the gift of a precious manuscript.

Neither the Pope nor anyone else at Rome appears to have been shocked by the new cosmological system. It is possible that the use of the word *hypothesis* in the title itself of this small treatise, as well as the word 'assumption' throughout the text, may have given the impression that it was only a matter of a new mathematical construction which made no claims to describe the true universe. Perhaps this was the interpretation given to the new theory by Widmanstadt. Without doubt, it is also possible that neither the Pope, nor the Cardinals, realized the implications of the system, or did not attach too much importance to its lack of agreement with the letter of the Holy Scriptures.[63]

However that may be, no-one raised objections against the theory, or its author. On the contrary, some years later, one of the members of the Roman Curia, Nikolaus Schönberg, Cardinal Archbishop of Capua, who seems to have acquired further information on the work of Copernicus, wrote to him, 1 November 1536, inviting him to publish his discoveries and requesting him to have a copy of his work prepared by Dietrich von Reder, a representative of the Curia. In 1536, that meant, a copy of *De Revolutionibus Orbium Coelestium*.

Copernicus did not accede to the friendly advice of the Cardinal of Capua. As far as is known, he did not even arrange for the copy requested by Schönberg to be made. In fact, the Cardinal was not the first to urge him to publish his work. Copernicus told Pope Paul III that all his friends, foremost of whom was 'his most dear Tiedeman Giese, Bishop of Kulm', had long since urged him to publish this work as a duty towards learning and humanity.[64] Still, Copernicus could not make up his mind. As a staunch Pythagorean, he considered that the difficult and sublime teachings of philosophy should not be made indiscriminately available to the common people, but should be reserved for the *élite*. Furthermore, as he said to Pope Paul III, he feared *calumniatorum morsus*. At his age—he was over 60—he wanted peace more than anything else.[65]

II

G. J. Rheticus and the *Narratio Prima*

In 1539, Georg Joachim Rheticus,[1] a young professor from the University of Wittenberg, arrived at Frauenburg. He had heard about the new theories put forward by Copernicus, and wished to know more. It is strange, yet worthy of note, that neither the Roman Catholic authorities objected to the coming of the young protestant, nor did the Protestant authorities object to his journey. He was well received by Copernicus, who entrusted him with the manuscript of *De Revolutionibus Orbium Coelestium*. Rheticus was immediately conquered and convinced. Conquered and seduced by the lofty mind and personal charm of Copernicus; convinced and enchanted by the beauty of his work.[2] Henceforth, he declared himself to be the disciple of Copernicus, whom he always referred to as his Preceptor and Master, *Dominus Doctor Praeceptor*. Very shortly after his arrival, Rheticus prepared a summary of the manuscript, and sent it as a letter to the mathematician and astronomer Johann Schöner.[3] This letter informed[4] Schöner that his master [Copernicus] 'has written a work of six books, in which, in imitation of Ptolemy, he has embraced the whole of astronomy, stating and proving individual propositions mathematically and by the geometrical method.'

'The first book contains the general description of the Universe and the foundations by which he undertakes to save the appearances and the observations.' 'The second book contains the theory of the first motion and the statements concerning the fixed stars which he thought should be made in that place.' 'The third book treats of the [apparent] motion of the Sun. . . .' 'The fourth book treats of the motion of the Moon and eclipses; the fifth, the motion of the remaining planets; the sixth, latitudes.' Rheticus makes no claim to have completely mastered the Copernican doctrine.[5] He frankly admits that he has only 'mastered the first three books, grasped the general idea of the fourth, and begun to conceive the hypotheses of the last two.'

However, he was impatient. The light should not remain hidden under a bushel. His letter to Schöner should make the new doctrine of his master, Copernicus, known to the world. To be doubly sure, he immediately published it, though anonymously, at Danzig.[6]

In this account—the famous *Narratio Prima*—which subsequently was nearly always included with editions of *De Revolutionibus Orbium Coelestium*,[7] we find not only an excellent and very able introduction to Copernican astronomy[8] and some very important biographical information, but also a very curious astrological account of historical events consequent upon variations in the eccentricity of the Earth's circle.[9]

Georg Joachim Rheticus was, above all, bent on clearing Copernicus of any suspicion of wanting to make changes, or of striving for originality. Thus, he wrote[10]: '. . . concerning my learned teacher and master, I should like to know and be fully convinced that for him there is nothing better or more important than to walk in the footsteps of Ptolemy and to follow, as Ptolemy himself did, the Ancients and those who came before him. However, when he realized that the phenomena, which astronomers are obliged to recognize, and mathematics forced him to make certain assumptions, even against his wish, he thought that it was befitting if he shot his arrows by the same method and aiming at the same target as Ptolemy, even though he employed a bow and arrows made from different materials and of a different kind from those of Ptolemy. At this point we should recall the saying: "He who seeks understanding must be free in mind."[11] My teacher, however, abhors whatever is alien to the mind of any honest man, particularly to a philosophic nature; for he is very far from thinking that in the desire for novelty he should rashly depart from the sound opinions of the ancient philosophers, except it be for very good reasons and when the facts themselves render it necessary. Such is his time of life, such is the seriousness of his character and distinction in learning, such, in brief, the loftiness of his mind and the greatness of his spirit that no thought of this kind can take hold of him. It is rather the mark of youth and of those who, in the words of Aristotle[12] "pride themselves on some speculations of little importance, or of those passionate intellects that are stimulated and swayed by any breath of wind and by their own moods, so that, as though their pilot had been washed overboard by the waves, they seize hold of anything that falls to hand".'[13]

The *Narratio Prima*, even more firmly than Copernicus himself, stresses the fact that the new astronomy remains faithful to the principle of uniform circular motion of celestial bodies. Thus, when explaining the theory of the motions of the Moon, Rheticus writes[14]: 'here in the case of the Moon, as a result of admitting this theory, we are liberated from the equant; furthermore, deductions from this theory agree with experience and all the observations. Finally, my teacher's theory dispenses as well with the use of equants for the other planets'; hence[15]; 'only on the basis of this theory was it possible to make all the circles of the Universe revolve uniformly and regularly about their own centres, and not about other centres—which is the essential property of circular motion.'

This is not all: Copernican astronomy makes it possible to avoid the inequalities of planetary motions by reducing them to a mere semblance. In fact[16]:

> 'The planets are each year observed as direct [in motion], stationary, retrograde, near to and remote from the Earth. These phenomena, instead of being attributed to the planets, can be explained, as my teacher shows, by a uniform motion of the spherical Earth, that is to say, by allowing the Sun to occupy the centre of the Universe, whilst the Earth, instead of the Sun, revolves around it [the centre], on an eccentric, which it has pleased my teacher to call the great circle. Indeed, there is something divine in the circumstance that a sure understanding of celestial phenomena must depend on the regular and uniform motions of the terrestrial globe alone.'

Furthermore, the simplification and systematization introduced by Copernicus into the world-system conform in the highest degree with the general metaphysical principle that nature, or more precisely God, does nothing in vain[17]:

> 'Mathematicians, no less than physicians, must agree with Galen, who affirms on many occasions that Nature does nothing without a purpose, and that our Maker is so wise that each of his works has not one use, but two or three and often more. As we see that this one motion of the Earth suffices [to produce] an almost infinite number of phenomena [appearances], should we not attribute to God, the creator of nature, that skill which we observe in the common makers of clocks? For they carefully avoid inserting in the mechanism any unnecessary wheels, or any whose function could be served by another [wheel] with a slight

> change of position. What could dissuade my teacher, a mathematician, from adopting an appropriate theory of the motion of the terrestrial globe, when he saw that by adopting this hypothesis, there sufficed, for the construction of a sound science [theory] of celestial phenomena, one single eighth sphere, and that motionless, the Sun being motionless at the centre of the Universe, and for the motions of the other planets, epicycles on an eccentric [deferent], or epicycles on an epicycle? Moreover, the motion of the Earth in its circle produces the inequalities of all the planets, except those of the Moon; this one motion appears to be the cause of all the inequalities . . .':

these inequalities are for the most part merely apparent.

Indeed[18]:

> 'If our eye were at the centre of the great circle (*orbis magnus*), lines of sight drawn from it through the planets to the sphere of [fixed] stars would, as the lines of the true motions, be moved in the ecliptic by the planets exactly as the schemes of the aforesaid circles and motions require, so that they would reveal the true inequalities of these motions in the zodiac. However, we, dwellers upon Earth, observe the apparent motions in the heavens from the Earth. Consequently, we refer all the motions and all the phenomena to the centre of the Earth as the foundation and inmost part of our habitation, by drawing lines from this centre through the planets, as though our eye had moved from the centre of the great circle to the centre of the Earth. Clearly, therefore, it is from this latter point that the inequalities of all the phenomena, as they are seen by us, must be calculated. If our purpose be to deduce the true and real inequalities in the motion of the planets, we must use lines drawn from the centre of the great circle, as has been explained.'

As far as Rheticus was concerned, the new astronomy was a revival of a very ancient philosophic doctrine. According to him 'it was by following Plato and the Pythagoreans, the greatest mathematicians of that divine age, that my teacher thought that to determine the causes of the phenomena, circular motions must be ascribed to the spherical Earth.'[19] Furthermore, 'the Ancients, not to mention the Pythagoreans, were well aware of the fact that the planets evidently have the centres of their deferents in the Sun, which is the centre of the Universe.'

The *Narratio Prima* helps us to understand the mind of Copernicus, and I shall have occasion to quote it fairly frequently. Let us now return to its historical astrology. It is difficult to say if Copernicus shared the views of his young friend, or was merely indifferent to them, but Rheticus says, 'when the eccentricity of the Sun reached its maximum,[21] the Roman government became a monarchy; and when the eccentricity had reached its limit, the quadrant of the mean value, the Muhammedan faith was established; another great empire was created and rapidly increased with the change in eccentricity. A hundred years hence, when the eccentricity will be at its minimum, this empire too will come to an end. In our time, it is at its zenith, from which, God willing, it will fall just as rapidly [as it rose], and it will fall with a violent crash.[22] We await the coming of Our Lord Jesus Christ when the centre of the eccentric reaches the other limit of the mean value, for it was in that position at the creation of the world. This calculation gives a result which hardly differs from the saying of Elijah, who prophesied under divine inspiration that the world will endure only for 6000 years,[23] during which time nearly two revolutions are completed.[24] It appears, therefore, that this small circle is verily the Wheel of Fortune, in virtue of whose revolutions the kingdoms of this world have their beginnings and vicissitudes.'

This astrological exploitation of the new heliocentric astronomy may seem astonishing, and even ridiculous, to us. Nevertheless, we must remember that astrology was one of the most firmly established beliefs of the period, and that Kepler himself at a later date was adept in that pseudo-science.

III

De Revolutionibus Orbium Coelestium
Osiander's Preface and the Letter to Pope Paul III

The *Narratio Prima* was a great success. A second edition was published in 1541 at Bâle, *cura et studio* the physician Achilles Ganarus. In this way the learned world came into possession of the first elements of the new doctrine.

The first reaction was rather favourable. Erasmus Reinhold, professor at the University of Wittenberg, published a commentary on a work by Peurbach in 1542, in the preface to which he expressed the hope of seeing astronomy restored by 'a new Ptolemy'.[1] Once the *Narratio Prima* had been published, there was no reason to delay publication of *De Revolutionibus Orbium Coelestium*. So, there is no need to be surprised that Copernicus gave way to the importunities of his friends. The precious manuscript—at least, that is what Gassendi says—was entrusted to Tiedemann Giese, who in turn had it sent to Rheticus[2] at Wittenberg. The latter, who had already brought out the purely mathematical (trigonometrical) portion,[3] caused it, or, more precisely, the copy he himself had made, to be printed by the well-known Nürnberg printer, Johannes Petreius.[4] Tiedemann Giese records that Copernicus received the first printed copy of his work on his death-bed, 24 May 1543.[5]

All these facts are well known. We know, too, that Rheticus who should have supervised the printing of the work at Nürnberg, had to leave for Leipzig, and handed this task on to his friend Andreas Osiander, a well-known, though slightly heretical, Lutheran theologian.[6] Furthermore, we know that Osiander, who had personal experience of the *rabies theologorum*, fully realized that there was a certain danger in maintaining the thesis developed by the new astronomy; so, he decided to take some precautions, as he feared a violent reaction on the part of theologians and Aristotelians. In fact, he, too, was worried by the boldness of Copernicus. The new concept of the Universe was clearly contrary to the Scriptures, and

Osiander was too good a Lutheran, even if heretical, to doubt their divine inspiration.

For this reason, as early as 1541 (April 20), he suggested to Copernicus an elegant way out of this difficulty by using a phenomenalistic theory of knowledge. In the opinion of Osiander, knowledge, and astronomy in particular, has only one aim and object, which is 'to save the phenomena', *salvare apparentias*. The astronomer's task and aim is not to discover the hidden causes, or the true motions, of celestial bodies—in any case, he is incapable of doing so—but to assemble and arrange his observations by means of hypotheses which enable him to compute and hence foresee and predict planetary positions both visible and apparent. These hypotheses of Copernicus, like those of other astronomers, do not claim (or should not claim) to be true, or even probable; they are merely devices for computation, and the best of them are not the truest, but only the simplest and the ones best suited to computation. 'I have always believed', he wrote to Copernicus, 'that hypotheses are not articles of faith, but the bases of computation; so that, even if they be false, it is of no consequence, provided that they exactly reproduce the phenomena of the motions.' He also counselled Copernicus to say this in his Introduction. 'For, in this way, you will be able to appease the Peripatetics and the theologians whose opposition you fear.'

In a letter to Rheticus bearing the same date (20 April 1541), Osiander stresses the fact that the Peripatetics and the theologians will be easily pacified if they understand that various hypotheses are possible; and that the hypotheses in question are put forward, not because they are true in fact, but because they provide us with rules for computing apparent and compound motions in the most convenient manner; that other persons can put forward other hypotheses; that one man can establish a scheme of hypotheses—and another man something different—and yet the same phenomena of motion may be deduced from them; that anyone is at liberty to invent the most convenient hypotheses, and should be congratulated if he is successful. . . .

The reply from Copernicus is not known to have survived, but Kepler, to whom we are indebted for our knowledge of this correspondence,[7] tells us that Copernicus with stoical firmness declared the necessity of publishing his opinions openly and without any dissimulation whatsoever. Thus, instead of yielding to the behest of Osiander, he stressed in his letter of dedication to Pope Paul III (which replaced the *introduction* of the manuscript in the printed work) the rights of scientific thought, and the obligation imposed on astronomy to see a true representation of cosmic reality.[8]

The objection raised by Copernicus did not stop Osiander from printing at the beginning of *De Revolutionibus Orbium Coelestium* a preface *To the reader concerning the hypotheses of this work*, in which, following the plan outlined by him in his letters to Copernicus and Rheticus, he proclaims the strictly 'hypothetical' character of the new theory, as well as all astronomical theories in general.[9]

This is what Osiander wrote:

'To the Reader concerning the hypotheses of this work.

'Since the novelty of the hypotheses of this work has already been widely reported, I have no doubt that some learned men have taken serious offence because the book declares that the Earth moves, and that the Sun is at rest in the centre of the Universe; these men undoubtedly believe that the liberal arts, established long ago upon a correct basis, should not be thrown into confusion. But if they are willing to examine the matter closely, they will find that the author of this work has done nothing blameworthy. For it is the duty of an astronomer to compose the history of the celestial motions through careful and skilful observation. Then turning to the causes of these motions or hypotheses about them, he must conceive and devise, since he cannot in any way attain to the true causes, such hypotheses as, being assumed, enable the motions to be calculated correctly from the principles of geometry, for the future as well as for the past. The present author has performed both these duties excellently. For these hypotheses need not be true nor even probable; if they provide a calculus consistent with the observations, that alone is sufficient. Perhaps there is someone who is so ignorant of geometry and optics that he regards the epicycle of Venus as probable, or thinks that it is the reason why Venus sometimes precedes and sometimes follows the Sun by forty degrees and even more. Is there anyone who is not aware that from this assumption it necessarily follows that the diameter of the planet in the perigee should appear more than four times, and the body of the planet more than sixteen times, as great as in the apogee, a result contradicted by the experience of every age? In this study there are other no less important absurdities, which there is no need to set forth at the moment. For it is quite clear that the causes of the apparent unequal motions are completely and simply unknown to this art. And if any causes are devised by the imagination, as indeed very many are, they are not put forward to convince anyone that they are true, but merely to provide a correct basis for calculation. Now when from time to time there are offered for one and the same motion different hypotheses (as eccentricity and an epicycle for the Sun's motion), the astronomer will accept above all others the one which is the easiest to grasp. The philosopher will perhaps rather seek the semblance of the truth. But neither of them will understand or state anything certain, unless it has been divinely revealed to him. Let us therefore permit these new hypotheses to become known together with the ancient hypotheses, which are no more probable; let us do so especially because the new hypotheses are admirable and also simple, and bring with them a huge treasure of very skilful observations. So far as hypotheses are concerned, let no one expect anything certain from astronomy, which cannot furnish it, lest he accept as the truth ideas conceived for another purpose, and depart from this study a greater fool than when he entered it. Farewell.'†

This short treatise of positivist and pragmatic epistemology, which seems very modern and most interesting,[11] and which in any case is very carefully composed and full of meaning from the point of view of the history and philosophy of science, was nevertheless very severely criticized by the friends of Copernicus.

Tiedemann Giese attributed this to the envy and despicableness of a man who himself refused to relinquish the traditional views and beliefs, preferring to make false statements aggravated by a breach of trust. So he sent a letter (27 May 1523) to the magistrates of Nürnberg, in which he accused Johannes Petreius of a breach of trust, and demanded a reprinting of the first pages of *De Revolutionibus Orbium Coelestium* with the addition of an explanatory note. Giese sent this letter to Rheticus: no-one was in a better position, he thought, to take the matter in hand. Giese added, that it would provide a good opportunity to print at the same time the biography of Copernicus, written by Rheticus, as well as the small treatise proving that the new astronomy was in no way contrary to the Scriptures.[12]

It is highly probable, that Rheticus, being at Leipzig, viewed the matter in a quite different light from Giese, who was at Kulm. No doubt, he saw that the magistrates of Nürnberg received Giese's protest, but he seems to have taken no responsibility for it, nor to have intervened personally. Nevertheless, he exacted and obtained from Osiander a written acknowledgement of authorship of the preface in question.[13] As for the magistrates, they contented themselves with passing Giese's letter to Johannes Petreius, who gave an

† We have used here the version given in E. Rosen, *Three Copernican Treatises*, New York, 1959, pp. 24–25. (Translator's note.)

insolent reply. The matter was then shelved. Consequently, most readers of *De Revolutionibus Orbium Coelestium* believed that Copernicus himself had provided the preface.[14] Some even accused him of hypocrisy; others thought that he was being cautious.

In fact, he had not been careful enough (assuming that he had even given attention to the matter), for he added to the book the letter he had received some years earlier from Nicolaus Schönberg, Cardinal of Capua.[15] He dedicated his work to Pope Paul III.[16] However, in the very fine letter of dedication in which he admits his fear of being attacked and condemned by 'certain persons' for having put forward novel, and apparently absurd, notions concerning the constitution of the Universe and the motions of the celestial orbs, and in which he appeals to the example and authority of the Pythagoreans, and discloses to the Pope his hesitation in publishing his work,[17] Copernicus then stoutly and proudly defends the position he has adopted, and the decision he has taken as a result of the entreaties of numerous philosophers and learned men. It was really for them that he wrote his book. *Mathemata mathematicis scribuntur*, he proclaimed; the ignorant had better remain silent. The reference to Lactantius, who made himself ridiculous by refusing to believe that the Earth is round, is obvious: it is not enough to be a good Christian, or even a good theologian, or indeed a Father of the Church, when it is a question of discussing astronomical matters.

In his introductory letter, Copernicus explains why he endeavoured to elaborate a new theory of planetary motions. The disagreement amongst mathematicians, the variety and multiplicity of astronomical systems (Copernicus cites the systems of homocentric spheres, and of epicycles and eccentrics, but contrary to what he had done in the *Commentariolus* he mentions no names), also the inability of all these systems to reproduce the apparent motions exactly whilst remaining faithful to the principle of uniform circular motion of celestial bodies, these facts caused him to conclude that the 'mathematicians' either had overlooked some basic principles, or, on the other hand, had introduced some false assumption into their systems. It was precisely for this reason, that 'as regards the most important matter, namely, the form of the Universe and the exact symmetry of its parts, they were unable to discover it, or reconstruct it.'[18] This would explain their failure: but where had they gone wrong?

Copernicus informs us that in order to find the answer he read all the books of the ancient philosophers dealing with this question on which he was able to lay his hands. He discovered in the works of Cicero that Nicetus thought that the Earth was endowed with motion. 'Later on, I discovered that Plutarch mentioned several others who held the same opinion; and so that they may be clear to all, I transcribe his words here: "Others, however, believe that the Earth is endowed with motion; thus, Philolaos the Pythagorean says that it moves around a central fire in an oblique circle as do the Sun and the Moon. Heraclides of Pontos and Ecphantos the Pythagorean do not ascribe to the Earth any motion of translation, but believe that it moves about its own centre in the manner of a wheel between sunset and sunrise."'

This view culled from the ancient philosophers encouraged him to examine this hypothesis for himself, notwithstanding its apparent absurdity, and urged him to try and discover a better explanation of celestial phenomena by means of it. In fact, he found it was possible to do so, and thereby arrived at a perfectly ordered Universe in which, 'if the motions of the moving celestial bodies were referred to the [orbital] motion of the Earth, and the latter were taken as the base for the revolution of each of the celestial bodies, it was possible to derive not only their apparent motions, but also the order and size of all the celestial bodies and spheres; and that in the firmament itself the bonds were such that nothing could be changed in any one of its parts without producing confusion in all the others and throughout the whole Universe.'

Evidently the mathematicians made the mistake of taking the Earth as the centre of the Universe and of celestial motions.

J. L. E. Dreyer, the author of one of the best histories of astronomy ever written, comments in this connection as follows[20]: 'According to this statement, Copernicus first noticed how great was the difference of opinion among learned men as to the planetary motions; next he noticed that some had even attributed some motion to the Earth, and finally he considered whether any assumption of that kind would help matters. We might have guessed as much, even if he had not told us.'

Dreyer, who seems to have little faith in the sincerity and candour of Copernicus in his letter to the Pope, considers therefore that the information provided by it does not allow us to answer the questions: Why did Copernicus feel obliged to place the Sun at the centre of the Universe? Was he influenced by reading the ancient philosophers?[21] Did he obtain the first idea of his system from Aristarchos of

Samos?[22] Or, did he first of all work out a heliocentric astronomical system, and subsequently find encouragement and confirmation for his discovery in the writings of the Ancients?

It is true that the account given by Copernicus is rather reserved and reticent. However, if he does not give details of the development of his thought, he does give some valuable indications which complement and confirm those that are found in the *Commentariolus.* Copernicus tells us quite plainly what it is he takes exception to in all the astronomical systems he criticizes: In the first place, they are unable to maintain the principle of uniform circular motion of celestial bodies; so that they are *physically impossible*, as far as Copernicus is concerned. In the second place, they give us an incoherent and irrational picture of the Universe.[23] Let us admit it, he was quite right.

Possibly, it was lunar theory, one of the weak points in the Ptolemaic system,[24] that Copernicus tackled in the first place; and his success in this direction, namely, the possibility of explaining motion without recourse to equants, may have encouraged him to do the same in respect of the other planets. This is what Rheticus seems to imply when he writes[25]:

> 'In his *Epitome*, Book V, Prop. 22, Regiomontanus says: "But it is noteworthy that the Moon does not appear so large at quadrature, when it is in the perigee of the epicycle, whereas if the entire disc were visible, it should appear four times its apparent size at opposition, when it is in the apogee of the epicycle. This difficulty was noticed by Timocharis and Menelaus who in their observations of the stars always use the same diameter of the Moon. However, experience has shown my teacher that the parallax and size of the Moon differ little, or not at all, from those that are observed at conjunction and opposition, whence it clearly follows that the traditional eccentric cannot be assigned to the Moon. Therefore, he supposes that the lunar sphere encloses the Earth together with its adjacent elements, and that the centre of the deferent is the centre of the Earth, about which the deferent revolves in a uniform manner, carrying the centre of the lunar epicycle." '

In Book I of *De Revolutionibus Orbium Coelestium*, when outlining the difficulties inherent in any theory of the motion of Venus and Mercury, Copernicus recalls a concept, mentioned by Martianus Capella and certain Latin writers, according to which the two planets have their centre of motion in the Sun, or, expressed more simply, revolve around it. He adds, that if anyone wished to develop this concept he should place the Sun at the centre of motion of Saturn, Jupiter and Mars, and having done so he would discover the true explanation of their motion.[26] Have we here a recollection of the path traversed by his own mind?

Rheticus mentions in the *Narratio Prima*,[27] that, apart from difficulties inherent in the theories of Venus and Mercury, there are large variations in brightness of the planet Mars, which are unexplainable by any geocentric theory: he says, that this fact convinced Copernicus that the planet did not revolve round the Earth, but that the Sun was at the centre of its motion. This seems to indicate the same line of thought. Yet, if Copernicus had continued this line of reasoning he would have developed Tycho Brahe's system and not his own.[28]

It is highly probable, as suggested by Apelt and Dreyer,[29] that Copernicus was struck by the fact that the motion of the Sun through the Zodiac played a very important part in Ptolemaic astronomy[30]; and especially that the period of revolution of the deferent of the interior planets was equal to one tropical year, and that of their epicycles, as well as that of the epicycles of the exterior planets, was equal to their synodic period. Rheticus seems to confirm this assumption. Indeed, he says that 'in the hypotheses of the three superior planets, the great circle takes the place of the epicycle attributed to each of the planets by the Ancients', and more generally[31] that 'it could be seen, from the commonly accepted principles of astronomy, that all the celestial phenomena conform to the mean motion of the Sun and that the entire harmony of the celestial motions is established and preserved under its control. Hence, the Ancients called the Sun, the leader, the governor of nature, and king.'

It would seem rather natural to deduce from this statement that the above mentioned circles are nothing more than projections of the Earth's orbital motion round the Sun; and furthermore, to support this discovery by the consideration that as 'the Emperor does not hasten from one place to another in order to impose his laws and authority upon his empire', so the Sun has no need to travel through the Universe in order to exert its power; and finally, to aver that 'God placed in the centre of the stage, His governor of nature, king of the entire Universe, resplendent by its divine glory, the Sun

> To whose ryhthm the gods move, and the world
> receives its laws and keeps the pacts ordained.'[32]

It is all very probable, and even quite true. Nevertheless, it is still a fact that the curious relationship between the motions of the planets and the Sun must have intrigued many philosophers, as Dreyer himself admits,[33] but no-one except Copernicus produced a heliocentric astronomy. Why? It is an idle question: because no-one before Copernicus had his genius, or his courage. Perhaps, it was because no-one between Ptolemy and Copernicus had been both an inspired astronomer and a convinced Pythagorean.

IV

The Cosmic Doctrine

Copernicus took exception to the great complexity of contemporary astronomy. In his opinion it was better to assume that the Earth has motion, however absurd that might seem, rather than let the mind be bewildered and disturbed by the almost infinite multitude of circles and spheres present in geocentric astronomy. Indeed, when we see a model of his Universe, we are seduced by its beauty and apparent simplicity. However, this impression is not quite correct. The number of circles in Ptolemaic astronomy is not so large as Copernicus says, and his own system is very liberally supplied with them.[1] Thus, the Earth alone has eight; and the other planets, too, have a considerable number. This fact is easily understood: there is no advantage simply by transferring the centre of motion from the Earth to the Sun, when considering one planet only: the orbital motion of the Earth replaces the motion of the planet on its epicycle, and that is all (Fig. 1). The gain, or economy in motion and in 'circles', becomes evident only when the system is considered as a whole; then, the motion of the Earth alone replaces all the epicyclic motions, or, otherwise expressed, the circle of the Earth is substituted for the epicycles of the planets. Now, as the Earth's motion is counterbalanced by the Sun's immobility, there is a net gain of five epicycles or motions. One could also stress the fact, that by restricting himself to uniform circular motion, Copernicus, as he loudly proclaimed, got rid of the equant circles. That is perfectly true; but this very refusal to accept the equant, *i.e.*, to allow the planets to move on their deferents with non-uniform orbital velocity, forced Copernicus to replace it by a supplementary epicycle.[2] Furthermore, from a general point of view, it could be maintained that the Copernican system is more complicated than the Ptolemaic, not necessarily in itself, but *for the observer*. In fact, the Ptolemaic system exemplifies the location of the planets (with the firmament as background) *as seen from the Earth*, and the

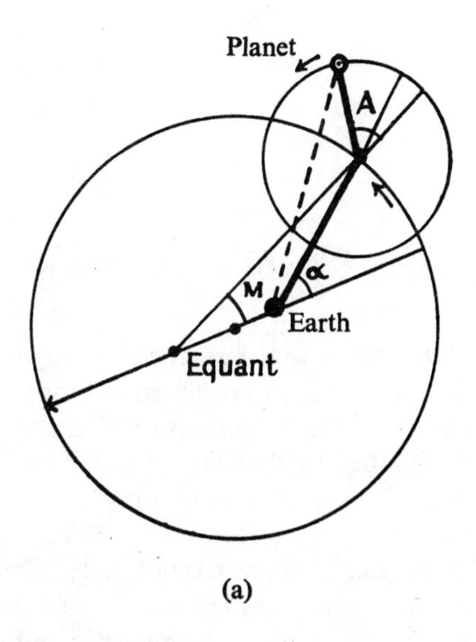

(a)

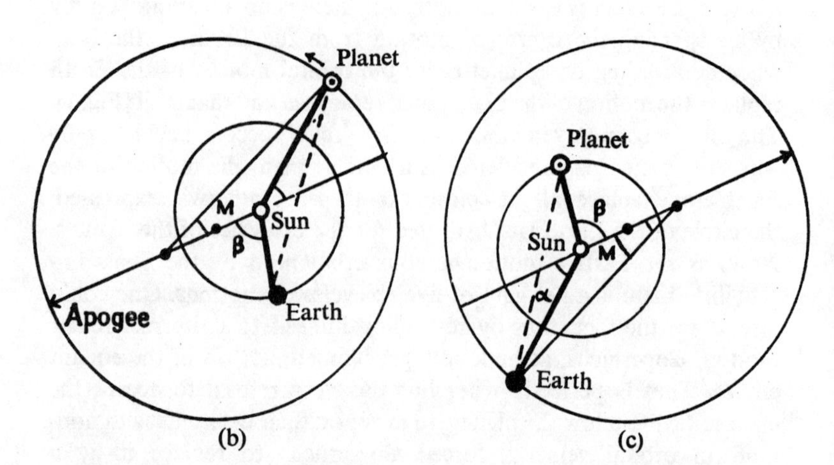

(b) (c)

Figure 1. Equivalence of the Ptolemaic and Copernican hypotheses. The figure shows the two possible cases: (a) Ptolemy's construction; (b) Heliocentric construction for Mars, Jupiter and Saturn; (c) Heliocentric construction for Mercury and Venus.

(From D. J. de S. Price, 'Contra Copernicus', *Critical Problems in the History of Science*, Madison, 1959, p. 202).

theory of their apparent motion (angular displacement) provides direct inter-relation between their apparent (geocentric) position; in the Copernican system, on the other hand, in order to determine the apparent (geocentric) position of a planet, it is first of all necessary to determine its heliocentric position, and then the heliocentric position of the Earth, from which the apparent position can be deduced, knowing the respective distances between the Sun–Earth and the Sun–planet. Galileo commented on the difficulty of understanding the Copernican system.[3] Paradoxically, it is the theories concerning the inferior planets—Venus and Mercury—which should have been simplified most by adopting heliocentrism: (What could be more simple, broadly speaking, than an explanation of their motions, once they had been referred to the Sun?)—and yet they are the most complicated. In Dreyer's opinion they are just as complicated as Ptolemy's theories; or even more so according to Neugebauer.[4]

The model for the motions of Venus is as follows.[5] The planet moves on a circle (revolving eccentric) A whose centre α is on the small circle (deferent) D which moves in the same direction as the Earth, but with double the angular velocity. Let c be the centre of the Earth's circle C, then ab will be the line of apsides produced of Venus; d the centre of the small circle (deferent D) whose radius (dm or dn) is equal to one-third of the distance cd separating the centre of the Earth's circle from the centre of motion of Venus (Fig. 2).

The motion of the circles is regulated in such a manner that when the Earth is at a or b, the centre α of the revolving eccentric A (circle of Venus) is at m.

The model for Mercury is even more complicated.[6] Again, let c be the centre of the Earth's circle C, ab the line of apsides produced of Mercury, A the revolving eccentric (circle of Mercury) whose centre α is on the small circle (deferent D) with its centre in d (Fig. 3).

The centre α will be at n when the Earth is at a or b; furthermore, the planet oscillates on the line kl, always directed towards the centre of this eccentric circle, or, otherwise expressed, on the radius of the latter from l to k, and from k to l. This oscillatory motion (produced by combining two circular motions[7]) is regulated in such a manner that Mercury returns to k whenever the Earth is at a or b (that is to say, every six months), and returns to l when the mean (heliocentric) longitude of the Earth differs by 90° from that of the apsides of the

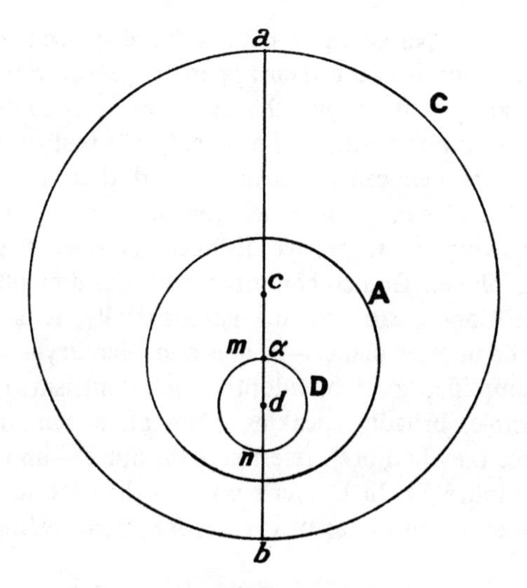

Figure 2.

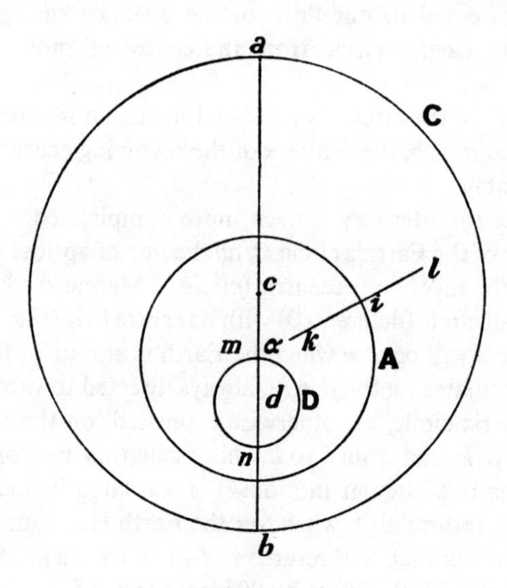

Figure 3.

planet, whilst the point *i* revolves about α on the eccentric *A* in 88 days.

> 'For this reason, Mercury by its proper motion does not always describe the same circle, but describes different circles depending on the distance from the centre; the smallest circle is described when Mercury is at *k*, the largest when it is at *l*, and the mean when it is at *i*. This motion is performed in almost the same manner as in the lunar epicycle, but in the case of the Moon it takes place on the circumference, whilst in the case of Mercury it is accomplished by reciprocal motion on the diameter, compounded of equal motions.[7] Now, the manner in which that is performed, has already been described in treating of the precession of the equinoxes.'

Namely, by combining two circular motions in such a way that the resultant motion shall be linear.

The models of the motion of the Moon, and of the superior planets are simpler. That for the Moon consists of two epicycles placed one on the other, as shown on the accompanying sketch, where *d* is the centre of the Moon's circle (*i.e.*, the Earth), *c* is the centre of the first epicycle, and *a* is the centre of the second epicycle. The radius *ae* of the second epicycle is equal to one-third that of the first radius *ca*.

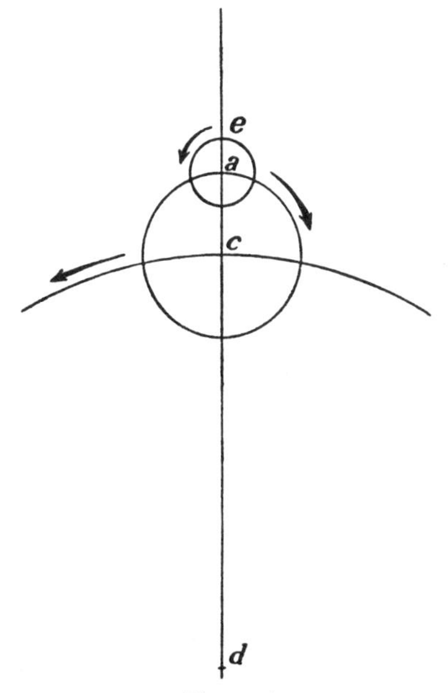

Figure 4.

The first epicycle moves in the opposite direction to the motion of the circle of the Moon (the deferent); the second epicycle moves in the opposite direction to the first (and hence in the same direction as the deferent) with an angular velocity equal to that of the first (Fig. 4).[8]

The model of the motion of the superior planets employs one epicycle only, and may be described approximately as follows[9] (Fig. 5):

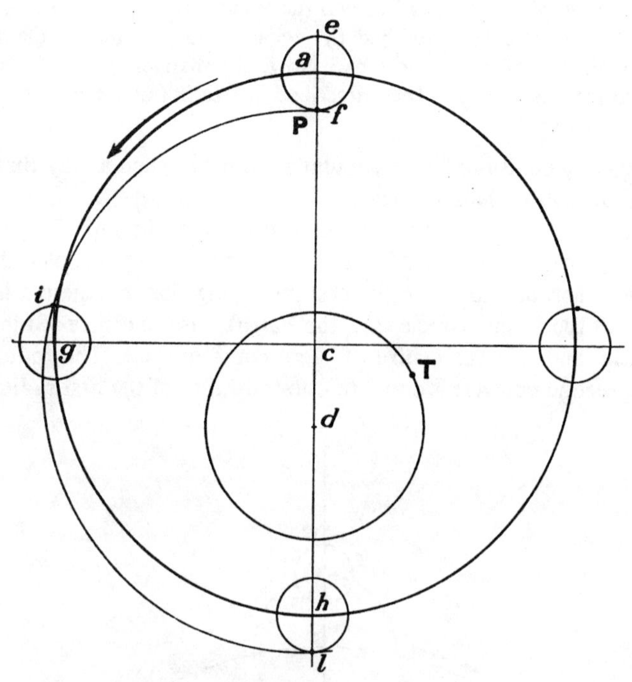

Figure 5.

d is the centre of the Earth's circle; *c* is the centre of the planet's circle (eccentric with respect to *d*). *cd* (the eccentricity) is three times *ae* (the radius of the epicycle) and their sum (*cd* + *ae*) is equal to the Ptolemaic eccentricity of the planet (distance between the Earth and the *punctum aequans*). *P* is the planet, *T* is the Earth; the epicycle and the deferent revolve in the same direction, and with the same angular velocity. Consequently, when the planet is at *f* the centre of its epicycle is at *a*; at *i*, when the latter is at *g*; and at *l*, when the latter is at *h*: the true path of the planet is then almost circular.

To summarize, the *orbis magnus* (great circle) by itself replaces the arrangement of Ptolemaic epicycles, as has been admirably shown by Kepler[10]; furthermore, it replaces the special models needed for the motion of the planets in latitude, explained at least partially[11] as projections of the Earth's orbital motion. The total number of motions and 'circles' is therefore reduced. However, there is no great gain—Kepler's estimate is ten circles—and one cannot criticize those who found that this was insufficient to justify the change, especially as it was necessary to enlarge the confines of the Universe almost infinitely in order to accommodate the great circle of Copernicus,[12] namely, to explain why a parallax of the fixed stars is never observed,[13] in spite of the annual motion of the Earth on this circle.

The great superiority of the Copernican system over the Ptolemaic does not depend, however, on the decrease in the number of celestial motions (and the number of their corresponding circles), but on the fact that they were standardized, regularized and systematized; it explained the inequality of apparent motions, with their retardations, stations, retrogressions and advances, by an effect of perspective resulting from the motion of the observer himself; it relegated these inequalities to the unreality of optical illusions; and substituted a much more systematic and regulated state of affairs for the incoherent universes of Aristotle and Ptolemy.[14]

As we have already seen, this is the feature that is stressed by Rheticus, who emphasizes even more than his teacher how the hypothesis of the Earth's motion simplifies and systematizes the explanation of celestial phenomena. Copernicus, on the other hand, does not rely on the theoretical and practical advantages of his doctrine. As far as he is concerned, they are unquestionably by-products of the truth. The great, and real, superiority of his system does not lie therein; it lies in safeguarding the essential principle of uniformity of motion of the circles (elimination of the equants), just as much as explaining the orderly principle governing the Universe, that is to say, agreement of the distance of the planets from the Sun with their period of revolution.

There is strict adherence to the principle of uniform circular motion for the planets, which Copernicus regards as one of the essential advantages of his system, but which is not the case in the Ptolemaic system, as we have already seen. The principle is, however, by no means bound up with heliocentrism, as Copernicus believed—if indeed he did believe it. In fact, as was brilliantly shown by Kepler, it is always possible to replace the 'equant circle' by an epicycle;

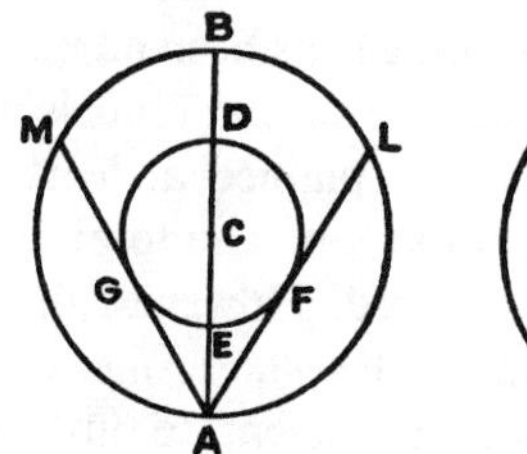

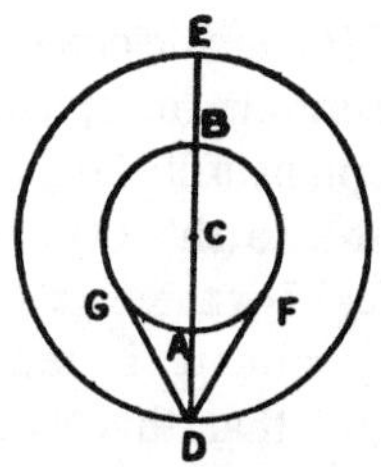

Figure 6a. General demonstration of the apparent inequality of planetary motion. (*De Revolutionibus Orbium Coelestium*, Lib. V, cap. III; TH., p. 322–323.)

AB = Circle of the Earth.
DE = Circle of Venus or Mercury.
A = Earth.
EGF = Planet.
The helio- and geocentric longitudes coincide when the planet is at *E* or *D*, and differ when it is at *G* or *F*.

AB = Circle of the Earth.
DE = Circle of a superior planet.
A = Earth (equinox).
GF = Earth.
D = Planet: the helio- and geocentric longitudes coincide when the Earth is at *A*, and differ when it is at *G* or *F*.

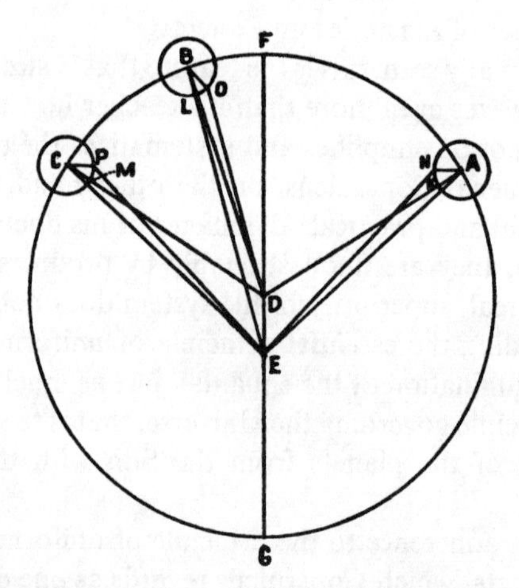

Figure 6b. Model of Saturn's motion (*De Revolutionibus Orbium Coelestium*, Lib. V, cap. V; TH., p. 329; *Opera Omnia*, p. 297.)

A, *B*, *C* = Positions of centre of epicycle.
AFBCG = Circle of planet.
N, *O*, *P* = Planet

FG = Line of apsides.
D = Centre of planetary circle.
E = Centre of Earth's circle.
The Sun is not shown.

similarly, it is always permissible[15] to substitute a concentric circle with an extra epicycle for an eccentric circle, as Copernicus shows on several occasions, following the example of Ptolemy. Inversely, it is always possible to substitute an eccentric for a concentric circle with an epicycle, with the saving of one circle, or one motion. It was precisely for this reason that Ptolemy opted for eccentrics; and that Copernicus, in *De Revolutionibus Orbium Coelestium*, followed his example by replacing two epicycles in the *Commentariolus* by an eccentric circle. Now, the same state of affairs exists with regard to the equant circle (or the equant point): its introduction results in the economy of one epicycle: and that is why Ptolemy the *mathematician* introduced it into his system and dropped the *physical* principle of uniform circular motion. Inversely, the Ptolemaic equant was rejected by Copernicus for *physical* reasons; and consequently, he was obliged to re-introduce epicycles peculiar to each celestial body (except the Earth) into his model of planetary motions, thereby losing part of the formal advantages of his system,[16] namely, elimination of Ptolemaic epicycles. In any case, the epicycles of Copernicus are not exactly equivalent to those of Ptolemy: in the first place, they are much smaller[17]; secondly, and this is important, their period of revolution is not equal to the apparent motion of the Sun (or real motion of the Earth), but equals that of the planetary circle on which they are carried. Apart from the resulting advantages for computation, the true path of the planets is always a *smooth* curve according to Copernicus, whereas according to Ptolemy the path bristles with nodes; furthermore, it is approximately circular,[18] at least for the superior planets.

Let us now consider the second fundamental principle of the Copernican system, namely, the agreement between the distance of the planets from the Sun and their period of revolution.[19] In this respect the superiority of Copernican astronomy over the Ptolemaic is obvious. The latter, too, certainly applies the principle stated by Aristotle,[20] according to which a longer time of revolution corresponds to a greater distance: but it can be applied only in the case of the superior planets, and not for the entirety of celestial bodies.[21] Although Ptolemy does indeed put forward determinations of distance, he gives them, so to speak, only by way of extra information.[22] In fact, Ptolemy's technical astronomy is based on the calculation of angles (and not of distances); consequently, the order of the celestial bodies could be reversed, and they could be placed anywhere or at any distance in the firmament[23] (always provided that the relationship between the length of the radii of the deferents and the epicycles be maintained). On the other hand, according to Copernicus, this order is found to be fixed in a quite definite manner, and, as he proudly informed Pope Paul III, it is not possible to modify this order in one respect without upsetting the remainder.[24] He was right, when he declared[25]:

> '... For this reason, the first law being admitted—and no-one can propose a more suitable one—that the size of the spheres is proportional to their periodic time, then the order of the spheres results; and starting with the highest is as follows.
>
> 'The first and highest of all is the sphere of fixed stars, which contains all things and contains itself, and is therefore motionless. It is, assuredly, that place in the Universe to which the motion and position of all other celestial bodies must be referred. For, if some believe that it, too, has motion of some kind, we for our part do not admit it, and shall show reason for this apparent motion in deducing the Earth's motion. Next comes the first of the planets, Saturn, which revolves in thirty years; next comes Jupiter whose motion of revolution takes twelve years; then Mars which takes two years. The fourth place is occupied by the annual revolution of the sphere which contains the Earth together with the sphere of the Moon. In the fifth place is Venus whose period is nine months. Finally, the sixth place is occupied by Mercury which revolves in the space of eighty days. In the midst of all sits the Sun....'

As for Rheticus, he goes even further. Compared with the Ptolemaic system, where there is no fundamental order, he not only notes the 'remarkable symmetry and interconnection of the motions and the spheres which result from the foregoing hypotheses, and which are not unworthy of God, nor unsuited to these divine bodies',[26] he sees also a profound meaning and confirmation of Copernican astronomy not only (as Copernicus saw it) in the fact that the speed—or slowness—of motion of the celestial spheres corresponds to their distance from the Sun, but also in the fact that this astronomy reduces their number from seven to six. In the Copernican universe 'there are only six moving spheres which revolve about the Sun, the centre of the Universe'. Rheticus continues[27];

> 'Who could have chosen a more suitable and more appropriate number than six? By what other number could mankind be more easily persuaded that the whole Universe was divided into spheres by God, the Author and Creator of the Universe? For the number six is exalted

above all others in the sacred prophecies of God, as well as by the Pythagoreans and other philosophers. What is more consistent with God's handiwork than the fact that the first and grandest of His works should be summed up in the first and most perfect numbers?'

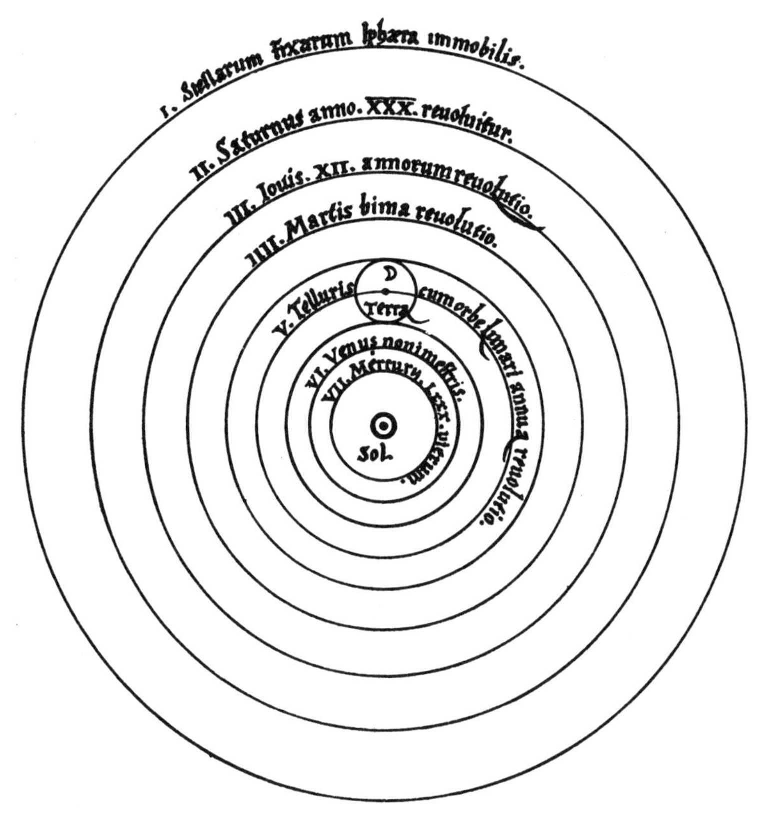

Figure 7.

The arguments put forward by Copernicus in favour of the new astronomy are of two kinds: (a) it gives a better explanation of certain phenomena (the apparent motion of Mercury, Venus and Mars) than the old astronomy; (b) it provides a much more systematic and ordered picture of the Universe than do either the Ptolemaic or Aristotelian systems. He says[28]:

'... So we find in this admirable arrangement a harmony of the Universe, as well as a certain relationship between the motion and the size of the spheres, such as can be discovered in no other way. For here the close observer may see why the progression and retrogression appear greater for Jupiter than for Saturn, and less than for Mars; and yet again, greater for Venus than for Mercury; and why these progressions and retrogressions appear more frequently in the case of Saturn than with Mars and Venus; moreover, he may see why Saturn, Jupiter and Mars are nearer to the Earth when they rise in the evening than when they disappear in, or emerge from, the Sun's rays. Mars in particular when it shines throughout the night, seems to rival Jupiter in magnitude, being distinguishable only by its reddish colour; otherwise, it is difficult to find amongst stars of the second magnitude except by close observation of its motion. All of these phenomena proceed from the same cause, namely, the Earth's motion. That nothing of the kind is observed in the case of the fixed stars proves them to be at an immense distance, which renders the annual motion of the sphere imperceptible even to the eye: for every visible object there is a certain distance beyond which it can no longer be seen, as is proved in optics. Indeed, the twinkling of the stars shows that there is still a vast space between Saturn, the highest of the planets, and the sphere of the fixed stars. It is by this behaviour that they are chiefly distinguished from the planets, for it is proper that there should be a great difference between moving and non-moving bodies. Verily, so perfect is this divine work of the Great and Supreme Architect. . . .'

A modern reader could express surprise at the complete absence of physical arguments in favour of the Earth's motion. He would be mistaken. Copernicus certainly gives none. In fact, he could not give any.[29] We must admire—as did Galileo[30]—the power and boldness of this mind, which by pure intellectual intuition was able to overcome all the obstacles that had impeded the advance of ideas.

V

The Physical Problem

The controversy between Copernicus and the supporters of the traditional conception of the Universe is very curious. From the point of view of an Aristotelian, his criticisms—basically, there is only one—do not carry weight, any more than do his replies to the old objections of Aristotelism against the Earth's motion. All the same, they are extremely important; they contain the germ of those which Galileo was to formulate at a later date on a totally fresh basis.

The fundamental objection of Copernicus to any supporter of geocentrism—and it is directed against Aristotle rather than Ptolemy—seems at first sight to be an extremely strong one. Copernicus says that it is absurd to want to impart motion to the *locus* (place) and not to the *locatum* (that which is set in a place); and so the starry firmament, which according to Aristotle is the *locus* of the Universe, must be considered to be without motion.

This seems to be a quite reasonable argument. Indeed, we believe, with Newton who used the same argument, and even with Aristotle, that it is absurd to impart motion to the very *places* in which objects are placed; similarly, it is absurd to allow the whole vast Universe (infinite for Newtonian astronomy) to revolve about a small grain of dust: we are persuaded. However, the Aristotelian (or the Ptolemaean) for whom, moreover, the argument of Copernicus is only a revival of the old objection made by Aristotle's opponents against his conception of the Universe, is not an argument at all. His Universe is not infinite, nor even *vast*, that is to say, beyond all measure, as it is for Copernicus. It is finite, though of considerable size, and consequently measurable.[1] Moreover, there is a qualitative difference between the Earth and the heavens: the Earth is heavy and inert, the heavens lack all heaviness. Therefore, though on the whole it is better to move the contents rather than the container, it is not so in the present instance: in order to give motion to the Earth (even if it be only rotation in a given position) an external, physical driving force of formidable power would be required. Nothing of the kind is needed for the motion of the heavens: they revolve in virtue of their own nature and their own perfection; in other words, they are moved by a spiritual driving force.[2]

Obviously, it would be even more difficult to confer orbital motion on the Earth about the centre of the Universe. The Earth, we repeat, is heavy, and therefore like all heavy bodies has a natural tendency towards the centre of the Universe (the stars not being heavy, do not possess this tendency). How could it be dislodged from that position once it was there? Or how prevented from returning thither, if by some miracle it were displaced? Is it not far more reasonable to accept that it is at rest at the centre of the Universe, as is confirmed by all experience? Far from introducing more order into the Universe, does not the Copernican conception introduce, on the contrary, perpetual disorder by wanting to keep the Earth out of its natural place?[3]

Copernicus replied by rejecting the cosmic concept of gravity which is that of Aristotelian physics. Heaviness is not the tendency of heavy bodies towards their natural place in the Universe; quite the contrary, it is the tendency peculiar to every celestial body, and to the Earth also, to form a whole. He says[4]:

> '. . . I consider that gravity is nothing more than a certain natural desire given by divine providence of the Architect of the Universe to all parts to recover their unity and wholeness by coming together again in the form of a globe. We may believe that this tendency is shared also by the Sun, the Moon and other wandering stars in such a manner that through its power they retain that roundness in which they appear, although they accomplish their course in divers ways. . . .'

This natural tendency of like things to come together and form a whole—a resuscitation of the ancient doctrine of Empedocles or of Plato—is certainly very different from universal gravitation.[5] Yet it opens the path to it, and moreover, contains an implicit negation of the concept of 'natural place',[6] which is the first step towards geometrization of space, one of the bases of modern physics.

The remaining counter-arguments put forward by Copernicus are hardly any better from the Aristotelian point of view. By quoting the famous line from Virgil—

Provehimur portu terraeque urbesque recedunt

—Copernicus shows that from the point of view of optics it is impossible to say if it is the observer, or what he observes, which is in motion. Very true, the Ptolemaean would retort; the well-known relativity of motion[7] certainly implies that terrestrial motion is optically *possible*; but at the same time it implies—always from the point of view of optics—that it presents no advantage with respect to the Earth's immobility.[8]

To the physical objection that terrestrial rotation would generate an enormous centrifugal force which would cause the Earth to shatter into pieces, Copernicus replied that the same objection could be made against the motion of the heavens, particularly as their speed is infinitely greater than that of the Earth.[9]

In fact, from the point of view of Ptolemy—or Aristotle, if we accept that the 'spheres' have no physical reality for Ptolemy—Copernicus is wrong, because the motion of the celestial spheres, which are considered to be imponderable, is of a quite different kind from that of the Earth, and could not give rise to a centrifugal force. Furthermore, the rotation of the heavens, which is their natural motion, cannot be the cause of an effect that would endanger their conservation and existence.

However, we must not deride these arguments. From the point of view of Aristotelian physics they are perfectly valid. Moreover, Copernicus used them himself in order to show that if the Earth turned on its axis, then this motion would be a *natural* motion, and consequently none of the effects deduced (hypothetically) by Ptolemy, on the supposition *that this motion would be violent* and *contra naturam*, could occur.

To the old argument that bodies not in contact with the Earth—clouds, and birds—should 'remain behind' in the case of a rotating Earth, or that a body thrown vertically upwards would never fall again to the place whence it was thrown, Copernicus replied that these bodies, being 'terrestrial' and consequently sharing the nature of the Earth, share also its 'natural' motion of rotation which coexists in them together with their own proper motion.[10]

We see that Copernicus, as did everyone before Galileo and Descartes,[11] accepted the distinction between *natural* motion and *violent* motion. On the other hand, he asserts that the same laws apply to the heavens as to the Earth,[12] and in so doing laid the basis of a profound change in human thought, to which History has given the name *Copernican Revolution*. We shall be able to understand this a little more clearly by considering the great astronomer's dynamics.

The dynamics of Copernicus is definitely not 'modern'. It differs, however, from that of his contemporaries because Copernicus geometrizes nature, being probably inspired by Nicholas of Cusa, whom he greatly surpasses.[13] He saw the Universe from the point of view of a 'mathematician' and geometer. Even his perception is geometrical: it is aesthetics of geometry, or more precisely of geometrical optics, and his physics—though he never said so *expressis verbis,* and perhaps was not even fully aware of the fact—is geometrical physics. Yet, not entirely! Copernicus, as we have just seen, does not abandon the idea of 'nature'. On the contrary, he attributes to the Earth and to each of the celestial bodies[14] their own particular 'natures'—and yet the geometrization of his thought is sufficiently strong and deep-seated to bring about a noticeable change in the Aristotelian concept of form. Thus, when medieval and classical physics speaks of 'forms', it usually has 'substantial forms' in mind. Copernicus, on the other hand, thinks in terms of *geometrical* form.

The implications of this conception, or fresh outlook, are numerous and far-reaching. For example, in the physics of antiquity, the kind of *natural* motion possessed by a body (rectilinear for sublunary bodies, circular for celestial bodies) was determined by the specific nature, and definite substantial form of the corresponding matter, whilst it was the geometrical form that was responsible in the view of Copernicus. If celestial bodies move about themselves, it is not because they possess a specific nature, but simply because they are spherical.

Copernicus seems to believe that a spherical shape, the most perfect geometrical form, and the one that all natural bodies endeavour to assume because of this very perfection, is not only more suitable for motion (as is universally admitted[15]), but is sufficient cause thereof, and naturally engenders *the most perfect and most natural* of motions, namely *circular* motion.[16]

This being so, it follows that: (a) the same reasoning allows and even obliges us to attribute the same circular motion to the Earth as to the planets—undoubtedly, we have here the reason why Copernicus dealt at such length and with such a profusion of argument with the sphericity of the Earth, a fact that was not in dispute in his time, and which could have been treated with less insistence had it not been so important for him to do so; (b) the same laws of motion which are applicable to the planets are equally applicable to the Earth; (c) the Earth, participating of the same circular shape and obeying the same laws of motion, as celestial bodies (the planets), is not set as a world apart—the sub-lunary world, the unworthy world and cesspool of corruption—but constitutes with them one single and unique Universe.

Geometrization of the concept of form places the Earth amongst the stars, and raises it, so to speak, into the heavens.[17]

We are now able to understand why Copernicus attached such great importance to the rule, or principle, of uniform circular motion, and why he regarded it as the basis of his celestial mechanics. It was the only means at his disposal to set the *machina mundi* in motion. In fact, in the dynamics of Copernicus, motion (circular) is caused by (or results from) the shape (spherical) of bodies. Bodies revolve because they are round, and to do so have no need of an external driving force, nor even an internal one. Put a round body (a sphere) in space—it will revolve. Put a heavenly body there—it, too, will revolve about itself without need of driving force to maintain the motion, nor will it need a *physical centre*, which Aristotle found it impossible to do without. Still, it will revolve only with a perfectly uniform motion, and for this reason the concept of 'equant' must, at all costs, be removed[18]: even if it be necessary to increase the number of motions, as we have seen already. Hence, the celestial mechanics of Copernicus is, so to speak, a mid-path between pure kinematics and dynamics, and that is why (like Ptolemy) he had no need to put anything whatever at the centre of his celestial bodies or spheres; not even the Sun (Figs. 8a, b, c, d).

Indeed, we have already seen that when Copernicus places the Sun at the centre of the Universe, he does not place it at the centre of the celestial motions: neither the centre of the Earth's sphere nor that of the planetary spheres is placed *in* the Sun, but only *near* to it,[19] and the planetary motions[21] are referred not to the Sun, but to the centre of the Earth's sphere[20]—eccentric with respect to the Sun. The centre of the terrestrial sphere certainly revolves about the Sun[22]; it is placed on a small epicycle whose deferent has the Sun for centre, but its motion is so slow—the epicycle makes one revolution in 3434 years and the deferent in 53 000 years—that, for practical purposes, it does not enter into the calculations.[23] As a result, we have the paradox, that in the celestial *mechanics* of Copernicus the Sun plays a very unobtrusive part. It is so unobtrusive that

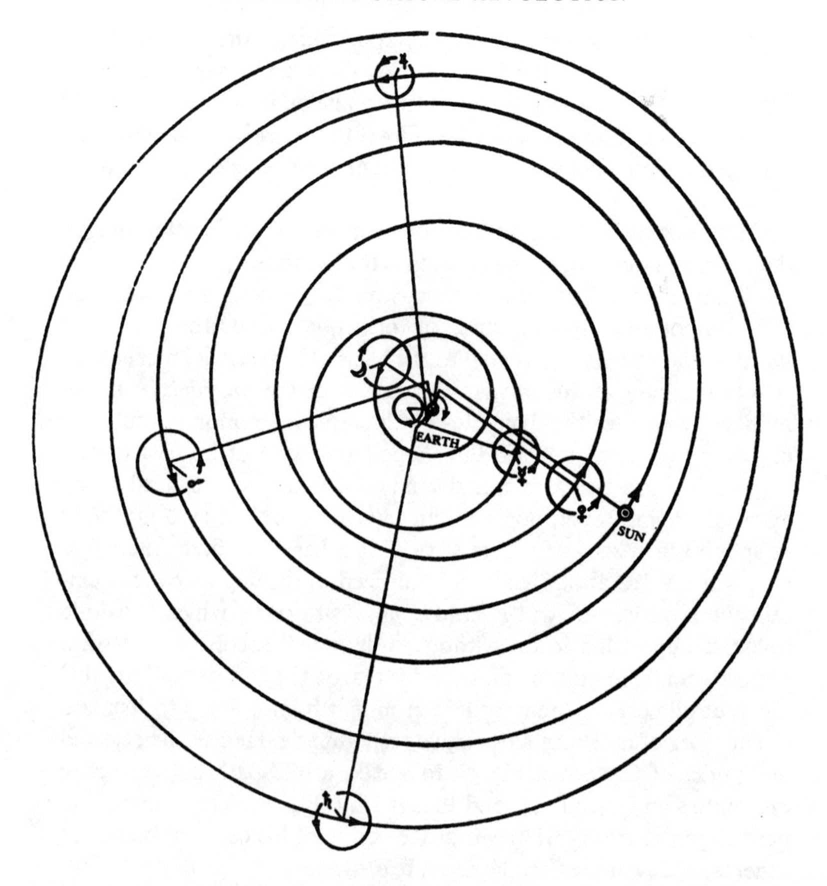

Figure 8a. The Ptolemaic system.

These diagrams show the relative complexity of the Ptolemaic and Copernican systems.

The radial dimensions in these diagrams cannot be exact, but the centres of the planetary orbits have been very carefully orientated with respect to the Zodiac. For example, in the Ptolemaic diagram, if a radial line be drawn from the Sun to the near point of the Earth, which is the centre of the solar orbit, it will be seen that it (the Earth) is located between the centres of rotation of Venus and Mars, exactly where the Ptolemaic geocentric theory requires.

The relative directions of rotation of the epicycles on the circumferences of the deferents, and of the planets on the epicycles are shown by arrows.

The planetary distances are arbitrary, as distinct from the Copernican system.

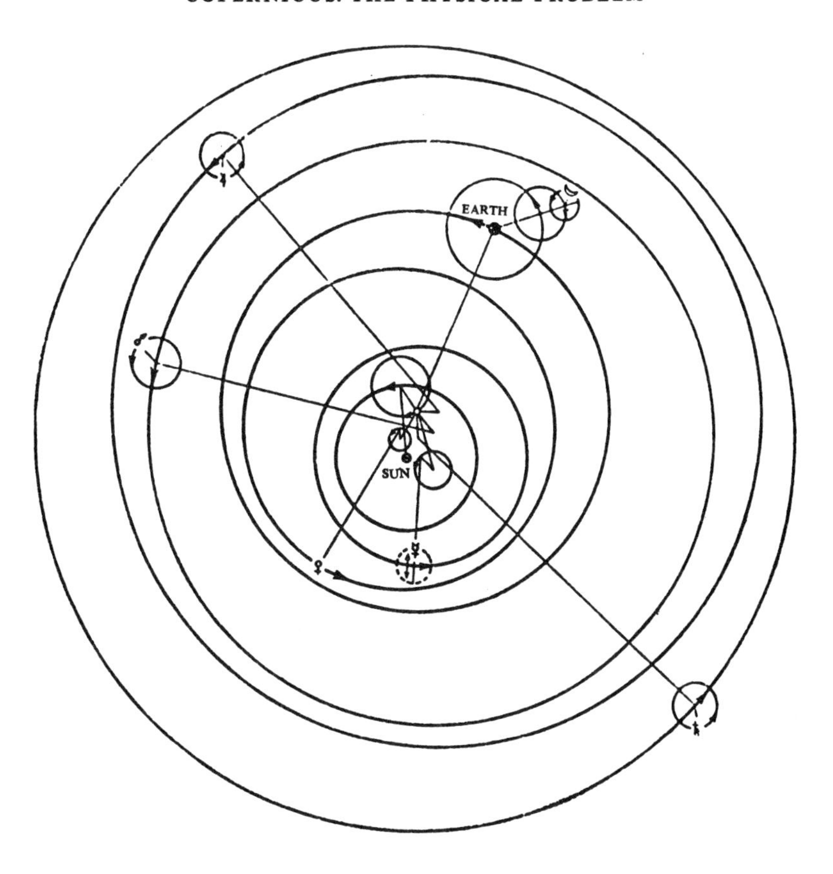

Figure 8b. New system according to Copernicus.

In the Copernican system, the Sun is at the centre.

The true centres of rotation of the planets at any moment are grouped about the centre of the terrestrial orbit at any moment.

In this system Mercury is given special consideration: the planet oscillates about the centre of an epicycle in preference to moving on the epicycle itself.

The planetary symbols are as follows:

☉	Sun	☾	Moon	♃	Jupiter
☿	Mercury	♁	Earth	♄	Saturn
♀	Venus	♂	Mars		

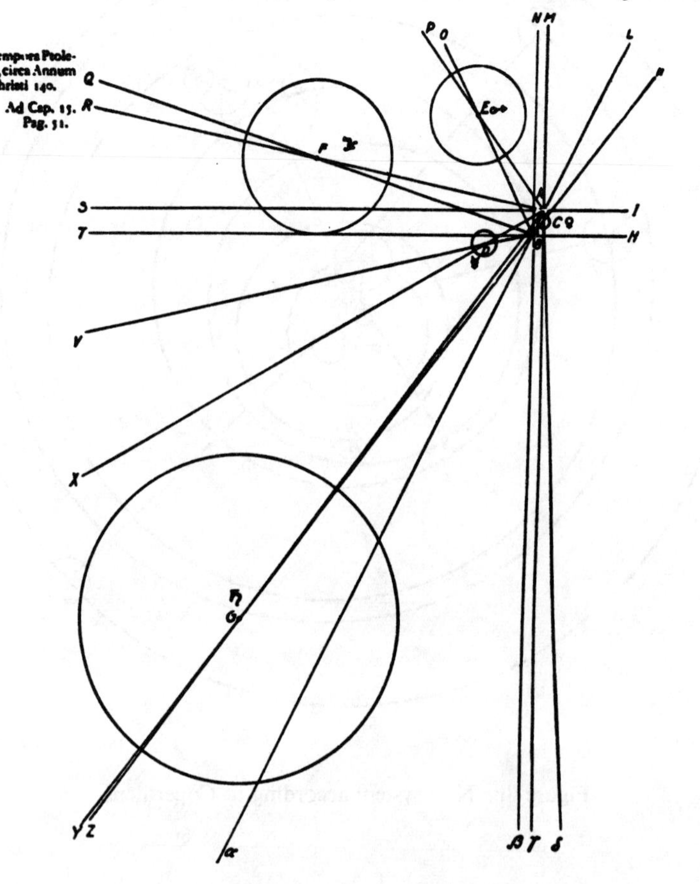

Figure 8c (see p. 149).

Diagram drawn for Kepler by M. Maestlin.

mundi, secundum sententiam Copernici, et numeros Tabularum Prutenicarum.

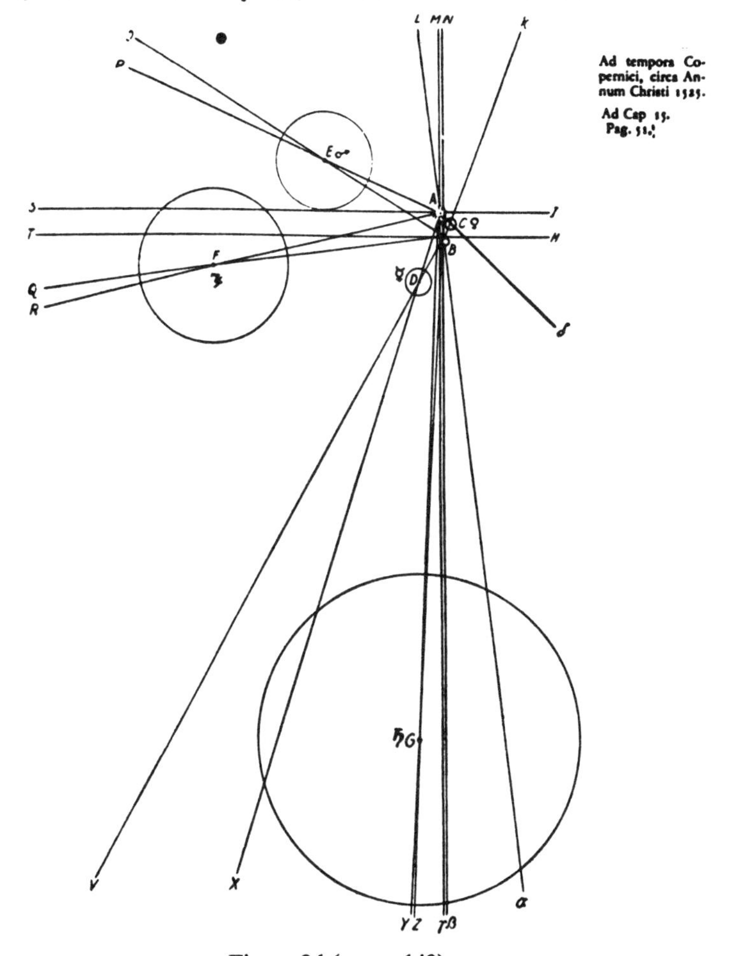

Figure 8d (see p. 149).

Diagram drawn for Kepler by M. Maestlin.

Figure 8c. Ad A Sol, centrum Mundi est: Circellus paruus ad B, est circulus eccentricitatis Orbis Magni Telluris, in huius fastigio seu loco remotiore à Sole, eccentrici orbis magni centrum consistebat tempore Ptolemaei, sed tempore Copernici in loco propiore. Hoc est, eccentricitas Orbis Magni erat illic propè maxima, hic ferè minima. Horum illud priore, siue sinistro, hoc posteriore, siue dextro schemate videre licet.

A B priore schemate est 4170. qualium semidiameter orbis magni est 100 000. Hinc maxima Terrae à Sole remotio est 104 170, et minima 95 830. Sed in altero schemate illa eccentricitas propè minima, est 32 195.

A C est circulus paruus eccentricitatis ♀. Huius semidiameter (qualium orbis magni semidiameter est 100 000) est 1040, et B C (dextrae figurae) eccentricitas centri parui circuli à centro orbis magni B, est 3120. Sed A C, eiusdem eccentricitas à Sole A, est 1262. Hinc maxima Veneris à ☉ distantia 74 232, et minima 69 628.

D centrum est circelli eccentricitatis ☿. Huius semidiameter est earundem, qua suprà, partium 2114½. eiusque eccentricitas à centro orbis magni D B 7345½ sed D A, eccentricitas eius à Sole 10 270. Vnde maxima Mercurij distantia à ☉ inuenitur 48 114½. et minima 23 345½.

E centrum est parui circuli eccentricitatis ♂. Huius semidiameter est 7602½. et B E eccentricitas ab orbis magni centro 22 807½. Sed A E eccentricitas à Sole 20 342. Vnde distantia ♂ à ☉ maxima 164 780. minima 139 300.

F centrum est parui circuli eccentricitas ♃. Huius semidiameter est 12 000. et B F eccentricitas à B 36 000. Sed A F à ☉ 36 656. Iouis maxima distantia à ☉ 549 256. minima 499 944.

Figure 8d. G centrum est parui circuli eccentricitatis ♄. Huius semidiameter est 26 075. B G est 78 225. et A G eccentricitas à ☉ 82 290. Saturni maxima remotio à Sole est 998 740. et minima 834 160.

Recta H B T est linea aequinoctialis respectu Terrae. Sed I A S, respectu Solis. Sic recta N B β est linea solstitialis respectu Terrae, et MA γ respectu Solis.

	tempore		Ptolemaei.	Copernici.
	♄	BGY	23 ♏	27 42 ♐
	♃	BFQ	11 ♍	6 21 ♎
ἀπόγειον	♂	BEO	25 30 ♋	27 ♌
	♀	BCK	25 ♉	15 44 ♊
	☿	BDV	10 ♎	28 30 ♏
	☉	BAL	6 8 ♋	6 50 ♋

	tempore		Ptolemaei.	Copernici.
	♄	AGZ	23 40 ♏	28 3 ♐
	♃	AFR	17 31 ♍	11 30 ♎
ἀφήλιον	♂	AEP	4 27 ♌	3 21 ♍
	♀	ACδ	4 39 ♑	19 48 ♒
	☿	ADX	29 42 ♎	13 40 ♐
	Terrae	ABα	6 8 ♐	6 40 ♑

we might say that it plays no part whatsoever. Its purpose is quite different: it gives light to the Universe, and that is all.

Perhaps I am wrong in saying that that is all; for the function accorded to the Sun, namely to give light to the Universe, is of supreme importance for Copernicus. It is this function which explains and assures the position it holds in the Universe—foremost in dignity and central in position.[24] The Sun, this *lampada pulcherrima*, is placed at the centre of the Universe in order to give it light, and hence life and motion. The central position is obviously the most suitable for this purpose: 'Indeed, in this splendid temple [the Cosmos] who could place this luminary in any better position than that from which it can illuminate the whole at once? Verily some have rightly called it the Lamp, the Mind, the Ruler of the Universe. Hermes Trismegistus calls it the Visible God, and Sophocles in *Electra* calls it the All-seeing. So the Sun, seated as it were upon a royal throne, rules the surrounding family of heavenly bodies . . .' There we have the motive—the real motive—which inspired the mind and soul of Copernicus.[25] It is not a purely scientific motive; it is much more than that.

The old traditions, the tradition of Metaphysics of Light (which throughout the Middle Ages maintained and accompanied the study of optics), Platonic memories, neo-Platonic and neo-Pythagorean renaissance (the visible Sun representing the invisible Sun; the Sun, master and king of the visible Universe, and hence symbolic of God —this is the conception perfectly expressed by Marsilio Ficino in his Hymn to the Sun), these traditions alone are capable of explaining the emotion with which Copernicus speaks of the Sun.[26] He adores it and almost deifies it.

Those who, like Digby and Kepler and many others, have associated Copernican astronomy with a kind of Sun-worship, linking it moreover with Christianity, were by no means disloyal to the inspiration of the great Polish thinker.

Copernicus, as I have often said[27] and as others have said long before me, is not a Copernican.[28] He is not 'modern'. His Universe is not the infinite space of classical physics. It has limits, as does the Universe of Aristotle. It is certainly much larger, very much larger, so large that it is *immeasurable* (*immensum*), and yet it is finite, being contained within—and limited by—the sphere of fixed stars.[29] The Sun is at its centre; and round the Sun are set the spheres (which sustain and carry the planets), these spheres being just as real as the crystal spheres of medieval cosmology. The spheres revolve because of their shape and carry along with them the wandering planets which are set in them like jewels. It is splendid orderliness, luminous astro-geometry, magnificent cosmo-optics which replace the astro-biology of Aristotle.

The ways of the Mind like the ways of God are mysterious and remarkable. Nothing seems more remote from our own knowledge than the vision of the Universe held by Nikolaus Copernicus. Without it, however, our knowledge would not have had its being.

Appendix†

'. . . In particular, I marvel at the kindness of two most distinguished men towards me, seeing that I readily recognize how slight is my scholarly attainment when measured according to my own abilities.

'One of them is the most illustrious prelate . . . the Most Reverend Tiedemann Giese, Bishop of Kulm. His Reverence acquired with deep devotion all the Virtues and doctrinal knowledge required of a bishop by Paul. He realized that it would of no little importance for the glory of Christ to have a proper calendar of the feasts[1] of the Church and a correct theory explaining the motions [of the heavenly bodies]. He did not cease urging my teacher, whose accomplishments and knowledge he had known for many years, to take up this problem, until he had persuaded him to do so. Now, as my teacher was of a sociable nature and saw that the world of learning also had need of an improvement of the motions, he willingly yielded to the entreaties of his friend, the reverend prelate. He promised that he would prepare astronomical tables with new rules, and [said] that if his work had any value he would not keep it from the world, as was done by Johann Angelus,[2] amongst others. He had long been aware, however, that the observations, taken by themselves, required [for a correct interpretation] hypotheses that would overthrow the ideas concerning the order of the motions and the spheres which had hitherto been discussed and promulgated, and which were commonly accepted and believed to be true; furthermore, the hypotheses in question contradicted our senses.

† *Narratio Prima . . . Borussiae Encomium.* E. Rosen has pointed out in his *Three Copernican Treatises* (New York, 1939, p. 188, n. 243; 2nd edition, 1960) that this *Encomium*, though published in the two original editions of the *Narratio* of 1540 and 1541, disappeared from the editions issued as appendix to *De Revolutionibus Orbium Coelestium*; it is not present in the Bâle edition (1566), nor in those of Warsaw (1854) and Toruń (1873). On the other hand, it is present in the edition of the *Narratio Prima* appended by M. Maestlin to Keplers' *Mysterium Cosmographicum* (Tübingen, 1596, and Frankfurt, 1621). It is also included in the re-issue of the *Mysterium Cosmographicum* by W. von Dick and Max Caspar in Johann Kepler, *Gesammelte Werke*, München, 1938, vol. I, pp. 125 f. The passage quoted above occurs on pp. 129 f.

'He decided, therefore, to imitate the *Alfonsine Tables* rather than Ptolemy, and to prepare tables with accurate rules, but without proofs. In that way he would provoke no dispute among philosophers: ordinary mathematicians would have an accurate [method] for computing the motions; but true scholars, on whom Jupiter had bestowed a particularly favourable look, would easily arrive, from the numbers set down [in the Tables], at the sources and principles from which all was deduced. In the same way, that hitherto, scholars had to deduce the hypotheses of the true motions of the starry sphere from the Alfonsine doctrine, so the whole system would be as clear as crystal to learned men. Yet the ordinary astronomer would not be deprived of the use of the tables which he seeks and desires apart from all theory. So, the Pythagorean principle would be observed that philosophy must be practised in such a way that its intimate secrets are reserved to the learned, well versed in mathematics, etc.

'His Reverence then remarked that such a work would be an incomplete gift to the world, unless my teacher set forth the reasons for his tables, and in imitation of Ptolemy included also the system or theory as well as the foundations and proofs on which he relied when investigating the mean motions and prosthaphaereses, and establishing the epochs as initial points in the computation of time. The bishop, furthermore, stressed the fact that the procedure of the *Alfonsine Tables* had produced great inconvenience and many errors, seeing that we were obliged to assume and approve their ideas in virtue of the principle that (as the Pythagoreans were wont to say) "The Master said so": a principle which has no place whatsoever in mathematics.

'Moreover, said the bishop, as the principles and hypotheses are diametrically opposed to the hypotheses of the ancients, it will be difficult to find among the learned hereafter anyone who [when the tables have been published] will undertake to examine their principles and to publish them after the tables have been accepted as being in agreement with the truth. There was no place in matters of learning, he asserted, for the practice frequently adopted in kingdoms, conferences and public affairs, where for some time plans are kept secret until the subjects see the beneficial results, and then remove from uncertainty the hope that they will approve the plans.

'As far as philosophers are concerned, he continued, those who are possessed of a more penetrating judgement and greater knowledge would carefully study Aristotle's lengthy discussions, and would note that, after having satisfied himself of the immobility of the Earth by numerous proofs, Aristotle finally takes refuge in the following argument[3]: "We have evidence for our view in what the mathematicians say about astronomy. For the phenomena observed as changes take place in the figures by which the arrangement of the stars is marked out occur as they would on the assumption that the Earth is situated at the centre."

'Accordingly the philosophers would then say: "If this concluding statement by Aristotle cannot be linked with his previous discussion, we shall be compelled, unless we are to waste the time and effort we have invested, rather to assume the true basis of astronomy. Moreover, we must work out appropriate solutions for the remaining problems under discussion. By returning to the principles with greater care and equal assiduity we must determine whether it has been proved that the centre of the Earth is also the centre of the Universe. If the Earth were raised to the lunar sphere, would loose fragments of earth seek, not the centre of the Earth's globe, but the centre of the Universe, inasmuch as they all fall at right angles to the surface of the Earth's globe? Again, since we see that the magnet by its natural motion turns north, would the motion of the daily rotation or the circular motions attributed to the Earth necessarily be violent motions? Further, can the three motions, away from the centre, toward the centre, and about the centre, be in fact separated? We must analyze other views which Aristotle used as fundamental propositions with which to refute the opinions of the *Timaeus* and the Pythagoreans."[4] If they [the philosophers] desire to understand the principal end of astronomy and the power and efficacy of God and Nature they will meditate on these questions and on others of the same kind.

'If the intention and decision of scholars everywhere be to hold passionately and insistently to their own principles, His Reverence said, then my teacher should not expect a more fortunate fate than that of Ptolemy, the king of this science. Averroes, who was on the whole a philosopher of the first rank, concluded that epicycles and eccentrics could not possibly exist in nature, and that Ptolemy did not know why the Ancients had introduced motions of rotation. His final judgement is: "Ptolemaic astronomy is nothing, as far as existence is concerned; but it is convenient for computing the non-existent."[5] As for those ignorant of learning, whom the Greeks call "those who know not theory, music, philosophy and geometry",[6] their cries should be ignored, because men of good-will do not undertake any labours for their sake.

'By such arguments, and many others, as I learned from friends well acquainted with this whole affair, the learned prelate drew from my teacher a promise to allow scholars and posterity to pass judgement on his work. Consequently, men of good-will and students of mathematics will be deeply grateful with me to His Reverence, the Bishop of Kulm, for having given this masterpiece to the world.'

NOTES

1. Rheticus alludes here to the scheme for reforming the calendar. Copernicus, at the end of his letter of dedication to Pope Paul III, writes: "Not long since, under Leo X, the question of reforming the ecclesiastical calendar was debated

in the Lateran Council [1512–17]. It was left undecided solely because the lengths of the year and the months, and the motions of the Sun and the Moon were held not to have been determined with sufficient accuracy. From that time, I have been minded to study these matters in a more exact manner, being encouraged by the most eminent Paul, Bishop of Fossombrone, who was sometime in charge of these deliberations." (*De Revolutionibus Orbium Coelestium*, TH., p. 11; P., pp. 48 f.) Copernicus, as we know, had not waited for the Lateran Council and Bishop Paul's exhortation, but, no doubt, he thought it was the proper thing to say to Pope Paul III.

2. For Johann Angelus, cf. Gessner, *Bibliotheca Universalis*, pp. 382 f., and *Allgemeine Deutsche Biographie*, vol. I, p. 457.

3. *De Coelo*, II, 14. The translation is from Thomas L. Heath, *Greek Astronomy*, London, 1932, p. 91.

4. The doctrine of the Earth's rotation has frequently been attributed to Plato. It may be mentioned in passing that it is rather strange that Copernicus never mentioned him as one of his precursors, whereas Rheticus does (cf. *Narratio Prima*, p. 469). It could be explained, however, if we believe that in his view the Earth's orbital motion was more important than its axial rotation. Furthermore, it can be accepted that he associated Platonism especially with the Florentine Academy.

5. Averroes, *Commentarium supra Metaphysicam Aristotelis*, Lib. XII, cap. iv, no. 45, Padua, 1473. Cf. E. Rosen, *Three Copernican Treatises*, p. 195, n. 251.

6. Aulus Gellius, *Noctes Atticae*, I, ix, 8. Cf. E. Rosen, *op. cit.*, p. 195, n. 252.

Notes

INTRODUCTION

1. Nicolai Copernici Torinensis, *De Revolutionibvs Orbium Coelestium, Libri VI.* Norimbergae apud Ioh. Petreium, Anno M.D.XLIII. A list of various editions of *De Revolutionibus* is given in note 4, p. 95. Quotations from that of Thorn (1873) are designated by TH.; those from my edition of Book I of *De Revolutionibus*, (*Des révolutions des orbes célestes*, Paris, 1933), are designated by P.

2. It is interesting to note that *De humana corporis fabrica* by Andreas Vesalius was published in this same year, 1543.

3. Not unexpectedly we find that the heliocentric system had already been expounded by the Greeks. According to some historians this was the first system to be put forward by Greek astronomy. (Cf. G. V. Schiaparelli, *I precursori di Copernico nell' Antiquita, La sfere omocentrice di Eudosso. . .*; *Origine dell sistema planetario eliocentrico presso i Greci. . . .* These studies date from 1873 and were published with later ones in *Scritti sulla storia dell'astronomia antica*, 3 vols., Bologna, 1925–27; P. Tannery, *Recherches sur l'histoire de l'astronomie ancienne*, Paris, 1893; J. L. E. Dreyer, *History of the Planetary Systems, from Thales to Kepler*, Cambridge, 1916; Sir Thomas L. Heath, *Aristarchos of Samos*, London, 1913; Pierre Duhem, *Le Système du Monde*, tome I, cap. I and cap. VII, Paris, 1913; P. Brunet and A. Mieli, *Histoire des Sciences: Antiquité*, Paris, 1935.) Nevertheless, the system was never accepted by the science of antiquity, (a) because it was never developed and elaborated sufficiently to become an astronomical theory capable of practicable application, and (b) because it was never able to answer the physical objections against the motion of the Earth. On the importance of part played by these two factors see my *Études Galiléennes* (Paris, 1940) and *Galilée et la révolution scientifique du XVIIe siècle* (Conf. du Palais de la Découverte, 1953). Cf. *infra*, p. 161.

4. It is rather strange that Arthur Koestler in his book *The Sleepwalkers: A history of man's changing vision of the universe* (London, 1959) reproaches Copernicus, whom he calls *the timid canon*, with lack of courage and intellectual boldness!

5. Galileo praises him also (*Dialogo sopra i due massimi sistemi del Mondo.* Edizione Nazionale, vol. VIII, p. 355; cf. *infra*, p. 109, n. 30) for

having (together with Aristarchus of Samos) overcome the evidence of his senses to the advantage of his faculty of reasoning.

6. Biblical authority was a powerful obstacle to the diffusion of the heliocentric theory, and prevented its general acceptance (at least in teaching) until the beginning of the eighteenth century. Nevertheless, it would be wrong to regard it as the only, or even the chief, obstacle as does, for example, Andrew White in his well-known *History of the Warfare of Science with Theology* (New York, 1895). In fact, at the time of Copernicus, and down to the time of Descartes and Galileo, that is to say, until the establishment of the new science, the opposition of Aristotelian philosophy and physics constituted an equally, if not more, powerful obstacle.

7. No doubt it could be objected that a century before Copernicus, in 1440, Nicholas of Cusa in *De docta ignorantia* (II, 17) had already proclaimed 'the Earth is a noble star' (*terra est stella nobilis*), and had removed it from the centre of the Universe, declaring, moreover, that this centre has no existence, seeing that the Universe is 'an infinite sphere having its centre everywhere and its circumference nowhere'; and it could be maintained that the work in question was probably known to Copernicus, whose mind could have been influenced by it (R. Klibansky, 'Copernic et Nicolas de Cues' in *Léonard de Vinci et l'expérience scientifique du XVIe siècle*, Paris, 1953). I do not dispute it. Yet, it is nonetheless true that the metaphysically very bold concept of Nicholas of Cusa, namely that of an undefined, if not infinite, Universe, was not accepted by Copernicus, nor by anyone else before Giordano Bruno; that his cosmology, scientifically speaking, is non-existent; and if he attributed any motion to the Earth, he does not endow it with any motion round the Sun. On the whole, his astronomical notions are so vague, and often so erroneous (for example, he endows both the Moon and the Earth each with its own proper light) that Nicholas of Cusa cannot by any means be ranked among the forerunners of Copernicus (except in dynamics); nor can he claim a place in the history of astronomy. Cf. Ernst Hoffmann, 'Das Universum des Nicolaus von Cues' (Cusanus-Studien I, *Sitzungsberichte der Heidelberger Akademie der Wissenschaften*, Philosophish-historische Klasse, 3 Abt., Heidelberg, 1929–30), and especially the *Textbeilage* by R. Klibansky (p. 41); cf. also my *From the Closed World to the Infinite Universe*, Baltimore, 1957. As for Nicole Oresme, whom P. Duhem wished to make 'a forerunner of Copernicus' (cf. *infra*, p. 110, n. 4), he undoubtedly admitted the possibility of the Earth's rotation on its axis (without believing it to be a fact), but he never envisaged the possibility of the Earth's orbital motion. Now, it is precisely this assertion of orbital motion that constitutes the Copernican revolution.

8. It is to be noted that geocentrism in no way implies an anthropocentric concept of the Universe; neither Aristotle nor the Stoics held that view. It is only in the Christian tradition that a connection between these

two is established and asserted, for in this tradition Earth is the place where the divine-cosmic drama—the Fall, Incarnation, Redemption—unfolds itself, and which alone gives meaning to the creation of the Universe.

9. Georg Joachim Rheticus, the only follower of Copernicus during his lifetime, emphasizes the importance of mathematics or geometry for the astronomer in his *Narratio Prima* (*De Libris Revolutionum Eruditissimi Viri et Mathematici excellentissimi reverendi D. Doctoris Nicolai Copernici Torunaei Canonici Vuarmaciencis. . . .*, *Narratio Prima*, Gedani, 1540; 2nd edition, Basileae, 1541); and, no doubt with the approval of his master, he put on the title-page of *De Revolutionibus* the famous adage which, according to tradition, was placed above the portal of the Academy —'Let no-one ignorant of geometry enter here' (*ἀγεωμετρητος ὀυδεὶς ἐισίτω*). This was not an innovation: astronomy has always been considered as auxiliary to mathematics. It may be recalled that Ptolemy's great work, known under the Arabic title of *Almagest*, has for its proper title: *ΜΑΘΗΜΑΤΙΚΗ ΣΥΝΤΑΞΙΣ* (*The Mathematical Composition*). The *Narratio Prima* is a very clever introduction to the work of Copernicus, and has accompanied most editions of *De Revolutionibus Orbium Coelestium*. There is an excellent English translation by E. Rosen in *Three Copernican Treatises*, New York, 1939; 2nd edition, 1960.

10. For the history of the spread of Copernicanism, see E. Zinner, 'Entstehung und Ausbreitung der Coppernicanischen Lehre' (*Sitzungsberichte der Physikalisch-medizinischen Sozietät zu Erlangen*, vol. LXXIV, 1943).

11. In passing, it may be mentioned that this was the idea Cardinal Bellarmino had formed of Copernicanism. Cf. 'Roberto Bellarmino a Paolo Antonio Foscarini', 12 IV 1615 (Galileo Galilei, *Opere*, Ed. Nazionale, vol. XII, p. 171):

> '*Dico che mi pare che V. P. et il Sig*[r]. *Galileo facciano prudentemente a contentarsi di parlare* ex suppositione *e non assolutamente, come io ho sempre creduto che habbia parlato il Copernico. Perchè il dire, che supposto che la terra si muova et il sole stia fermo si salvano tutte l'apparenze meglio che non porre gli eccentrici e epicicli, è benissimo detto, e non ha pericolo nessuno; e questo basta al mathematico: ma volere affermare che realmente il sole stia nel centro del mondo, e sole si rivolti in sè stesso senza correre dall'oriente all'occidente, e che la terra stia nel 3*[e] *cielo e giri con somma velocità intorno al sole, è cosa multo pericolosa non solo d'irritare tutti i filosofi e theologi scholastici, ma anco di nuocere alla Sante Fede con rendere falsa le Sante Scritture.*'

It is interesting to note that Bellarmino, as well as Kepler, considered the Copernican doctrine to imply rotation of the Sun.

12. E. F. Apelt in *Die Reformation der Sternkunde* (Jena, 1852, p. 166), which is still full of interest though published so long ago, writes:

> '*Denn die neue Lehre des katholischen Domherrn, ein echt deutsches Gewächs wie der Protestantismus selbst, ist vorzugsweise von Protestanten gepflegt und ausgebildet worden. . . . Das Schicksal der neuen Astronomie war von da an gewissermassen an das Schicksal des Protestantismus gefesselt.*'

Nothing is further from the truth than this assertion which only shows to what extent national and religious prejudices can distort the judgement of a very experienced historian. In fact, although Rheticus and Kepler were Protestants (and so was Tycho Brahe), protestantism had as much as germanism to do with the new astronomy, and that was *absolutely nothing*. One might even claim that protestantism, far more than catholicism, was hostile to astronomy and Copernican cosmology. Indeed, whilst Copernicus was encouraged by high dignitaries of the Roman Catholic Church, such as Tiedemann Giese, Bishop of Kulm (Chełmno), Cardinal Nicolaus Schönberg, Bishop of Fossombrone, and even Pope Paul III who accepted dedication of the *De Revolutionibus Orbium Coelestium*, on the other hand, Luther and Melanchthon reacted even before the book was published. According to a famous passage in the *Tischreden* (Weimar edition, vol. I, p. 419), Luther (as related by Aurifaber) said on 4 June 1539:

> '*Es ward gedacht eines neuen Astrologi, der wollte beweisen dass die Erde bewegt würde und umginge, nicht der Himmel oder das Firmament, Sonne und Monde; gleich als wenn einer auf einem Wagen oder in einem Schiffe sitzt und bewegt wird, meinte er sässe still und ruhete, das Erdreich aber und die Bäume gehen um und bewegen sich. Aber es gehet jtzt also: wer da will klug sein, der soll ihm nichts lassen gefallen, was Andere machen, er muss ihm etwas Eigens machen, das muss das Allerbeste seyn wie ers machet. Der Narr will die ganze Kunst Astronomiae umkehren, aber wie die heilige Schrift anzeiget, so hiess Josua die Sonne still stehen and nicht das Erdreich.*'

Melanchthon in a letter, dated 16 October 1541, to Bernardus Mithobius declared that the attempt of Copernicus (whom he calls *Sarmaticus Astronomus*) 'to give motion to the Earth and bring the Sun to a standstill' is absurd, and that the propagation of such notions should not be tolerated by a wise government. Some years later, in his *Initia doctrinae physicae* (Wittenberg, 1549, pp. 60 f., 99 f.; 2nd edition, 1550) he took a definite stand against Copernicus, and used all the arguments based on common sense, Aristotelian physics, and the authority of the Scriptures. Once more, he demands drastic measures against this impious doctrine. Calvin, to be sure, never mentions Copernicus, but anti-Copernicanism was the general attitude of Protestant theologians throughout the sixteenth century. Opposition was not restricted to theologians: the *Synopsis Physicae* of Comenius, which was very popular in the seventeenth century, is violently antagonistic to the new astronomy.

There has been an attempt (Wilhelm Norlind, 'Copernicus and Luther, A Critical Study', *Isis*, vol. XLIV, fasc. 3, 1953, p. 273) to 'save' the reputation of Luther, Norlind stresses the lack of authenticity of the *Tischreden* text, which was edited by Aurifaber, who was quite capable of making his own additions to the notes of the participators; Norlind compares the quotation given above with one from Lauterbach's *Journal*, where under the date 4 June 1539 is another version of the same conversation (*Tischreden*, vol. IV, p. 142):

> '*De novo quodam astrologo fiebat mentio, qui probaret terram moveri et non coelum, solem et lunam, ac si quis in curru aut navi moveratur, putaret se quiescere et terram et arbores moveri. Aber es gehet jtzunder also: wer da will klug sein, der soll ihme nichts lassen gefallen, das andere achten; er muss ihme etwas eigens machen, sicut ille facit qui totam astrologiam invertere vult. Etiam illa confusa tamen ego credo sacrae scriptorae, nam Josua iussit solem stare, non terram.*'

In this Lauterbach version, as Norlind points out, the contemptuous term '*Narr*' is missing. So, Norlind assumes that it was interpolated by Aurifaber. This is a possibility, however, it seems more likely that the latter gave an accurate account—the grossness of Luther's language is well known—and that it was Lauterbach who dropped the term, when translating Luther's remarks into Latin. This is all the more likely as the preceding phrase, '*er muss ihm etwas eigens machen*, etc.', which is reproduced by Lauterbach, is not to be regarded as *praise* of Copernicus (p. 275), as Norlind believes, but as *censure*. Luther takes Copernicus to task for his inordinate desire for originality; Melanchthon did the same, and also accused him of copying the Ancients, especially Aristarchus of Samos. Cf. Emil Wohlwill, 'Melanchton und Copernicus', *Mitteilungen zur Geschichte der Medizin und Naturwissenschaften*, vol. III, p. 260, 1904. As far as the Roman Catholic Church is concerned, mention must be made of the rejection of Copernicanism by Christoforus Clavius (in 1570 in his *In Sphaeram Joannis de Sacro Bosco Commentarius*) and its violent condemnation by F. Maurolico (*Computus Ecclesiasticus*, 1575); cf. Edward Rosen, 'Galileo's Mis-statements about Copernicus', *Isis*, vol. 49, 1958.

13. On the condemnation of heliocentrism see the excellent work of G. de Santillana, *Le Procès de Galilée*, Paris, 1955.

THE ASTRONOMICAL REVOLUTION

CHAPTER I. FIRST OUTLINE: THE *Commentariolus*

1. Nicolai Copernici, *De hypothesibus motuum coelestium a se constitutis Commentariolus*, the first outline of the heliocentric system (cf. *infra*, n. 52) written by Copernicus some years after his return from Italy. Mentioned by Tycho Brahe (*Astronomiae Instauratae Progymnasmata, Tychonis Brahe Dani Opera Omnia*, edited by Dreyer, Copenhagen, 1913–1925, vol. II, pp. 428 f.) who received a copy from his friend the Czech physician Thaddeus Hagecius (probably in 1575) and communicated it to several mathematicians in Germany, this work remained unknown and unpublished until 1854, when it was published (very badly) in the Warsaw edition of *De Revolutionibus Orbium Coelestium* (1854). In 1877, Maximilian Curtze found a copy at Vienna, and published it in *Mittheilungen des Copernicus-Vereins für Wissenschaft und Kunst zu Thorn*, fasc. I, *Inedita Coppernicana*, Leipzig, 1878. Another copy was found in 1881 in the library of *Kungliga Svenska Vetenskaps Akademien* at Stockholm by Arvid Lindhagen, and published by him in *Bihang till K. Svenska Vet. Akad. Handlinger*, vol. VI, no. 12, 1881. A fourth, critical, edition was prepared by Maximilian Curtze (*Mittheilungen des Copernicus-Vereins*, vol. IV, 1882) and published by Leopold Prowe as an appendix to his biography of Copernicus (*Nicolaus Coppernicus*, vol. II, pp. 184–202, Berlin, 1884). A. Müller translated it into German in 1899: 'Coppernicus, Einführung in sein astronomisches Hauptwerk', *Zeitschrift für die Geschichte und Altertumskunde Ermlands*, vol. XII, 1899. An excellent English translation was published by Edward Rosen in *Osiris*, vol. III, 1937, and re-issued with informative introduction and notes in *Three Copernican Treatises* (New York, 1939). Finally, in 1949, Fritz Rossmann published it together with an introduction, translation and notes, to which was added Kepler's translation of Aristotle's *De Coelo*, Book II, cap. XIII and XIV: *Nikolaus Kopernikus, Erster Entwurf seines Weltsystems*, München, 1949. I have used this edition when quoting from the *Commentariolus*.

2. With regard to the difficulty of avoiding modification to the thought of Copernicus, and introducing ideas which are foreign to it, see the illuminating remarks of Émile Meyerson, *Identité et Réalité*, Appendix III, 'Les Coperniciens et le principe d'inertie', Paris, 1908; 5th edition, Paris, 1956. A good example of this difficulty is provided by the translation of the work of Copernicus published by the Academy of Toruń in 1877. The title of the work of the great astronomer is *De Revolutionibus Orbium Coelestium*, which means *On the revolutions of the celestial spheres*. Now, the translator of this work, C. L. Menzzer, writes: *Ueber die Kreisbewegungen der Welt-Körper*, which means *On the circular motions of the bodies of the universe*, and is not at all the same thing: in fact, Menzzer's translation implies a denial of the real existence of the planetary 'spheres', which nevertheless had been accepted by Copernicus. It is rather odd, that no-one

not even Moritz Cantor who added a preface to Menzzer's work, noticed the contradiction perpetrated by the latter. Cf. my note 'Traduttore-traditore' in *Isis* (1942) and my edition of Book I of *De Revolutionibus Orbium Coelestium, De la révolution des orbes célestes*, p. 24, n. 2, Paris, 1933. E. Rosen in his *Three Copernican Treatises* (p. 19, n. 50) points out that H. O. Taylor in his *Thought and Expression in the Sixteenth Century*, New York, 1920, writes: *The Revolution of the Celestial Bodies*, and that Flammarion in his *Vie de Copernic* does the same: *Les Révolutions des corps célestes*.

3. The notion of a 'forerunner' is a very dangerous one for the historian. It is no doubt true that ideas have a *quasi* independent development, that is to say, they are born in one mind, and reach maturity to bear fruit in another; consequently, the history of problems and their solutions can be traced. It is equally true that the historical importance of a doctrine is measured by its fruitfulness, and that later generations are not concerned with those that precede them except in so far as they see in them their 'ancestors' or 'forerunners'. It is quite obvious (or should be) that no-one has ever regarded himself as the 'forerunner' of someone else, nor been able to do so. Consequently, to regard anyone in this light is the best way of preventing oneself from understanding him.

4. The best are those of J. B. Delambre, *Histoire de l'Astronomie Moderne*, vol. I, pp. 85–142, Paris, 1821, and of J. L. E. Dreyer, *History of the Planetary Systems, from Thales to Kepler*, Cambridge, 1906; 2nd edition under the title, *A History of Astronomy*, pp. 305–344, New York, 1952, and Thomas S. Kuhn, *The Copernican Revolution*, Cambridge (Mass.), 1957; cf. also Arthur Koestler, *The Sleepwalkers, A History of Man's Changing Vision of the Universe*, London, 1959, which contains a long chapter (part III, pp. 119–219), 'The Timid Canon', on Copernicus—interesting but malicious. A bibliography of Copernicus has been published by the Polish Academy of Sciences: Henryk Baranowski, *Bibliografia Kopernikowska*, Warsaw, 1958.

The fundamental work by L. A. Birkenmajer, *Mikolaj Kopernic*, Cracow, 1900, has never been translated, unfortunately.

In my own account I have disregarded certain technical matters (observational and computational methods), as well as the problem of the latitude of planets, which at the present time even more so than in the days of Copernicus would be understood only by 'mathematicians'.

5. The best biographies of Copernicus are still those of Leopold Prowe, *Nicolaus Coppernicus*, in 2 vols., Berlin, 1883–84; and, of course, the work of L. A. Birkenmajer cited in n. 4. Cf. also the *Mikolaj Kopernic, jako uczony i tworca*, Cracow, 1923, as well as his *Stromata Copernicana*, Cracow, 1924.

6. An exception must be made in respect of the works by L. A. Birkenmajer cited above in nn. 4 and 5; Pierre Duhem, *La Théorie physique, son*

objet, sa structure, Paris, 1906; 2nd edition, Paris, 1914; *Études sur Léonard de Vinci*, 3 vols., Paris, 1909–13; as well as that by T. S. Kuhn cited in n. 4.

7. G. V. Schiaparelli, *I precursori di Copernico nell'antichita*, Bologna, 1873; German translation, *Die Vorläufer des Copernicus im Altertum*, Leipzig, 1878, p. 85: 'The great conflict between the Ptolemaic and Copernican systems revolved round the same physical and cosmological principles [as in Antiquity]. Both systems are capable of being used to represent the phenomena, the one as well as the other. From the geometrical point of view they are equivalent; they agree with each other just as well as with the eclectic system of Tycho Brahe'.

8. Much discussion has taken place on the name of Copernicus; the most fantastic etymologies have been put forward with the sole object of proving either his Germanic, or his Polish, character. In fact, as Stanislas Kot points out in 'Nationalité et culture polonaise de Copernic' (*Le Monde*, 7 April 1954), this name is truly Polish and derives from the village of Kopernic in Silesia, where the family of the great astronomer originated. The name itself comes from *koper*, meaning fennel. From this village of Kopernic, an ancestor of Copernicus migrated to Cracow, where he founded a family. Cf. 'Ille Sarmaticus Astronomus' in *Études Coperniciennes* (Académie Polonaise des Sciences et des Lettres, Centre Polonais de Recherches Scientifiques de Paris, *Bulletin*, no. 13–16, 1955–1957), pp. 233–254, and 'Le Nom de Copernic', *ibid.*, pp. 255–258.

9. The Watzelrodes—or Watzenrodes—in spite of their rather Germanic name seemed to have been good Poles (enemies of the Teutonic Order). The mother of Barbara and Lucas was a Modlibog; one of her sisters had 'married a Konopacki, who belonged to one of the most important Polish families in Prussia' (S. Kot, *ibid.*). However, cf. Johannes Papritz, 'Die Nachfahrentafel des Lukas Watzenrode', *Kopernicus-Forschungen* (Deutschland und der Osten, 1322), Leipzig, 1943.

10. National antagonism and national hatred are not entirely modern phenomena. We find them at their height during the Middle Ages in Spain, France and Bohemia. Their intensity was greatest where they depended on religious antagonism as in the Balkans, or between Poland and Russia; or where they were associated with oppositional religious movements such as that of the Hussites. Nothing of the kind is to be found in Poland; opposition to Prussia of the Teutonic Order was a political opposition, not a national opposition.

11. Particularly, if that meant that he belonged to the German *Volkstum*, not only objectively but also subjectively (*sich gefühlsmässig dem deutschen Volkstum verbunden fühlte*); and was conscious of *völkischen Gegensatzes zum Polentum*, cf. Hans Schmauch, 'Nikolaus Kopernikus ein Deutscher', *Kopernikus-Forschungen*, pp. 13, 15, 30.

12. Consequently, Copernicus had not the slightest difficulty in com-

mencing his studies at Cracow and continuing them at Bologna and Padua; it was quite the normal thing to do.

13. He wrote in German to Duke Albert of Prussia. To conclude that German was his *Umgangssprache* as is done by Hans Schmauch (n. 11, *supra, op. cit.*, p. 29) is rather absurd.

14. F. and K. Zeller, who edited *De Revolutionibus Orbium Coelestium*, note Germanisms in the Latin as written by Copernicus, and so deduce that his mother tongue was German, and conclude that Copernicus belonged to the German nation (Nicolai Copernici Thorunensis, *De Revolutionibus Orbium Coelestium libri sex*, München, 1949, p. 428); it seems to me that the conclusion rests on a very dubious foundation.

15. Giordano Bruno, however, calls him 'German'. Cf. E. Cassirer, *Individuum und Cosmos in der Philosophie der Renaissance*, Leipzig, 1927, p. 197.

16. The writers in *Kopernikus-Forschungen* (cited *supra*, n. 9) claim even Albert de Brudzewo for the German *Volkstum.*

17. *Commentariolum super theoricas novas planetarum Georgii Peurbacchii* per Mag. Albertum de Brudzewo, Mediolanum, 1494 and 1495; the *Commentariolum* was re-edited by L. A. Birkenmajer, Cracow, 1900.

18. The biographies of Copernicus agree in saying that Albert de Brudzewo gave him private lessons. It was possible, of course; cf. E. Brachvogel, 'Nikolaus Kopernikus in der Entwicklung des deutschen Geisteslebens' in *Kopernikus-Forschungen*, p. 93. In spite of some nationalistic exaggerations, which are accounted for by the date—1943!—Brachvogel's work is a serious and learned contribution.

19. The Diocese of Warmia (Ermland), first of all under the protectorate of the Teutonic Order, and afterwards (from 1466) under that of the kings of Poland, was a practically autonomous entity, whose bishop was both the temporal and spiritual head.

20. The Statutes of Frauenburg required Canons to have studied theology, law and medicine. Cf. F. Hipler, *Spicilegium Copernicanum*, Braunsberg, 1873, p. 261.

21. As Hans Schmauch does (n. 11, *supra, op. cit.*, p. 13).

22. Amongst the four 'nations' of the University of Paris we find the *nation picarde* and the *nation burgonde*. At the University of Bologna, which was differently constituted from that of Paris (the former was a *Universitas Scholarum* and the latter a *Universitas Magistrorum*), there were fourteen 'nations', namely, Gallic, Portuguese, Provençal, Burgundian, Savoyard, Aragonese (which included Catalonian also), Navarrese, German, Hungarian, Polish, Bohemian and Flemish; cf. H. Rashdall, *The Universities of Europe in the Middle Ages*, Oxford, 1895, vol. I, pp. 158 f.

23. G. J. Rheticus says so (*Narratio Prima*, Thorn edition of *De Revolutionibus Orbium Coelestium*, 1873, p. 448). What is certain, is that he made astronomical observations.

24. He even translated into Latin the Greek *Epistles* of Theophylactus Simocatta, which he dedicated to his uncle, and had printed in 1509.

25. The Neo-Platonic and Pythagorean inspiration—the Renaissance mentality—breaks forth even in the style of Copernicus, who speaks of the 'revolutions of the divine Universe', and refers to Pythagoras at every turn.

26. Although he had studied medicine, he did not receive his doctorate.

27. The Bishopric of Warmia (Ermland) occupied a very important political, as well as military, position; and its bishops were obliged on occasions to become soldiers. Copernicus had perforce to concern himself with putting the towns of his diocese into a proper state of defence, especially Allenstein (Olsztyn) which was attacked by the troops of Albert of Hohenzollern in 1620.

28. In 1519 Copernicus drew up a report (presented to the Diet of Graudenz in 1622) on means of restoring the value of currency greatly debased by minting coins in inferior alloy. This report, enlarged and carefully elaborated, became the small treatise *De Monetae cudendae ratio* (1628). There is a French translation: N. Copernic, 'Discours sur le frappe des monnaies . . .' in *Écrits notables sur la monnaie: XVIe siècle*, vol. I, Paris, 1934.

29. Copernicus the physician seems to have enjoyed a high reputation. For example, in 1541, Albert of Hohenzollern asked him to treat one of his counsellors, Georg von Kunheim. Copernicus went to Königsberg, and stayed there until the patient had recovered.

30. His astronomical instruments included a quadrant, an armillary sphere and a *triquetrum*: this last named instrument eventually came into the possession of Tycho Brahe.

31. E. Brachvogel, 'Nikolaus Koppernicus und Aristarch von Samos' in *Zeitschrift für die Geschichte Ermlands*, Braunsberg, 1935, vol. XXV, p. 67, which refers to the work of L. Birkenmajer; see also E. Zinner, *Entstehung und Ausbreitung*. . . , cited on p. 73, n. 10.

32. On the other hand, he made use of certain observations of Bernhard Walther, friend and pupil of Regiomontanus, as well as some observations of Johann Schöner.

33. E. Brachvogel, *op. cit.*, pp. 66 f., and E. Zinner, *op. cit.*, p. 218. Cf. Appendix.

34. *Narratio Prima*, p. 485. Quotations from the *Narratio Prima* are from the edition of *De Revolutionibus Orbium Coelestium* published at Thorn; concerning G. J. Rheticus, cf. *infra*, p. 90, n. 1. The text of Rheticus is so characteristic that I shall quote it *in extenso*: 'The divine Plato, master of wisdom as Pliny styles him, affirms not indistinctly in the *Epinomis* that astronomy was discovered under the guidance of God. Others perhaps interpret this opinion of Plato's otherwise. But when I see that my teacher always has before his eyes the observations of all ages together with his

own, assembled in order as in catalogues; then when some conclusion must be drawn or contribution made to the science and its principles, he proceeds from the earliest observations to his own, seeking the mutual relationship which harmonizes them all; the results thus obtained by correct inference under the guidance of Urania he then compares with the hypotheses of Ptolemy and the Ancients; and having made a most careful examination of these hypotheses, he finds that astronomical proof requires their rejection; he assumes other hypotheses, not indeed without divine inspiration and the favour of the gods; by the application of mathematics he geometrically establishes the conclusion which can be drawn therefrom by correct inference; then he harmonizes the ancient observations and his own with the hypotheses which he has adopted; and when he has completed all these operations, he finally sets down the laws of astronomy.'

35. E. F. Apelt (*Die Reformation der Sternkunde*, Jena, 1852, p. 150), indeed, exaggerates when he says:

> 'If it be asked: What practical advantages has astronomy derived from the Copernican system? we should reply: Absolutely none, at the time! The Copernican system, in the form in which it left the hands of its author, is no more in agreement with the heavens than that of Ptolemy.'

In fact, the theory of Copernicus simplified calculation by eliminating a certain number of useless circles, and in particular greatly improved lunar theory, though to a much less degree that of Mars. His methods were used by Erasmus Reinhold in his *Tabulae Prutenicae* (1551). Furthermore, by combining his own observations with those of his predecessors, Copernicus was able to give far greater accuracy to certain astronomical constants. However, 'the Copernican revolution' did not depend on perfecting *astronomical methods*, but on establishing a new cosmology.

36. The *Tabulae Prutenicae* of Erasmus Reinhold (1551) have errors amounting to 4 or 5 degrees, cf. J. Kepler, *Astronomia Nova* . . . ., ed. C. Frisch (*Opera Omnia*, Frankfurt-am-Main, vol. III, 1860), pp. 146 f.

37. In a famous passage in his *Ephemerides* (1551), G. J. Rheticus records that Copernicus told him that he would be as happy as Pythagoras must have been after discovering the theorem bearing his name could he attain an accuracy of ten minutes in his (Copernicus) astronomical observations. On this matter it must be mentioned that E. Zinner (*Entstehung und Ausbreitung* . . ., pp. 241 f.) does not give much credence to the accuracy of this report by Rheticus.

38. The superposition of circles is equivalent to combining trigonometrical series. Cf. J. L. E. Dreyer. *A History of Astronomy*, New York, 1952, pp. 196 f., 201.

39. In the Ptolemaic system the centres of the planetary epicycles move with constant angular velocity not with respect to the centre of the deferent,

but with respect to an eccentric point, *punctum aequans*, as if they had uniform motion on a circle with its centre at the equant. (Cf. *infra*, n. 54.) It was precisely this violation of the principle of uniformity of celestial motion (it was in effect an abandonment of the principle) against which Copernicus protested.

40. Cf. J. L. E. Dreyer, *A History of Astronomy*, New York, 1952, pp. 201 f.

41. E. F. Apelt, *Johann Keplers Astronomische Weltansicht*, Leipzig, 1852, p. 11: 'Dieser Teil seines Lehrgebandes [first inequality] ist im Grunde weiter nichts als eine Uebertragung der ptolomäischen in die veränderte Konstruktionsweise der heliozentrischen Hypothese.'

42. In fact, since the establishment of mathematical astronomy by Hipparchos and Apollonius. Cf. P. Duhem, *Le Système du Monde*, Paris, 1913, vol. I, pp. 434 f.

43. The first attempt to do this was made by Ptolemy himself in *Ὑποθέσεις τῶν πλανωμένων* (French translation by N. Halma, *Hypothèses des Planètes*, Paris, 1820), where he adopts a system of *real* spheres enclosed one within another, thus providing a physical basis for planetary motion. The Peripatetic, Adrastos of Aphrodisias, and the Platonists, Derkyllides and Theon of Smyrna, had previously done the same thing for the astronomy of Hipparchos simply by using concentric spheres. Cf. P. Duhem, *Le Système du Monde*, Paris, 1915, vol. II, pp. 80, 87 f. The last astronomers (and the fact is perhaps not without importance) who, by inspiration from Arab sources, replaced the abstract circles of the *Almagest* by solid spheres were Peuerbach and Regiomontanus, whose works Copernicus possessed, and with which he was familiar.

44. [The passages dealing with Ptolemy's views on the nature of astronomical hypotheses, and on the neo-Aristotelian view of astronomy, would undoubtedly have been revised by the author had he lived a few years longer. The medieval system of planetary spheres, nested so as to fit exactly between the fixed stars and the highest sub-lunary element (fire), has often been called the 'Ptolemaic System', but until recently the earliest known reference to the system was by Proclos (fifth century), who did not mention Ptolemy by name. In an article included in a volume dedicated to Koyré himself,* W. Hartner argued ingeniously that Heiberg's edition of Ptolemy's *Planetary Hypotheses* was incomplete, and that it lacked a passage on planetary sizes and distances. B. R. Goldstein soon afterwards re-examined the Arabic manuscripts which Heiberg's collaborators had used as the only source of much of the text, and found that Hartner's predictions were remarkably accurate. An edition and translation of the missing part has been published by Goldstein,† and this contains con-

* *Mélanges Alexandre Koyré*, Paris, 1964, pp. 254–282.

† 'The Arabic version of Ptolemy's *Planetary Hypotheses*', *Transactions of the American Philosophical Society* (N.S.), vol. LVII, part 4, 1967.

clusive evidence that the system of nesting spheres was indeed Ptolemy's. Ptolemy's sentiments on the reality of the spheres is as much a matter for discussion as ever, of course, but it is necessary to point out that the evidence available to Koyré was incomplete. (Editoral addition.)]

In Ptolemy's *Almagest* the reality of the circles is never positively stated; they are no more than mathematical expedients. The contrary views of mathematicians and philosophers ended finally in purely pragmatic and phenomenalistic epistemology, which, despairing of being able to ascertain the true motions of celestial bodies, assigned to astronomy the sole task of establishing a system of computation for predicting and methodizing phenomena. The famous Platonic injunction *σώξειν τά φαινόμενα*, *salvare apparentias*, which meant initially: expose the true nature of what appears, changed its meaning and became the motto of a science which renounced knowledge of reality and contented itself with appearances. For the history of this discussion between 'realists' and 'positivists', which began with Proclos and Simplicios, continued in Arabian science and Latin scholasticism, and involved some surprising changes and alliances (for example, the Averroists who resolutely denied the reality of the astronomical circles, and through their devotion to Aristotle consequently adopted a phenomenalistic epistemology *for astronomers*, found themselves allied with the nominalist Occamites who denied the objective value of natural science as such in favour of revelation, and not Aristotle), cf. P. Duhem, *La Théorie Physique*, Paris, 1908; *Le Système du Monde*, Paris, 6 vols, 1913–17, 2nd edition, 1955; F. Boll, 'Die Entwicklung des astronomischen Weltbildes im Zusammenhang mit Religion und Philosophie' in Paul Hinneberg, *Die Kulter der Gegenwart*, Teil III, Abt. III, Leipzig, 1921; J. L. Heiberg, 'Geschichte der Mathematik und Naturwissenschaften in Altertum' in W. Otto, *Handbuch der Altertumswissenschaft*, München, 1925; Étienne Gilson, *La Philosophie au Moyen Age*, 2nd edition, Paris, 1944; and A. C. Crombie, *Robert Grosseteste and the Origins of Experimental Science*, Oxford, 1953.

Copernicus, naturally, could not fail to have been aware of these discussions, seeing that Padua was the centre of Averroism. It is quite possible that the attitude of the Averroists strengthened his desire to reform astronomy by basing it on the discovery of the true laws of real motions. In fact, the attitude of the ancient and medieval 'positivists', whose doctrine is usually modernized and misinterpreted by modern positivist historians, does not depend on the *adoption* of a new scientific ideal, but in *despair* or *renunciation* of the possibility of attaining to truth, whether it be in a particular field of reality such as astronomy, or (in the case of sceptics and nominalists) in all branches of natural knowledge; this attitude resulted from the fact that the essence of things and their real causative relationships remain inaccessible to our understanding. Ancient and medieval positivism always entails devaluation of any branch of knowledge which deals only with phenomena (appearances) as compared with one that treats of reality. Consequently, it is the opposite of modern positivism which denies, not the cognizability, but the very existence of a universe of realities underlying appearances, and which glories in its lack of realism.

45. This reference to the Pythagoreans is very curious and most significant; cf. *infra*, p. 38 and n. 17, and Appendix, p. 68.

46. *De Revolutionibus Orbium Coelestium*, TH., p. 4, P., p. 38. If we take Copernicus literally, that would bring us to about the year 1506.

47. In his *Commentariolus de hypothesibus motuum coelestium a se constitutis*, edited by Rossmann, pp. 11 f., Copernicus writes: 'Let no-one suppose that I have gratuitously asserted with the Pythagoreans the motion of the Earth; strong proof will be found in my exposition of the circles.'

48. The *De Revolutionibus Orbium Coelestium* is modelled on the *Almagest*. Rheticus says so, and repeats it. Thus, *Narratio Prima*, p. 447: '*D. Doctor Praeceptor meus sex libros conscripsit, in quibus ad imitationem Ptolemaei singula mathematicos, & Geometrica methodo, docendo & demonstrando, totam Astronomian complexus est.*'; and *ibid.*, p. 458: '*Cum in principio nostrae Narrationis praemiserim D. D. Praeceptorem suum opus ad Ptolemaei imitationem instituere . . .*'; cf. *supra*, p. 29. It was intended originally that *De Revolutionibus Orbium Coelestium* should contain eight books, but in the course of preparing it, Copernicus reduced the number to six. Book I contains a general exposition of the system of the Universe, accompanied by a treatise on trigonometry; Book II contains an exposition of spherical astronomy, accompanied by a star catalogue in the preparation of which Copernicus used ancient data as well as more recent observations, and recalculated the fundamental elements of motion (length of year, precession of the equinoxes, etc.); the remaining Books III to VI give detailed theories of the apparent and real motion of the Sun, Earth, Moon and the planets.

Paradoxically, Copernicus did not spend about twenty years in the elaboration of definite theories of planetary motion, but in the simple assertion of heliocentrism, which has played a revolutionary rôle in the history of thought, but only at the cost of abandoning the said definite theories. No doubt, heliocentrism as revived by Copernicus would have had no more success than the heliocentrism of the Ancients had it not been accompanied by proof (by practical demonstration) of the possibility of making it the basis of a system for computation.

49. The observations of 1532 relate to a determination of the apogee of Venus. Copernicus noted them on a piece of paper which he inserted in his copy of *Tabula Directionum* by Regiomontanus; cf. preface to the Toruń edition of *De Revolutionibus Orbium Coelestium*, p. XVII, and M. Curtze, *Reliquae Copernicanae*, p. 29. Cf. also J. L. E. Dreyer, *A History of Astronomy*, p. 310. However, a subsequent 'brushing up' of the manuscript could have lasted right up to the time of publication. One could even go further

and say, with E. Zinner (*Entstehung und Ausbreitung* . . ., p. 239), that the work was never really finished, seeing that it lacks the general conclusion to which reference is made in the body of the text. According to A. Koestler, the book was not finished because the result of his work, namely, the great complexity of the system disappointed and discouraged its author. As a result, *De Revolutionibus Orbium Coelestium* is 'an unreadable and unread book' (*The Sleepwalkers*, p. 215); unread even by the historians of Copernicus (*ibid.*).

50. Cf. *supra*, n. 1.

51. The discovery and publication of the *Commentariolus* have given rise to numerous discussions concerning its title and date of composition. As regards the date, M. Curtze (*Mittheilungen* . . ., etc.) considers it to have been written *after De Revolutionibus Orbium Coelestium*, between 1533 and 1539; and J. L. E. Dreyer (*A History of Astronomy*, p. 316) is in close agreement (after *De Revolutionibus Orbium Coelestium*, but before 1533). On the other hand, L. Birkenmajer considers it to have been written *before De Revolutionibus Orbium Coelestium*, and even gives an exact date—1512—(*Mikolaj Kopernic*, p. 79). The question has now been decided in favour of L. Birkenmajer who discovered in the catalogue of his (Copernicus) library, made, 1 May 1514, by the historian Matthias de Miechow (of Cracow), the following notice: '*Item sexternus Theorice asserentis terram moveri, solem vero quiescere*', which can only refer to the *Commentariolus* (*Stromata Copernicanam*, Cracow, 1924; and A. Birkenmajer, 'Le premier système héliocentrique imaginé par Nicolas Copernic', in *La Pologne au VIIe Congrès International des Sciences Historiques*, Warsaw, 1944, vol. I, pp. 95–96). With regard to the title: Nicolai Copernici, *De hypothesibus coelestium a se constitutis Commentariolus* (identical in the two known manuscripts), M. Curtze has drawn attention to the term *hypothesis*, which is used there, and concludes that Copernicus regarded his system as 'hypothetical'. On the other hand, Leopold Prowe (*Nicolaus Coppernicus*, vol. I, p. 288; vol. II, p. 185) very rightly stresses the fact that Copernicus had never done so, and concludes that the title is not authentic; which opinion was accepted on the whole by historians of Copernicus, including Dreyer and Birkenmajer. More recently, A. Birkenmajer (*op. cit.*) has rather come round to Curtze's opinion; he defends (rightly, I believe) the authenticity of the title, but concludes (wrongly, in my opinion) that Copernicus, at the time of the *Commentariolus* (1510–15), assigned no more than an 'hypothetical value' to heliocentrism, and only at a later date came to admit its truth. Eugen Brachvogel considers this to be most unlikely ('Nikolaus Koppernicus und Aristarch von Samos', *Zeitschrift für die Geschichte und Alterthumskunde Ermlands*, 1935, vol. XXV, pp. 41 f.), the contents of the *Commentariolus* being no more 'hypothetical' than those of *De Revolutionibus Orbium Coelestium*. Hence, E. Brachvogel, relying on the aversion Copernicus had from the term 'hypothesis', proven,

as he says, by Adolf Müller, S.J., in the introduction to his translation of the *Commentariolus* (*Zeitschrift für die Geschichte und Alterthumskunde Ermlands*, 1899, vol. XII, p. 360), concludes that the title has been superadded? By whom? F. Rossmann (*Nikolaus Kopernikus, Erster Entwurf seines Weltsystems*, Munich, 1948, p. 34) gives the answer: it was Tycho Brahe! The whole of this discussion is fundamentally unsound: the term 'hypothesis' has been construed in its modern meaning, and not as it was understood by Copernicus. It has been admirably shown by E. Rosen (*Three Copernican Treatises*, pp. 22, 29, 30 f.) that Copernicus used the term as the equivalent of *principium* and *assumptio*, that is to say, in the sense of fundamental propositions of the system; consequently, the term in no way implies attribution of a 'purely hypothetical' value to the proposition in question. So contrary to the assertions of the above mentioned writers, when Copernicus speaks of the hypothesis of the Earth's motion, as he does in both the *Commentariolus* and *De Revolutionibus Orbium Coelestium*, he definitely does not mean to cast doubt on the matter (E. Rosen, *ibid.*, p. 30), any more than he casts doubt on the principle or hypothesis of circular and uniform planetary motion.

52. L. Birkenmajer (*op. cit.*, in n. 51, *supra*) has the credit of pointing out that the system described in the *Commentariolus* is 'concentro-bi-epicyclic', whilst that of *De Revolutionibus Orbium Coelestium* is 'eccentro-epicyclic'; that is to say, in the former work, planetary motions are performed by means of concentric circles each carrying two epicycles (one on the other), whilst in the latter work the concentric circle and the first of the epicycles are replaced by an eccentric circle (the deferent) which carries only one epicycle. In this way, there is a saving of one circle per planet (Fig. 9). It must be pointed out, that in both the *Commentariolus* and *De Revolutionibus Orbium Coelestium* the Sun is not at the centre of the circles.

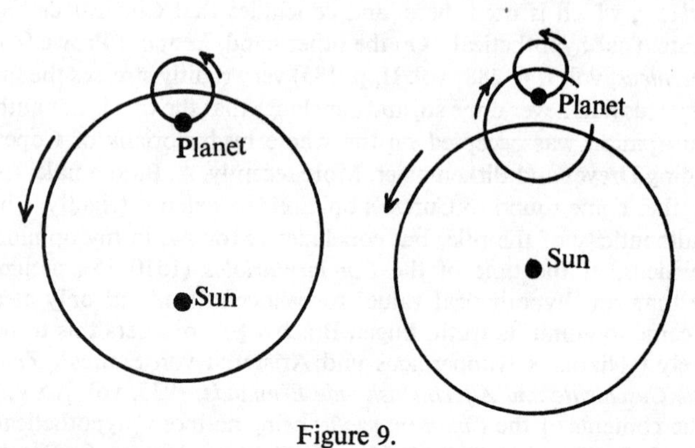

Figure 9.

Scheme in *De Revolutionibus Orbium Coelestium.* Scheme in *Commentariolus*

53. *Commentariolus*, p. 9. One might ask why Copernicus felt obliged to mention the out-of-date system of homocentric spheres used by Callippos and Eudoxos (he mentions it also in his letter to Pope Paul III). Was it because that is the system on which Aristotle modelled his own, and therefore, when speaking of 'those only use homocentric spheres', Copernicus is really aiming at the followers of Aristotle?

54. *Ibid.*, 9–10:

> '*Non enim sufficiebat nisi etiam aequantes circulos imaginarentur, quibus apparebit neque in orbe suo deferente, neque in centro proprio aequali semper velocitate sidus moveri. Quapropter non satis absoluta huiusmodi speculatio neque rationi satis concinna.*'

Greek astronomy was always faithful to a set system; namely, to represent celestial motions by a combination of uniform circular motions. However, whilst adhering to this principle, Ptolemy, in fact, considerably distorted it by considering as uniform certain motions which are hardly so, and even not so at all. These motions, though endowed with variable orbital velocity, and hence with a likewise variable angular velocity with respect to the centre of the eccentric circle on which the motion takes place, nevertheless had a constant *angular* velocity with respect to a specific point within the circumference in question, this point being called the 'point of equality'—*punctum aequans*. In the case of the exterior planets and Venus, Ptolemy placed this centre of the equant on the line passing through the Earth and the centre of the deferent circle (line of apsides) at a distance from the latter equal to its distance from the Earth (eccentricity), and hence symmetrically with respect to the former: this procedure is known technically as 'bisection of eccentricity'. This point is regarded as being the centre of a fictitious circle (the equant), equal to the deferent, on which in imagination the centre of the epicycle is placed, the arcs of which, swept out with fictitious uniform motion, provide a measure of its motion. The mechanism is more complicated for Mercury and the Moon. In the case of Mercury, the *punctum aequans* is placed between the Earth and the centre of the hypocycle of the mobile deferent of the planet; in the case of the Moon, the *punctum aequans* is itself mobile (Fig. 10).

55. Cf. *Commentariolus*, pp. 9 f.

56. This statement does not contradict by any means the third 'petition' according to which the circles (deferents) of the planets complete their revolution round the Sun *tanquam in medio omnium existentem, ideoque circa solem esse centrum mundi*, but it is contrary to the Aristotelian–Ptolemaic dogma according to which all the 'wanderers' revolve round one single centre, namely, the Earth. According to Copernicus, one only of the 'wanderers', namely, the Moon, does in effect revolve round the Earth, whilst the remainder revolve round the Sun; this means that there is not one single centre of motion, but two.

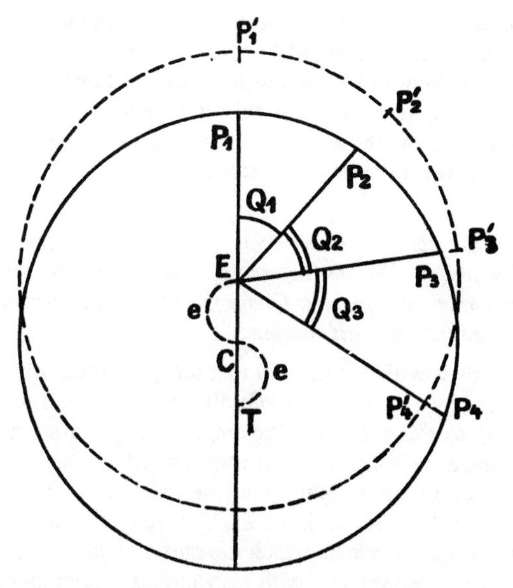

Figure 10.

T—Earth; *C*—centre of circular orbit (eccentric); *P*—centre of epicycle of planet; *E—punctum aequans; $P_1P_2P_3$*—deferent circle; $P'_1P'_2P'_3$—equant circle. Ptolemy regards the motion of the centre of the epicycle of the celestial body on the deferent as uniform when the segments P_1–P_2, P_2–P_3, P_3–P_4, etc., swept out in equal times are such that the angles Q_1, Q_2, Q_3, etc., are equal also; or more exactly, when the segments P'_1–P'_2, P'_2–P_3, P'_3P'–$_4$, etc., of the imaginary circle $P'_1P'_2P'_3$... (the equant circle), by which they are measured, are equal. Copernicus objected to this extension to the concept of uniformity in circular motion, because he considered it to be unwarranted and even slightly dishonest; he was unwilling to admit into his celestial mechanics anything other than *truly* uniform motions.

57. Taken literally this statement is incorrect. In the system described in the *Commentariolus*, the centre of the terrestrial circle is the common centre of the planetary circles. Now, the centre of the *orbis magnus* does not coincide with the Sun, but is displaced from it by an amount equal to one-twenty-fifth of its radius. The eccentricity was inevitable (it corresponds to the eccentricity of the circle of the Sun in the systems of Hipparchos and Ptolemy) in view of the non-uniformity of the Sun's apparent motion. It cannot be avoided except by endowing the Earth (or the Sun) with an epicycle. As Kepler was to remark at a later date, this was something that no-one could make up his mind to do. Cf. chap. V, n. 19.

58. Besides orbital motion and diurnal rotation about its axis, the Earth,

according to Copernicus, has a third motion, the purpose of which is to maintain the Earth's axis parallel to itself in spite of its motion of translation round the Sun. The Earth's axis is inclined to the ecliptic (its circle), and Copernicus was of the opinion that were it not constantly corrected, it would describe a truncated conical surface in space, and would point to different regions of the celestial vault. This notion is understandable only if we admit the material nature of the spheres in which the planets (and the Earth) are enshrined. The needlessness of this third motion very soon became apparent, with the discovery of the law of inertia, and even before that, by Kepler. Copernicus himself could not dispense with it; he used it most ingeniously for two purposes: (a) to explain how the Earth maintained the direction of its axis, and (b) by assuming a quite small difference in velocity of the third motion and that of orbital motion, he found an explanation for the precession of the equinoxes. Unfortunately, he placed too much reliance on the data and teachings of the 'Ancients', and assumed a variation in the rate of this precession throughout the centuries; in order to explain this, he added a fourth motion to the three motions of the Earth, this fourth motion being the libration of the terrestrial axis, produced by two supplementary circles revolving in 3434 and 1717 years respectively.

59. *Commentariolus*, p. 28:

> '*Sicque septum omnino circulis Mercurius currit, Venus quinque, Tellus tribus et circa eam Luna quatuor. Mars demum Juppiter et Saturnus singuli quinque. Sic igitur in Universum 34 circuli sufficiunt, quibus tota mundi fabrica totaque siderum chorea.*'

E. Zinner has pointed out (*Entstehung und Ausbreitung*, p. 187), that Copernicus neglects the motions of the planets in latitude, the displacement of the apsides, as well as that of the nodes of the Moon. In order to take account of these, he would have required (according to Zinner) another four circles, making 38 in all, and not 34. Zinner is in error: at least seven, not four, more circles would be needed.

60. It may be assumed—no doubt gratuitously, but it seems likely to me—that Watzelrode counselled prudence, at least, initially. At a later date, when Copernicus had modified his astronomical system by replacing the epicyclo-epicycles by eccentrics, he could not have wanted diffusion of a work which gave a false idea of his theories, as well as premature and incorrect determinations of fundamental constants.

61. It is indeed inconceivable that the universities would not have reacted, at least in their lecture courses.

62. On this, see Max Müller, *Johann Albrecht von Widmanstetter*, Bamberg, 1908. The fact had already been mentioned by Tiraboschi (*Storia di Letteratura Italiana*, vol. VII, p. 706).

63. After all, Tiedemann Giese, Bishop of Kulm, saw nothing contrary

to the Roman Catholic faith in heliocentric astronomy, any more than did Copernicus himself. Assuredly, they believed, as had already been explained by St Thomas, that the Scriptures spoke the language of common sense, and that their purpose was to teach us the *via in coelum* and not the *viae coelorum*, cf. *infra*, p. 379, n. 5.

64. *De Revolutionibus Orbium Coelestium*, TH., p. 4; P., p. 38. Tiedemann Giese (1480–1550) of Danzig, the great friend of Copernicus, was Canon of Frauenburg for thirty years, a position held by Copernicus himself. Giese became Bishop of Kulm (Chełmno) in 1538, and succeeded to the See of Warmia in 1548 after the death of Johann Dantiscus.

65. *Ibid.* According to A. Koestler (*The Sleepwalkers*, p. 153), it was not fear of the theologians, but of the mob, the common herd; it was fear of ridicule that inspired the 'timid canon' to refuse publication of his work. It is possible . . . but it seems very unlikely to me: in fact, Osiander in his letter to Copernicus speaks of ways of pacifying the theologians and peripatetics (cf. *supra*, p. 35).

CHAPTER II. G. J. RHETICUS AND THE *Narratio Prima*

1. Georg Joachim Rheticus (his true name was Lauschen) was born 16 February 1516 at Feldkirch in the ancient Roman province of Rhaetia (whence his adopted name in accordance with the custom of the times). After studying at Zürich, he went to Wittenberg in 1532, where he was made professor of mathematics in 1536. He seems to have been well received during his stay in Prussia (1538–40), even though he was a Protestant; he travelled the country, of which he prepared a map (now lost); besides the *Narratio Prima*, he wrote a highly laudatory *Encomium Prussiae*, a biography of Copernicus (lost), and a small treatise aiming to show that the new astronomy does not conflict with the Holy Scriptures. Tiedemann Giese would have liked to have seen this treatise published also, but it is lost. He returned to Wittenberg at the beginning of 1540 in order to take up teaching again. His lectures dealt with the astronomy of Alfraganus and Ptolemy. The lectures of the second half of the year were announced by him (1 May 1540) in the following terms: 'I have been ordered to give once more a course on the *Sphaera* of Joannes de Sacrobosco.' It is clear that the atmosphere at Wittenberg, where in the meantime Melanchthon had received the *Narratio Prima* sent him by Rheticus from Danzig, was not favourable to a course on Copernican astronomy. In the summer of 1541, Rheticus returned to Frauenburg, and it was undoubtedly then that he made the copy of *De Revolutionibus Orbium Coelestium*, which at a later date served as the basis for the printed version of the work. He returned once more to Wittenberg at the end of 1541, and was even made dean of the Faculty of Arts; but he did not stay there for long. In 1542, he accepted a chair at the University of Leipzig, which he occupied,

with rather extended absences, until 1551. In 1557, he was at Cracow. Rheticus seems to have been something of a wandering spirit. Twice, in 1551 and 1567, he announced the publication, or the preparation, of many works among which, curiously enough, there is a 'German astronomy' that was never written. He did write, however, the most important work on trigonometry in the sixteenth century—the *Opus Palatinum de Triangulis* with trigonometrical tables calculated to the base (*sinus totus* or length of the radius) of 1 000 000 000 000 000. He did not finish this work, which was completed and published long after his death by his pupil, L. V. Otto, in 1596. Rheticus died in 1574.

2. Rheticus compares Copernicus with Regiomontanus and even with Ptolemy, and considers that his teacher is inferior to neither of them. The comparison with Ptolemy, and the designation—the New Ptolemy—was usual, in the sequel.

3. Johann Schöner (1477–1547) was born 16 January 1477 at Karlstadt, near Nürnberg; he was minister of the Jakobskirche at Bamberg, and then professor of mathematics at the *Gymnasium* at Nürnberg from 1526 to 1546. He edited the works of Regiomontanus and of Werner; he wrote a series of works on the preparation of the ecclesiastical calendar, the use of the astronomical globe, etc.... an *Opus Astrologicum*, Nürnberg, 1539, and *De Judiciis Astrologicis*, Nürnberg, 1543. His complete works (*Opera Mathematica*) were edited by his son, Andreas Schöner, at Nürnberg in 1561.

4. Cf. *supra*, p. 84, n. 48.

5. Rheticus tells us that he had barely ten weeks in which to study the work of Copernicus. He had been ill, and after recovering, went for several weeks to rest at Löbau, whither he had been invited by Tiedemann Giese. It was during this time, that he had extremely interesting conversations with Giese, which throw an astonishing light on the mind of Copernicus (and of Giese, too), and which he records in his *Encomium Prussiae*; cf. Appendix, pp. 67 f.

6. *De Libris Revolutionum Eruditissimi Viri et Mathematici excellentissimi reuerendi D. Doctoris Nicolai Copernici Torunnaei Cannonici Vuarmaciensis Narratio Prima ad clarissimum Virum D. Joan. Schonerum per quandem Juvenem Mathematicae Studiosum*, Gedani, 1540. The second edition (Bâle, 1541) has the author's name: per M. Georgium Joachimum Rheticum. Rheticus had the intention to follow this work with a *Narratio Secunda*, but he never did so, although publication of *De Revolutionibus Orbium Coelestium* (a book which is just as difficult as the *Almagest*) made a second narration particularly desirable, and not, as has been said, unnecessary. It is quite likely that Rheticus was scared by the reaction of Luther, and more especially by that of Melanchthon, and decided that discretion was the better part of valour. A. Koestler suggests (*The Sleepwalkers*, pp. 172 f.) that Rheticus must have been deeply wounded by what he (Koestler) calls 'the master's betrayal', for Copernicus, in his

preface, had omitted all mention of Rheticus and the part he had played in bringing about publication of *De Revolutionibus Orbium Coelestium*. To me, this hardly seems probable: it is quite obvious, that Copernicus could not mention collaboration with the protestant Rheticus, when he was dedicating his book to Pope Paul III.

7. Only the last (critical) edition by F. and K. Zeller, *Gesamtausgabe*, München, 1949, vol. II, is an exception.

8. So as not to frighten off his reader, Rheticus does not start his account with cosmology—he says that there is very little new in the first two books, and therefore starts with the third. He starts with the technical aspect by showing first of all the practical advantages of the new astronomy. Thus, from the very outset he speaks of the *motion of the Sun* and not of the *motion of the Earth round the Sun*, because it comes to the same thing from the point of view of computation, when it is a matter of calculating the *apparent* motion of the Sun on the celestial vault. Only after he has shown the advantages of the mathematical theory elaborated by Copernicus, does he pass on to an account of the cosmology and shows how it systematizes and unifies planetary theory.

9. Cf. *supra*, p. 33.

10. *Narratio Prima*, p. 490; cf. *supra*, chap. 1, n. 47.

11. *Ibid.*, p. 445. This sentence serves as *motto* for the *Narratio Prima* as well as for the *Dissertatio cum Nuntio Sidereo* . . . of Kepler. It comes from the *Didaskalikos*, a manual of Platonic philosophy by Albinus. Cf. E. Rosen, *Three Copernican Treatises*, p. 187, n. 241.

12. *De Mundo*, 391 a 23–24.

13. One wonders if the insistence of Rheticus on the 'Ptolemaism' of Copernicus might not have been for 'tactical' reasons, namely, from a desire to protect his teacher from the reproach, or accusation, of wanting to make changes, or of striving for originality (cf. p. 31). No doubt, that was partly the case; the desire for change was a great sin in a period devoted to tradition and respect for authority, when any step forward had to be presented as a movement backwards (return to origins). In this instance, however, there is also a statement of fact: in spite of his break with Ptolemy in two respects (rejection of geocentrism and equants), Copernicus remained deeply imbued with Ptolemaic concepts, and could even be regarded as a follower—the greatest follower—of Greek astronomy. This was the verdict of Kepler, who, in the *Mysterium Cosmographicum* (cf. *infra*, p. 137), defends Copernicus against the charge of having abandoned the ancient truth without justification; on the other hand, in the *Astronomia Nova*, he reproaches him with having followed Ptolemy too slavishly by agreeing with Ptolemy in preference to nature. Cf. *Astronomia Nova*, chap. XIV, p. 234:

'*Copernicus divitiarum suarum ipse ignarus Ptolemaeum sibi exprimen-*

dum omnino sumsit, non rerum naturam, ad quam tamen omnium proxime accesserat.'

Furthermore, as Kepler points out, Copernicus having decided not to depart too radically from Ptolemy (cf., *infra*, p. 154), and to refer planetary motions to the centre of the Earth's orbit (mean Sun) and not to the true Sun (p. 59), he deprived himself of the opportunity to explain variations in the longitude of planets by the Earth's orbital motion, and consequently had to introduce an 'adjustment' into the planetary circles.

14. *Narratio Prima*, p. 460:

> '*Porrò, doctissime D. Schonere, quemadmodum nos hinc in Luna ab aequante liberatos esse uides, & tali insuper theoria assumpta, quae experientiae, & omnibus obseruationibus correspondet, ita etiam in reliquis planetis aequantes tollit, tribuens cuilibet trium superiorum unum solummodo epicyclum, & eccentricum, quorum uterque super centro aequaliter moueatur, & pares planeta in epicyclo cum eccentrico reuolutiones faciat. Veneri autem, & Mercurio eccentri cum eccentrici.*'

15. *Ibid.*, p. 461.

16. *Ibid.*, p. 460. The Earth's orbital motion provides an explanation of (and renders unnecessary) the irregular movements of the planets (stations, retrogradations, forward motions), which in Ptolemaic astronomy were the result of their proper motion on their epicycles, and were designated 'second inequality'. It could be said that Copernicus eliminated these two 'second inequalities', or rather that he transferred them all to the Earth.

17. *Ibid.*, p. 461. Comparison of the work of the Creator with that of a clockmaker, first met with in the works of Jean Buridan, was extended by Kepler to assimilate the mechanism of the *Machina Mundi* to that of a clock. Cf. *infra*, p. 377, n. 8.

18. *Ibid.*, p. 481. The motions of the planets should be referred to the centre of the Earth's orbit, and not to the Sun.

19. *Ibid.*, p. 468.

> '*Cum D. Praeceptor meus Platonem, & Pythagoreos summos Diuini illius seculi Mathematicos sequens sphaerico terrae corpori circulares lationes ad τῶν φαινομένων causas assignandas, tribuendas censeret, uideretque (quemadmodum Aristoteles quoque testatur) uno attributo terrae motu, & alias item lationes ipsi ad stellarum imitationem competere, tribus eam principio ut maximè praecipuis moueri motibus, assumendum iudicauit.*'

20. *Ibid.*, 452. I agree with the opinion expressed by J. L. E. Dryer (*A History of Astronomy*, p. 333), that the views expressed by Rheticus coincided with those of his teacher under whose supervision he wrote the

book. At that period, such views were quite normal; from the time of Ptolemy to that of Campanella there was a firm alliance between astronomy and astrology. E. Rosen (*Three Copernican Treatises*, p. 122, n. 57) points out that if Rheticus were an ardent adept of astrology, as revealed by the Preface to Werner's *De Triangulis Sphaericis*, and his other works, then Copernicus never showed any inclination whatsoever for it. Rosen is right: compared with his predecessors and successors—Peurbach and Regiomontanus, Tycho Brahe and Kepler—Copernicus seems never to have engaged in astrological predictions. Are we therefore justified in concluding that he did not believe in them? Possibly, but not certainly.

21. In fact, it is not the eccentricity of the Sun, but of the Earth. Cf. *infra*, p. 114, n. 22.

22. Cf. *Narratio Prima*, p. 453. It is the Turkish Empire whose destruction is predicted by Rheticus.

23. This prophecy does not occur in the *Bible*, but in the *Talmud*. (E. Rosen, *Three Copernican Treatises*, p. 122, n. 54.)

24. The small circle on which the centre of the Earth's circle moves makes one revolution in 3434 years. Cf. pp. 59 and 115, fig. 12.

CHAPTER III *DE REVOLUTIONIBUS ORBIUM COELESTIUM* OSIANDER'S PREFACE AND THE LETTER TO POPE PAUL III

1. *Theoricae novae planetarum Georgio Purbacho ab Erasmo Reinholdo . . . auctae*, Nürnberg, 1542, Preface, fol. 4. Reinhold, indeed, does not name Copernicus, but the reference to him is sufficiently clear for no-one to be mistaken.

In the Preface to his *Tabulae Prutenicae*, Nürnberg, 1551, which is frankly and explicitly based on the work of Copernicus, Reinhold renews his eulogy, and this time mentions him by name. However, it should be noted that Reinhold borrows from Copernicus only his methods of computation, and in no-wise shares his cosmological concept. This fact was suspected by P. Duhem and E. Wohlwill, and finally proved by A. Birkenmajer; cf. A. Birkenmajer, 'Le commentaire inédit de E. Reinhold au *De Revolutionibus de Copernic*', *La Science au XVIe siècle*, Colloque de l'Union Internationale d'Histoire des Sciences, Royaumont, Juillet 1957; Paris, 1960.

2. M. Zinner (*Entstehung und Ausbreitung* . . ., p. 244), on the other hand, thinks that Rheticus took it with him in 1541.

However that may be, it was not the autograph manuscript of Copernicus, which belonged to G. J. Rheticus and is now preserved in the Nostitz Library at Prague, and which was published in photographic facsimile in 1945 (Nikolaus Kopernikus, *Gesamtausgabe*, München, 1945, vol. I), that was used by the printer, but the copy made by him. Cf. on this point,

Ryszard Gansiniec, 'Rheticus—éditeur de Copernic', Académie Polonaise des Sciences et des Lettres . . ., *Bulletin*, nos. 13–16, Paris, 1955–57.

3. *De lateribus et angulis triangulorum tum planorum rectilineorum angulorum, tum sphaericorum libellus eruditimus et utilissimus, cum ad pleroque Ptolemaei demonstrationes intelligendas, tum vero ad alia multa, scriptus a Clarissimo et doctissimo viro, D. Nicolai Copernico Toronensi. Additus est canon semissium subtensarum linearum in circulo.* Excusum Wittembergae per Joannem Lufft. Anno MDLXII.

4. Nicolai Copernici Torinensis, *De Revolutionibus Orbium Coelestium libri VI*, Nürnberg, 1543. The printer has placed the following appeal to the reader on the title-page:

> *Habes in hoc opere iam recens nato, & ædito, studiose lector, Motus stellarum, tam fixarum, quàm erraticarum, cum ex ueteribus, tum etiam ex recentibus obseruationibus restitutos; & nouis insuper ac admirabilibus hypothesibus ornatos. Habes enim Tabulas expeditissimas, ex quibus eosdem ad quoduis tempus quàm facillime calculare poteris. Igitur eme, lege, fruere.*

A second edition (together with the *Narratio Prima*) appeared at Frankfurt in 1566; a third edition, bearing the title *Astronomia Instaurata*, was published with explanatory notes by Nicolaus Mullerus at Amsterdam in 1617; a fourth edition with a Polish translation on the opposite page was published at Warsaw in 1854; a fifth edition, *ex autoris autographo*, appeared at Toruń in 1873. A facsimile of the first edition of *De Revolutionibus Orbium Coelestium* was published in 1927 by Hermann at Paris; a photographic edition of the autograph manuscript of Copernicus appeared in Vol. I of the *Gesamtausgabe* of his works (München, 1945), and Vol. II contains a critical edition of the work (München, 1949). In addition, Book I (*Cosmology*) of *De Revolutionibus Orbium Coelestium*, together with a French translation, was published by me at Paris in 1933; and in 1953 it was issued in a Polish translation with profuse and valuable notes (regrettably in Polish) by the Polish Academy of Sciences, Warsaw.

5. Cf. L. Prowe, *Nicolaus Coppernicus*, vol. II, p. 420.

6. Andreas Osiander (1498–1552), preacher of the Lorenzkirche, Nürnberg, in 1522, was one of Luther's first partisans; though he differed from him in the interpretation of 'justification by faith'. In 1548, he was obliged to leave Nürnberg, first for Breslau and then for Königsberg, where he became a professor at the university founded by Duke Albert of Prussia.

7. Johannes Kepler, *Apologia Tychonis contra . . . Ursum* (*Opera Omnia*, edited by Frisch, vol. I, pp. 246 f.). This defence of Tycho against the attacks of the ultra-positivist Nicolaus Reymarus Ursus remained unfinished and unpublished. Ursus, who had been accused by Tycho Brahe of plagiarizing his world-system, published a small book, *De Hypothesibus Astronomicis*, at Prague in 1597, in which he rejects Tycho Brahe's charge

of plagiary, and taunts him into the bargain with having stolen his world-system from Apollonius. Kepler, who had undertaken the task at the request of Tycho Brahe, no doubt thought that the death of the latter put an end to the polemic. However, he was anxious to re-establish the truth and defend Copernicus against the charge made by Ramus of having not only formulated hypotheses, which in the opinion of Ramus were unnecessary, but also of having used hypotheses which he knew to be false in order to deduce planetary motions and positions from them (a most reprehensible business)—instead of elaborating an astronomy without hypotheses, which was something that Ramus sincerely prayed for. Kepler, therefore, reproduced on the verso of the title-page of the *Astronomia Nova* the passage in which Ramus makes the charge in question, together with the reply in which he (Kepler) reveals the name of the writer of the famous preface to the reader (*Opera Omnia*, edited by Frisch, vol. III, p. 136; *Gesammelte Werke*, München, 1937, vol. III, p. 6). Here are the passages in question from Ramus and Kepler. P. Ramus, *Scholarum Mathematicarum*, Lib. II, p. 50:

> '*Commentum igitur Hypothesium absurdum est: Sed tamen commentum in Eudoxo, Arisoteles, Callipo simplicius, qui veras hypotheses arbitrati sunt: imo tanquam Deos ἀνάστρων orbium sunt venerati. At in posteris fabula est longe absurdissima, naturalium rerum veritatem per falsas causas demonstrare. Quapropter Logica primum, deinde Mathematica Arithmeticae et Geometricae elementa, ad amplissimae artis puritatem et dignitatem constituendam adjumenti plurimum conferent. Atque utinam Copernicus in istam Astrologiae sine Hypothesibus constituendae cogitationem potius incubuisset. Longe enim facilius ei fuisset, Astrologiam astrorum suorum veritati respondentem describere, quam gigantei cujusdam laboris instar, Terram movere, ut ad Terrae motum quietas stellas specularemur. Quin potius e tot nobilibus Germaniae scholis exoriare Philosophus idem et Mathematicus aliquis, qui positam in medio sempiternae laudis palman assequare. Ac si quis caducae utiltatis fructus tantae virtutis praemio proponi possit, regiam Lutetia professionem praemiam conformatae absque hypothesibus Astrologiae tibi spondebo; sponsionem hanc equidem lubentissime, vel nostrae professionis cessione, praestabo.*'

To which Kepler replied:

> '*Fabula est absurdissima, fateor, Naturalia per falsas demonstrare causas: sed fabula hoc non est in Copernico: quippe qui veras et ipse arbitratus est, Hypotheses suas, non minus quam illi tui veteres suas: nempe tantum est arbitratus, sed et demonstrat veras; testem do hoc Opus. Vin'tu vero scire fabulae hujus, cui tantopere irasceris, architectum?* Andreas Osiander *annotatus est in meo exemplari, manu* Hieronimi Schreiber *Noribergensis. Hic igitur* Andreas, *cum editioni Copernici praeesset, praefationem illam,*

quam tu dicis absurdissimam, ipse (quantum ex ejus literis ad Copernicum colligi potest) censuit prodentissimam, posuit in frontispicio libri; Copernico *ipso aut jam mortuo, aut certe ignaro. Non igitur μυθολογεεῖ Copernicus, sed serio παραδοξολογεῖ, hoc est φιλοσοφεῖ quod tu in Astronomo desiderabas.*'

Cf. *infra*, p. 381, n. 12. As a matter of fact, it could be that Ramus, whilst rejecting 'hypotheses', was not pursuing the ideal of a purely computational astronomy, as understood by Tycho Brahe (and myself), but objected only to the preconceived dogma of circular motion. In this case, Kepler's claim to have realized this wish would be more justifiable than it appears at first sight. Yet Kepler had already raised this claim in his *Mysterium Cosmographicum* . . .; cf. *infra*, p. 388, n. 4.

8. Was the Letter of Dedication to Pope Paul III substituted for the introduction of the manuscript by Copernicus himself, or by Osiander? This question has engaged the attention of historians of Copernicus who are accustomed to accuse Osiander of all possible and imaginable misdeeds, including that of having altered even the title of the work written by Copernicus. According to the evidence of Johannes Praetorius (Letter to Herwart von Hohenburg in 1609; published by E. Zinner in *Entstehung und Ausbreitung der Coppernicanischen Lehre*, Erlangen, 1943, p. 454) the original title was *De Revolutionibus Orbium Mundi*, or simply *De Revolutionibus*, but it was changed to *De Revolutionibus Orbium Coelestium*. E. Rosen has demolished this charge by showing that neither the replacement of *mundi* by *coelestium*, nor the addition of *orbium coelestium* to *de revolutionibus*, makes any difference, seeing that they are in perfect agreement with the thought of Copernicus, and do not lend any particular support to the phenomenalistic interpretation that Osiander wished to give it (cf. E. Rosen. 'The Authentic Title of Copernicus' Major Work', *Journal of the History of Ideas*, 1943, vol. IV). The same argument is applicable to the present problem: it is difficult to see what Osiander would have gained by the substitution. E. Zinner, who charges Osiander with having substituted *coelestium* for *mundi*, considers that the substitution of the Letter to Pope Paul III for the original introduction was made either by Copernicus himself, or by Rheticus with the consent of Copernicus (E. Zinner, *Entstehung und Ausbreitung* . . ., p. 266). However, see also R. Gansiniec, 'Quel est le véritable titre du chef-d'œuvre de Copernic', Académie Polonaise des Sciences et des Lettres . . ., *Bulletin*, nos. 13–16, Paris, 1955–57.

9. *De Revolutionibus* . . ., TH., pp. 1 f.; P., pp. 27 f. In the opinion of A. Koestler (*The Sleepwalkers*, p. 567), historians of Copernicus have misunderstood the import of Osiander's preface: both of them support the heliocentric theory—otherwise, why did Osiander bother himself about publication of *De Revolutionibus Orbium Coelestium*? Osiander's criticisms

are directed only against certain details of the system, which are just as unlikely and unconformable to empirical reality as the corresponding details in Ptolemy's astronomy: *e.g.*, the non-variation in brightness of Venus. Furthermore, the arrangement of circles seems, in Osiander's view to be just as 'hypothetical' in the system of Copernicus as in that of Ptolemy: in both cases it is merely a question of computational procedures without any claim to reality. Koestler believes that Copernicus himself shared this point of view. Copernicus, being convinced of the positive value of heliocentrism, was not, and could not be, sure of having discovered the 'true' mechanism of the spheres, the reality of which he could not seriously accept, seeing that it would imply the necessity for endowing the Earth also with the same mechanism; in which case its non-existence (non-reality) is easily perceived. For Copernicus, therefore, the concrete arrangement of circular planetary motions (the entire assembly of circles) would have only an 'hypothetical' value, as it did for Osiander and Ptolemy, and would be no more than a mathematical device. I must admit, that Koestler's most ingenious explanation does not convince me; it seems to me that Osiander's 'positivism' extends to include heliocentrism. Moreover, that was how it (the preface) was understood by his contemporaries, as well as by Kepler; and it is only in these circumstances that its adoption, which Osiander suggested to Copernicus, could have served as a protective screen for his doctrine and protected it from the hostility of theologians and Peripatetics. As for the attitude of Copernicus, his rejection of Osiander's strategem shows quite clearly that he was not in agreement. It is possible, and even highly probable, that Copernicus, whilst not regarding his circles and spheres as mathematical fictions, did not feel so sure of having succeeded in determining the true structure of the cosmic model in all its details, as he was of having discovered its fundamental nature.

10. Osiander's objection is specious and is just as valid against Ptolemy as against Copernicus. In fact, the visual size of Venus should be much larger at apogee than at perigee . . . if the difference were perceptible. With the naked eye, it is not. As regards its brightness, that is a function of its phases: the surface illuminated by the Sun is a *maximum* at apogee and a *minimum* at perigee. The fact that Venus exhibits phases is in agreement with Copernican astronomy, but not with Ptolemaic astronomy. Their discovery (theoretical) was credited to Copernicus by Galileo.

11. It represents a point of view which can be traced back to the Greeks, and which was most clearly expressed by Simplicius. Osiander has been highly praised (wrongly in my opinion) by many modern historians for having upheld it; including P. Duhem, *La Théorie Physique*, Paris, 1908, pp. 58 f., and more recently E. J. Dijksterhuis, *Die Mechanisierung des Weltbildes*, Leipzig, 1956, pp. 330 f., Lynn Thorndike (*A History of Magic and Experimental Science*, New York, vol. V, 1941, p. 413, n. 33) goes even further and wonders if the preface in question 'was not by Rheticus

rather than Osiander, or perhaps their joint work'; this seems to be a somewhat gratuitous hypothesis in view of the agreement amongst contemporary evidence, and the admitted Copernicanism of Rheticus.

12. T. Giese's letter to G. J. Rheticus was published at Cracow in 1615; republished in the Warsaw edition of *De Revolutionibus Orbium Coelestium* in 1854; in Hipler's *Spicelegium Copernicanum*; and in L. Prowe's *Nicolaus Copernicus*, vol. II, p. 420. German translations will be found in Prowe, *ibid.*, vol. I, p. 339; and in the German edition of *De Revolutionibus Orbium Coelestium* by C. Menzzer, *Ueber die Kreisbewegungen*, etc., Toruń, 1879.

13. This disavowal was not made public by Rheticus; nor did he keep it secret. Even before it was made public by Kepler, the fact that Osiander, and not Copernicus, was the author of the preface was not entirely unknown in the sixteenth century. Hieronimus Schreiber, successor to Rheticus at the University of Wittenberg, knew of it, and, on the evidence of Kepler, noted the fact in his copy of *De Revolutionibus Orbium Coelestium*; Petrus Apianus also knew about it, and he too did not keep the information to himself. A note on this matter, making express mention of Apianus (*Apianus mihi dixit*) is to be found in the copy which belonged to Johann Jacob Fugger, as well as in that which belonged to Maestlin. Cf. E. Zinner, *Entstehung und Ausbreitung . . .*, Beilage E, pp. 452 f.

14. Kepler's remarks quoted in n. 7 above ought to have put an end to this misunderstanding. Unfortunately, few persons seem to have read Kepler, not even historians of astronomy. Consequently, some of them, amongst whom I am sorry to name Montucla, Laplace and Delambre, have ascribed Osiander's preface to Copernicus.

15. The Dominican Nikolaus Schönberg (1472–1537); Archbishop of Capua, 1520; created cardinal, 1535.

16. Alessandro Farnese (1468–1549), elected Pope in 1537 and took the name of Paul III. A great patron of the arts and sciences, he was also one of the promoters of the Counter-Reformation; during his pontificate the Society of Jesus (1540) and the Holy Office (1542) were established.

17. *De Revolutionibus Orbium Coelestium*, TH., p. 5; P., pp. 36 f.:

> 'Therefore, when I considered how my contrary assertion that the Earth has motion would seem an absurd *ἄχρόαμα* to those who know the judgement of centuries confirms the opinion that the Earth is placed immovably at the centre of the Universe, I hesitated long whether I should bring to the light my Commentaries written to prove its motion, or whether on the other hand it were better to follow the example of the Pythagoreans and others who . . . were accustomed to impart the mystery of philosophy only to their friends and intimates, and then not in writing but only by word of mouth . . . so that noble discoveries made by the diligence of great men should not be despised by the foolish and ignorant.'

18. *Ibid.*, TH., p. 6; P., pp. 43 f.

19. *Ibid.*, TH., p. 6; P., pp. 43 f. According to Diogenes Laertius, the real name of the Nicetus in question was Hicetas. The passage in the works of Cicero to which Copernicus refers is to be found in the *Academica*, I, iv, chap. 39; that of the [pseudo] Plutarch is taken from *De Placitis Philosophorum*, I, iii, chap. 13, which Copernicus quotes in Greek from the Bâle edition of 1531. Concerning the familiarity of Copernicus with the writers of Antiquity, cf. L. Ideler, *Ueber das Verhältnis des Copernicus zum Altertum*, 1876; G. Schiaparelli, *I precursi di Copernico nell'antichità*, Bologna, 1873 (German translation, *Die Verläufer des Copernicus im Altertum*, Leipzig, 1876); and especially the excellent study by E. Brachvogel, *Nikolaus Koppernicus und Aristarch von Samos*, Braunsberg, 1935.

20. Cf. Dreyer, *History of the Planetary Systems*, p. 312.

21. The texts of the ancient philosophers to which Copernicus refers are not unknown texts, recently discovered; everyone had read them, but no-one had had the inspiration to substitute a heliocentric system for that of Ptolemy (or Aristotle). As for the more restricted principle of the Earth's diurnal motion (rotation on its axis), this had been known from remote times, if only because it had been stated, discussed and rejected both by Ptolemy and Aristotle.

22. The name of Aristarchos of Samos does not occur in the printed version of *De Revolutionibus Orbium Coelestium*, but in the manuscript it follows that of Philolaus (Book I, chap. xi, pp. 129 f.):

> '*Et si fateamur Solis Lunaeque cursum in immobilitate quoque terrae demonstrari posse, in caeteris vero errantibus minus congruit. Credibile est hisce similibusque causis Philolaum mobilitatem terrae sensisse, quod etiam nonnulli Aristarchum Samium ferunt in eadem fuisse sententia, non illa ratione moti, quam allegat reprobatque Aristoteles.*'

The work to which Copernicus refers is Aristotle's *De Coelo*, II, 14. The similarity between the systems of Copernicus and Aristarchos was noticed at an early date by Melanchthon, who charged Copernicus in his *Initiae Doctrinae Physicae* (Wittenberg, 1549, and 1550) with wanting to create something novel at all costs in order to satisfy his vanity, and with having done nothing but copy Aristarchos. Erasmus Reinhold in his *Hypotyposes Orbium Coelestium* (Strasbourg, 1568) supported his esteemed colleague, asserting the dependence of Copernicus on his Greek predecessor whose propositions he adapted to his own use (L. A. Birkenmajer, *Mikolaj Kopernik*, p. 639; Brachvogel, *Nikolaus Koppernicus und Aristarch von Samos*, p. 16). Since then, nearly everyone has accepted as a fact that the Copernican doctrine originated with Aristarchos. It has been objected that this was an historical impossibility (Brachvogel, *ibid.*, pp. 8–17; and 'Nikolaus Kopernicus in der Entwicklung des deutschen Geisteslebens' in *Kopernikus-Forschungen*, Deutschland und der Osten, Leipzig, 1943, vol.

22, pp. 40 f.): the *Arenarius* of Archimedes is the only place where an account of the heliocentric astronomical doctrine of Aristarchos of Samos is to be found, and it was not published until 1544. To this fact, it may very well be countered that Copernicus had seen a manuscript of it, and may even have possessed one: indeed, Rheticus, in his *Narratio Prima* (p. 461) mentions one.

23. *De Revolutionibus Orbium Coelestium*, TH., p. 10; P., pp. 45 f. In fact Ptolemy's astronomical universe does not constitute a system: the mechanisms of planetary motion, though they are similar (at least, for the superior planets), are independent of each other and are not connected except through their common centre, the Earth. It was otherwise with Copernicus. Thus, his insistence on a rational order of the Universe—'the best work of the supreme architect'—is most characteristic; note also his revival of the principle of uniform circular motion; cf. *supra*, p. 26, and p. 58. In Book V, 2, Copernicus repeats the charge against Ptolemy of having abandoned the principle of uniform motion by introducing equants.

24. In the Ptolemaic system the variation in distance between the Earth and the Moon is such that there should be a fourfold change in the apparent size of the latter. Furthermore, the Ptolemaic lunar theory is extremely complicated; it even includes one mobile equant.

25. *Narratio Prima*, p. 459; cf. *supra*, p. 32. This is also the opinion of Neugebauer, who, having shown how Copernicus in his theory of the Moon had succeeded in replacing the Ptolemaic equant by a double epicycle, writes: 'This obvious advantage of the use of secondary epicycles induced Copernicus to apply the same construction also to the planetary motion and thus to initiate complications which destroyed the inherent elegance and simplicity of the Ptolemaic model'; cf. O. Neugebauer, *The Exact Sciences in Antiquity*, 2nd edition, Providence, R.I., 1957, p. 197. However, it should be noted that as the motion of the Earth plays no part in lunar theory, the success of this theory must have encouraged Copernicus to dispense with Ptolemaic equants in planetary theory by replacing them by a second epicycle, and not to transfer the centre of the motions from the Earth to the Sun.

26. *De Revolutionibus Orbium Coelestium*, TH., p. 27; P., p. 110. The Encyclopaedia of Martianus Capella, *De nuptiis Philologiae et Mercurii libri duo, de grammatica, de dialectica, de rhetorica, de geometria, de arithmetica, de astronomia, de musica libri septem*, which comprises the *trivium* and the *quadrivium* of the seven liberal arts, was *the manual* throughout the Middle Ages; it was translated into German in the eleventh century by Notker Labeo, and published in Latin at Vicenza in 1499. Consequently, everybody was familiar with the theory which made Venus and Mercury satellites of the Sun; Martianus Capella called it the 'Egyptian theory', though it derives in fact from Heraclides of Pontos. It is recorded not only

by Martianus Capella, but also by Macrobius in his commentary on Cicero's *Somnium Scipionis*; by Chalcidius in his commentary on *Timaeus;* as well as by many other authors amongst whom mention may be made of Nicholas of Cusa and Pico della Mirandola. (Cf. L. Birkenmajer, *Mikolaj Kopernik*, pp. 162, 194, 245; and E. Brachvogel, *Nikolaus Koppernicus und Aristarch von Samos*, p. 29, n. 66.) With the exception of Johannes Scotus Erigena (cf. P. Duhem, *Le Système du Monde*, vol. III, p. 61), no-one before Copernicus had given any serious attention to it, or tried to extend it to the other planets. Copernicus seems not to have done so; he went straight from the Ptolemaic system to his own without the intermediate Tychonian stage.

27. Cf. *Narratio Prima*, p. 461.

28. In the Tychonian system the planets revolve round the Sun, and the Sun (together with all the planets and satellites) revolves round the stationary Earth. From the point of view of computation, which takes account only of the relative positions and motions of the bodies in the solar system, this conception is identical with that of Copernicus; and as observers, both before and after Copernicus, observe *celestial motions* from the Earth, the whole Universe is, practically speaking, Tychonian. The reason why Tycho Brahe developed his own system seems to have been because of the conflict of Copernicanism with literal interpretation of the Bible on the one hand, and especially the inability of Copernicans to give a satisfactory answer to Aristotle's objections against the motion of the Earth, on the other hand. Cf. my *Études Galiléennes*, Paris, 1940, fasc. III.

29. J. L. E. Dreyer, *History of the Planetary Systems*, p. 312; 'It must have struck him as a strange coincidence that the revolution of the Sun round the zodiac and the revolution of the epicycle-centres of Mercury and Venus round the zodiac should take the same period, a year, while the period of the three outer planets in their epicycles was the synodic period, *i.e.*, the time between two successive oppositions to the Sun.'

30. E. Zinner, prompted, no doubt, by his desire to 'Germanize' Copernicus, and at all costs to count him among the German precursors, attaches great importance to the fact that Peurbach in his *Theoricae Novae Planetarum* directs attention to the relationship between the motion of the three superior planets on their epicycle and the motion of the Sun, as well as the agreement between the motion of the Sun and Venus (*Entstehung und Ausbreitung*, p. 97). More important still, in his opinion, is the similar remark made by Regiomontanus, who, when he was 20 and studying Al-Biṭrūjī, noted that Venus in the course of its zodiacal motion is linked with the Sun; whereas this is not so in the case of the superior planets, which, in contradistinction to Venus, are linked in their motion to the epicycle (*ibid.*, p. 129). If to this we add the fact that Regiomontanus asserts that the Sun, being the source of heat and light, should be in the midst of the planets like a king in his kingdom, or the heart in a living

being; and that Georg Hartmann (1489–1554), a maker of sun-dials at Nürnberg, and a great admirer of Regiomontanus, was in possession of a fragment of a letter from the latter in which is written (if the statement be true) that 'the motion of the stars must be slightly modified on account of the motion of the Earth' (*ibid.*, p. 135), then it is obvious (at least it was so to Zinner) that Regiomontanus had pointed out the motion of the Earth in his non-surviving letters, and that it was these ideas, transmitted through Brianchini and Novara, which influenced the young Copernicus. I doubt if anyone, unless he were blinded by nationalistic passion, would accept Zinner's argument. Obviously, we do not know what was in the non-surviving letters of Regiomontanus: yet, on the other hand, we do know that neither Bianchini, nor Novara, ever showed the slightest leanings towards heliocentrism. As for that sentence about the rôle of the Sun, Zinner must know as well as anybody that it was a commonplace remark. Cf. E. Brachvogel, 'Nikolaus Kopernikus in der Entwicklung der deutschen Geisteslebens', *Kopernikus-Forschungen*, pp. 90 f., where short shrift is given to Zinner's arguments.

31. Cf. *Narratio Prima*, p. 462.

32. *Ibid.*, p. 465:

'*Deinde, quod his quidem consentaneum est, Deum, in huius theatri medium Solem, suum in natura administratorem totiusque uniuersi Regem, Diuina maiestate conspicuum collacasse:*

> *Ad cuius numeros & Dij moueantur, & orbis*
> *Accipiat leges, praescriptaque foedera seruet.*'

The verses quoted by Rheticus come from *Urania sive De Stellis* by Giovanni Giovano Pontano, published at Florence, 1514. Copernicus possessed a selection of poems by Pontano, printed at Venice, 1501.

33. As, for example, Peurbach in his *Theoricae Novae*. In fact, this relationship could not have escaped the notice of any reader of the works of Ptolemy, seeing that it was definitely postulated by him, and indeed in a way that reveals the dominant rôle of the Sun in a far more striking manner than that used by Dreyer. According to Ptolemy, the centres of the rotating eccentrics of the inferior planets are always on the straight line joining the Earth to the Sun (the radius vector of the latter). It follows that these eccentrics could rotate about one and the same centre, which could be placed in the Sun, as Heraclides of Pontos clearly understood. The epicyclic radii-vectores of the superior planets are always parallel to the radius-vector of the Sun, and consequently not only complete one revolution in one synodic period, but also move with the same angular velocity as the Sun, in exactly the same way as the deferents of the eccentrics of the inferior planets. Whence it follows, that if the deferent-epicyclic mechanisms be replaced by an exactly equivalent mechanism of rotating eccentrics, then the eccentrics of the three superior planets would have their

centres on the Sun's radius-vector; the centre would naturally be placed in the Sun exactly as in the case of the centre of the epicycles of Venus and Mercury. Thus, Tycho Brahe's system is present, as it were, in embryonic form in Ptolemy's system, and in such a clear manner that it is inconceivable to me that Ptolemy should not have noticed it. Why, then, did he not develop it? Probably because he did not wish to; but preferred to make the planets revolve round the Earth; and wished to endow each with its own independent arrangement of proper motions. In other words, he wanted to do exactly the opposite of what Copernicus wanted to do.

CHAPTER IV. THE COSMIC DOCTRINE

1. A. Koestler (*The Sleepwalkers*, p. 572) has counted 48.
2. Kepler, too, lost no time in resorting to non-uniform motions.
3. G. Galilei, *Dialogo soprai due massimi sistemi del mondo* (*Opere*, Ed. Naz, vol. VII, p. 416): '*Sistema Copernicano difficile a intendersi e facile ad effectuarsi.*' Cf. E. F. Apelt, *Johann Keplers Astronomische Weltansicht*, p. 10.

> '*Das Kopernikanische System stellte eine ganz neue und höchst schwierige Aufgabe an die Geometer: es verlangte neue Regeln für die Berechnung der Planetenörter, d.i., eine neue Theorie. Im Ptolemäischen System gibt die Theorie unmittelbar den geometrischen Ort eines Planeten, d.h., seinen Ort an der Himmelskugel, im Kopernikanischen dagegen nur den heliocentrischen Ort desselben, und aus diesem und dem jedesmaligen Ort der Erde im Raume muss dann erst der geocentrische Ort des Planeten berechnet werden.*'

4. O. Neugebauer, *The Exact Sciences in Antiquity*, pp. 202 f.; p. 204: 'The popular belief that Copernicus' heliocentric system constitutes a significant simplification of the Ptolemaic system is obviously wrong. The choice of the reference system has no effect whatever on the structure of the model, and the Copernican models require about twice as many circles as the Ptolemaic models and are far less elegant and adaptable.'
5. *De Revolutionibus Orbium Coelestium*, Lib. V, cap. xxii, pp. 368 f.; Dreyer, *History of the Planetary Systems*, pp. 336 f.
6. *De Revolutionibus Orbium Coelestium*, Lib. V, cap. xxv, pp. 376 f.; Dreyer, *op. cit.*, p. 337.
7. In *De Revolutionibus Orbium Coelestium*, Lib. IV, cap. iv, p. 166, Copernicus explains how a vibratory, or oscillatory, motion is compounded from two circular motions. Two circles, whose diameters are in the ratio of 1 to 2, are required. If the smaller circle be made to roll on the circumference inside the larger circle, then any point on the circumference of the smaller will describe a straight line passing through the centre of the larger

diameter. The same result can be obtained if this point on the circumference of circle *FCE* be made the centre of another circle *GMC* equal to the former, and both circles be made to rotate with the same speed in opposite directions. In the manuscript of *De Revolutionibus Orbium Coelestium*, Copernicus had added that if the circles were unequal they would not describe a straight line, but a conic section known as an ellipse by mathematicians (Fig. 11).

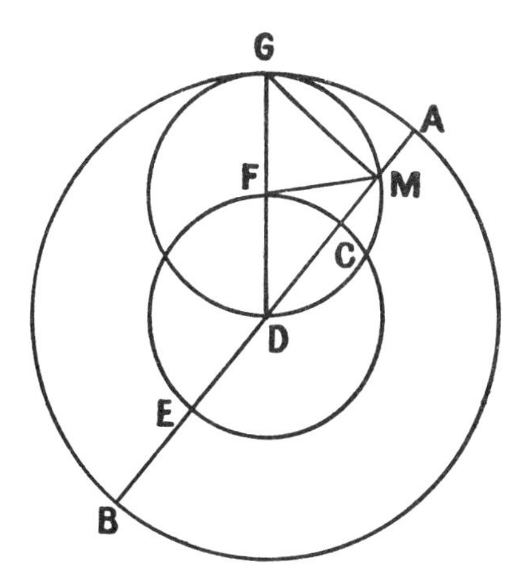

Figure 11.

Dreyer assumes (*op. cit.*, p. 331) that Copernicus crossed out this passage because he realized that the line in question would not be a true ellipse, but a hypocycloid which would only bear resemblance to it. He adds that the theory of producing a straight line by compounding two circular motions was already known to Naṣīr al-din al-Tūsī.

8. *De Revolutionibus Orbium Coelestium*, Lib. IV, cap. xvii, p. 298; Dreyer, *History of the Planetary Systems*, p. 334. Ch. Victor Roberts has found that the lunar theory of Copernicus, save for some numerical differences, is identical with that of the Syrian astronomer Ibn al-Shāṭir (1306–75), who moreover used the same scheme (concentro-bi-epicyclic) for his theory of the Sun (this is rather surprising for he was able to dispense with Ptolemy's eccentric circle for the Sun by using only one epicycle). Did Copernicus know of Ibn al-Shāṭir's work? Roberts was unable to answer this question, noting that no Latin translation of this work has ever been reported. Cf. C. V. Roberts, 'The Solar and Lunar Theory of Ibn al-Shāṭir' *Isis*, 1957, vol. XLVII, pp. 128 f.

9. *De Revolutionibus Orbium Coelestium*, Lib. V, cap. iv, pp. 325 f. Dreyer, *op. cit.*, pp. 335 f.

10. *Mysterium Cosmographicum*, p. 175; *v. infra*, p. 131. Kepler does not take into account the reintroduction of epicycles compensating for the equant. The *Mysterium Cosmographicum* is quoted from M. Caspar's edition, *Gesammelte Werke*, München, 1938, Vol. I.

11. Partly, but not entirely, because Copernicus refers the planetary motions to the centre of the Earth's circle (mean Sun) and not to the true Sun; cf. *supra*, p. 92, n. 13.

12. The radius of the *orbis magnus* (distance of the Earth from the Sun) being more than one thousand times the Earth's diameter, the linear dimensions of the Universe must be increased at least 2000 times and its volume 2000^3 times. The presence of the solar system at the centre of this vast and empty sphere must have seemed absurd to astronomers and philosophers of the sixteenth century; and, contrary to the assertions of Copernicus and Rheticus, incompatible with divine wisdom. Tycho Brahe considered that one of the greatest advantages of his system was the possibility it offered of *decreasing* the dimensions of the Universe.

13. The absence of parallax of the fixed stars has always been the strongest astronomical argument against heliocentrism. Astronomers considered the explanation given by Aristarchos and Copernicus to be mere evasion, contravening the strict rules of logic. It must be admitted that they were right, from the point of view of a strictly experiential science.

14. As mentioned above (p. 101, n. 23), the Ptolemaic universe does not constitute a system; the planetary models are independent of each other, and apart from a curious, but quite unjustifiable linking of planetary motions with those of the Sun, the celestial motions are not interconnected by any intelligible relationship.

15. Johann Kepler, *Astronomia Nova*, Lib, I, cap. iv; *De Revolutionibus Orbium Coelestium*, Lib. III, cap. xv; Lib. IV, cap. vii; and especially Lib. V, cap. i.

16. The motion of the planet on its circle is consequently just as complicated according to Copernicus as according to Ptolemy.

17. The epicycles used by Copernicus are so much smaller than those used by Ptolemy—it was pointed out on p. 48 that their radii equal one quarter of the Ptolemaic eccentricity—that Derek J. Price ('Contra Copernicus' in Marshall Claggett (editor), *Critical Problems in the History of Science*, Madison, Wisconsin, 1959) very justifiably calls them 'epicyclets'. Consequently, the Copernican 'circles' are much closer than those of Ptolemy, and, as Kepler rightly observed, they do not touch, but have empty spaces between them; (cf. figure on p. 148).

18. Cf. *supra*, Fig. 5.

19. Kepler's ideas developed from this principle. Cf. *infra*, pp. 139, 151, 185.

20. Aristotle, *De Coelo*, II, 10.

21. The period of revolution of the deferents of the interior planets is one year. The sphere of the fixed stars has the fastest motion.

22. The dimensions of the Cosmos, in the case of an orderly separation of the planets, can only be determined if it be assumed also that the corresponding spheres in which their epicycles are located succeed each other without loss of continuity, *i.e.*, without leaving empty space between them.

23. E. F. Apelt, *Die Reformation der Sternkunde*, Jena, 1852:

> '*In dem ptolemaischen System hat zwar jeder Epicykel zu seinem Deferenten ein durch synodischen und den siderischen Umlauf gegebenes Verhältnis, aber das Verhältnis der deferirenden Kreise zu einander bleibt unbestimmt.*'

24. The only distances that could be determined directly by pre-Copernican, or Copernican, astronomy were those of the Moon and the Sun. The others had to be computed. Such computation was impossible in the Ptolemaic system, but immediately became so in the Copernican system. In the case of the inferior planets, it results from the fact that, at their *maximum* elongation, they are at the apex of the right angle of the right-angled triangle formed by the planet, the Earth and the Sun. As the side, Earth–Sun (hypotenuse), is known, and the angle at the apex occupied by the Earth is provided by observation, it follows that the distances Earth–planet and Sun–planet can be calculated. We can summarize as follows: at conjunction, the Sun, the Earth and the planet are in a straight line; at a time *t*, later, the Earth and the planet will have traversed arcs corresponding to the time required for one revolution divided by *t*; the difference between these two arcs gives the angle at the apex occupied by the Sun in the triangle formed by the Sun, the Earth and the planet; the angle at the apex of the angle occupied by the Earth is provided by observation; the side, Earth–Sun, of the triangle is known; the rest follows immediately.

It must not be forgotten that all these calculations assume that the distance from the Earth to the Sun is known, and that all the others are determined as a function of this one distance. Consequently, they are too small to the extent that the distance, Earth–Sun, accepted by Copernicus (1142 terrestrial semi-diameters) is smaller than the true distance, which is about twenty times greater. On the other hand, in ratio to this fundamental distance, they provide a very good, close approximation, as pointed out by Dreyer (*History of the Planetary Systems*, p. 339), to the results given by the ratio of deferent-radius to epicycle-radius found by Ptolemy (*De Revolutionibus Orbium Coelestium*, Lib. V, cap. 5, 14, 19, 21, 27).

Seeing that Copernicus used the same value for the distance Earth–Sun as given by Ptolemy, namely 1142 terrestrial semi-diameters, it follows (a) that all the Copernican distances are just as incorrect as those of Ptolemy;

	Distance from Sun	
	According to Copernicus	According to modern astronomy
Mercury	0·3763	0·3871
Venus	0·7193	0·7233
Earth	1·0000	1·0000
Mars	1·5198	1·5237
Jupiter	5·2192	5·2028
Saturn	9·1743	9·5389

and (b) that the dimensions of the Copernican universe are enormously increased compared with those of the Ptolemaic universe, though the dimensions of the planetary system are not. Consequently, there is a vast empty space between the sphere of Saturn and the sphere of the fixed stars, which was so scandalous in the view of the contemporaries—and successors—of Copernicus.

25. *De Revolutionibus Orbium Coelestium*, TH., pp. 28 f.; P., pp. 113–118; *Narratio Prima*, p. 467.

26. *Narratio Prima*, pp. 466 f.

27. *Ibid.*, p. 467. Six is the first perfect number. Kepler, although as much a Pythagorean as Rheticus, was of the opinion that the perfection of a *number* does not suffice to explain a tangible structure. Cf. *Mysterium Cosmographicum*, edited by Max Caspar, *Gesammelte Werke*, München, 1938, vol. I, p. 13, and *infra*, p. 139. On the other hand, he fully accepted the Pythagorean concept of a *harmony of the spheres*, which is evoked by Rheticus. Concerning this concept, which was just as popular in Antiquity as in the Middle Ages and Renaissance, cf. F. Piper, *Mythologie der christlichen Kunst*, Weimar, 1847–51; Th. Reinach, 'La Musique des Sphères', *Revue des Études Grecques*, 1900.

28. *De Revolutionibus Orbium Coelestium*, Lib. I, cap. x, TH., p. 29; P., pp. 116 f. Consequently, the great advantage of the system from the point of view of Copernicus lies in the revelation of its systematic structure; not in the best agreement with observational data and ease of computation. History has proved him to be right.

29. The first physical proof of the Earth's motion (rotation) was suggested by Galileo who saw in this rotation combined with orbital motion the cause of tides; unfortunately, his theory was wrong. The first *correct* physical proof was not forthcoming until 1672 when Richer, who had gone to Cayenne to make astronomical observations, noticed that his clock, which had been adjusted at Paris, lost 2′ 28″ daily. This phenomenon could only be explained by a decrease in the force of gravity resulting from

the greater centrifugal force at the equator than at the latitude of Paris. The first experiment to demonstrate the Earth's rotation by a physical effect, namely, the deviation towards the East of a freely falling body, was suggested by Newton, and performed by Hooke, in 1679 (cf. my 'An unpublished letter of I. Newton to R. Hooke', *Isis*, 1953). The positive result was undoubtedly caused by experimental errors. Foucault's experiment was made in 1851. The discovery of the parallex of fixed stars, proof of orbital motion, was made by Bessel in 1837.

30. Galileo Galilei, *Dialogo sopra i due massimi sistemi del mondo*, Edit. Nazionale, vol. VII, p. 355:

> '. . . *nè posso a bastanza ammirare l'eminenza dell'ingegno di quelli che l'hanno ricevuto e stimata vera, ed hanno con la vivacità dell'intelletto loro fatto forza tale a i proprii sensi, che abbiano possuto antepor quello che il discorso gli dettava, a quello che le sensate esperienze gli mostravano apertissimamente in contrario. Che le ragioni contro alla vertigine diurna della Terra, già esaminate da voi, abbiano grandissima apparenza, già l'abbiamo veduto, e l'averle ricevute per concludentissime i Tolemaici, gli Aristotelici e tutti i lor seguaci, è ben grandissimo argomento della loro efficacia; ma quelle esperienze che apertamente contrariano al movimento annuo, son ben di tanto più apparente repugnanza, che (lo torno a dire) non posso trovar termine all' ammirazion mia, come abbia possuto in Aristarco e nel Copernico far la ragion tanta violenza al senso, che contro a questo ella si sia fatta padrona della loro credulità.*'

CHAPTER V. THE PHYSICAL PROBLEM. CONCLUSION

1. The Universe of Aristotle and of Ptolemy (the Universe of pre-Copernican astronomy) is obviously extremely small when compared with the infinity of pre-relativity cosmology as well as of present-day cosmology; it is also very small compared with the Universe of Copernicus. All the same, it is sufficiently large not to be constructed on a 'human scale', as is sometimes imagined. Indeed, though 20 000 terrestrial semi-diameters (240 million km) are as nothing when compared not only with the four light years which separate us from the nearest star (Prox. Cent.), but even with the true dimensions of the solar system, it was nevertheless a sufficiently large distance for Ptolemy (and Euclid, Cleomedes, Geminos and others) to declare that the Earth is as nothing, or like a dot, with respect to the celestial vault (G. V. Schiaparelli, *I precursi di Copernico nell'antichita*, Bologna, 1873, German translation: *Die Verläufer des Copernicus im Altertum*, Leipzig, 1876, pp. 73 f.). Finiteness of the Cosmos gives much satisfaction to metaphysical intellect, but not to the emotions: the gods of Antiquity inhabited the Earth, not the heavens. The religions that involve worship of heavenly bodies are based on a misunderstanding: the celestial bodies worshipped by people—the Sun, Moon and planets—are not those

of astronomy; one must be a philosopher to worship the astronomical celestial bodies.

2. They could even revolve, without need of a spiritual driving force, in consequence of the *impetus* imparted to them by God at the Creation, as explained by Jean Buridan (P. Duhem, *Études sur Léonard de Vinci*, Vol. III, pp. 52 f; *Le système du Monde*, vol. VIII, pp. 328 f.) and Nicole Oresme (P. Duhem, *Études*, etc., vol. III, pp. 354 f.; *Le Système du Monde*, vol. VIII, pp. 341 f.); the latter gave as the reason, that the celestial spheres being imponderable offer no sensible resistance to motion. Naturally, it would be quite different in the case of the Earth. Cf. Marshall Clagett, *Science of Mechanics in the Middle Ages*, Madison, Wisconsin, 1959, and W. Hartner, 'Remarques sur l'historiographie et l'histoire des sciences du moyen âge' in *IXe Congrès international d'histoire des sciences, Barcelona-Madrid, 1959*, Paris, 1960.

3. The famous verses of John Donne admirably express the hopeless feeling of cosmic incoherence with which Copernicanism was received:

> . . . New philosophy puts all in doubt,
> The element of fire is quite put out,
> The Sun is lost, and the Earth, and no man's wit
> Can well direct him, where to look for it . . .

Furthermore, Copernican astronomy upset the theological order of the Universe by placing the Earth in the heavens and the Sun in the lowest position in the Universe, where Hell should be. Cf. J. Donne, *Conclave Ignatii* (1611) quoted by Margorie Nicolson, 'The New Astronomy and English Imagination' in *Science and Imagination*, Ithaca, 1956.

4. *De Revolutionibus Orbium Coelestium*, Lib. I, cap. ix, TH., p. 44, P., pp. 101 f. As pointed out by Max Jammer (*Concept of Force*, Cambridge, Mass., 1957, pp. 72 f.), similar ideas, originating from the Stoics and transmitted by Galen, reappear in Girolamo Fracastoro (*De sympathia et antipathia rerum*, Venice, 1546) as well as in Antonius Ludovicus (*De occultis proprietatibus, libri quinque*, Lisbon, 1540); prior to them, similar ideas occur in Marsilio Ficino and in the heart of the Middle Ages in the School of Chartres (Raymond Klibansky, *The Continuity of the Platonic Tradition during the Middle Ages*, London, 1950). It is interesting to note that Nicole Oresme, when dealing with the theories concerning the motion *deorsum* of heavy bodies, says amongst other things (*Quodlibeta*, qu. XXII) that the cause is possibly '*Similitudo illius ad quod movetur, quia omnia similia habent quandam inclinationem naturalem ad invicem, sicut contraria habent se fugere et corrumpere naturaliter, nec de hoc potest deddi bene causa.*' In fact, this concept of gravity, together with that of the theory of *impetus* allowed Oresme to adduce classical arguments against the motion of the Earth (objections based on a stone thrown upwards in the air, clouds and birds); his arguments are very similar to those of

Copernicus, and perhaps even better. Oresme considered that motion (rotation) of the Earth was quite possible. In his view, it would be far more rational to impress a diurnal rotational motion on the Earth than to make the immense sphere of the Universe revolve about it. However, we should not be encouraged thereby to reject the traditional doctrine and to assert terrestrial motion on the pretext that it is more reasonable. Do not religious dogmas include things absurd from the point of view of reason, and true for all that? (*Le Livre du Ciel et du Monde*, edited by A. D. Menut and A. J. Denomy, Toronto, 1943, Lib. IV, pp. 243 f.; and P. Duhem, 'Un Précurseur Français de Copernic, Nicole Oresme', *Revue Générale des Sciences pures et appliquées*, November, 1909). Did Copernicus know of Nicole Oresme? It is not impossible, but seeing that *Le Livre du Ciel et du Monde* was written in French, it seems very improbable to me. It is interesting to note that exactly similar arguments to those of Oresme had already appeared in Arabian astronomy. S. Pines. 'La Théorie de la Rotation de la terre à l'Époque d'Al-Bîrûni' in *Journal Asiatique*, 1956, vol. 249, pp. 301 f.

5. To arrive at this, the idea of qualitatively specific 'natures' must be abandoned, and it must be accepted that the whole Universe, or rather, that all the bodies of which it is composed, have a 'similar' nature. Cf., my study, *La Gravitation Universelle de Kepler à Newton*, Edit. du Palais de la Découverte, Paris, 1955.

6. Copernicus did not abandon entirely the idea of the natural place of bodies. Thus, having explained the nature of gravity, he then tells us in chapter VIII that the rectilinear motion of bodies (the fall of heavy bodies, the rise of light bodies) '*supervenit jis quae a loco suo naturali perigrinantur vel extruduntur vel quomodolibet extra ipsum sunt . . .*' and only occurs to things *non recte se habentibus.* Rectilinear motion, being a motion for the restitution of order and adjustment of previous disturbance, appears therefore less natural than circular motion; furthermore, it is not uniform, but accelerated and has necessarily a limit, whilst circular motion may be prolonged indefinitely.

One could say that these are the classical ideas of Aristotelian physics, save for the essential difference that they are applicable only to the Earth (or to each of the planets taken individually), but not to the Cosmos. The Earth, as such, is not attracted towards the centre of the Universe, as towards its natural place; nor are the Moon and other planets similarly attracted. Expressed in another way, its natural place is not its position in space, but the region of the Universe occupied by its sphere.

7. Concerning the concept of relative motion, cf., P. Duhem, *Le Mouvement relatif et le Mouvement absolu*, Paris, 1902.

8. O. Neugebauer, *The Exact Sciences in Antiquity*, p. 204, considers that the same applies to orbital motion:

'It is, of course, of no interest whether we say that *O* [the observer]

> rotates about S [the physical Sun] or $\bar{S}$ [the mean Sun] about O . . . Thus it is evident that cinematically the two models are hardly different except for Copernicus's insistence on using circles for every partial motion where Ptolemy had already much greater freedom of approach.'

The re-evaluation of Copernicus seems very much a question of the day. For example, D. J. Price in his article 'Contra Copernicus' (Marshall Clagett (editor), *Critical Problems in the History of Science*, Madison, Wisconsin, 1959) considers the general admiration of Copernicus as a mathematician to be a dangerous myth, for he is much inferior to Ptolemy. According to Price (*op. cit.*, p. 198), 'Copernicus himself did not understand what he was doing. He believed that the change from geostatic to heliostatic would make the theory more accurate, whereas it leaves it precisely unchanged.' The mathematical technique of Copernicus is certainly that of Ptolemy, and the two systems are equivalent mathematically, so it is difficult to accept that 'he did not realize that he was the first inventor of a mathematical planetary system as distinct from a mathematical theory of the individual planets' (*op. cit.*, p. 199). In fact, Copernicus (as well as Rheticus) continually emphasized that the new astronomy reveals the inherent order of the Cosmos, whereas the old astronomy does not. Finally, it was not in perfection of mathematical theory, but in the agreement of heliocentrism with reality, that Copernicus, as well as his successors, saw the superiority of his system over that of the Ancients.

9. *De Revolutionibus Orbium Coelestium*, Lib. I, cap. vii; TH., p. 20; P., p. 87: 'If the Earth revolved, said Ptolemy of Alexandria, that is to say made one revolution in a day . . . this motion, which in twenty-four hours passes over the whole circumference of the Earth, would be extremely violent and of unsurpassable speed. Now, things moved by a violent motion seem to be totally unable to gather together, but must rather disperse. And, he continued, the dispersed Earth would long since have passed beyond the sky itself (which is quite absurd).' Copernicus is apparently referring to the *Almagest*, Book I, chap. vi. However, Ptolemy does not speak there of the destructive action of *centrifugal force*, but only of the disruptive force of motion as such. Had Copernicus found some precise mention of centrifugal force? It is possible, but it is also likely that he himself put this interpretation on Ptolemy's text, either because he believed that this disclosed the true meaning of the great astronomer, or because he felt that he ought to make his objection in a better, *i.e.*, stronger, way.

10. Jean Buridan and Nicole Oresme had already said almost the same; see nn. 2 and 4.

11. In fact, even for Galileo, circular (orbital) motion of the planets remains a natural motion and does not produce centrifugal forces. Descartes was the first to introduce them into the heavens; and Borelli, into astronomy.

12. It might be objected that this had already been done by Epicuros. Epicurianism, however, never inspired a scientific theory, neither in antiquity, nor after its revival by Gassendi.

13. The similarity between the dynamics of Nicholas of Cusa and of Copernicus has frequently been noted, and rightly so (P. Duhem, *Études sur Léonard de Vinci*, vol. II, pp. 186, 201). Both Copernicus and Nicholas of Cusa assert that roundness is the cause of motion; on the other hand, the theory of *impetus* is not found in the writings of Copernicus, neither in the form given to it by the Parisian nominalists (Buridan and Oresme), nor in the form given to it by Nicholas of Cusa.

14. From this point of view, a fragment of the planet Earth (a stone), if transported near to the Moon or to Mars, would fall back again to Earth, and not on to the Moon or on to Mars; the same would be true in respect of fragments of the Moon or of Mars, which would fall back again to *that* planet with which they are related by *their nature*, and not on to that one which, by chance, would be the nearest. On the other hand, according to Aristotelian physics, if the Earth by some remote chance were transported to the Moon, and a fragment of the Earth were detached, this fragment would not return to the Earth, but would immediately fly away to the centre of the Universe.

15. Aristotle, *De Coelo*, I, 2. After all, it is superfluous to quote authorities: everyone believes and admits that a round ball is made to roll. For example, Nicholas of Cusa (*De Ludo Globi*) explains that a perfectly round body, placed on a perfectly smooth surface and then set in motion, would never be able to stop, because it would be impossible for it to remain at rest on an atom (a sphere touches a plane only in one single point); furthermore, being round, it reproduces, whilst turning, the same position with respect to the supporting surface, and consequently has no reason for stopping. In short, it turns because it is round. Similarly with the spheres and circles of Copernicus: they revolve because they are round, and because their very nature, their (geometric) shape, expresses itself in their circular motion. P. Duhem, *Études sur Léonard de Vinci*, vol. II, pp. 181–201; E. Cassirer, *Das Erkenntnisproblem in der Philosophie und Wissenschaft der Neueren Zeit*, Berlin, 1906, vol. I, pp. 284 f.; Annelise Maier, *Die Vorläufer Galileis im XIV Jahrhundert*, Rome, 1949, pp. 151 f.

16. E. Brachvogel (*Nikolaus Kopernikus in der Entwicklung des deutschen Geisteslebens*, pp. 49 f.) does not agree with my interpretation, and cites the works of Copernicus in which the latter describes the Sun as 'governing the family' of surrounding celestial bodies; and he quotes Rheticus who finds the 'principle of motion and light' in the Sun. E. Brachvogel interprets the doctrine of Copernicus as an expression of the conception of Antiquity (especially of the Stoics) according to which the Sun is the source of heat and of life for the Universe, and considers that Copernicus did not see in the roundness of the spheres and circles the source of their

motion, but only its 'natural' character. For my part, I think I have adequately shown the Neo-Platonic and Pythagorean inspiration received by Copernicus, and that we can accept that, for him, the Sun is in the last analysis the source of life and, consequently, of motion, because it is the source of light. Nevertheless, I believe that my dynamic interpretation of the geometrization of the shape of 'divine bodies' is correct.

17. The reasons for raising the Earth to the heavens are clearly very different for Copernicus than for Nicholas of Cusa. Cf., *supra*, chap. I, n. 7.

18. Kepler saw this quite clearly. Uniform circular motion does not need a driving force; whereas non-uniform motion demands the action of some driving force, even non-uniform action of this force. Cf. *infra*, p. 192 f.

19. Similarly for the other planets: their centre of motion, *i.e.*, the centres of their deferent circles are placed *near* to the Sun, but not *in* the Sun. It was Kepler who transferred the centre of planetary motions into the Sun. Cf. *infra*, p. 154.

Figs. 8a and 8b make use of the excellent diagram prepared by William D. Stahlman for G. de Santillana's translation of Galileo's *Dialogo sopra i due Massimi Sistemi del Mondo* (Chicago, 1953), and reproduced in his *Le Procès de Galilée*, Paris, 1955.

Figs. 8c and 8d are the diagrams prepared by Michael Maestlin for Kepler's *Mysterium Cosmographicum*.

20. In other words, the 'mean Sun' and not the 'true Sun'.

21. This was done because Copernicus accepts that the Earth moves with uniform motion on, or rather, with its circle—as does the Sun according to Ptolemy. Kepler considered that Copernicus did so 'in order not to depart too far from Ptolemy'. As Kepler pointed out (p. 172) it was precisely on that account that the orbital motions of the Earth and planets do not provide an adequate explanation of the apparent changes in latitude of the latter; consequently, Copernicus was under the necessity of admitting their reality and explaining them by an oscillatory motion of the planetary circles as a function of the Earth's motion. It was, as Kepler remarked, quite irrational.

22. *De Revolutionibus Orbium Coelestium*, Lib. I, cap. xxii, p. 222; *Narratio Prima*, p. 453, J. L. E. Dreyer, *History of the Planetary Systems*, p. 332, from which Fig. 12 (p. 115) is taken.

23. Kepler distinguished between 'Copernicus who speculates' and 'Copernicus who calculates'. Cf. *infra*, p. 190.

24. It is not always, or perhaps not sufficiently, appreciated that by placing the Sun at the centre of the Universe in virtue of its dignity, Copernicus returned to the Pythagorean conception and completely overthrew the hierarchy of positions in the ancient and medieval Cosmos, in which the central position was not the most honourable, but, on the contrary, the most unworthy. It was, in effect, the *lowest*, and consequently

appropriate to the Earth's imperfection. Perfection was located *above* in the celestial vault, above which were 'the heavens' (Paradise), whilst Hell was deservedly placed beneath the surface of the Earth.

25. Lynn Thorndike (*History of Magic and Experimental Science*, New York, 1941, vol. V, p. 425) treats this passage from *De Revolutionibus Orbium Coelestium* as a 'rhapsodical lapse'. Personally, I feel that Thorndike has failed to recognize, under the cloak of rhetoric, the deep emotion with which Copernicus was imbued.

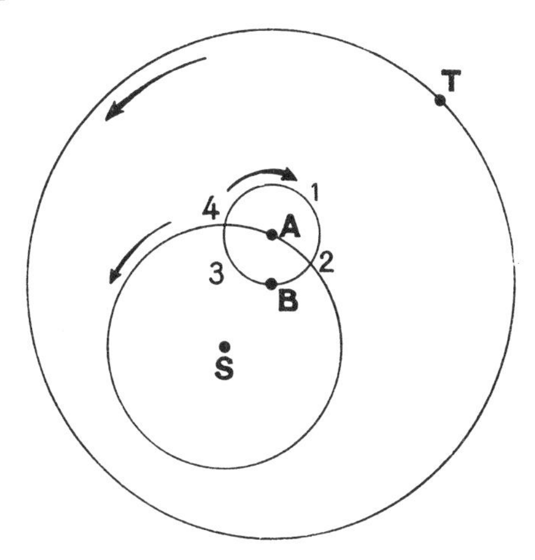

Figure 12.

S = Sun. *T* = Earth, whose circle has its centre at *B*, located on a peri-circle (epicycle) whose centre *A* is in its turn located on the deferent *D*, which has its centre in the Sun.

26. *De Revolutionibus Orbium Coelestium*, Lib. I, cap. x; TH., p. . . . ; P., p. 115. E. Zinner (*Entstehung und Ausbreitung* . . ., pp. 130 f.) has collected numerous passages which illustrate this tradition. Others may be found in E. Brachvogel, *Nikolaus Kopernikus in der Geschichte des deutschen Geisteslebens*, pp. 61 f. Cf., also, Nicolaus Copernicus, *De Revolutionibus Orbium Coelestium*, Warsaw, 1953, Book I, notes.

27. In my 'Nicolaus Copernicus', *Quarterly Bulletin of the Polish Institute of Arts and Sciences*, New York, 1943; and elsewhere.

28. If Copernicus had been able to foresee the evolution of heliocentric astronomy, it would no doubt have made him frightened and indignant. Heliocentric astronomy started by rejecting the fundamental principle of uniform circular motion, and then even rejected the principle of the circular motion of celestial bodies. It suppressed the planetary spheres and even the celestial vault. . . . One might say that it developed in spite of itself.

29. The infinite nature of the Universe of Copernicus has been asserted by G. McColley ('Nicolaus Copernicus and an Infinite Universe', *Popular Astronomy*, 1936, vol. XLIX), but his interpretation is untenable, as has been shown by F. Johnson (*Astronomical Thought in Renaissance England*, Baltimore, 1937, p. 107). Cf. my *From the Closed World to the Infinite Universe*, Baltimore, 1955.

II

KEPLER AND THE NEW ASTRONOMY

Introduction

'Mihi non minus admirandae videntur occasiones quibus homines in cognitionem rerum coelestium deveniunt, quam ipsa Natura rerum coelestium.'

In the galaxy of great minds whose combined efforts have produced that which it is customary to call the 'scientific revolution of the seventeenth century', Johann Kepler occupies a unique position. It could be said, that he was both in advance of, and behind, his contemporaries. He was, in fact, the first to overcome, at least partially, the obsession with circularity from which Galileo was never able to free himself, and to divest astronomy of the cumbrous apparatus of spheres and circles which Copernicus had rigidly maintained. Consequently, from the purely scientific viewpoint, he, far more even than Copernicus who technically was a Ptolemaean or even a rather strict Hipparchan, was the true founder of the New Astronomy. Yet, it was he who, with all the power at his command, opposed Giordano Bruno's attempt to make the Universe infinite[1]; and believed with the strength of iron—or rather with the strength of ice, for he thought it was composed of ice—in the existence of a celestial vault enclosing the Universe and containing the fixed stars.

He, too, was the first to conceive and conscientiously follow, if not realize, a programme directed to the scientific unification of the starry Universe and the sublunary Universe, and substitute celestial dynamics for the kinematics of circles and the metaphysics of spheres in traditional astronomy. He it was who based this dynamics on solar astro-biology by ascribing to the Sun a spiritual, perhaps even intellective, driving force, and to the Earth not only a spiritual, but also a sensory driving force. He based his defence of heliocentrism on a conception of the Universe which, by an inspiration both

Pythagorean and Christian, saw therein an expression of the divine Trinity, and caused the concept of the Cosmos, hereafter condemned, to shine with supreme brilliance. . . .

Johann Kepler is a veritable Janus; and that is why a study of his ideas is so attractive, yet so difficult. If we may judge by the bibliography of Kepler,[2] it is more difficult than attractive. It is well-known that we lack an exhaustive study of Kepler.[3] This is a rather serious deficiency, for Kepler was not only a genius of the first rank, who has left a deep impression on all the branches of knowledge he touched—mathematics, optics, astronomy—he was also an influence who played a decisive part in the development of science. Without Kepler, without the strenuous work he undertook, without his years of laborious and wearisome calculations, the progress of astronomy would have been delayed for a century; without Kepler, there would have been no Newton.[4]

Now, as regards these calculations, this work which no-one would have done in his stead, for no-one amongst his contemporaries understood its importance,[5] would Kepler have undertaken it in order to sustain him in his isolation had not his scientific passion as an astronomer been exceeded by his ardour as a metaphysician and even his faith as a believer?

Naturally, I shall not attempt a detailed consideration of Kepler's scientific work; I shall restrict myself to that of Kepler the astronomer.[6] The direction and deep inspiration of this work consist, according to Kepler, in the substitution of a 'philosophy' or 'celestial physics' for the 'theology' or 'metaphysics' of Aristotle[7]; in the demonstration that 'celestial physics' is to be elaborated by geometry and computation; in the proof that 'celestial physics' and 'terrestrial physics' do not constitute two 'physics', but one only.[8]

Clearly, this represents a complete overthrow of the structure and meaning traditionally ascribed to the science of celestial bodies, and the question naturally arises: How did Kepler come to adopt such a revolutionary attitude? By way of answer we could suggest: Kepler asked himself—*a quo moventur planetae* (what is it that makes the planets move)? He gave this question a dynamic significance, which was something no-one had done before.

In astronomy of the Middle Ages and Antiquity, this problem, which seems quite natural, does not arise; or, if one prefers, it is resolved before it arises. In the traditional cosmological model, the planets as such do not move, but are moved by, or with, the spheres and celestial orbs in which they are set. The spheres, being made of a very special, imponderable material and offering no resistance to motion (particularly circular motion), are moved in virtue of their own proper natures, or even by 'spirits' or 'intelligences' which are attached to them. The causes of motion in this instance are trans-physical.[9]

On the other hand, the problem of the causes of planetary motion has no meaning for mathematical astronomy: the circles and spheres which appear in this astronomy are mere adjuncts to computation, and there is no claim for their physical reality. Celestial motions belong to the realm of pure kinematics: astronomy is one thing, physics is something quite different.

The advent of Copernican Astronomy changed nothing, or practically nothing, in this state of affairs. There was scarcely any dynamic problem for Copernicus, because in his view the planets were still transported by the celestial spheres, which, like the Earth, revolved simply as a result of their spherical shape. Even Tycho Brahe and Giordano Bruno, who no longer believed in planetary spheres, did not raise the decisive question. Tycho Brahe wavered between belief in planetary spirits and a purely computational attitude; Giordano Bruno, on the other hand, frankly adopted an animistic conception. Kepler was the first, and the only one up to the time of Descartes, to ask for a physical explanation.

The reply that I have outlined above proves therefore to be right, but the question then arises: Why did Kepler ask for this explanation?

I believe that he did so for a highly metaphysical reason, namely, that he divided the visible Universe quite differently from the way in which his predecessors had done. He did not counter-balance the Earth by the heavens, nor the latter by the planets (notwithstanding his veneration for the Sun), but, pursuing a suggestion of Copernicus,[10] he counter-balanced the motionless Universe (the Sun,[11] space and the fixed stars), *en bloc*, by the moving Universe, which comprises the planets and Earth, and which thereby acquired in his mind a unity and similarity in nature and structure that rendered it subject to the same physical laws.

Furthermore—but here we come to the psychological aspect, the very structure of Kepler's genius, a structure that we can try to understand, though it would be in vain to try and explain it—Kepler always raised questions that nobody else raised, and sought answers on matters where certainly nobody else saw any problem. He was firmly convinced that an answer must be forthcoming to every reasonable question.

Whilst his predecessors accepted the cosmological state of affairs as the ultimate fact beyond which it was impossible to enquire, and at most believed (as did Copernicus) that they had discovered its specific law of order,[12] Kepler boldly sought the underlying reason. The Cosmos was not formed by chance; it was created by God; and God, assuredly, did not create it *temere*, haphazardly, but on the contrary was guided by rational considerations and followed a perfect architectural plan.[13] Consequently, Kepler considered it perfectly reasonable to seek the architectonic principles which determined the structure and composition of the Cosmos; and not to limit himself, as did Copernicus, to exposing the principle on which the solar system is arranged, namely, that the planets complete their course round the Sun in a time which increases the further they are away. Kepler asked himself the further questions: *Why* are the superior planets slower in their motion than the inferior planets? and *why* is there a fixed number of planets, namely, six and not four, eleven or fourteen?

Let us be quite clear on the matter! These questions are not concerned with God's purpose in creating the Universe. There is no intention of developing a cosmo-theology centered on the needs of Man, of which there are many examples in history. Although it was quite clear to Kepler that the Universe was created for man, and that man forms the axiological centre of the world and so determines to a very great extent its architecture,[14] nevertheless, Kepler was looking for something quite different from a teleological explanation. In the first place, he was concerned with finding the constructional laws (he called them the 'archetypal' laws) which, in the mind of the Creator, directed the creation of the Universe. In Kepler's view, these laws could only be mathematical, or indeed, to be more precise, geometrical ones. In the next place, he was concerned to find the physical (dynamic) means used by the Divine Architect, or Engineer, to keep his construction together, or to set it in motion.

This was by no means a preposterous undertaking in Kepler's estimation. Indeed, why should not an earthly mathematician be capable of penetrating, at least partially, the mind of the Divine Mathematician?[15] Was not Man, that is to say his soul, created in the image and likeness of God?

It is this attitude which is revealed in the *Mysterium Cosmographicum*, Kepler's first writing; and this very same attitude is found again in his last great works, the *Harmonice Mundi* and the *Epitome Astronomia Copernicanae.*[16] It is practically certain that this attitude inspired all his other works.

I. THE BEGINNINGS

I

Mysterium Cosmographicum

Johann Kepler was born in 1571 at Weil (now called Weil der Stadt) in Württemberg, and studied at the University of Tübingen. It is worth noting that, although he had developed a keen interest in astronomy through the influence of his teacher, Michael Maestlin, and had zealously studied the subject, it was not to this science but to theology that he originally intended devoting himself: in fact, he wanted to become a pastor. It so happened, however, that the combined post of mathematician to the Province of Styria and teacher at the Protestant seminary (*Stiftschule*) at Graz fell vacant by the death of the occupant, Georg Stadius (1550–93). Early in 1594 the provincial authorities in their search for a candidate enquired of the University of Tübingen, the senate of which recommended Kepler,[1] who naturally could not refuse. He accepted, but made it clear that, as far as he was concerned, it would be only a temporary position, and that he hoped after some years to embrace an ecclesiastical career, for which he was particularly well suited in the opinion of his teachers, and to which he himself aspired with all his heart and soul. After all, had not he received, two years before, from his native town of Weil a bursary on the recommendation of the Faculty of Theology? Fate—or Providence—decided otherwise.[2]

The rather low level of teaching at Graz bored him; moreover, he had hardly any pupils. His official duties as mathematician to the Province involved the preparation of calendars and making 'prognostications',[3] and were scarcely more heartening. They did leave him, however, with leisure for his private investigations and rapturous meditation on the structure of the Universe. The *Mysterium Cosmographicum*,[4] which he wrote in 1595, and published, thanks to the help of Maestlin,[5] in 1596, was the fruit of these meditations.

The *Mysterium Cosmographicum*, which contains the germ of Kepler's future great discoveries, has all the spontaneity of youth.

With touching and naive confidence, Kepler reveals himself to us completely: the preface to the work explains its purpose, and at the same time traces the development of his ideas:[6]

> 'Dear Reader, (he says), it is my intention in this small treatise to show that the almighty and infinitely merciful God, when he created our moving world and determined the order of the celestial bodies, took as the basis for his construction the five regular bodies which have enjoyed such great distinction from the time of Pythagoras and Plato down to our own days; and that he co-ordinated in accordance with their properties the number and proportion of the celestial bodies, as well as the relationships between the various celestial motions.
>
> 'However, before giving proof of this, I wish to state the reasons which have led me to write this small treatise, as well as the manner in which I have proceeded; in this way I shall be better known, and better understood.
>
> 'Already six years ago, when I benefitted from the company of the most famous Michael Maestlin, I was aware of the very unsatisfactory nature of the then accepted conception of the structure of the Universe. Moreover, I developed so much enthusiasm for Copernicus, whom my teacher had frequently mentioned in his lectures, that on many occasions I defended his opinions in disputations with candidates [for a degree], and I even prepared a whole disputation to defend the thesis that the 'prime motion' had its origin in the rotation of the Earth [and not in that of the firmament]. I was already trying to ascribe the Sun's motion to the Earth by virtue of physical, or if preferred, metaphysical reasons, as Copernicus had done by virtue of mathematical reasons.[7] To this end I have, little by little, collected and put together all the mathematical advantages which [the system of] Copernicus possesses in comparison with [that of] Ptolemy, partly drawing them from Maestlin's lectures, and partly discovering them for myself.'

Kepler's 'disputation' has not survived,[8] but it is highly probable that the arguments used by Kepler the student to defend Copernican astronomy were already the same as those which appear in the masterly first chapter of the *Mysterium Cosmographicum*, where Kepler insists to a much greater degree than did Copernicus on its systematic, explanatory and rational character. Indeed, in Kepler's view, the great superiority of the Copernican system compared with the Ptolemaic does not depend on the fact that it did away with some few circles, and hence facilitated computation, even though this was by no means negligible, especially as regards the saving in the number of circles and consequently of motions. That is not the main point.

What made the Copernican system vastly superior to the traditional teachings was its ability to explain matters which for Ptolemy were no more than bare observational data. Thus, it was able to explain (and similarly eliminate) certain apparent inequalities in planetary motions, such as stations, and retrogradations, which Ptolemy was obliged to accept, but which Copernicus explained. He explained them all, simultaneously, by one single factor, namely the Earth's motion and the change in relative positions occupied by the Earth in its orbit and by the other planets in theirs. He explained what Ptolemy was unable to do—why the inferior planets can never move far enough away from the Sun to be in opposition to it as in the case of the superior planets; he explained also—and this was something Ptolemy was obliged to accept as a plain fact—why, in the case of the superior planets, apogee is always at conjunction, and perigee is always at opposition to the Sun. He explained why, in the system of circles on which the planets have their motion, there is always one of them—the deferent of inferior planets and the epicycle of superior planets—which completes its course in the same time as the Sun completes its own; in other words, he explained why the Sun plays such an important part in the planetary system. The part it plays in the Ptolemaic Universe is just as important, even more so perhaps, than in the Copernican, as Kepler showed with admirable insight by presenting his readers with geometrical diagrams of the two rival systems. There is this difference, however; with Ptolemy, this pre-eminent rôle of the Sun is based on nothing, whilst with Copernicus it forms part of the very structure of the system, which replaces the empirical disorder of his predecessor by a rational arrangement.

Kepler himself said[9]:

> 'My confidence [in Copernicus] was upheld in the first place by the admirable agreement between his conceptions and all [the objects] which are visible in the sky; an agreement which not only enabled him to establish earlier motions going back to remote antiquity, but also to predict future [phenomena], certainly not with absolute accuracy, but in any case much more exactly than Ptolemy, Alfonso and other astronomers. Furthermore, and this is much more important, things which arouse our astonishment in the case of other(s) [astronomers] are given a reasonable explanation by Copernicus, and thereby he destroys the source of our astonishment which lies in the ignorance of causes.[10] The easiest way to convince oneself of this is to read the *Narratio* by Rheticus, for it is not given to everyone to understand the writings of Copernicus on *the Revolutions*.'[11]

The system of Copernicus was the *true* one for Kepler. So, taking a resolute stand against the 'hypothetical' interpretations of Copernicanism suggested by Osiander in his famous preface to *De Revolutionibus Orbium Coelestium*, Kepler went on to say that he had never been able to agree with those who, pleading examples of proofs by which a true statement is deduced from false premises, maintain that it is possible for the hypotheses put forward by Copernicus to be false, but nevertheless that real phenomena can be deduced from them.[12]

> 'The analogy [between a true conclusion derived from false premises and the theory of Copernicus] is not valid, for a [true] conclusion from false premises is accidental; and its inherent falsity betrays itself as soon as it is applied to something other than that for which it was deduced. . . . The point in question is quite otherwise for him who places the Sun at the centre; for once this hypothesis has been put forward he will be able to demonstrate any of the phenomena which indeed appear in the sky; he can go forwards or backwards in time, deduce one phenomenon from another, thus showing how they are intimately connected together; and the most complicated demonstrations will always bring us back to the same initial hypotheses.'

Isn't the possibility of drawing correct conclusions from false premises an established fact in astronomy? Can't we deduce the same observable phenomena starting from different 'hypotheses'? Ought not the Copernican, more than anyone, to accept this possibility because he is obliged to recognize the fact that the Ptolemaic system, which he considers to be wrong, nevertheless enables him to draw up tables and to compute the motions of celestial bodies?

Kepler tried to dispel the illusion in logic which is the basis of this objection[13]:

> 'It will perhaps be objected that the same can also be said, or at least could be said formerly, about the tables and hypotheses of the Ancients, namely, that they too agree with the phenomena [and allow them to be deduced]; and that Copernicus, however, rejected them as being false. It would seem, then, that the same can be said about Copernicus, namely, that he deceived himself by his hypothesis, although he gave a satisfactory account of the phenomena. At the outset, I shall say that the hypotheses of the Ancients do not give answers to certain very important questions. For example, they take no account of the cause of the number, the amount and the duration of retrogradations [of planets]; nor do they explain, why these retrogradations agree so well with the position and

the motion of the mean Sun. Now, all these things have of necessity some underlying cause.'

which was revealed by Copernicus when he demonstrated the wonderful orderliness governing all these phenomena.

'Furthermore, Copernicus contested none of the hypotheses which agree with observation and account satisfactorily for appearances; if anything, he accepted and explained them. [At first sight], it seems that he changed many things in the accepted hypotheses, but in fact it is not so. Indeed, it may be that the same conclusion results from two pre-suppositions, different in species, because these two [pre-suppositions] fall in the same genus, and it is in virtue of the genus [of cause and not of specific nature] that the result in question is produced. Thus, Ptolemy had shown [the reason for] the rising and setting of celestial bodies, but [he did not prove it] with respect to the nearest middle term, namely, with respect to the fact of the Earth's immobility at the centre [of the Universe]. Neither did Copernicus make this demonstration with respect to the middle term [which in his system corresponds to the Earth's immobility in Ptolemy's], namely, the fact that the Earth completes one revolution about the centre of the Universe whilst remaining at a certain distance from it.

'It is sufficient for both of them to say [as in fact they do] that the phenomena in question occur because there is a certain antithesis between the Earth and the firmament as regards motions, and that the distance of the Earth from the centre [of the Universe] is not significant with respect to the fixed stars.

'Consequently, Ptolemy has not demonstrated the phenomena which he examined with respect to a false or accidental middle term, but has only sinned against the law χατ' ἀυτὸ [of logic] by considering that [the things] which happened because of the genus, happened because of the species. So, it becomes evident, why, starting from a false arrangement of the Universe, he was nevertheless able to demonstrate things which are true and conformable with the sky as well as our eyes; and it is clear also that this does not authorize us in any way to assume anything of a like nature in regard to the Copernican hypotheses.

'On the contrary, the result is rather similar to what I said at the beginning, namely, that the principles of Copernicus, which reveal the constant cause unknown to the Ancients of a large number of phenomena, cannot be false. This was fully realized by the celebrated Tycho Brahe. Although he did not agree with Copernicus regarding the position of the Earth, nevertheless he admitted its competency to account for hitherto unknown facts: for example, the fact that the Sun is the centre [of motion] of the five planets. In fact, the proposition that the Sun is motionless at the centre [of the Universe] is a much too restricted mean

TABELLA I. Exhibens ordinem sphaerarum coelestium mobilium: simulque veram proportionem magnitudinis earum iuxta medias suas distantias: item angulos prosthaphaereseon earundem in orbe Magno Telluris, secundum sententiam Copernici.

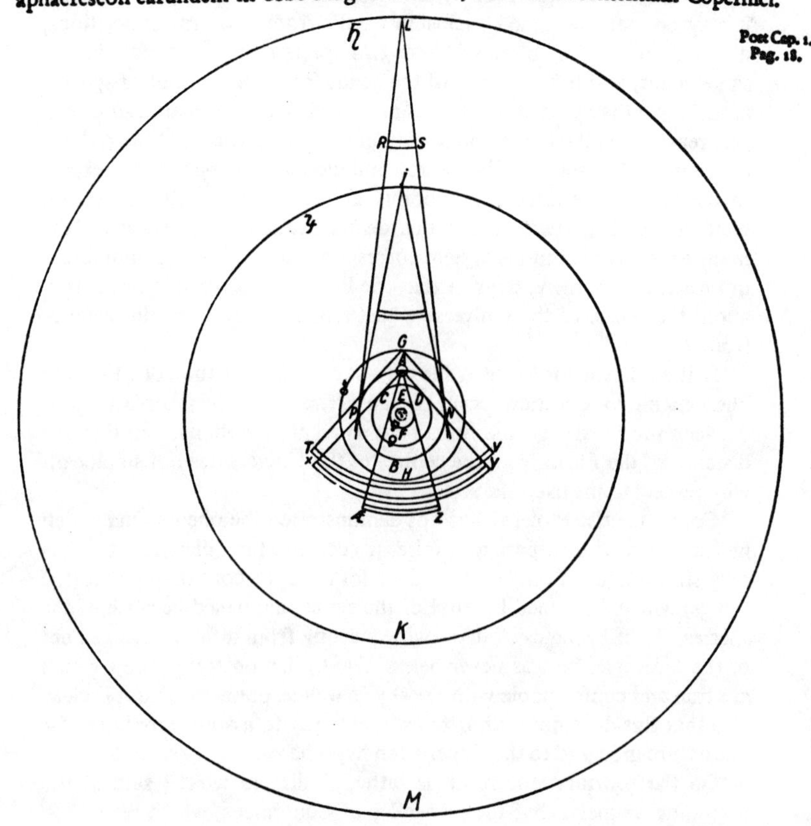

Figure 1a (see p. 135).

term for the demonstration of retrograde motions. The general [proposition] that the Sun is at the centre of the five planets is sufficient.'

Having thus shown that the foundation for truth can be nothing but the truth, and having at the same time very skilfully shown the structure of the *real* logical foundations (purified by formal analysis) of Ptolemy's and Tycho Brahe's arguments (which analysis, moreover, reveals their complete agreement with Copernicus), Kepler thereafter felt justified in passing from 'astronomy', *i.e.*, the purely computational study of celestial phenomena, to 'physics and cosmography', *i.e.*, the study of *reality*[14]:

> 'Now, in order to pass from astronomy to physics or cosmography, the hypotheses of Copernicus are not only contrary to the nature of things, but on the whole they agree with it. Nature likes simplicity and unity. Nothing trifling or superfluous has ever existed: and very often, one single cause is destined by itself to [produce] several effects.[15] Now, with the traditional hypotheses there is no end to the invention of circles; with Copernicus, on the other hand, a large number of motions is derived from a small number of circles.[16] Consequently, Copernicus has not only

---

In centro, vel propre est SOL immobilis.

E F minimus circa Solem circulus est MERCVRII, qui restituitur diebus 88 ferè.

Hunc sequitur VENERIS C D, cuius reuolutio circa eundem Solem est dierum 224. cum besse.

Qui hunc sequitur A B, TELLVRIS est, cuius reuolutio dierum 365. et quadrantis. Dicitur ORBIS MAGNVS, propter vsum multiplicem.

Circa Tellurem est orbiculus velut epicyclus, SPHAERAE LVNARIS, ad A, eodem motu per anni spacium cum tellure ad eandem stellam fixam rediens. Sed eius propria reuolutio ad Solem habet dies 29. cum dimidio.

Post hunc est Orbis MASTIS G H, qui cursum vnum sub fixis stellis, siue ad Solem, absoluit diebus 687.

Hunc excipit post magnum interuallum, Sphaera IOVIS I K, habens ambitum dierum 4332. cum quinque octauis ferè.

L M vltimus et maximus, est SATVRNI, eius tempus periodicum dierum 10 759. cum quinta.

FIXAE veró STELLAE adhuc tam inæstimabilli interuallo altiores sunt, vt ad eam, quæ est inter Solem et Terram intercapedo sensibilis non sit Et eæ sunt in extremo, sicut Sol in centro, penitus immobiles.

Angulus T G V, vel Arcus T V, prosthaphæresis est, siue parallaxis, quam Orbis Magnus Teliuris ad Sphæram Martis habet.

Sic P I N est eiusdem Orbis Magni parallaxis ad Sphæram Iouis: et P L N, siue R L S vel R S arcus ad sphæram Saturni.

Ita X A Y, vel X Y arcus est parallaxis sphæræ Veneris: ut et Z A Æ, vel Z Æ sphæræ Mercurij parallaxis, ad Orbem Magnum.

TABELLA II. Exhibens ordinem sphaerarum coelestium, et utcunque proportionem orbium et epicyclorum, atque angulos vel arcus prosthaphaereseon eorundem, iuxta medias distantias, secundum Veterum sententiam.

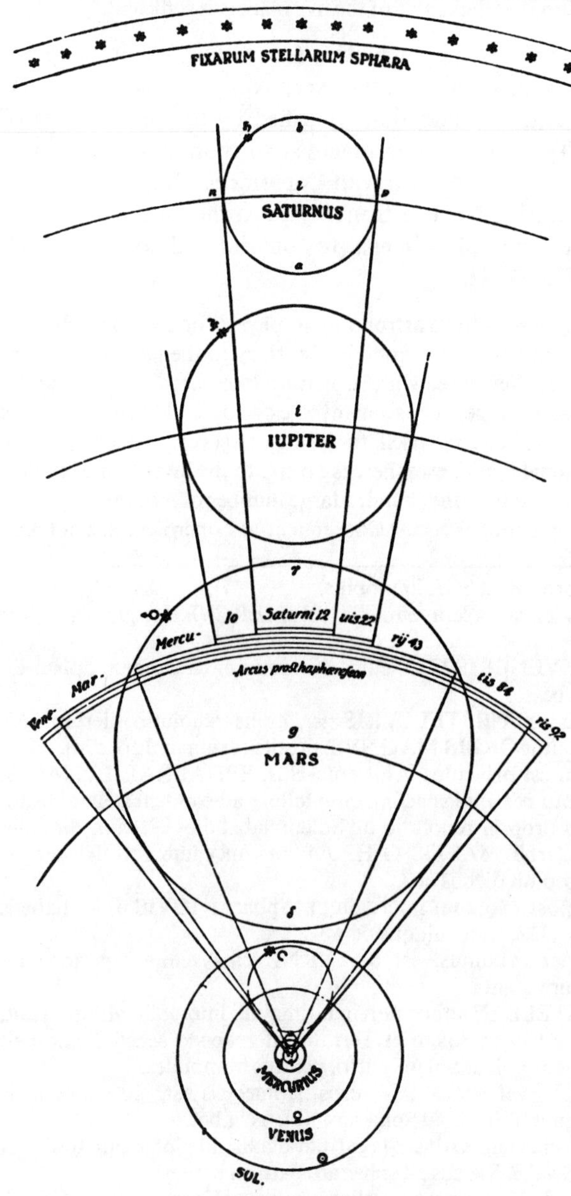

In centro TERRA est, sola immobilis. Intimus circa Terram orbiculus LVNAE Sphaeram repraesentat, cuius motus menstruus est. Hunc proximè MERCVRII orbis circumdat: quem sequitur VENERIS, et postea SOLIS Sphaera, annua omnes conuersione volubiles. Reliquorum trium superiorum MARTIS, IOVIS et SATVRNI orbes, FIXARVM quoque STELLARVM Sphaeram, arcus, quos circa terram, ceu centrum integros describere, et complere quisque potest, indicant. Martis orbis biennio conuertitur. Iouis 12. annos quàm proximè, requirit. Et Saturni ferè 30. annos. Fixae Stellae 49000. annis, iuxta Alphonsinorum placita, periodum restituunt. Quantas singulorum (praeter ☽) epicycli in concentrico circulo prosthaphaereses, in medijs distantijs faciant, arcus, rectis ex terra ductis, et epicyclos singulos tangentibus intercepti, additi graduum numeris, monstrant.

Post Cap. 1. Pag. 18.

Figure 1b (see p. 135).

freed nature from a laborious and useless accumulation of so vast an array of circles, but he has furthermore opened to us an inexhaustible treasure of truly divine considerations concerning the marvellous orderliness of the Universe.

'For I have no hesitation in asserting that everything, that Copernicus has demonstrated *a posteriori* and on the basis of observations interpreted geometrically, may be demonstrated *a priori* without any subtlety of logic.'

Kepler added that Aristotle himself, were he still alive,—a wish frequently voiced by Rheticus—would certainly have borne out this assertion. He felt that he could not embark on a full account of Copernican theory relating to planetary motion—he recommended the reader to refer to Rheticus, or Copernicus himself. He expressed his intention of limiting himself to a brief sketch of it accompanied by two diagrams which would enable the reader to understand how Copernican *reality* could give rise to Ptolemaic *appearance*.[17]

'In order to become acquainted with the order of the Universe according to Copernicus, consider figure I at the end of this chapter together with what is written thereon. Copernicus ascribed four motions to the Earth under various conditions (for brevity, Copernicus said three, but there are really four[18]). As a result of these motions, certain apparent inequalities in the motion of the other planets, as seen by us, become evident.'[19]

'The first motion is that of the sphere, or circle, which in one year carries the Earth as a celestial body around the Sun. Now, this circle, being eccentric and furthermore of variable eccentricity,[20] must be considered in three different ways. In the first place, we shall take no account of its eccentricity. The motion of this circle, or of the Earth, enables us to make an economy of three eccentrics in the usual hypotheses, namely those of the Sun, Venus and Mercury. Because the Earth revolves round these three planets, the inhabitants of the Earth are led to believe that these three [planets] revolve round a stationary Earth. Consequently, they make three motions out of one. If there were other celestial bodies within the Earth's orbit, they would ascribe motion to them also. Furthermore, this circle [of the Earth] being admitted, the three large epicycles of Saturn, Jupiter and Mars disappear also. It will be seen from the accompanying drawings that the Earth, viewed from Saturn (which is almost motionless, because it has the slowest motion), moves on its circle, first approaching and then receding from Saturn; but the [Earth's] inhabitants believe that it is Saturn which moves on an epicycle with a backward and forward motion, whilst they themselves [the Earth] are motionless at the centre of its circle. Therefore, they project the circle *AB* [on to the sky] and believe it to be the epicycles *g*, *i*, *l*. In the same way, the latitudes of the five planets appear to undergo changes, because the Earth during the course of its motion on its circle approaches and moves away from the planets. Now, in order to 'save' these librations, Ptolemy was obliged to introduce five more motions, all of which can be eliminated by the motion of the Earth by itself.

'Consequently, all these motions, eleven in number, are eliminated from the Universe, and one single motion, that of the Earth, replaces them all[21]; furthermore, explanations are found for the causes of many other phenomena, which Ptolemy could not account for by means of his more numerous motions.

'1. In the first place, Ptolemy could be asked: How is it, that three of the eccentrics, namely those of the Sun, Venus and Mercury, have the same [period] of revolution ?

'This is a question to which Ptolemy can give no answer. On the other hand, it is easy to understand why the apparent periods of revolution are the same if it is accepted that the eccentrics in question, namely the circles which move the Sun, Mercury and Venus round the Earth, are nothing more than projections of the Earth's proper motion on to the firmament.

'2. Why do the five planets show retrograde motion, whilst the luminaries [the Sun and the Moon] do not?

'First of all, with regard to the Sun, the answer is that it is motionless. Consequently, the motion of the Earth, which is always direct, is referred purely and simply to the Sun, except that it appears to be directed towards the opposite part of the sky. With regard to the Moon, the answer is that the annual motion of the Earth's circle is common to both the Earth and the Moon; and two bodies which have the same motion throughout the whole of their course appear to be at rest with respect to each other. Consequently the Earth's motion is not revealed by the appearance of motion by the Moon as is the case with respect to the other planets. With regard to the superior planets, Saturn, Jupiter and Mars, it must be said that they are slower than the Earth, and consequently the Earth's motion is ascribed to them [by the inhabitants of the Earth]. Therefore, to observers on the globe of Saturn, the Earth seems to go forward when it moves on the semi-circle *PBN* above [beyond] the Sun, but backward when it moves on [the semi-circle] *NAP*, and to be stationary [when it is] at *N* and *P*: by the same consideration it is inevitable that Saturn, as seen by us from the Earth, will appear to move towards the immediately opposite parts of the sky. Thus, when the Earth is at *BNA*, Saturn appears to be at *bna* of the other figure (II). The inferior [planets], Venus and Mercury, appear to have retrograde motion because they are faster than the Earth; [consequently] Venus, if the Earth be stationary, describes on the most distant part of its circle the opposite path to that which it describes on that part [of its circle] nearest to the Earth.

'In the third place, one might ask [and Ptolemy would be unable to give any answer] why the epicycles of the large [planetary] circles are so small, and why those of the small ones are so large; in other words, why the prosthaphaeresis (προσθαφαίρεσις) of Mars is greater than that of Jupiter, and that of Jupiter is greater than that of Saturn.[22] Also, why is it that Mercury does not have a greater one than Venus, seeing that it is in a lower sphere than Venus; whereas, in the case of the other four planets, it is always greater for the lower [planet]? The answer is simple: the Ancients thought that the true circles of Mercury and Venus were their eccentrics. Now, the circle of Mercury, which moves the fastest, is the smallest. As for the [prosthaphaeresis of the] superior [planets], it is proportionately greater, the nearer the [planets] are to the Earth; the prosthaphaereses therefore appear that much greater. This is the reason why the prosthaphaeresis of Mars, the nearest, is the greatest, whilst it is least for Saturn, which is the highest [planet]. For, if the eye were placed at *G*, the circle *PN* would subtend the angle *TGV*; but if it were at *L*, the same circle would subtend the angle *RLS*.

'4. The Ancients also wondered, not unreasonably, why the three superior [planets], when in opposition to the Sun, are always at the lowest point of their epicycle [nearest to the Earth], whilst at conjunction they are at the highest point; in other words, why, if the Earth, Sun and [Mars] *G* be in the same straight line, Mars can only be at γ on the epicycle. There is a simple explanation according to the system of Copernicus. It is not Mars on its epicycle, but the Earth on its circle which causes this change. Consequently, when the Earth moves from *A* to *B*, the Sun is then between [Mars] *G* and the Earth, *B*; and Mars appears to come back on its epicycle from γ to δ. When the Earth is at *A*, which is the point nearest to *G* [Mars], *G* and the Sun, as viewed from *A*, appear to be in opposition to each other. All this is immediately clear from the figures.'

In fact, Keplers' diagrams show better than any descriptive argument both the true structure of the solar system and its apparent structure as it appears to an observer bound to the Earth in motion. They reveal the (Copernican) reality underlying the (Ptolemaic) appearances; they explain the latter and substitute comprehension for astonishment.

However, there is much more in the Copernican system than this mere substitution of reality for appearances. There is the revelation of a rational order, deducible *a priori* from the structure of the Universe, or, at least, from the solar system—a fact of which Copernicus was quite unaware. This was Kepler's great discovery to which we shall now turn our attention.

Kepler's main preoccupation was a search for this rational order; so we are not unduly surprised to learn[23]:

> 'That there were three things above all others the cause of which I sought without wearying, namely, the number, size and motion of the orbits. I was induced to try and discover them because of the wonderful resemblance between motionless objects, namely the Sun, the fixed stars and intermediate space, and God the Father, God the Son, and God the Holy Ghost; this analogy I shall develop further in my Cosmography. Now, seeing that motionless objects behave as they do, I had no doubt but that [objects] in motion were governed by a similar harmony.'

It is an odd argument that bases the search for a mathematical law governing planetary motion on the discovery of a wonderful resemblance between the Cosmos and the Holy Trinity. However, we must not be too surprised thereat: the progress of human thought—even of scientific thought—rarely follows the laws prescribed by pure logic, and great discoveries are frequently made as a result of flights of fancy.

Kepler continued[24]:

> 'First of all, I tried numbers; that is to say, I tried to see if by chance one orbit were twice, three times, or four times another; and in the absence of any relationship between the orbits themselves to see if there were not some relationship between their differences.'

It was in vain. Kepler tells us that he

> 'lost much time in this endeavour, this play with numbers. For I found no regularity either in the relationship between the orbits themselves, nor between their differences.'

Kepler was not at all discouraged, for

> 'apart from other reasons, of which I shall speak later, I always derived hope and consolation from the fact that the motion [period of revolution] seemed always to depend on the distance, and wherever there was a large interval between the orbits, there was a similar difference between the motions. For, I said to myself, if God has related the motion to the distance in an orbit, he must certainly have related the distances themselves to something else.'
>
> 'Now, as I did not succeed in my expectations in this way, I chanced

to try another which was exceedingly daring. Between Jupiter and Mars, as well as between Venus and Mercury, I introduced two new planets, invisible on account of their small size, to which I ascribed certain periods of revolution.'

It was, indeed, exceedingly daring, but it did not yield the desired result; for

'even though, by this means, one succeeded in establishing a proportion [between the dimension of the orbits], the calculation provided no certain conclusion. The result gave no definite number of moving celestial bodies, neither in the direction of the fixed stars, nor in the direction of the Sun.'

In fact, as the space could be infinitely subdivided so as to fill all the positions corresponding to the terms of the relationship, it was necessary to insert an infinite number of increasingly smaller orbits between Mercury and the Sun; furthermore, one could, and perhaps should, add a series of trans-Saturnian planets to the solar system.

One could, doubtless, forgo arrangements of this kind, and try to find the number of planets on the basis of the properties of pure numbers: but Kepler, however Pythagorean he might be, could not accept a solution of that kind. The key to the structure of the Universe was not to be found, as far as he was concerned, in the ratios between pure numbers, but in the relationships between solid figures. One might say that his God is a geometer, and not an arithmetician. Consequently, he was of the opinion[25]:

'That there is no number whose perfection is such that one can deduce from it that the number of planets must be exactly equal to it and to no other, or that the planets are not infinite in number. Although Rheticus in his *Narratio* deduced the number of spheres from the sacred character of the number six, this conclusion seems unlikely to me.[26] Indeed, when speaking of the structure of the Universe, one's demonstration should not be based on numbers which have acquired a particular significance only in relation to things engendered since the creation of the world.'

This rather cryptic phrase on the part of Kepler means that Rheticus was wrong to account for the number of planets through the perfection of the number six; for, though the number six has without question a certain perfection, because it is a perfect number and also the first perfect number, this asset is not enough to serve as a basis for the divine decision to create only six celestial bodies. On the contrary, the number six acquires its particular and unique significance as a result of this creation. Therefore, God must have used something else as a basis, namely, structural relationships existing before the creation, and they cannot be numerical relationships. Kepler did not accept the pre-eminence of number *qua* number; numbers, for him, always meant the number of something—they were 'denominate numbers' and not 'abstract numbers'. Before the creation of the Universe there was nothing to be numbered except for the persons of the Holy Trinity.

Nor was he any more successful in using trigonometrical relationships to express the connection between a planet's distance (from the Sun) and its motion (period of revolution), or, more exactly, as we shall see later, between its distance and its motive power. The constructional method devised by Kepler was based on a square with an inscribed quadrant of a circle; on one side of the square (that which represented the radius of the Universe from the Sun to the fixed stars) the positions of the planets were marked; lines were drawn from *AC* parallel to the base *CB* so as to cut the quadrant, and these lines represented the motive power ascribed to each of the celestial bodies, the Sun, planets and fixed stars. However, Kepler lacked one essential piece of information, namely, the value of the radius. Moreover, the construction, though it gave a zero value to the motive power of the fixed stars, gave an infinite value to that of the Sun[27] (Fig. 2).

Kepler finally realized that he was on the wrong track. Nevertheless, we, who know the sequel, cannot but recognize in his fruitless and persistent efforts the inspiration which led him to the solutions which he discovered ten and twenty years later. There was his investigation of the change in motive power acting on celestial bodies as a function of their distance from the Sun; and the search for a definite mathematical relationship between distance and period.

It cost Kepler much time and trouble in exploring these paths that led to nothing. Once again, however, the effort was not entirely useless. By making mistakes the mind is at length able to approach the truth.[28]

> 'I lost nearly the whole summer in this difficult work (said Kepler). Finally, I came near to the truth of the matter on the occasion of a trivial event. I believe that it was by a gift of Divine Providence that I suddenly received what I had never been able to obtain hitherto through all my labour; I believe it all the more seeing that I never ceased praying

to God to let me succeed in my undertaking, especially, should it prove that Copernicus had proclaimed the truth.

'Now, it happened on the 9th or 19th July 1595,[29] that I wished to show my students how the great conjunctions jump through eight signs [of the Zodiac] and pass successively from one trigon to another. I drew

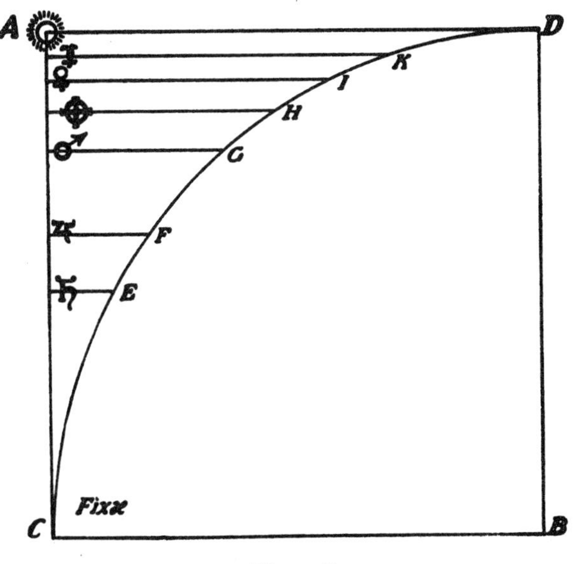

Figure 2.

a large number of triangles (if they may be so called) in a circle in such a manner that the end of one always formed the beginning of the next (it is shown on the accompanying diagram of the conjunctions of Jupiter and Saturn). Now, the points in which the triangles cut each other form a small circle; for the radius of a circle inscribed in such a triangle is one half the radius of the circumscribed circle. The ratio between the two circles was, to the eye, exactly the same as that which is found between [the orbits of] Saturn and [of] Jupiter, and the triangle is the first of the geometrical figures, just as Saturn and Jupiter are the first planets. I immediately tried [to determine] the second distance, [which exists] between Mars and Jupiter, by means of a square, the third by means of a pentagon, the fourth by means of a hexagon.' (Fig. 3.)

Once more, he was unsuccessful. In spite of all his efforts to discover them in this manner—and Kepler said, 'I should never finish were I to describe everything in detail'— the distances between the planets bore no relationship to the geometrical figures hierarchically arranged according to the number of their angles. The mind of the Divine Geometer, obviously, had not followed such a simple principle as that.

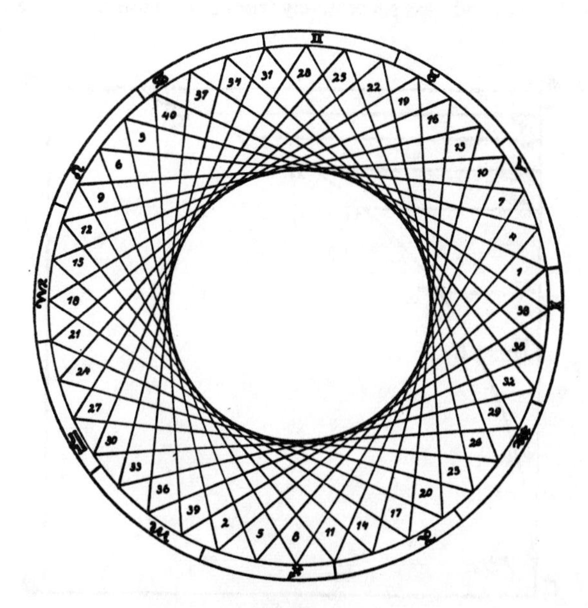

Figure 3.

'The end of this unfortunate attempt was also the beginning of a final, successful [attempt]. I realized, in fact, that if I wanted to continue the arrangement of the figures in this manner, I should never come to the Sun, and that I should never find the reason why there must be six planets rather than twenty or one hundred. Yet, figures appealed to me, for they are certainly quantities, and therefore something that existed before the firmament, for quantity was created in the beginning together with the [heavenly] bodies, and the Sun on the second day [only [30]]. I then said to myself, I should obtain what I am seeking, if five [particular] figures, among the infinitely many, can be found for the size and relationships of the six spheres assumed by Copernicus. Then, I went even further. What have plane figures to do with corporeal [planetary] spheres? Clearly, one should resort to solid bodies.

'Here then, Dear Reader, is my discovery and the content of my short treatise. If this be told to anyone with the least knowledge of geometry, he will immediately think of the five regular solids and their relationship to their circumscribed and inscribed spheres.'

Henceforth, Kepler no longer doubted his success. Is anyone ignorant of the fact that there are only five regular solids, and that there cannot be any more? Five solid bodies—and that means six spheres at the very most. Much more work would be needed in order to determine, from a comparison of calculated with real (observed) distances, the order in which these solids should be enclosed one within the other; and having done so, to explain the appropriateness of this order.[31] The solution to the problem of the fixed number of planets made Kepler confident that he was at last on the right track. Moreover, the very principle of the solution seemed eminently reasonable and worthy of having been adopted by the Divine Architect. What, indeed, have abstract numbers, or lines, or plane figures to do with the construction of a corporeal Universe? Isn't it obvious that only solid bodies are involved?

God undoubtedly created solid matter in the very beginning. Kepler even says that if we had a precise statement of its essential nature, it would not be difficult to understand why God did create solid matter, and not something else, in the first place. Even without fully understanding, one has an inkling of the reasons for doing so: the fact is, that God intended to create quantity; and to do this, He had need to bring into being all the characteristics which belong to the essence of the body: for the quantity of a body *qua* body is in some measure the origin of form and quantity. Now, if God had decided to create quantity, it was because He wanted to produce the contrast between a curve and a straight line. The divine Nicholas of Cusa, and some others, realized this, and had ascribed a primordial importance to the difference between the two; and they had the boldness to assign the curve to God, and the straight line to that which was created.

Although this connection is sufficient in the sight of God to serve as a basis for the rôle of quantity and the need to create matter and bodies, there is finally another, more important reason. God's image is revealed in the sphere by God the Father at the centre, God the Son on the surface, God the Holy Ghost mid-way between the central point and the circumference.[32]

> 'For, that which Cusa attributes to the circle, and others to the globe, I, (said Kepler), attribute only to the surface of the sphere. Nor am I convinced that there can be any curve more perfect than a spherical surface. A globe [solid body] is more than a spherical surface; it is mingled with rectitude which alone fills its interior [volume].'

According to Kepler, then, a globe consists of a spherical surface inseparably linked with all the diameters which fill its interior. In consequence of the connection between the *curvature* of the surface and the *rectitude* of the diameter, certain properties of the cube, formed by straight lines, are nevertheless reflected in the globe; similarly for the square in relation to the circle which is only a plane section of a globe.

> 'Yet again! Why were the difference between the curve and the straight line and the nobility of the curve taken by God as the bases of the creation of the Universe? Why? indeed! if it were not because it was absolutely necessary that the most perfect Creator should produce the most beautiful handiwork. For, "it is not fitting (as Cicero says in his book *de Universitate*, when quoting Plato's *Timaeus*), and has never been fitting, that he who is supreme should do anything except it transcend in excellency[33]".'

Therefore, God wished to create the most beautiful Universe, and to express his own Being, as far as possible, in this Universe; and that was the very reason why he had to use the curve and straight line in its construction, and so establish grandeur based on material substance.[34]

> 'Now, therefore, it has already been shown by Aristotle that the Universe is spherical (curvilinear), and Copernicus agrees with him in asserting the sphericity of the starry vault, although contrary to Aristotle he denies that it possesses mobility any more than does the Sun which he places at the centre of the Universe.'

In short, with the Sun corresponding to the Father, the sphere corresponding to the Son, and between them the space filled with celestial aura corresponding to the Holy Spirit, with these three the creative Trinity reveals its own image in the Universe.

Having explained the framework of the Universe, it now remains to examine the contents, namely the fixed stars and the planets, in order to determine their number and order relying on the principle of rectilinearity inherent in the creatural being.

Now, if Kepler were firmly convinced that God created (and distributed) the fixed stars not *temere*, but in accordance with the necessity for perfection, he nevertheless considered that the discovery of the reasons by which the Creator was guided in His action were perhaps beyond human comprehension.[35] Were not they so numerous that, in the words of the Psalmist, God alone telleth the number of the stars? Consequently, were not they in some way related to infinity? Kepler therefore invites us to give our attention to the planets, which are closer to us, and much fewer in number.

The principle of rectilinearity as such offers no immediate help: in fact, the number of straight lines is infinite; so is the number of plane surfaces bounded by straight lines, as well as the number of solid bodies limited by planes. However, the situation is entirely changed when we consider the regular solids, that is to say, those formed by regular and equal polygons having equal angles. Furthermore, these solid bodies are related to the spheres, because they can be inscribed in the spheres, and conversely, the spheres can be inscribed in them. There is a fixed number of such solid bodies; there are five only, as proved by Euclid.

Now, seeing that we have, on the one hand, an infinite number of irregular bodies, and on the other, a fixed number of regular solids, it is obviously fitting that there should be two kinds of celestial bodies in the Universe, these bodies being distinctly distinguished one from the other (as motion is from rest); one kind (the fixed stars) is similar to the infinite, the other (the planets) is similar to the finite.[36]

> 'We have, therefore, a globe, which is spherical according to both Aristotle and Copernicus, and which is dependent on motion; and we have the solids which depend on number and size; what more is left except to say with Plato *Θεὸν ἀεὶ γεωμετρεῖν*, and in the structure of this mobile Universe to inscribe the solids in the spheres, and the spheres in the solids until no solid remains to be placed within or beyond the moving spheres. . . . Therefore, if the five solids with their corresponding spheres be inserted one within the other, the number of spheres will be equal to six.
>
> 'Now, therefore, if some era of the world's existence conceived the idea of arranging the Universe in such a way as to place six moving spheres round the Sun, then that era has imparted the true Astronomy to us. It happens that Copernicus has precisely six spheres of this kind, and taken two by two they are in the required proportion for being placed between the five solids in question."[37]

Consequently, Copernican astronomy, which not only reduces the number of planets from seven to six—Kepler does not use this argument *expressis verbis*, though it is clearly implied—but also sets them at theoretically justifiable distances from the Sun, must be considered to be true until someone puts forward a better hypothesis, or rather explains how it is that something, which is deducible in this manner using the soundest arguments based on the principles of nature, can be produced by chance, and how it is that the agreement between facts and their explanation is purely fortuitous. Kepler considered this to be quite impossible. Nothing is to be more admired, and nothing is more likely to be convincing than the fact that Copernicus discovered his system and determined the distances by groping and without any preconceived idea.

Kepler continued[38]:

> 'The Earth [the sphere of the Earth] is the measure for all the other spheres. Circumscribe a Dodecahedron about it, then the surrounding sphere will be that of Mars; circumscribe a Tetrahedron about the sphere of Mars, then the surrounding sphere will be that of Jupiter; circumscribe a Cube about the sphere of Jupiter, then the surrounding sphere will be that of Saturn. Now place an Icosahedron within the sphere of the Earth, then the sphere which is inscribed is that of Venus; place an Octahedron within the sphere of Venus, and the sphere which is inscribed is that of Mercury.[39]
>
> 'There you have the reason for the number of the planets."[40] (Fig. 4.)

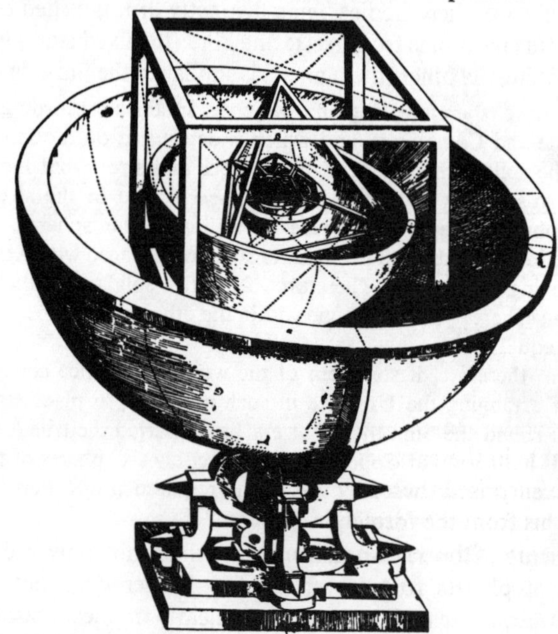

Figure 4. α Sphere of Saturn. β Cube. γ Sphere of Jupiter. δ Tetrahedron. ε Sphere of Mars. ζ Dodecahedron. η Sphere of the Earth. θ Icosahedron. ι Sphere of Venus. κ Octahedron. λ Sphere of Mercury. μ Sun, mid-point, stationary centre.

Kepler certainly did not overlook the fact that the agreement between his construction and astronomical data, *i.e.*, the agreement between distances calculated by means of his system and those provided by observation, was only approximate, and very approximate at that, especially for the extreme members of the series. In fact, if the solids be enclosed one within the other in the manner described by Kepler, then the planetary spheres are envelopes having significant thickness, and in every case a distinction must be made between the external sphere and the internal sphere. The former is inscribed in the next higher regular solid, the latter is circumscribed about the next lower regular solid. Taking 1000 as the radius of any internal planetary sphere, then the value for the next planet is given in the following table:

					By calculation	According to Copernicus
Relationship	♄	Saturn	1000	Jupiter	= 577	635
between	♃	Jupiter	1000	Mars	= 333	333
the radii	♂	Mars	1000	Earth	= 795	757
of the	♁	Earth	1000	Venus	= 795	794
spheres	♀	Venus	1000	Mercury	= 577[41] or 707	723

If, however, the thickness of the Earth's sphere be considered to enclose that of the Moon as well, then the ratio of its radius to that of Venus will be as 1000 to 847; on the other hand, it will be 0·801 that of Mars (Fig. 5).

The agreement is decidedly not perfect, but it is too significant to be accidental. The discrepancy between data and calculation is considerable in the cases of Jupiter and Mercury, though we ought not to be surprised thereat in view of the enormous distance which separates us from the former, and the difficulty in providing a satisfactory theory of the motion of Mercury.[42]

Furthermore, the data given by Copernicus are not absolutely correct in that he has referred the planetary motions not to the true Sun, even though he had placed it at the centre of the Universe, but to the centre of the terrestrial sphere[43] (mean Sun); he had done so 'in order to shorten the calculations' and 'in order not to alarm his readers by departing too much from Ptolemy.' Neglecting other

TABELLA IIII. Ostendens veram amplitudinem orbium coelestium, et interstitiorum, secundum numeros et sententiam Copernici.

Ad Cap. 14.
Pag. 49.

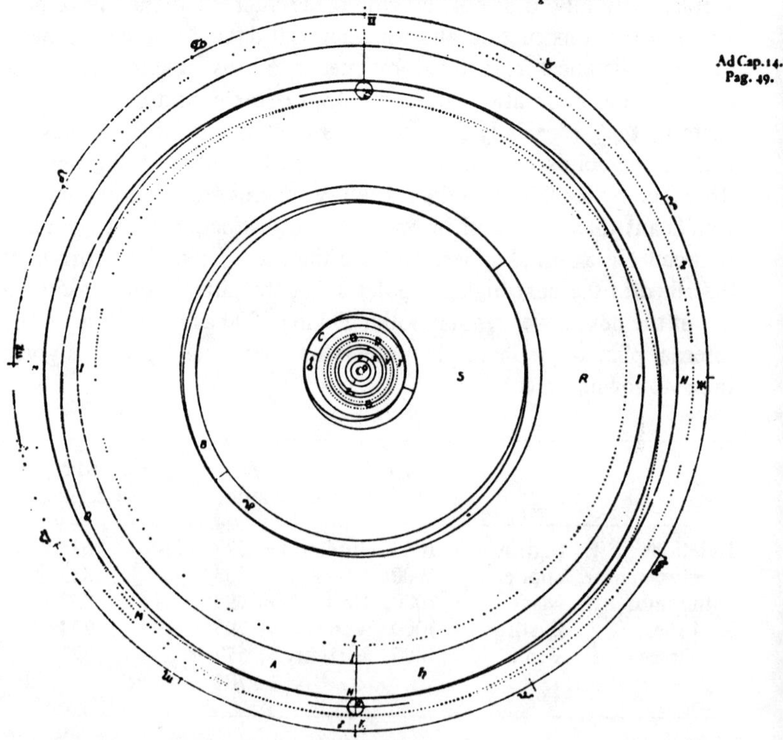

Figure 5.

Extremus circulus Zodiacum refert in Orbe stellato, descriptus ex centro Mundi vel Orbis Magni, vel etiam ex globo Terreno, quia totus Orbis Magnus ad eum insensibilis est.

A Saturni systema, concentricum ex G centro Orbis Magni. B Systema Iouis. C Martis.

D Circulus siue via centri globi terreni concentrica ex centro G, eum sphærula Lunari duobus locis appicta. Duæ coecæ lineæ circulares orbis terræ cum inserta Luna crassitiem denotant.

E Duo circelli delineantes crassitiem systematis Venerij, intra quam omnis eius motuum varietas perficitur.

F Spacium inter duos circellos, in quo omnis motuum stellæ Mercurij varietas perficitur.

G Centrum omnium et prope ipsum corpus Solare.

Circulus per O et P transiens (cuius hic tantùm duo arcus comparent) eccentrepicyclus Saturni est.

Linea curua per Q, atque per perigaeum epicycli in O apogæo eccentrici positi, et per apogæum eiusdem in P perigæo eccentrici, est via planetæ eccentrica. Circulus quidem non est, sed tamen à circulari linea sensibiliter non differt. H I Crassities duobus circulis concentricis inclusa, quam via Saturni eccentrica sibi vindicat.

Linea curua, vel quasi circulus per M, et per apogaeum epicycli in O, atque per perigæum eiusdem in P transiens, eccentricus est, quem Ptolemæus Aequantem vocat.

K L Crassities duobos coecis circulus concentricis intercepta, quam totus epicyclus, et æquans ille requirunt.

Planeta vero ultra H nunquam ascendit, nec infra I descendit.

Similibus particularibus orbibus cæteræ sphæræ etiam distinctæ intelligantur, qui tamen, ne multitudo linearum negotium potiûs obscuraret, qupm declararet, hic omittuntur. Ideo in Ioue et Marte via eorum eccentrica, duoque eam continentes circuli concentrici, in cæteris soli concentrici descripti sufficiunt.

Spatia intermedia. R locus Cubi. S Tetrædri. T Dodecaedri. V Icosaedri.
X Octaedri. Z est spacium inter Saturnum et fixes, infinito simile.

difficulties, such as the absence of any thickness for the Earth's sphere, which arise, it is clear that the centre of the terrestrial sphere cannot have any decisive rôle in the structure of the Universe, and that all the distances must be recalculated taking into account the eccentricity of the Earth (or the Sun) which had been ignored by Copernicus.

Furthermore, it is not enough to calculate the mean distances of the planets from the Sun: their paths, or orbits, if we keep to the conception of Copernicus, are eccentric with respect to the Sun. Therefore, the thickness of the spheres should be sufficient to take this eccentricity into account: at aphelion the planetary globe should touch (but not go beyond) the outer surface of its sphere, and at perihelion it should touch its inner surface. Even this is not enough, for the eccentricity of the planetary globes in itself presents a problem: indeed, why are they eccentric? and not centred on the Sun? It is a difficult question which Kepler did not tackle directly; and he was not to give the answer till much later in his *Harmonice Mundi.* For the time being, he was fully occupied in trying to discover if the planetary globes could be fitted into the 'spheres' as determined by the five regular solids.

No doubt Kepler thought that the distances referred to the Sun would conform better with his theory than would those given by Copernicus. Unfortunately, the long and difficult calculations which Maestlin carried out for him did not confirm his hopes; the new figures did not agree any better with his conclusions than did the old ones. After all, it was not surprising. The astronomical data themselves were only approximate: the astronomical tables were anything but accurate, and the differences between them were very considerable, being no less so than those revealed by comparing them with present-day observations.[44] Hence, Kepler hoped that the advancements in astronomy, particularly the work of Tycho Brahe, would show better agreement between his theory and reality.

I hope I shall not be reproached for making long quotations from the *Mysterium Cosmographicum* which illustrates Kepler's mentality so well; Kepler's works have never been translated in their entirety into English, and an analysis of them is no substitute for the impact of the author's own words. Kepler's mentality seems very strange to us, and the reasoning inspired by it seems fantastic or even harebrained. Nevertheless, ignoring his answer, was not Kepler's *question* a legitimate one? Was it possible, in the long run, to be satisfied with *ascertaining* the composition of the planetary Universe and accepting the structure of the solar system as a bare fact? Was not it necessary to seek out the laws determining this structure and, it may be, explaining its stability? Was not it increasingly probable that the solar system was not constituted *temere*, just anyhow, and that the planets cannot have been placed just anywhere?[45] In this connection, mention may be made of the confirmation provided by the curious Law of Bode and the presence of the asteroids or minor planets between Mars and Jupiter.

Let us now return to the *Mysterium Cosmographicum*, which has still much of interest to yield. We have already seen that Kepler from the very beginning, from the time when as a pupil of Maestlin he defended Copernican astronomy in public disputations, had striven to substitute a dynamic concept for the purely, or almost purely kinematic concept of tradition (and of Copernicus), and to demonstrate the superiority of the new concepts both from the mathematical and physical points of view; also, we have seen him try to determine the variation in motive force of the planets as a function of distance. In the *Mysterium Cosmographicum*, he does not restrict himself to seeking the structural laws, the 'archetypical' laws of the Cosmos, he endeavours also to find the physical and transphysical causes responsible for its motions, and which explain why the planets move more slowly the farther they are away from the Sun.

The answer to the question Kepler asked himself seems very simple at first sight: the planets move more slowly, *i.e.*, complete their paths in longer periods of time the more distant they are from the centre of the Universe, because they have of necessity a much longer path to travel. Aristotle had already stated this in *De Coelo*, Lib. II, cap. 10; and Ptolemy agreed. Now, it happens that ancient astronomy, whilst admitting the principle of increase in the period of revolution with the distance from the centre, was obliged to contradict it by ascribing an identical period to the three inferior planets (Venus, Mercury and the Sun)[46]; on the other hand, the system of Copernicus was in perfect agreement with the said principle. However, the principle in question is incomplete, and the proposed explanation, which implies that the planets move with equal (orbital) velocities, and consequently that the motive forces responsible for their motions are similarly equal, does not agree with the facts: the periods of revolution are not directly proportional to the distances, but differ considerably therefrom, as may be seen from the following table[47]:

	Saturn ♄ Days	Jupiter ♃ Days	Mars ♂ Days	Earth ♁ Days	Venus ♀ Days	Mercury ☿ Days
♄	10 759 12					
♃	6 159	4332 37				
♂	1 785	1282	686 59			
Earth	1 174	843	452	365 15		
♀	844	606	325	262 30	224 42	
☿	436	312	167	135	115	87 58

The first line in each column of the above table shows the number of days and sixtieth parts of a day in which the planet whose sign is at the head of the column completes one revolution; the succeeding numbers give the period of revolution for the other planets on the assumption that they all have the same orbital velocity, *i.e.*, the periods of revolution if they were directly proportional to the distances. It follows that not only the path to be traversed varies with the distance from the centre, but also the speed with which it is covered: the former increases, the latter decreases; and this behaviour can be explained only by a corresponding variation in the motive force acting on the planetary spheres, or on the planets.

In Kepler's time the only 'motive forces' that could be accepted as operating in the heavens were 'animal' forces and not 'material' forces. Kepler did not dispute this; at least, not yet (later on he was to say[48] that he had adopted this doctrine as a result of the influence of Joseph Scaliger); but it was not the nature of the driving forces in question which interested him, it was their power. So, in that chapter of the *Mysterium Cosmographicum* particularly devoted to the relationships between the motions and the spheres, he says[49]:

> 'But if we hope to come closer to the truth and find some law (equality) in these relationships, we must accept one of the following two assertions: either, the motive souls are weaker the farther they are away from the Sun; or, there is only one motive soul at the centre of all the spheres; that is to say, it is in the Sun, and this soul moves the planets which are near to it with more vigour than it does those which are farther away, because at a great distance the force involved is weakened. In the same way, therefore, that the source of light is found in the Sun, and that the origin [the common point] of the spheres is in the same position as the Sun, *i.e.*, at the centre of the Universe, so do life, motion and the soul of the Universe originate in the Sun. In this ordering of the Universe the

> fixed stars are endowed with rest, the planets with moderate activity, but the Sun is endowed with the primary and inherent activity, which is incomparably greater than the much smaller activities of all other things to the extent that the Sun by the splendour of its appearance, the efficiency of its power, and the glory of its light far surpasses everything else.'

Kepler obviously inclined towards the second of these two possibilities: there is only one motive soul, that of the Sun, and its action is attenuated by distance. In what proportion? He considered that the attenuation was most probably the same as in the case of light the value for which is known from optics.

It seemed very probable[50] that this attenuation did take place in the same proportion as in the case of light. At a later date, Ismaël Boulliau took exception to this very point when Kepler once again used this analogy in his *Astronomia Nova*, and emphasized the difference between the method of propagation of motive force (or *species*) and that of light. This by no means implied any fundamental change in his ideas, but was simply the result of correcting a mistake, or rather a double error which he had made in the *Mysterium Cosmographicum*.

In fact, by the laws of optics, the attenuation of light is proportional to the square of the distance. Kepler, however, did not understand the matter in this way, and instead of considering the propagation of light as taking place in *space*, he supposed that it was propagated in a plane. For, he said[52];

> 'there is as much light or as many solar rays in a small circle as in a large one; and as it is denser in the small circle and more extended in the large one, the measure of this attenuation, both for light and the motive power, must be sought in the proportion of the circles themselves.'

which means that it is directly proportional to the distance, and not to its square.

Two factors are therefore implied in the solution of the problem of planetary motion and the period of revolution. The planets move more slowly the farther away they are from the Sun, because the force which moves them is that much weaker. Yet, at the same time, they have to cover a proportionately longer course; 'accordingly the greatest distance from the Sun acts, so to speak, twice over in order to prolong the period of revolution, and consequently the increase in the period of revolution is double with respect to that of the distance.' From this it should follow that the ratio of the periods are proportional to the square of the distances; but, making yet another mistake, Kepler added instead of multiplying and so arrived at the following law: the ratio of the distances of two planets [from the Sun] is proportional to that between the period of the faster moving planet and the arithmetical mean of the periods of the faster and slower moving planets. For example, the periods of Mercury and Venus are 88 and $224\frac{2}{3}$ days respectively; one half the difference of these values is $68\frac{1}{3}$; hence, the distances of Mercury and Venus (from the Sun) are in the ratio of 88 to $156\frac{1}{3}$ ($88 + 68\frac{1}{3}$): and similarly for other planets.[53]

One is inclined to say that Kepler's conception of the Universe was settled from this time onwards.[54] Undoubtedly, there were to be changes, some of them quite important: the spheres and the circles disappeared from the structure of the Cosmos; the temporal notion of musical harmony completed the purely spatial harmony resulting from enclosing the regular solids one within the other[55]; 'animal' force was replaced by 'corporeal' forces[56]; the Sun ceased to be completely motionless and acquired motion about its axis. Nevertheless, Kepler's conception remained in its broad outlines exactly as we have just described it. Kepler's letter (3 October 1595)[57] to his teacher Maestlin provides a striking summary of it:

> 'God created a fixed number of bodies in the Universe [planets]; but this number is an accident [property] of quantity [geometrical], by which I mean, that the numbers are in [related to] the Universe. For, before the Universe existed, there was no number, except [that of] the Trinity which is God Himself. Now, if the Universe has been created in conformity with the measure of numbers, then it has been done in conformity with the measure of quantities. There is no number, but only infinity, to be found in a line or a surface. Therefore, as regards bodies, the irregular bodies are to be put aside, for it is a question of the foundation of the best arrangement for the creation. Six bodies are then left. The globe, or more precisely, the hollow sphere and the five regular solids. The spherical shape befits the celestial vault. In effect, the Universe has a double nature: mobile and immobile. The one is in the image of the divine essence considered by itself, the other is in the image of God in so far as He creates, and therefore has that much less significance. A curve has a natural relationship with God; the straight line has relationship with that which has been created. The sphere has a

threefold quality: surface, central point, intervening space. The same is true of the motionless Universe: the fixed stars, the Sun, and the *aura* or intermediate aether; and it is true of the Trinity: the Father, Son, and Holy Ghost.

'The mobile Universe is to be related to the regular solid bodies, of which there are five. If they be regarded as the boundaries or walls (for which I provide ample reasons), they cannot give rise to more than six things. Consequently, there are six bodies in motion round the Sun. Therefore, the Sun which keeps its place, motionless, in the midst of the planets, and which is nevertheless the source of all motion, provides the image of God the Father, the Creator, for creation is to God, as motion is to the Sun. As the Father creates through the Son, so the Sun gives motion in the midst of the fixed stars; for if the fixed stars did not create a void through their lack of motion, nothing could be given motion. . . . But the Sun diffuses and bestows motive power across the *intermedium*, in which the planets are placed, in the same way that God the Father, considered as Creator, acts through the Holy Ghost, or in virtue of the Holy Ghost. Consequently, it necessarily follows that motion is proportional to the distances.'

Once more, Kepler's ideas may strike us as being odd and absurd. Nevertheless, it cannot be denied that this curious assimilation—inspired by Nicholas of Cusa[58]—of the sphere, the Universe and the creative Trinity guided his mind, and that these mystical speculations led him to make the Sun the dynamic, as well as the architectonic, centre of the Cosmos, and thereby to introduce the first, extremely important, modification to the Copernican system of astronomy.[59]

In the Copernican astronomical system the Sun is, indeed, situated at the centre of the Universe, but from the practical, or at least technical, point of view does not fulfil any function there. It gives light to the Universe, and that is all. The centres of planetary motion are not located in the Sun, but only in the proximity; and these motions are not referred to the Sun, but to the centre of the Earth's sphere, 'so as not to differ too much from Ptolemy's concept', as was pointed out by Kepler. It is certainly true that this centre, which is eccentric with respect to the Sun, is itself situated on an epicycle whose centre in its turn executes a circular motion round the Sun; but these two motions are much too slow to be of significance for computation, or to have any effect on the mechanism of planetary motions.[61] (The epicycle of the terrestrial circle makes one revolution in 3434 years, and its deferent does the same in 53 000 years; moreover, they were only introduced by Copernicus for the purpose of explaining certain discrepencies between the old and new tables.) Though the Universe of Copernicus is heliocentric, his astronomy is not; it is merely heliostatic: and if, as Kepler has shown, the Sun plays just as important a rôle in the Universe of Ptolemy as it does in that of Copernicus, then the Earth, on the other hand, plays a scarcely less important rôle in the astronomical system of Copernicus as it does in that of Ptolemy. It was Kepler who made both the Universe and its astronomy heliocentric, by relating planetary motions to the Sun instead of to the centre of the terrestrial sphere.

Now, if he decided to transfer the origin to 'the body of the Sun', *i.e.*, the common point of the planetary orbits, he did so because distances reckoned *from the Sun* (and not from some imaginary point) were able to play a rational part in the architectonic design of the Universe; for the origin or source of the motive power which would confer dynamic unity upon the whole could only be in 'the body of the Sun', and not in the centre of the terrestrial sphere, where there is no body at all.

II. CELESTIAL PHYSICS OR *ASTRONOMIA NOVA*

I

Kepler and Tycho Brahe

Kepler devoted his *Astronomia Nova* to defining the force which moves the planets, or, more precisely, substituting a quasi-magnetic, physical force for the 'animal' force of the *Mysterium Cosmographicum*; and to ascertaining the strict mathematical laws governing its action, as well as elaborating a new theory of planetary motion based on observational data provided by the labours of Tycho Brahe.

There was an interval of ten years between the *Astronomia Nova* (published in 1609 though completed in 1607) and the *Mysterium Cosmographicum*. They were ten eventful years full of discovery.[1] The *Mysterium Cosmographicum* was not a success, notwithstanding the enthusiastic support of Maestlin,[2] who took a definite stand in favour of Copernicus against Tycho Brahe[3] in the excellent preface to the new edition of the *Narratio Prima* of Georg Joachim Rheticus, which, together with an appendix containing the calculations he had made for Kepler, was added to the printed copy of the *Mysterium Cosmographicum*. The work did not arouse the enthusiastic reception Kepler expected.[4] Nevertheless, it drew the attention of the great Danish astronomer to Kepler, and the sequel was to be of great importance. Tycho Brahe whilst at Wandsbeck, where he had taken refuge after having been obliged to quit Denmark, received a letter (dated 13 December 1597[5]) from Kepler in 1598 in which the latter asked his opinion on the *Mysterium Cosmographicum*. Tycho Brahe made a very gracious, though slightly ironic, reply, saying that it had given him much pleasure to receive and peruse the *Mysterium Cosmographicum*. Whilst he criticized the work—he expressed the opinion that astronomy should proceed *a posteriori*, and try to find concordant relationships only after having established the facts—he nonetheless recognized the usefulness of Kepler's speculations as well as the cleverness of the author whom he invited to pay him a visit for the purpose of discussing these matters and comparing the theories with his accumulated observational data.[6] He wrote as follows:

> 'Your book entitled *Prodromus Dissertationum Cosmographicarum*, I have already seen and perused as far as my other occasions permitted me to do. It has really given me more than an ordinary measure of pleasure. Your shrewd intelligence and keen mind shine clearly therein; it was an original and ingenious idea to relate, as you have done, the distances and periods of the planets to the symmetrical properties of the regular solids. There seems to be a reasonably good agreement on the whole, and the slight discrepencies with respect to the proportions given by Copernicus are of no great importance, for they themselves disagree considerably with the phenomena. I heartily commend the ardour you have shown in making these enquiries. Yet I should not care to say that you are right in everything. By using the true values for the eccentricities of the planets as obtained by myself over many years, it would be possible to make more accurate verification; but as I am at present much occupied in preparing and publishing my astronomical works, which I was unable to finish in Denmark, I have not the time to make such a comparison. Perhaps it will be possible some other time.'

Tycho Brahe sent Kepler some information on the values for eccentricities as found by Ptolemy, Copernicus and himself, which values did not fit in at all well with Kepler's ideas and were, moreover, vitiated by his adherence to Copernicus, which Tycho Brahe considered absurd, for it did not bring about an elimination of the Ptolemaic epicycles as was claimed by Copernicus, though it was accomplished in Tycho Brahe's own system. So, he continued (p. 199):

> 'This and many other things, which I have not the time to treat at length, give me doubts about your discovery, which otherwise is most ingenious.'

Tycho Brahe was of the opinion that it would have been better if Kepler had tried to find analogies, *i.e.*, numerical relationships, in his system. It should be possible.

> 'For there is no doubt but that everything in the universe is in keeping and has been ordered by God in accordance with fixed harmony and proportion, in such a manner that it may be represented just as well by numbers as by forms, as was foreseen formerly to some extent by the Pythagoreans and the Platonists. Therefore, direct the power of your mind to this matter, and if you find perfect agreement without the least defect or deficiency, then you will be in my view a great Apollo.'[7]

At the same time, in another letter to Maestlin, Tycho Brahe expressed himself with far greater severity on the subject of Kepler's speculations[8]:

> 'A short while ago, I received a learned work from the excellent mathematician J. Kepler of Styria in which he attempts in an ingenious manner to establish a progression between the planetary spheres, as arranged by Copernicus, and the five regular solids. He wrote to me recently on this subject, and, as far as the burden of my work allowed me, I conveyed to him my opinion of his speculations.
>
> 'If improvement in astronomy must be made *a priori*, by means of these regular solids, rather than *a posteriori* on the basis of a knowledge of facts obtained by observation, as you suggested,[9] we shall assuredly have to wait too long, if not for ever and in vain, until someone does it. Assuming that the use of the proportions of the regular solids must depend on previous observations and be confirmed by theory, it follows that, apart from general relationships, whatever they may be, one cannot deduce particular details from them with the required accuracy; this fact, undoubtedly, will not have escaped your notice.'

Kepler did not receive Tycho Brahe's letter till much later on 18 February 1899, eight days after he had received the copy that the latter had sent to Maestlin. There was no question of making the journey to Wandsbeck; it was too far, too difficult, and too expensive. Moreover, Kepler had been offended by Tycho Brahe's criticisms, and more so by those which had been sent to Maestlin.[10] In addition, the political situation in Styria, where the Archduke Ferdinand had initiated a policy of anti-protestant repression, made it impossible for him to leave. However, he did not forget the invitation, and when towards the end of 1599 he learned that Tycho Brahe, who, in the meantime had been named 'Imperial Mathematician' by Rudolf II, had arrived at Prague, he decided to visit him at the castle of Benatky which the Emperor had placed at his disposal.

In fact, some months after arriving in Bohemia and settling at Benatky, Tycho Brahe had sent Kepler a very cordial and generous letter (it arrived at Graz only after Kepler had left) in which he offered his friendship and support,[11] though at the same time he repeated his criticisms of the *Mysterium Cosmographicum*:

> 'Since it came to my notice, (he wrote), I have always had a deservedly high opinion of your book and of its author, too; and I have never refused—nor shall I ever refuse—my commendation, in any way whatsoever, of yourself or others who have acquired the credit of concerning

themselves with these sublime subjects so far removed from general understanding, and who earnestly seek therein the kernal of truth. I have, on the contrary, always commended them, and shall ever praise them. There is one thing in particular of which I do not approve in your ingenious work; it is the error that you make, in company with many others, of attributing a certain reality to the celestial spheres in order to facilitate the path of Copernican concepts and [to be able] to endorse them more easily. That the celestial motions conform to a certain symmetry, and that there be reasons why the planets perform their circuits about one centre or another and at different distances from the Earth or the Sun, I do not deny. However, the harmony and proportion of this arrangement must be [sought for] *a posteriori*, where the motions and the circumstances of the motions have been definitely established, and must not be determined *a priori* as you and Maestlin would do; even then they are very difficult to find. Should anyone succeed in this task, he would surpass Pythagoras who had felt the existence of a beautiful harmony between celestial objects and even throughout the whole Universe. If the circular motions in the heavens by the manner of their arrangement sometimes seem to produce—to anyone who has the vain curiosity to note such oddities—divers angular figures mostly of oblong shape, then that can only be by accident, and the mind recoils with horror from such a supposition. The orbits of celestial bodies must be composed entirely of circular motions, otherwise they would not return perpetually and uniformly on their courses, and would be deprived of their perennity. This does not take account of the fact that [their courses] would be less simple and more irregular, and so would be less suitable for scientific study and computation.

'I shall willingly and with great pleasure speak with you at greater length on these and other things, and impart to you many of my observations, if you visit me one day as you promise. It will be less difficult for you than heretofore, because I have fixed upon a fresh site for Urania in Bohemia not far from you, and I am living in the Imperial castle of Benatky, five leagues from Prague. I am sure that you must already have learned from others that I was graciously summoned here from Germany by His Imperial Majesty, who received me with much kindness and liberality. Nevertheless, I should not wish you to be forced to come here through the afflictions of fate, but that you would come of your own accord, as well as from love and liking for studies of interest to us both. Whatever may happen, you will find in me not a friend of fortune, such as it may be, but your friend, who will not refuse you his support, even in adversity.'

On arriving at Benatky, Kepler was received with open arms by Tycho Brahe, who kept him there four months, and suggested that he join his staff. This offer was finally accepted by Kepler after some hesitation[12]; both men were proud and plain-speaking, and unable to get on with each other at first; there were clashes, a quarrel and reconciliation. In 1601, on Tycho Brahe's suggestion, Kepler was officially appointed as his assistant, and when Tycho Brahe died in this very same year, Kepler succeeded him as Imperial Mathematician.[13]

Tycho Brahe's aim throughout the whole of his life, and for which purpose he had brought together an enormous mass of observations of hitherto unknown accuracy, was to prepare new celestial tables to replace the *Alfonsine Tables*, which were several centuries old, as well as the *Tabulae Prutenicae*, which though quite recent, were just as inaccurate as the others.[15] The task assigned to Kepler when he joined the staff was to work out the theory for the motion of the planet Mars.[16] At a later date, Kepler was to regard this decision as the intervention of Providence, or even, more precisely, the final, supreme act of Providence which had inexorably guided him to meet Tycho Brahe and to study the motion of Mars.

When we consider that the outcome of this meeting and this study was the birth of the new astronomy, we are inclined to agree with Kepler.[17] We might, indeed, be surprised that Providence should have followed such devious paths and used such drastic means as the catastrophes of Hveen and Graz[18] in order to achieve its purpose. However, it is well known that the ways of Providence are obscure and disconcerting to the human mind, at least for the sceptic. For a believer, such as Kepler, they were on the contrary glaringly obvious; so, we shall do well to listen to and learn from him[19]:

'*The circumstances which led me to* [*concern myself with*] *the theory of Mars*. 'It is true that the divine voice, which commands men to learn astronomy, expresses itself in the world, not in words and syllables, but through things themselves and through the agreement of the human intellect and senses with the entirety of celestial bodies and phenomena. Nevertheless, there is also a sure destiny which secretly urges men towards certain arts and gives them the certitude that they participate also in the divine providence, as well as being part of creation.

'As soon as I was able in my youth to taste the sweets of philosophy, I embraced it wholeheartedly and with extreme passion, though not devoting myself especially to astronomy. It is true, that I had an adequate knowledge of it, and that I had no difficulty in understanding the course of studies prescribed for geometry and astronomy, namely, the study of figures, numbers and proportions. These studies were prescribed, and in no way something that might indicate a particular inclination on my part for astronomy. As holder of a bursary from the Duke of Württemberg, when I saw my fellow students, whom my prince, at the request of foreign nations wished to send them, steal away on divers pretexts, but in fact because they refused to leave their native land, I, being of a tougher fibre, decided quite early to go willingly wherever it might be.

'Now, in the first place, an astronomical post was offered to me, and to tell the truth I accepted it only because I was compelled to do so in obedience to my superiors. It was not that I was alarmed by the remoteness of the place [to which I was summoned], a fear which I had condemned in others (as I have already said), but that I hesitated on account of the unforeseen and lowly character of the post, as well as my limited knowledge of that branch of philosophy. Nevertheless, I accepted it, being richer in intellect than in knowledge, and loudly protesting that I was in no way abandoning my right to another kind of life [in an ecclesiastical post] which seemed much more attractive to me. The result of my studies during the first two years may be seen in my *Mysterium Cosmographicum.* Furthermore, the way in which my teacher Maestlin urged me to apply myself to the other parts of astronomy may be read in this small book in his letter printed in front of the *Narratio Prima* of Rheticus. I thought very highly of my discovery, the more so as I saw that Maestlin was well pleased with it also. Yet, I was less stimulated by the unreasonable promise he made to my readers of a general work from my pen—he called it *opus cosmicum seu uranicum*—than by my own eagerness to learn from [Tycho Brahe's] astronomy restored,* if my discovery could withstand examination in conjunction with accurate observations, for it had been shown in that book that they agreed with the less precise observations of popular astronomy. From that time onwards, therefore, I started to think seriously about obtaining observational data for myself. In 1597, I wrote to Tycho Brahe asking him what he thought of my little book; in his reply, he mentioned among other things certain of his observations, and in consequence I had a burning desire to see them. Now, Tycho Brahe, who played an important part in my destiny, never ceased urging me to visit him; but, as the distance between us prevented me from doing so, I am firmly of the opinion that it was by the grace of divine Providence that he came to Bohemia. I went to him then at the beginning of 1600 in the hope of being informed about the corrected eccentricities of planetary orbits. In the first week, I noticed that in common with Ptolemy and Copernicus he confined himself to the mean motion of the Sun,[20] whereas the [motion of the] visible [Sun] was more in agreement with my little book (as may be seen in the book itself); so I asked Tycho Brahe's permission

* *Astronomiae instauratae mechanica,* Nürnberg, 1602.

to use his observations in my own way. At that time, his personal assistant, Christian Severinus [Longomontanus] was working on the theory of the planet Mars, which opportunity itself had put into his hands seeing that they were then concerned with [carrying out] observations on the acronical position of Mars, that is to say, the opposition of Mars to the Sun in the 9° of ♌. If Christian had been busy with some other planet, I, too, should have been obliged to work on the same one.

'I regard it as yet another sign of Divine Providence that I came [to Benatky] precisely at the time when he [Longomontanus] was engaged on Mars; for, to be able to arrive at an understanding of the arcana of astronomy, it was absolutely necessary to take the motion of Mars as the basis, otherwise these secrets would have remained eternally hidden from us.'

Kepler was absolutely right. Even if the task of working out a theory of the motion of Mars, which fell to his lot after the departure of Longomontanus, were most laborious and most difficult—the motion of Mars has always been the despair of astronomers[21]—it was also the one which alone was able to lead him to his great discovery, namely, the ellipticity of planetary orbits. It so happens, that the orbit of Mars is not the most eccentric—that distinction belongs to Mercury—but it is the only one whose eccentricity is sufficiently large[22] to be apparent in the observational data of astronomy before the time of Galileo, or even Tycho Brahe. This was the very reason why it was so difficult for Ptolemy, as well as for Tycho Brahe and Copernicus, to account for the orbit in terms of circular motions.

Whilst he assigned Kepler the task of studying the motion of this planet, Tycho Brahe, nevertheless, did not give him a free hand. He asked—and he renewed the request on his deathbed—that the motion should be treated according to his (Tycho Brahe's) principles, and not according to those of Kepler, or of Copernicus. Kepler fulfilled this request, without conforming to it exactly.

This partly explains the unusual character and extreme difficulty of the *Astronomia Nova*, which are responsible for the exceptional interest of the work. Indeed, in this book, which is unique among the great classics of science, and in which all astronomical problems are treated three, and even four, times after the manner of Ptolemy, Tycho Brahe, Copernicus and finally Kepler himself, Kepler does not restrict himself to setting forth the results, as did Copernicus and Newton: he relates at the time, intentionally[23] as he did in the *Mysterium Cosmographicum*—the development of his thought, his efforts, and his setbacks,[24] Kepler's mind was so constituted that he was unable to find the way to truth without first having explored all the paths leading into error[25]—but perhaps the mind of man in general is naturally framed in this manner.

Johann Kepler's *Astronomia Nova* is indeed a new astronomy; in a certain sense it is even more so than that of Copernicus. In this work, for the first time, we find that the principle which Plato had laid down as fundamental to astronomy, and which had dominated the science for two thousand years, namely, the principle according to which the motion of celestial bodies is necessarily a circular motion (or is composed of circular motions) is abandoned. For the first time, too, this motion is explained by the action of a physical force. '*Astronomia Nova ΑΙΤΙΟΛΟΓΗΤΟΣ seu Physica Coelestis tradita commentariis de motibus stellae Martis*,'[26] that is to say '*New Astronomy, based on causes, or Celestial Physics, expounded in Commentaries on the motion of the planet Mars*': the very title of Kepler's work proclaims, rather than foretells, a revolution.

Once more, let Kepler speak for himself[27]:

> 'At the present time (he wrote in the preface to the *Astronomia Nova*), it is a very hard lot to have to write mathematical works, especially when they deal with astronomy. If proper accuracy be not observed in the propositions, explanations, demonstrations and conclusions, then the work is not mathematical. On the other hand, if attention be paid thereto, then the work becomes very difficult to read, especially in Latin, which language has no articles and lacks the felicity of Greek. Is not this the reason why there are so few good readers today, and the others, for the most part, decline to read? How many mathematicians take the trouble to read the *Conics* of Apollonius in its entirety? Yet, the subject is one that is better adapted to explanation by lines and diagrams than is astronomy.
>
> 'I, who am considered to be a mathematician, become tired on reading my own work when I try to have a clear understanding of the proofs which I myself have incorporated in the diagrams and text. If I obviate the difficulty of understanding the subject by introducing descriptive explanations here and there, then I seem to be prolix . . . and that is just as bad a fault; for prolixity of expression makes understanding just as difficult as does brevity of explanation. The latter fails to be taken in by the intellect, the former repels it. The one lacks illumination, the other disturbs us by its undue brilliance. In the former case the eye is not stimulated, in the latter it is dazzled.'

To tell the truth, any historian of Kepler finds himself in a similar position. If he follows, step by step, the reasoning of the author of the *Astronomia Nova*, which reasoning is far more difficult for us than it was for Kepler, or his contemporaries, it will become very prolix. On the other hand, if he neglects to do so, it runs the risk of becoming incomprehensible, or superficial; or, at least, it will fail to reveal the decisive steps in Kepler's thought. Thus, *mutatis mutandis*, he will say, in sympathy with the hero: 'At the present time, it is a very hard lot to have to write historical works on scientific thought. . . .'

However that may be, Kepler says[28]:

> 'it is my intention above all in this work to promote astronomical doctrine (particularly with respect to the motion of Mars) in all its three forms,[29] and in such a way that one can compute by means of the tables corresponding to the celestial phenomena, that which it has not been possible to do hitherto with sufficient accuracy. For example, in August 1608 the planet Mars was almost 4 degrees away from the position assigned to it by the Prussian computation [that of E. Reinhold]. In August or September 1593, this error, which has been completely eliminated in my computations, amounted to nearly 5 degrees.
>
> 'Now, while pursuing this purpose, which I fortunately achieved, I came also upon Aristotle's metaphysics, or more precisely, *Celestial Physics*, and *I studied the natural causes of motion*.[30] This consideration provided very clear reasons which show the Copernican doctrine to be true, and the other two [those of Ptolemy and Tycho Brahe] to be false.'

In fact, Kepler was not greatly concerned with proving the falsity of Ptolemaic astronomy. It seemed out-dated to him, and rightly condemned by the very existence of Tycho Brahe's system. Furthermore, it was totally irrelevant to any dynamic interpretation of celestial motions, and even incompatible therewith. Astronomically speaking, the discussion is centred on Copernicus and Tycho Brahe, or, more precisely, on Copernicus modified, or even utterly overthrown by Kepler, and Tycho Brahe. Paradoxically enough, it is Aristotle who looms on the fringe of the discussion behind Tycho Brahe. For, in the last analysis, the latter uses Aristotle's arguments against the motion of the Earth. Therefore, Aristotle's physics, or metaphysics, must be discussed.

Although the subjects treated by Kepler in the various chapters of the *Astronomia Nova* are, strictly speaking, quite different, they are in fact very closely connected. Consequently, the treatment of these subjects, and the various parts of the books devoted to their explanation are 'intermixed and intertwined'. Kepler adds, that in order to achieve his purpose, *i.e.*, arrive at an improvement in the methods of astronomical computation, he 'tried several paths, namely those taken by the Ancients, as well as those discovered by himself through following their example'. It happened that 'not one of them led to the desired end except the one that is subjected to the physical causes set forth in this work'.[31]

> 'Now, the first step towards determining the physical causes [of planetary motion] consists in proving that the common point of the eccentrics [that is to say, the point to which the motions of the eccentric planetary orbits must be referred] is not some point or other in the vicinity of the Sun, as believed by Copernicus and Tycho Brahe, but is the centre of the solar body itself.'

In fact, as had already been hinted in the *Mysterium Cosmographicum*, physical causes can have their origin only in a real (physical) body. They cannot emanate from a mathematical point, where there is nothing at all. The geometrical mechanics of Copernicus, based on a belief in the real existence of corporeal spheres carrying the planets, just as much as the concept which explains their motions through the action of motive souls (a concept occasionally favoured by Tycho Brahe, and even considered as a possibility by Kepler in his *Mysterium Cosmographicum*), seem to be adaptable to the absence of any physical centre of motion.

The necessity to find a physical explanation of this, and therefore to transfer the common point of their orbits into the body of the Sun, introduced rather serious consequences. Kepler had been convinced of this necessity long before his meeting with Tycho Brahe, whom he tried to convert to his viewpoint. In the first place, shifting the common point of the orbits to the Sun involves a displacement of the line of apsides and the positions of the planets when at opposition: in the case of Mars the difference in longitude can amount to 5°, which is no small matter.

Furthermore, as mentioned above,[32] it leads to the abandonment of one of the most essential theses of Copernicanism, namely, that of the absolute uniformity of rotation of the planetary spheres, and an unexpected return to Ptolemy's concepts, especially to a resurrection of the equant, which Copernicus prided himself on having eliminated from astronomy, and which Tycho Brahe, who was a most strict observer of Copernicus in this respect, only reluctantly accepted.

In fact, if we assume, as Kepler had done in his *Mysterium Cosmo-*

graphicum, that the various planets move more slowly the farther they are from the Sun, and that they do so for a physical and not only for an architectonic reason, it is natural enough to extend this notion to the motion of individual planets in their orbit, and to expect that in their courses about the Sun they will move more slowly the farther they are from the Sun; and more quickly the nearer they are to it.[33] Also, we may conclude that the inequality of their motion (the first inequality—that of motion on the deferent) is not only apparent, being explained by the eccentricity of the planetary circle (as required by Copernicus), but *real* (as Ptolemy had recognized it to be); and, once more contrary to Copernicus, that their motion is uniform only with respect to a *punctum aequans* situated on the other side of the centre of the circle in question.

It is even quite natural to apply this notion to the Earth itself, and to postulate that in its motion round the real Sun (and not round the mean Sun = centre of the Earth's circle), it moves on its circle with truly—and not apparently—variable velocity; or, in other words the inequality of the Sun's apparent motion round the Earth cannot be accounted for by a simple eccentricity, and the theory of the Earth's motion round the Sun, or of the Sun round the Earth, must include, exactly as in the case of the other planets, either a Ptolemaic equant, or the epicycle introduced by Copernicus in his theory of motion to replace the equant for the definite purpose of preserving absolute uniformity of circular motion, this being the only motion legitimately accepted by him for astronomical usage. Now, if we abandon the Copernican principle of uniform circular motion, and if—with Ptolemy—we assume true non-uniformity of planetary motions, it then becomes possible by reintroducing the equant to get rid of the Copernican epicycle; and, wonderful to relate, we thereby attain the traditional objective of astronomy and ascribe *truly circular* orbits to the planets.

Thus, paradoxically, the Keplerian innovations, the introduction of dynamical concepts into the theory of planetary motion, the transfer of the origin of the orbits into the real Sun, made possible, or implied, a step backwards; and the complications arising from these innovations resulted in the long run in an unexpected and extraordinary simplification, as well as a surprising standardization of the kinetic structure of the astral universe, in which, for the first time in the history of astronomy, planetary motions were *in reality accomplished in circles*.[34]

It certainly seems—and Kepler positively says so in the *Astronomia Nova*[35]—that an idea of this kind entered his mind when, in chapter xxii of the *Mysterium Cosmographicum*, he raised against himself the objection that planetary motions according to his own dynamical explanation would be of a different kind from the orbital motion of the Earth (or the Sun), which has no equant, either in the Ptolemaic or in the Copernican system. It is clear that even his audacious mind recoiled before the idea of introducing complications of this kind into the theory of the Earth's (or Sun's) motion, which complications had been considered unnecessary by both Ptolemy and Copernicus.[36]

Six years later he no longer hesitated. To Tycho Brahe, who in his theory of the motion of Mars had assumed a simple eccentricity of the Sun's orbit (that meant the Earth's orbit for Kepler), *i.e.*, assumed uniformity of the Sun's (Earth's) motion on its eccentric circular orbit, he raised the objection, Copernicus and Ptolemy notwithstanding, that a simple eccentricity was not enough and that it must be completed by the introduction of an equant, or of an epicycle. However, undoubtedly made wise by his experience with the *Mysterium Cosmographicum*, and thereafter acknowledging the worth of Tycho Brahe's criticism of the respective parts played by *a priori* and *a posteriori* arguments in building astronomical theories, Kepler did not base his objections this time on dynamical considerations, but on an accurate study of observational data—those of Tycho Brahe himself. In fact, the latter had already discovered in 1591 that the Sun's orbit is not of a fixed dimension, but on the contrary undergoes periodic expansion and contraction; this fact naturally greatly complicated planetary theory. Now, this phenomenon, which is puzzling and totally inexplicable if we consider it to be real, is on the contrary quite simply explained, as Kepler pointed out, if we treat it as a mere appearance resulting from the fact that the Sun's (or the Earth's) motion is not uniform unless it is referred to a *punctum aequans*. As Dreyer says in his excellent summary of Kepler's reasoning in chapter xxii of the *Astronomia Nova*:

> 'In this case it is easy to see that the annual parallax or difference between the heliocentric and geocentric longitude of a planet will vary with its position with regard to the Earth's line of apsides. If Mars be in the prolongation of this line and be observed from two points at equal distances on both sides of the line, then the parallaxes will be equal, no matter where in that line the *punctum aequans* be situated. But if Mars

be about 90° from the Earth's apsides and be observed from the apsides or from two points in mean anomalies α and $(180° - \alpha)$, the parallaxes will not be equal unless the *punctum aequans* be in the centre of the orbit, but they will differ more or less according as the Earth is nearer to or farther from its apsides.'[37]

II

First Attack upon the Theory of Mars

Kepler always regarded his decision to transfer the centre of motions (intersection of orbital planes) into the true Sun (the 'body of the Sun') and to assign an equant to the Earth's motion[1] as being of prime importance. He was right, for from that starting point his study of the motion of Mars[2] progressed through a series of successes and failures; and, as we have already remarked, it was the necessary starting point for the new astronomy.

The theory of Mars worked out by Tycho Brahe (and Longomontanus), based on their observations of 1577, 1582, 1586 and checked by the calculation of this planet's position at ten successive oppositions[3] with the Sun (they ranged from 1580 to 1600), was able to give the longitude of these oppositions with considerable accuracy: the difference never exceeded 2′, and could be attributed therefore to observational error. However, the theory completely failed to give proper values for the latitude. It was quite natural then for Kepler, in his critical revision of Tycho Brahe's theory, to give attention to determining the inclination of the orbit of Mars to the ecliptic. Here it was, that he gained his first 'victory' in the conflict with Mars. This investigation was made by four different methods, one of which, that of Copernicus, was based on direct observation of latitude during opposition; and as a result he was able to establish, (a) that the value of the inclination is 1° 50′; (b) that the plane of the orbit passes through the true Sun; (c) that the inclination is constant, and not variable, as thought hitherto. Expressed in another way, he established that transferring the origin (common point) of the orbital planes into the true Sun provides a constant value for the inclination, and proves that its apparent variation derives from the fact that the orbital plane was made to pass, not through the true Sun, but either through the Earth (Ptolemy), or through the centre of its circle (Copernicus), or—what is the same thing—through the mean Sun (Tycho Brahe).[4]

Having in this manner amply confirmed the value of his innovation and his criticism of Tycho Brahe's theories (and those of Copernicus as well), Kepler then proceeded to a fresh determination of the orbit of Mars by referring it to the true Sun, and assuming, in the same way as Ptolemy did, an actual non-uniform motion on its circle; that is to say, he introduced the *punctum aequans* into his theory. In order to determine the orbit, or what comes to the same thing, the position of the line of apsides (longitude at aphelion), the eccentricity and the mean anomaly at any given date, Copernicus and Tycho Brahe, who made the *punctum aequans* coincide with the centre of the circle,[5] and Ptolemy, who placed it in a position symmetrical with respect to the observer (bisection of the eccentricity), could be satisfied with the data provided by observing the position of Mars at three oppositions.[6] Kepler declined to assume 'bisection of the eccentricity', and—here we see how much the impenitent apriorist of the *Mysterium Cosmographicum* had been influenced by Tycho Brahe—decided to try and determine the position of the equant by purely empirical means, *i.e.*, by depending entirely on observation. For this purpose it was necessary to make use of not three, but at least four, oppositions.

From the ten sets of observations by Tycho Brahe which were available to him, he chose four (those of 1587, 1591, 1593 and 1595) and deduced from them the times of the true oppositions. Then, he proceeded more or less as follows[7]: on the circle *HFEIDG* which represents the orbit of Mars, the points *D*, *G*, *F* and *E* indicate the observed positions of the planet; *C* is the centre of the circular orbit; *S* is the Sun; *A* is the *punctum aequans*; *HI* is the line of apsides (Fig. 6). For the position of this line and the value of the mean anomaly at the first opposition, namely *HSF* and *HAF*, Kepler took, in the first instance, those which had been found by Tycho Brahe. The heliocentric longitudes, *i.e.*, the direction of the lines *SF*, *SE*, *SD* and *SG*, as well as the size of the angles *FSE*, *ESD*, *DSG* and *GSF* formed between them, could be taken directly from observational data. The angles formed by the lines drawn from the points *F*, *E*, *D* and *G* to the *punctum aequans A* (the differences of the mean anomaly) could be calculated, as the period of revolution and, hence, the mean motion of Mars were known. In triangles *ASF*, *ASE*, *ASD* and *ASG* having the same base *AS* (distance between the Sun and the *punctum aequans*), the angles at the base are known; so, it is possible to calculate the sides *SF*, *SE*, *SD* and *SG*, namely, the respective distances of Mars from the Sun, and to express them in terms of *AS*.[8] Knowing the sides *SE* and *SF*, as well as the angle *FSE* of triangle *FSE*, the angles *EFS* and *FES* can be calculated; similarly, the angles *SFG* and *SGF* of triangle *FSG* can be calculated; and so on. In this way, the angles of the quadrilateral *FEDG* can be found. Now, as the points *F*, *E*, *D* and *G* must lie on the circumference of a circle, it follows that $\angle EFG + \angle EDG = \angle FGD + \angle FED = 180°$.

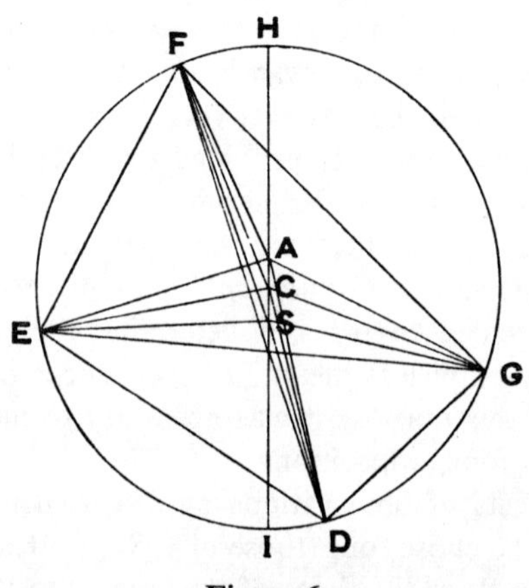

Figure 6.

It now remains to find the position of the point *C* (centre of the circle *FEDG*), and note, in particular, if it is located on the line *SA*, and, if so, its distance from points *S* and *A*. The length of the sides of the quadrilateral *FEDG* have been calculated; therefore, in each of the isosceles triangles formed by one side of the quadrilateral and the lines drawn from corners to the point *C*, the base and the angle at the apex are known. Consider, for example, the side *FG*: the angle *FGC* will be twice angle *FEG*, or twice the sum of angles *FES* and *SEG*; and each of the angles *CFG* and *CGF* will be equal to one half of (180° − *FCG*). Hence, the length of the radii *CF* and *CG* can be found in terms of *FG*; and *FG* itself being known in terms of *AS*, it is then possible to express *CF* and *CG* in terms of *AS*. Furthermore, in triangle *CFS*, the angle *CFS* equals the difference between *GFS* and *GFC*[9]; consequently, it is possible to determinate the side *CS* as well as angle *CSF* in triangle *CFS*. In order that the line *CS* shall fall on *AS*—that is the condition for *C* to be on *AS*—angle *CSF* must be equal to angle *HSF*. If this is the case, so much the better. If not, then the initial direction chosen for *HI* (the line of apsides), *i.e.*, the value of the angles *HSF* and *HAF*, must be altered, and the calculation outlined above must be repeated. This was what Kepler had to do until the two required conditions were satisfied by the solution, namely, the points *F*, *E*, *D* and *G* being on the circumference of a circle, its centre *C* falls on the line joining *S* (the Sun) and *A* (the *punctum aequans*).

Kepler, who devotes 15 pages *in folio* (Lib. II, cap. xvi) to a description of his method and the calculations related thereto, concludes by saying, that if the reader finds the chapter both boring and difficult, then he should take pity on the author, who had to do the same work sixty-six times before finally arriving at the solution[10] which gives for the longitude at aphelion a value of 28° 48′ 55″ (Leo), and for the distances of the Sun and the *punctum aequans* from the centre of the orbit, $CS = 11\,332$ and $AC = 7232$, the radius of the orbit being taken as 100 000 (or, taking the radius as unity, $AC = 0{\cdot}072\,32$ and $CS = 0{\cdot}113\,32$).

Kepler's result, which he finally realized to be inadequate, enables tolerably accurate calculations to be made; and Kepler used it to a considerable extent in the final working out of his theory of the motion of Mars. The theory which we have just outlined, and which Kepler subsequently called *hypothesis vicaria*—a vicarious hypothesis—was important to him as a working tool. For us, however, its importance is not technical, but psychological. It reveals one aspect of Kepler's mentality, an aspect that might be called 'Tychonian' or even 'empiric' (I have already referred to this[11]), and which gives us a better understanding of the limits, or preferably, the very structure of his *a priori* method. That God had created the Universe after due consideration and in accordance with a plan, which was architecturally perfect and expressible mathematically, was something Kepler never doubted; nor did he ever abandon his efforts to grasp the intention of the Divine Mathematician. Because of his deep conviction on this matter, he was not able to be satisfied with a mere approximation: *ubi materia, ibi geometria*, he had declared[12]; and, so, there could not be any laxity between facts and theory. The agreement must be rigorous, else it is worthless. Furthermore, there must be a clear distinction between legitimate *a priori* reasoning and pseudo-*a priori* assertions, which are without foundation. For example, the bisection of the eccentricity which Ptolemy adopted without any supporting proof. Undoubtedly, a symmetrical arrangement seems reasonable; but is it really necessary?[13] Kepler had his doubts—even at the time of his *Mysterium Cosmographicum*. In 1600, even before his association with Tycho Brahe, as he proudly informs us in the *Astronomia Nova*, he decided not to accept it, but to try and determine the distances between the Sun, the centre of the orbit and the equant point[14] empirically from observations. This decision was strengthened by the thorough realism of Kepler's epistemology: astronomy must harmonize with reality in every respect. It was a useful decision, for Kepler's results were capable of being interpreted quite differently: it was only necessary for him to share the total eccentricity indicated by his theory (11 332 + 7232 = 18 564) between the orbit of Mars and its epicycle in the Copernican (or Tychonian) manner; that is to say, to give the value of 14 948 to the distance between the Sun and the centre of the planet's deferent circle, and to place a small epicycle of radius 3616 on the deferent; or yet again, to place, not one, but two, epicycles of radii 14 948 and 3616 respectively on a circle concentric with the Sun.

We have already seen that when Copernicus decided to dispense with Ptolemy's equants and revert, as it were, to Hipparchos in order to maintain the principle of uniform circular planetary motion in all its rigour, he was obliged, in order to 'save' the phenomena, to reintroduce into the theory of planetary motion one epicycle, which was precisely what the Ptolemaic equant rendered unnecessary. He lost thereby even the advantage—the simplification of the system of motions—which he had gained through the orbital motion of the Earth. Nevertheless, and this is something of unquestionable importance, he was able to replace the enormous Ptolemaic epicycles by much smaller epicycles, which moreover agreed in their rotary motion with the period of the planet carried by them, and not with the period of the Sun as they did according to Ptolemy. Copernicus achieved this by ascribing three-quarters of Ptolemy's total eccentricity (distance between the Earth and the *punctum aequans*) to the eccentricity (distance between the centres of the circles on which move the Earth and the planet respectively). The true motion of the planet consequently ceased to be truly uniform and circular; but the path traced by the planet became smooth and was compounded of uniform circular motions,[15] which was something of prime importance from the point of view of Copernican dynamics.

Kepler went in the opposite direction[16]: from Copernicus (Hipparchos) he went back to Ptolemy; and it was undoubtedly for the purpose of justifying this step that he dealt at length in the early chapters of the first book of the *Astronomia Nova* (caps. III and IV) with the equivalence, admittedly approximate, of the Ptolemaic theory (eccentric and *punctum aequans*) and the Copernican (eccentric and epicycle, or concentric circle and two epicycles).

One might then ask, why, if the two theories are roughly equivalent, preference was given to that of Ptolemy? The reason is, that in Kepler's view, it is superior to that of Hipparchos–Copernicus, and for two reasons. Not only is it mathematically simpler and more elegant, but also it is physically 'nearer the truth', in that it freely recognizes the physical fact of the variation in speed of planetary motions, instead of trying to conceal it by an artificial device (superposition of uniform circular motions). This certainly does not explain the reason for the variation—that was to be Kepler's problem—but, at least, by doing away with useless epicycles, it enables us to glimpse the possibility of such an explanation. For—and here I am anticipating the development of Kepler's ideas just as he himself did in the *Astronomia Nova*—Kepler, whilst making use of the mechanism of epicycles, definitely seems to have become increasingly sceptical of the possibility of providing a physical explanation of epicyclic motion; and hence more and more inclined to deny its reality.[17] In fact, a combination of rotary motion of solid spheres encased one within another is quite conceivable; but it is a totally different matter when the planets move freely round a geometric point. Not only is it then necessary to ascribe to them intelligence, over and above purely animal or motive 'souls', to guide them on their courses, but it is also extremely doubtful if any intelligence whatsoever could be capable of making the necessary calculations and, for instance, of describing a circle without reference to a physically determined point as its centre.[18] Hence, the apparent equivalence of the theories in question does not stand up to close examination; at least, not as far as Kepler was concerned. Therefore, even if his *hypothesis vicaria* did not represent the phenomena with greater accuracy than did Tycho Brahe's, it seemed to him, nevertheless, to be *fundamentally* superior.

Let us now return to a consideration of this *hypothesis vicaria*.[19] At first sight, it seemed to provide an unassailable solution of the problem of the motion of Mars, and consequently to provide brilliant confirmation of the general views on the mechanism of planetary motions. Indeed, in the face of facts, namely, Tycho Brahe's observational data, it was possible to account for the heliocentric longitude of the twelve observed oppositions[20] with errors not exceeding 2′ 12″. Unfortunately, as regards latitude, the *hypothesis vicaria* was no better than Tycho Brahe's theory, which it was intended to replace. When Kepler made his calculations anew, using the observed latitudes at the oppositions of 1585 and 1593, he obtained results for the orbital eccentricity which differed greatly from the values obtained by him by dividing the total eccentricity into two unequal portions, they nevertheless agreed almost perfectly with one half of the total eccentricity:

$$\frac{0{\cdot}186\,54}{2} = 0{\cdot}092\,82.$$

Kepler concluded from this that Ptolemy had undoubtedly made similar trials, and that the 'bisection' of the eccentricity had not been based on reasons of pure symmetry, as he had at first thought, but, on the contrary, had been based on observational data; and he decided to repeat the calculations basing them on this bisection, the importance of which he had failed to recognize.[21] Unfortunately, the result was just as deplorable. He obtained quite satisfactory agreement between observed and calculated positions for the apsides and positions 90° away from them; but at intermediate positions, corresponding to displacements of 45° or 135°, the discrepancy between calculation and observation amounted to 8′.

Now, eight minutes of arc do not amount to much; and it was precisely the small value of this discrepancy, which was less than the limit of accuracy of astronomy in ancient times, (it was 10′ for Ptolemy), that explains, as Kepler pointed out, why Ptolemy was able to adopt bisection of the eccentricity without running into difficulties. However:

> 'Seeing that the Divine Goodness has given us in Tycho Brahe a most diligent observer whose observations have revealed the error of 8′ in Ptolemy's calculation, it is only right that we should thankfully accept this gift from God, and put it to good use. We must undertake to discover ultimately the true nature of celestial motions.'

The improvements in observational technique and accuracy of measurement introduced by Tycho Brahe had completely changed the whole situation: a difference of 8′ *compared with Tycho Brahe's data* could not be ignored, and these eight minutes 'provided the foundation, and pointed the way to the reformation of the whole of astronomy', which was inaugurated by Kepler. These eight minutes obliged him to reconsider, and finally reject, the fundamental axioms of the science of celestial bodies, which no-one had questioned for two thousand years, and which he himself had blindly accepted.[22]

Kepler had based his theory of the motion of Mars on the hypothesis of circular motion of the planet itself on its eccentric deferent, taking into consideration its distance from the Sun. The discrepancy mentioned above, showed him that Mars was not where it should have been, but was in fact nearer to the Sun than his theory required. In other words, its course was a curved one within his hypothetical circular orbit.

Furthermore, a close examination of the *hypothesis vicaria* and comparison with the determinations of eccentricity from the latitude of Mars (there is a fresh one in chapter xvii) led to some surprising facts (chap. xx), namely, 'that in the plane of the eccentric [orbit] there is no fixed point with respect to which the planet moves with a constant angular velocity; and this point must be moved backwards and forwards along the line of apsides'; which displacement cannot take place through natural causes.[23]

Kepler adds that this is a necessary consequence from the axioms of the circularity of planetary orbits, and that this axiom together with the one relating to the uniformity of convolute motion, even with respect to a *punctum aequans*, must be abandoned, as he will show later in his book.[24]

However, for the moment we have not yet reached that point: the collapse of the *hypothesis vicaria*, the setback over bisection, did not shatter Kepler's faith in the circularity of planetary orbits. It was only about a year later, in 1602, that he was seized by doubt.

III

Study of the Earth's Motion

The effort made by Kepler in developing and checking his *hypothesis vicaria* was not entirely wasted, even though it failed to provide the results he hoped to obtain. The hypothesis in question, whilst revealing its own inadequacy, was nevertheless a very valuable working tool; and calculation of the eccentricity starting from the latitudes, whilst showing that bisection of the eccentricity was not completely in agreement with fact, did show, however, that it was much nearer the truth than Kepler had at first thought. Consequently, it was not surprising that, in his subsequent investigations, he should use the *hypothesis vicaria* (for calculating the heliocentric longitude of Mars), and once again accept the Ptolemaic hypothesis of bisection, even for the Earth.

The setback in his attempts to formulate a correct theory of the motion of Mars convinced Kepler that he had to make a fresh beginning, and for a start direct his attention, not to the 'first' inequality, but to the 'second', namely, that one, which in the Ptolemaic system, derives from the motion of the planet on its epicycle, and in Copernican astronomy is explained by the projection of the Earth's orbital motion on to the sky. This particular motion, which is of prime importance because it is involved in all planetary motions, was therefore the one that ought to have been studied first of all. Now, this had never been done with sufficient accuracy. After all, this is understandable, because the motion did not exist for Ptolemy, and as far as Copernicus was concerned it was the simplest of all the motions,[1] as was that of the Sun for Ptolemy.

We know that this was not Kepler's view, and that he protested to Tycho Brahe that the Earth's motion is no more uniform than that of the other planets: 'inspired by his genius' he explained the fact by the apparent expansion and contraction of its orbit (for Tycho Brahe, this meant the Sun's orbit).

In order to make an exact determination of the Earth's orbit, Kepler used an extremely ingenious and original procedure, which consists in transporting ourselves to the planet Mars and observing the motion of the Earth from that standpoint in order, first of all, to ascertain several positions of the Earth on its orbit, and from them to find the orbit itself. In order to facilitate these determinations, and to avoid taking into account the proper motion of Mars (which was inadequately known) it would be desirable to immobilise the planet in the sky. Obviously, this could not be done; but an equivalent result is obtained by using only those observations made when Mars is in the same place on its orbit; that is to say, the individual observations are separated by an interval of 687 days (the sidereal period of Mars).

In Kepler's words[2]:

> 'I shall search [in Tycho Brahe's records of observations] for three, or any number, of observations made when the planet was at the same position on its eccentric, and from them, using the law of triangles [trigonometrical calculation], find the distance to the equant point from an equal number of points on the epicycle or annual circle [of the Earth]. As a circle is determined by three points, I shall determine from each [set of] three such observations the position of the circle and its apsides, which I have provisionally assumed to be known, as well as the eccentricity with respect to the equant point. If there be yet a fourth observation, it could serve as a check.'

Kepler's mathematical argument is rather complicated. It may be set forth schematically as follows.[3] From Tycho Brahe's observations on Mars, he selected three made at intervals of 687 days (the period of revolution of Mars round the Sun), when the planet was in exactly the same position with respect to the Sun. On these particular dates, the Earth was obviously at different positions on its path. The heliocentric longitude of Mars, *i.e.*, the direction of the line SM drawn from the Sun to the planet, or, more precisely, its projection on the plane of the ecliptic, could be easily calculated by means of the *hypothesis vicaria*, or even from the theory of the motion of Mars worked out by Tycho Brahe. The (heliocentric) longitude of the points E_1, E_2, E_3, being the positions of the Earth when Mars was at M, could be similarly determined from the Tychonian theory of the Sun. Hence, the angles E_1SM, E_2SM, E_3SM were known. Tycho Brahe's observations gave the angles SE_1M, SE_2M, SE_3M directly. Consequently, the angles SME_1, SME_2, SME_3 were known also. It was then quite simple to determine the distances SE_1, SE_2, SE_3 (Fig. 7), taking *SM* as base. As three points are sufficient to define a circle, it was then quite easy to calculate the dimensions of the (circular) orbit of the Earth and fix the position of its centre *O* with respect to *S*, *i.e.*, the direction of the apsides. By similar reasoning, namely, identifying *S* not with the true Sun, but with the mean Sun, *i.e.*, with the point from which the Earth's motion appears to be uniform (the *punctum aequans*), Kepler determined its distance with respect to the centre of the orbit. Now, if the theory of the Earth's motion adopted by Copernicus (or that of the Sun's motion adopted

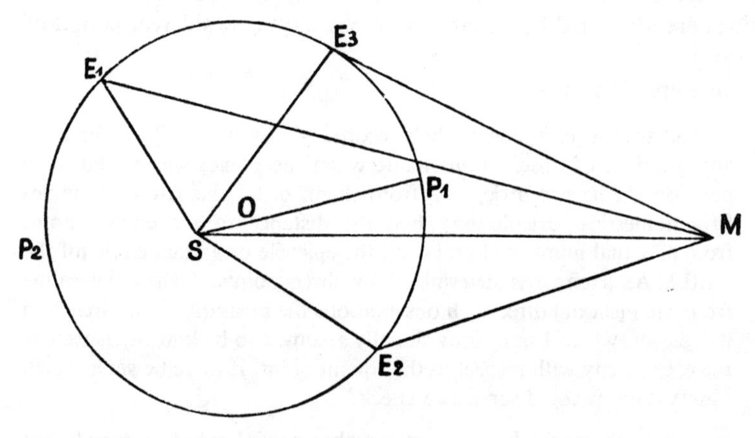

Figure 7.

by Tycho Brahe) were true, then the two points should coincide, and the radii vectores SE_1, SE_2, SE_3 should be equal. However, on the contrary, calculation confirmed Kepler's supposition that the Earth's eccentricity should be bisected, as in the case of the other planets: the *punctum aequans* and the Sun were found to be approximately at the same distance (0·018 of the radius taken as unity) from the centre of the orbit, that is to say, at a distance very nearly equal to one half of the eccentricity ascribed to it by Tycho Brahe (0·358 40).[4]

The result obtained by Kepler was so important, and from his point of view so satisfactory, seeing that it proved the structural uniformity of the planetary universe in which he had always believed,[5] that he did not restrict himself to the proof outlined above, but added two others. One of these uses the method of false position, starting with four (and not three) positions of Mars; the other computes, for five different dates (separated from each other by the orbital period of Mars), the respective positions of the Earth, Sun and the planet assuming bisection of the eccentricity as a basis of calculation.[6]

The method invented by Kepler for determining the terrestrial orbit can obviously be reversed. Taking the value of this orbit (now known), it is possible to calculate that of Mars. To do so, it was only necessary once more to delve into the treasury of Tycho Brahe's observations and this time calculate the distance of Mars from the Sun when the Earth was in the same position on its orbit in successive years,[7] instead of calculating, as formerly, the distance of the Earth from the Sun corresponding to the same position of Mars in the sky (at dates separated by the period of Mars).

We can go even further and use as a basis of calculation any position whatsoever of the Earth on its orbit (which is now completely known). Starting from one contemporary observation of Mars, which gives its geocentric longitude, it is possible to determine: (a) its heliocentric longitude; (b) the distance between the Earth and the Sun on that date; then the corresponding distance between the Sun and Mars can be determined trigonometrically. In the same way, the orbit of the planet can be found exactly.

We should have expected Kepler to do this, namely, continue—or resume—his study of the orbit and motion of Mars. It is somewhat surprising to note that he did nothing of the kind, but, on the contrary, turned to an investigation of the causes of planetary motion, a matter which he touched lightly upon in the *Mysterium Cosmographicum*. Most likely, he had been tired out by the innumerable calculations to which he had devoted himself—they cover more than 900 pages in his manuscripts preserved in the library at Pulkovo—and he had little inclination in consequence to embark on them again. So, we find that he abandoned his attack on Mars for more than a year, and turned to other work, such as chronology and optics.[8] It may be that he had been discouraged by the setback to his *hypothesis vicaria*; and that confirmation of his youthful intuition concerning dynamics had brought him back to the problems which then preoccupied him; or, it may be that his striking success in referring the non-uniform motion of the planets to the true Sun instead of to an abstract point (the mean Sun) had increased his aversion from using the *punctum aequans* (another abstract point) to describe these motions; it may also have increased his desire to give a mathematical form to the orderly rule of behaviour in his system—planets move slower the farther they are away from the Sun. This mathematical form would give direct expression to the dynamic structure of the motion in question, instead of employing an artificial 'trick' such as the equant, and consequently would enable its nature to be understood, in addition to allowing calculations to be performed. The situation must have been very much so, especially as he had recently discovered that in the theory of the motion of Mars it was impossible to assign this *punctum aequans* an unequivocal position on the line of apsides. Finally, his desire to substitute a fresh technique for Ptolemy's mathematical technique could be explained by the fact that, from the point of view of computation, Ptolemy's concepts were not entirely in agreement with his own.[9]

Whatever may be thought of my hypothetical reconstruction of the motives and stimuli which inspired and guided Kepler, the fact remains, that having shown that the Earth, just as much as the other planets, moves with a motion that is *truly* and not only *apparently* non-uniform, Kepler turned to an investigation of the causes and structure of planetary motion.

IV

A quo moventur planetae?

'The planets move faster the nearer they are to the Sun, and more slowly the farther they are away from it.' This fundamental axiom of Copernican cosmology, which Kepler had already adopted in the main in his *Mysterium Cosmographicum* for the whole planetary universe and had even extended to the motion of individual planets, may be formulated much more accurately, provided always that it is applied only to *each* individual planet, and not to them collectively. *The velocity of a planet in its orbit is inversely proportional to its distance from the body about which it revolves.* This, in Kepler's mind, is at first only another statement—yet, how much more reasonable!—of the Ptolemaic theory, which asserts that motion on a deferent is faster the *farther* the moving point (centre of the epicycle) is from the *punctum aequans*, and slower when it is *nearer*; and even that its speed is strictly proportional to its *distance* from that point.[1] Ptolemy, it is true, made this assertion only in respect of velocity at the *apsides*. Kepler, however, extended this relationship to the whole path of the orbit; and did so, as historians have not failed to remark, without justifying this extension by any kind of proof; nor was he even aware of the necessity for doing so. It seems to me, that this happened not only in virtue of the 'continuity principle', of which he made ample use in his mathematical work, but also, and perhaps especially, because Ptolemy in his view was providing a fact: at the apsides, that is to say under *optimum* observational conditions, the velocities of planets are proportional to their distance from the central body. This fact agreed so well with his dynamical principles (celestial physics) and Tycho Brahe's observations, that he could not regard it as anything but a general relationship; and if this relationship were not fully confirmed by Ptolemy for other points on the path, in other words, if the agreement between his theory and Ptolemy were only approximate, then the discrepancy was in fact too small to be used as a valid objection against him. There was nothing surprising about this discrepancy; it merely showed that Ptolemy's mathematical technique was not—how could it be?—adequate for the celestial dynamics he was then engaged in developing; and that it was therefore necessary to develop another, based not on abstract concepts, such as the equant, which are fundamentally spurious, but based on reality itself, whose structure must be reproduced.[2] Now, if we abandon the purely kinematic view of Ptolemy, as well as belief in the solid spheres of Copernicus, which Tycho Brahe rendered impossible—these are, in any case, adaptable only to *truly* uniform motion, and consequently imply the re-introduction of epicycles into the mechanism of planetary motion; and if, therefore, we must seek an explanation of the *truly* non-uniform motion of celestial bodies, then we must assume, because the *velocities* of the planets are inversely proportional to their distance from the Sun, that the same relationship holds in respect of the *motive forces*[3] which drive, or carry, them on their path. So, in Kepler's own words:

'I have already said that Ptolemy, being led thereto by observations, had bisected the eccentricity of the three superior planets, and that Copernicus copied him. As a result of Tycho Brahe's observations, bisection was shown to be probable also in the case of Mars, as we have described in chapters XIX and XX and as will appear more clearly in chapter XLII. Furthermore, in regard to the Moon, Tycho Brahe proceeded more or less in the same way. Bisection has just been proved in respect of the theory of the Sun (according to Tycho Brahe), or of the Earth (according to Copernicus). There is no objection to the same assumption in the case of Venus and Mercury. . . . All planets have this bisection. Eight years ago[6] (it is even more, now) in my *Mysterium Cosmographicum* I postponed [to a later date] a discussion of the cause of the Ptolemaic equant[7] only because it was impossible to decide, from the principles of astronomy as commonly accepted, if the Sun (or the Earth) required a *punctum aequans* as well as bisection of the eccentricity. This question must now be regarded as settled, particularly as we have confirmation through the evidence of improved astronomy [of the fact] that the Sun, or the Earth, have also a *punctum aequans*. This being so, I say that the cause of the Ptolemaic equant which I pointed out in my *Mysterium Cosmographicum* may be taken as true and correct, seeing that it is general and is the same for all the planets. In this part of my work, I shall explain the matter in greater detail.

'As the explanation is general, I use the term "planet" [in its most general sense]. (In this chapter and several following ones) the reader can take it to mean the Earth of Copernicus, or the Sun of Tycho Brahe.

In the first place, it should be understood that, according to the Ptolemaic hypothesis, the velocity at perihelion and the slower speed at aphelion are approximately proportional to [the length] of the line drawn from the centre of the Universe to the planet,[8] however great the eccentricity may be.' (Cf. Fig. 8.)

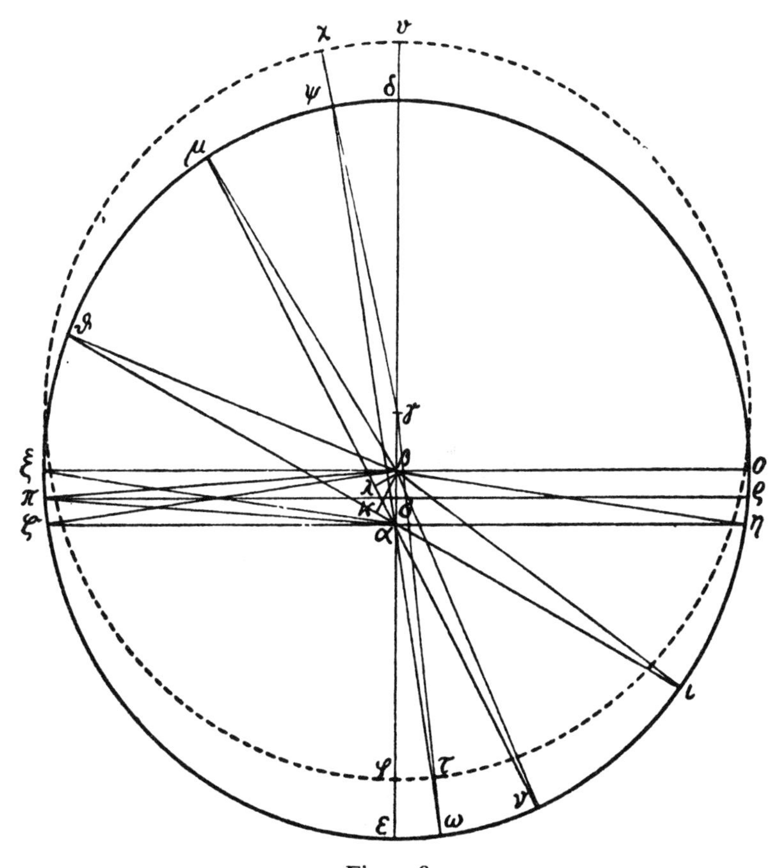

Figure 8.
α=centre of the Universe.
β=centre of deferent.
γ=centre of equant circle.

'The arc of the time $\upsilon\chi$, in the Ptolemaic sense, has nearly the same ratio to the arc of the path $\delta\psi$[9] as the distance $\alpha\delta$ of the arc $\delta\psi$ from the centre of the universe has to the mean distance $\delta\beta$ of the points π and ρ from α; and similarly the ratio of the arc of the time $\varphi\tau$ to the arc of the

> path $\varepsilon\omega$ is nearly equal to the ratio of the distance $\alpha\varepsilon$ of the arc $\omega\varepsilon$ from the centre of the universe α to the mean distance $\varepsilon\beta$ of the points π and ρ from α.'

From this, by a calculation which it is unnecessary to reproduce, Kepler arrived at the following conclusion: the time during which a planet remains on a given arc of its eccentric circle is greater the farther this arc is from the centre of the universe.[10] In other words, the planets move with non-uniform velocities; and this statement, in its turn, is explained by a corresponding variation in the force which is responsible for their motion. Now, if this force vary as a function of the distance between the planet and the centre of the Universe, it follows that the force is located at this centre, and seeing that the Sun is at the centre of the Universe, it follows that *the force which moves a planet is located in the Sun.* To continue in Kepler's words[11]:

> 'It has been shown in the previous chapter that the time during which a planet remains on equal portions of the eccentric circle (or traverses equal distances in space) is proportional to the distance of the path from the point from which the eccentricity is measured; or, more simply expressed, the planet in its motion round the point which is taken as the centre of the Universe is moved with less force the farther it is away from this point. Whence it necessarily follows that the cause of this diminution in strength must be either in the body of the planet itself and in its inherent motive force, or else at the assumed centre of the Universe.
>
> 'Now, a well-known axiom in natural philosophy states that phenomena produced at the same time and in the same way, and which are attentuated to the same degree, are [either] inter-related, [or] result from a common cause. In this instance, the increase or decrease in motion always coincides proportionally with an approach to or recession from the centre of the Universe. Consequently, either this diminution in force causes the celestial body to recede from the centre of the Universe, or the recession causes the diminution in force, or again, both effects result from a common cause. However, it is impossible to conceive a common cause for these two phenomena; and in the following chapters we shall show that there is no need for such an explanation. Now, it is not in accordance with [the laws of] nature for the distance from the centre to be dependent on the strength or weakness of the motion in longitude[12]; for distance, in one's mind and in [the order of] nature, comes before motion in longitude. In fact, there is never any motion in longitude without [the existence of] distance from the centre, because it requires space for its fulfilment; though a distance from the centre may certainly be conceived without motion.[13] Therefore,

distance is the cause of the degree of motion; a greater or less distance will result in a greater or shorter time for traversing the path.

'As [the idea of] distance is a relative concept whose meaning relates to concepts [placed] in relation to one another,[14] and which, in themselves (if we neglect the *relata*), have no meaning, the cause of the changes in motion must consequently be found in one of the *relata*.

'Would it be a change taking place in one of the bodies in question, for example, the moving body? The body of a planet does not of its own accord become heavier when it moves away from, nor lighter when it moves towards[15] [the centre of the Universe]. It would be a much too ridiculous assertion to say that any animal force residing in the body of the planet, and which imparts motion to the celestial body, increases and decreases in turn without becoming strained or weakened with age.[16] Moreover, it passes comprehension how this animal force could guide the planet through the space of the Universe, where there are no solid spheres, as Tycho Brahe has proved. Finally, a round body would lack members, such as feet or wings, by the movement of which, exerting a certain pressure and counter-pressure, the soul would be able to carry its body through the aether, as birds do through the air.'[17]

From this point onwards, Kepler thought it very unlikely that a soul could impart translatory motion to a body (it would be different were it only a question of rotation in a given position): the 'animistic' hypothesis consequently implies the existence of spheres and, in addition, leads to a very unlikely increase in the number of souls, for it is necessary to assume one of them is required for each motion performed by the planet.[18]

The non-existence of corporeal spheres, as proved by Tycho Brahe, and the possibility that the souls and intelligence in question are indeed inherent in the planets provide Kepler with the opportunity for a little pleasantry in the preface to the *Astronomia Nova*,[19] which preface, however, represents the end and not the beginning of his considerations on this matter:

'Conditions for the existence of the intelligence and motive souls become very difficult, because, in order to guide the planets with two motions compounded together [the motions on the deferent and the epicycle], they would be forced to take many things into account; at the very least, they would be forced to take into account, simultaneously, the origins, centres and periods of the two motions. If, however, the Earth were being moved, most of these things could be performed by corporeal, and not animal, faculties, for example, by magnetic [faculties], which are more general [than animal faculties].'

Let us now return to chapter xxxiii, where Kepler pursues his argument:

'Therefore, the only hypothesis that we can assume is the one which places the cause of the increase and decrease [in motion] in the origin of the *relata*, namely, in the point which we have taken as the centre of the Universe and from which the distances have been calculated.

'Consequently, if as a result of an increase, or decrease, in the distance of the body of the planet from the centre of the Universe the motion of the planet become slower, or faster, respectively, [then] the source of the motive force must necessarily be in the point which we have taken as the centre of the Universe.

'This being accepted, light is thrown on the way in which the cause functions; and, in particular, it will be seen that the planets move in a manner which conforms to the behaviour of a balance, or a lever.[20] To say that it is more difficult to impart motion to a planet the farther it is from the central point (hence, the motion is slower), is the same as saying that the heavy body becomes heavier the farther it moves away from the fulcrum of the supporting arm; the body itself does not become heavier, but merely exerts its effect over a greater distance. In both cases, either that of the balance and lever, or that of planetary motion, the decrease in power is proportional to the distance.[21]

'In the first part, I have begun a study based on physical arguments to discover what body is located at the centre; my intent is to learn if there be none at all, as is assumed for the purposes of calculation by Copernicus, and Tycho Brahe also at times; or if the Earth be at the centre as is assumed by Ptolemy, and Tycho Brahe also at times; or if it be the Sun, as I myself believe, and as Copernicus does, too, when he engages in speculation.[22] In particular, I have there assumed as a principle, that which was a fit subject for geometric proof in chapter xxxii, namely, the proposition: a planet moves less rapidly when it is distant from the point from which its eccentricity is calculated.

'Relying on this principle, I came to the conclusion that it was more likely for this point, the centre of the Universe, to be occupied by the Sun, or the Earth in Ptolemy's view, than by nothing at all. Having proved our principle, we shall, in this chapter, revert to this proof and show its probability. It should be remembered, that in the second part [of this work] I showed that [observed] phenomena during opposition agree satisfactorily [with theory] when we relate the oppositions of Mars to the observed position of the Sun; we then determined both the eccentricity and the distances from the centre of the body of the Sun itself, with the result that the Sun itself comes [to be] at the centre of the Universe (according to Copernicus), or, at least, at the centre of the planetary system (according to Tycho Brahe).

'Of these two possibilities, one is based on physical probability, and the other concludes in favour of reality from possibility. For this reason, I have delayed until chapter LII [the task of] proving, by means of observations, that we must necessarily refer the planet Mars to positions of the visible [true] Sun, and make the line of apsides, which bisects the eccentric, pass through the body of the Sun itself.... The impatient reader, who cannot wait, may read this chapter immediately.[23] When he has done so, let him then read the following....

'As, therefore, the Sun is situated at the centre of the system, it follows, from what has already been proved, that the source of the motive force also is situated in the Sun, seeing that it is placed exactly at the centre of the Universe.

'No doubt, I should have been entitled to attention with a sympathetic ear,[24] if that, which I have already proved *a posteriori* by a rather long investigation, had been attempted *a priori* (starting from the particular significance of the Sun), so as to show that the source of life in the Universe (by *life* I mean motion of celestial bodies) is identical with the source of light which is the ornament of the whole Universe, as well as the sourse of heat which invigorates everything.

'As for Tycho Brahe, or all those who have preferred to adopt his general hypothesis for a second inequality,[25] let them see for themselves how, by a semblance of truth, they once more subvert that physical uniformity which they have accepted for the greater part of the Universe[26]: in fact, for them, too, the Sun is situated at the centre of the planetary system.

'From the above, one of the following propositions necessarily follows: either, the motive force which is located in the Sun and which imparts motion to all the planets imparts motion to the Earth; or, the Sun and the planets which are linked to it by its motive force are guided round the Earth by a specific force which is located in the Earth.

'In fact, Tycho Brahe himself destroyed the notion of the real existence of the spheres. On the other hand, in the third part, I have indisputably proved the existence of an equant in the theory of the Sun, or of the Earth. Whence it follows that the Sun, if it moves, does so more quickly or more slowly according as it is nearer to, or farther from the Earth: in this case, the motion of the Sun would be controlled by the Earth. However, if the Earth move, it must do so [as do the other planets] through the influence of the Sun, and it will move faster, or slower, according as it is nearer to, or farther from, the Sun, whose force remains constant. Apart from these two possibilities, there is no other.

'I abide by Copernicus, and allow the Earth to be a planet. No doubt, one could raise against Copernicus the same objection with regard to the Moon, as that which I raised against Tycho Brahe with regard to the five planets, namely, that it seems absurd for the Moon to be given motion by the Earth, and furthermore be so closely linked with the latter as to be conveyed also round the Sun. Nevertheless, I prefer to make an exception only for the Moon, which, on account of its position, is linked with the Earth (as I have shown in *Astronomiae Pars Optica*); and to seek the cause only of the Moon's motion in a force located in the Earth yet extending to the Sun (as will be discoursed upon in chapter XXXIX), rather than to ascribe to the Earth the cause of motion of the Sun and all the planets linked to the latter.'

In the preface to *Astronomia Nova*—prefaces are always written after the books have been finished—Kepler summarized the discussion and expressed himself as follows[27]:

'If the Earth move round the Sun, then the law governing its speed, or its slowness, depends on the degree of its proximity to, or remoteness from, the Sun.

'Now, this phenomenon is observed also in the case of the other planets, namely, they are accelerated, or retarded, according as they are nearer to, or farther from, the Sun. The proof of this is a purely geometrical one. From this proof, which is quite sound, we deduce from a physical supposition that the source of motion of the five planets is located in the Sun. Consequently, it is highly probable that the source of the Earth's motion also is located in the same place as the source of motion of the five planets, that is to say, in the Sun. For this reason, it is most probable that the Earth also is endowed with motion, because a probable cause for its motion exists.

'On the other hand, that the Sun remains motionless in its place at the centre of the Universe is rendered plausible, apart from anything else, particularly by the fact that the source of [celestial] motions, at least [in the case] of the five planets, is situated in the Sun. For, whether one adopts the view of Copernicus or that of Tycho Brahe, in both cases the source of motion for the five planets is situated in the Sun; according to Copernicus, the same applies to the sixth planet, the Earth. Now, it is more likely that the source of motion is motionless where it is situated than that it has motion.[28]

'Yet, if we adopted the view of Tycho Brahe, and said that the Sun has motion, it would at least remain proven that the Sun moves slowly when it moves away from the Earth, and quickly when it approaches, not only in appearance, but also in reality.'

In this case, the small Earth would move the enormous Sun; and not only the Sun alone, but also the five planets linked to it. There is nothing absurd in this supposition so long as one considers (as did Tycho Brahe) the Sun, together with the planets, to be *celestial bodies*, whereas the Earth is not one; that is to say, it is not absurd so long as one assumes that they have quite different properties, especially in considering the Earth as a *heavy body* offering resistance to motion, whilst the celestial bodies are devoid of weight, and consequently offer no resistance. On the other hand, the assumption immediately becomes absurd, when with Kepler (and Copernicus before him) one considers them to be bodies of the same nature as the Earth, or at least of a similar nature; and especially as soon as one considers that they are just as difficult to move as the Earth.

This is precisely what was done by Kepler, who boldly and knowingly inaugurated the modern view of the essential identity of the elements making up the Universe, which is the basis for unifying 'terrestrial physics' and 'celestial physics' under one single set of principles. Kepler did not push this view to the limit; indeed, he made an exception of the fixed stars: and, in addition to a quantitative distinction,[29] maintained a qualitative distinction or one of kind between the planets. So, in a fine passage in the preface to his *Astronomia Nova*, he proclaimed the falsity of Aristotelian physics (and metaphysics) with its conception of space involving the doctrine of 'natural places', and the constitution of corporeal bodies through an association of qualities producing their 'natures' which allot them to fixed places throughout the Cosmos. Keplerian space, even though finite and enclosed within the celestial vault, is a perfectly homogeneous, fully geometrized, space in which the 'places' are strictly equivalent. Nor do bodies tend towards their 'natural places', because there are no such 'places'. Or, if preferred, all 'places' are equally 'natural' for bodies, all of which are equally capable of being in all 'places'; and, when there, they remain at rest without showing the slightest tendency to move. The Aristotelian concept, which distinguishes between 'natural' and 'violent' motions, being correlative with its general conception of the Cosmos and material existence, is therefore just as false.[30]

Expressed more precisely and positively, the Aristotelian doctrine which attributes a natural 'heaviness', or 'lightness', to bodies, and in consequence 'natural' motion towards a limiting 'height', or 'depth', (centre of the Universe and the periphery) is totally erroneous and indefensible. Weight, far from being a quality peculiar to 'heavy' bodies, or even a tendency to move towards the centre of the Universe, is, on the contrary, a relative property, the result of *attraction*, and this in the most emphatic meaning of the word: not an inherent tendency of like things, or the parts of a whole, to reunite (as Copernicus thought), but a real 'traction', *i.e.*, an action *ab extra.* A stone does not tend towards the Earth, but is attracted by it, and attracts it, too; the Earth attracts the Moon, which in turn attracts the Earth.[31] In fact:

> 'The true theory of gravity is based on the following axioms:
>
> 'Every corporeal substance as such is capable of remaining at rest in any place in which it is put by itself and outside the influence of a cognate body.
>
> 'Gravity is a mutual corporeal affection between cognate bodies and tends to unite them (which, in this order of things, the magnetic property is, too); so that *it is the Earth which attracts the stone, rather than the stone which tends towards the Earth.*[32]
>
> 'Heavy bodies (even if we place the Earth at the centre of the Universe) do not go towards the centre of the Universe as such, but towards the centre of the cognate body, *i.e.*, the Earth. For this reason, no matter where the Earth be placed, or whither it be transported by its animal faculty, heavy bodies will always go towards it.[33]
>
> 'If the Earth were not round, heavy bodies [coming] from various directions would not go straight towards the central point of the Earth, but [would go] to different places.
>
> 'If two stones were placed near to each other anywhere in the Universe, outside the influence of a third cognate body, these stones, after the manner of two magnetic bodies, would meet at some intermediate position, each drawing nearer to the other in proportion to the weight of the other.
>
> 'If the Earth were not held on its course by an animal, or some other, equivalent force,[34] it would rise towards the Moon by one fifty-fourth part of the distance [separating them], and the Moon would fall towards the Earth by fifty-three parts of this distance; and they would meet at this point, it being assumed that their substances are of the same density.[35]
>
> 'If the Earth ceased to attract [to itself] the waters of the sea, they would rise and pour themselves over the body of the Moon.
>
> 'The sphere of the Moon's drawing power extends as far as the Earth . . . and it follows . . . that the sphere of the Earth's drawing power extends . . . as far as the Moon and far beyond. . . .'

Keplerian attraction is exerted between bodies of identical, or similar, nature; but it is not a function of the 'nature' but of the 'size', the '*moles*',[36] of the bodies. 'Large' bodies act more powerfully than small ones, and in proportion to their 'size'. Their resistence to attractive action (or drawing power), as well as any other kind of motive action, is in the same proportion. If Keplerian bodies in themselves are no longer heavy and have lost their tendency to remain at rest in a fixed (natural) position, they have, nevertheless, retained—or acquired—a certain 'weight',[37] or something analogous to it, namely, a tendency to remain at rest in any 'place' whatsoever. (Would not these 'places', in effect, become 'natural' for them?) These bodies have retained a certain lack of power to stir themselves, a certain resistence to motion, in short, a certain *inertia*. This inertia is possessed by them all, small or large, heavy or light, celestial or terrestrial, and implies that, on Earth as in the heavens, all motion needs some motive power, cause or force without the constant application of which this motion would cease.[38]

If the idea of attraction allowed Kepler to answer the classical objections—adopted by Tycho Brahe—against the Earth's motion, namely, a stone thrown vertically upwards would never fall back to the place whence it was thrown if the Earth moved; and a cannon-ball aimed due North or South would never hit its target,[39] then it was the concept of *inertia*, the last vestige,[40] and last generalization,[41] of Aristotelian ontology, which allowed him to support the truth of Copernican astronomy by means of *physical* arguments.[42]

If all motion—and not simply the motion of some bodies—imply a motive power as the cause thereof, Kepler was right to conclude[43]:

'Therefore, let us consider two bodies, the Earth and the Sun: which of the two bodies is more likely to be the cause of motion in the other? Is it the Sun, which gives motion to the other five planets as well as the Earth, or is it the Earth which gives motion to the Sun, the motive power of the other five, and which is so much larger than the Earth? In order not to say that the Sun acquires motion from the Earth, which would be absurd, we must assume that the Sun is immobile, and that the Earth moves.

'What is there to be said about the period of revolution of 365 days? By its magnitude it comes between that of Mars, which is 687 days, and that of Venus, which is 225 days. Does not nature loudly proclaim that the course which requires 365 days depends on its position between the courses of Mars and Venus round the Sun, and that this course, therefore, is that of the Earth round the Sun, and not the Sun round the Earth? However, this is more the concern of my *Mysterium Cosmographicum* and here I must not adduce any proof other than those which are dealt with in that book. . . .

'Nevertheless, I think that the reader will pardon me for bringing forward some corrections to certain objections which prejudice the mind, and consequently rob the proofs of their power of conviction. Moreover, they are not too far removed from the subject of my work, especially in books III and IV which deal with the physical causes of planetary motion.

'However, that will suffice as far as concerns the truth of the Copernican hypothesis. Let us now revert to the matters on which I discoursed at the beginning of this introduction. I started by saying, that in this work I shall not base astronomy on fictitious hypotheses, but on physical causes; and to do so I have been constrained to pursue this end by stages. The first consisted in proving that the eccentrics of the planets coincide in the body of the Sun. Then, by deductive reasoning, seeing that the solid spheres do not exist (as was proved by Tycho Brahe), I proved, as a consequence, that the body of the Sun is the source and seat of the force which causes all the planets to revolve round it.'

The results obtained are obviously of prime importance. The modification introduced by Kepler into the structure of the Copernican Universe—transferring the origin of the planetary orbits into the real Sun—confirmed the truth of the Copernican doctrine by allowing him to base it on a dynamic conception of the astral Universe. Conversely, the dynamic conception confirmed and explained the necessity for making the transfer in question. Indeed, if the planetary motion be performed with variable velocity, and if the variation in the latter be a function of the distance between the planet and the Sun, then this motion can be explained only by a motive force, and this motive force cannot be located anywhere except in the body of the Sun.

V

The Motive Force

The planets move round the Sun only because there is some motive force which causes them to move. Now, what is this force? This is a particularly difficult question, which is made all the more difficult seeing that the location of this force in the Sun prevents any elucidation of its nature; and this location is not absolutely certain. In fact, the case of the Moon provides an objection by showing that the Earth is endowed with a similar force; so, we are compelled not to associate it too closely with the Sun.

Basically, all we know about this force, is its mode of action; this action, and therefore this force, gets weaker as the distance increases; to be precise, it varies inversely as the distance. This fact, undoubtedly, enables us to distinguish it from other forces that might, or even do, emanate from the source in question; but it does not enable us to discover its nature.

As a result, Kepler found himself in a difficult, and even rather paradoxical position. He raised the necessity for a dynamical explanation in order to deduce therefrom the mathematical law of planetary motions, and the form of the equations governing these motions. Now, this law (velocity is inversely proportional to distance) is the only definite piece of information; and the dynamical explanation, which should serve as its basis, does no more than agree with it, and therefore is dependent on it. Strictly speaking, the dynamical explanation seems to be quite unnecessary. Furthermore, in Kepler's chain of reasoning, which led to the law of areas from the law of velocities, the nature of the force does not even appear as a middle term.

Why, then, did Kepler place so much value on it? and why did he spend so much time on it? I feel that the answer can only be found in Kepler's fundamental aversion from any purely formal (positivist) attitude involving calculation. As I have said many times, in his case it was not a matter of knowing how to compute and predict the positions and motions of the planets, but of revealing the true structure of the Universe, and the true motions of celestial bodies by providing in addition a causal and structural (archetypical) explanation. It seems to me, that this enables us to understand why

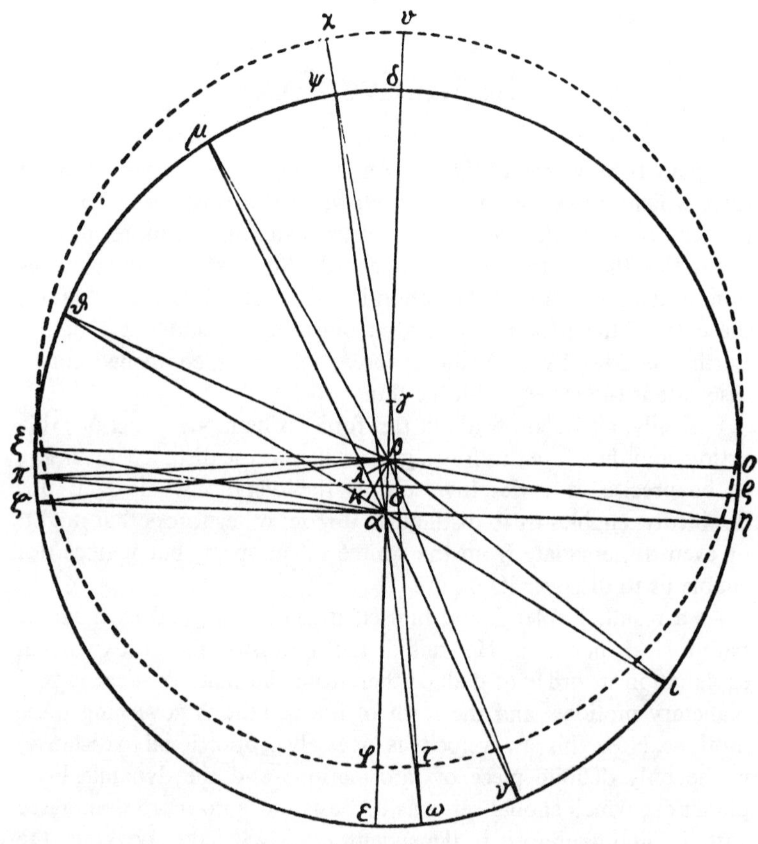

Figure 8a.

Kepler rejected the Ptolemaic formula and substituted his own: it was precisely because Ptolemy's formula was no more than a formula —velocity proportional to the distance from the equant; and because his own—velocity inversely proportional to the distance from the prime motive body—expressed causality. From this point of view, therefore, acknowledgement of the existence of a motive force was much more important than exact knowledge of its particular nature.

Now, in order to ascertain this particular nature, which is not directly accessible to us, we can only proceed by analogy with other forces and other more usual, better known, emanations.The knowledge to be acquired in this manner will, no doubt, be very vague and incomplete. Nevertheless, it will enable us to understand something of the kind of reality with which we are dealing: the motive force will reveal itself to be of an intangible nature, closely related to light and magnetic force[1]:

> 'Let us now pass to a consideration of the motive power which resides in the Sun[2]; we shall immediately become aware of its close relationship with light. Now, as the perimeters of similar regular solids, and hence also those of their circles, are to one another as their semi-diameters, it follows that the path described along the circle reckoned from α through δ is to the path described along the circle reckoned from α through ε as $\alpha\delta$ to $\alpha\varepsilon$ (Fig. 8a); but the strength of the motive power at ε is to the strength of the motive power at δ as $\alpha\delta$ is to $\alpha\varepsilon$, as has been proved in chapter XXXII.[3]
>
> 'Consequently, the power at ε to the power at δ is inversely as the circle δ to the smaller circle ε, that is to say, the more the power is dispersed (rarefied) the weaker it is; and conversely, the more it is concentrated (dense), the stronger it is. Thus, we clearly see that there is just as much power in the whole path of the circle through δ as in the shorter path through ε. This has been clearly proved as regards light in chapter I of *Astronomiae Pars Optica.*[4] Therefore, the motive power of the Sun and light agree entirely in all their properties.'

This claim on the part of Kepler, which we have already encountered in the *Mysterium Cosmographicum*, is a rather surprising one. All the more so, because on this occasion he expressly refers to the *Astronomiae Pars Optica*, where he had certainly shown that there is as much light at a distance *a* from a source as at a distance *b*; but in this instance, his argument applies to surfaces, and not to lines; and he had deduced that the intensity of light is inversely proportional to the *square* of the distance (and not inversely as the distance, as in the case of the motive force).

In Kepler's view, these were probably no more than secondary differences—differences in kind which prevent us from *identifying* the motive force with light, but do not prevent us from classing them as being of the same genus, and establishing their relationship to each other.

'Although the light of the Sun cannot itself be the motive power, we could nevertheless, ask if light might not play the part of an instrument or vehicle for use by this motive force.[5]

'It seems that we ought to take into account the following: *primo*, [the action of] light is impeded by opaque [bodies]; therefore, if light were the vehicle for the motive force, occultation [of one planet by another] would result in moving celestial bodies becoming motionless.[6] Furthermore, light is propagated orbicularly[7] in straight lines, whilst the [motive] force, to be sure, is propagated in straight lines, but circularly; that is to say, it acts only in one direction from West to East; it does not act in the reverse direction, nor towards the poles, etc. We shall possibly be able to answer these objections in the following chapters.

'Finally, seeing that there is as much motive force in a large and distant circle as in one which is smaller and nearer, it follows that none of the force is lost on the way from the source [to the object]—nothing is lost between the source [of the motive force] and the moving object. Therefore, the *effluxus* [of this motive force], like that of light, is intangible; it is not comparable with that of odours [which is effected] by a decrease in the substance [of the fragrant] body, nor to that of heat [issuing] from an incandescent furnace, etc.; these kinds of *effluxus* occupy and utilize the intermediate space. In the same way that light, which illuminates all things terrestrial, is an intangible *species* of fire which is located in the body of the Sun, so the motive force which embraces and moves the bodies of the planets is a non-material *species* of the motive force which resides in the Sun itself, [a motive force] of inestimable strength which is the prime source of all motion in the Universe. Now, seeing that the *species* of this motive force (like the *species* of light, (cf. *Astronomiae Pars Optica*, cap. I)) cannot be regarded as dispersed in the intermediate space between the source and the body which is moved, but [must be regarded] as received by the moving body in proportion to that part of the path which is occupied by the moving body[8]: it follows that this motive force (or *species*) will not be a geometrical body, but will be something like a surface, exactly like light.[9] In general: immaterial *species*, when they spread through space, do not do so according to corporeal dimensions, even though they have their origin in a body (for example, light in the body of the Sun). On the contrary, they [the *species*] behave according to their own law of propagation (*defluxus*), whose term and limit are not in itself, but only on the surface of illuminated things[10]; thus, light appears as though it were a surface because surfaces receive and limit its propagation (*defluxus*). Similarly, as bodies are moved [by the motive force], the result is that this force appears as though it were a geometric body because the [moving bodies] receive the motive *species* through their entire materiality, and set a limit to its propagation. Therefore, the motive *species* can never exist or subsist anywhere in the Universe except in moving bodies themselves; and, as in the case of light, it is never [in fact present] in the intermediate space between the source and the moving body, though in relation to this space it seems in some way to have been there.'[11]

Kepler's idea involves certain difficulties which he tried to dispel[12]:

'It has already been said above that the motive force is distributed throughout space, and that in certain places it is more dispersed (rarefied) and in others more concentrated (dense), and that this is the cause of the increase, or decrease, in the motion of the planets.[13] Now, it has just been said that this force is, in fact, an immaterial *species* of its source, and that it is never to be found anywhere save in the object which is moved, such as the body of a planet. However, to be without substance and yet subject to geometric dimensions; to propagate itself throughout the whole vastness of the Universe and yet never to be anywhere save where there is a moving body, seem to be quite incompatible [properties]. To this we reply: even though the motive force be something that is not material, nevertheless, it is co-ordinated with material substance, because it moves the latter, *i.e.*, the body of the planets; therefore, it is not free [from subjection] to geometrical laws, at least, in as much, that it exerts a material action, namely, vection.

'In fact, we see that planetary motion takes place in time and space, and that the motive force emanates from its source and propagates itself throughout the space of the Universe: these are matters of geometry. Why, then, should this force be contrary to other geometrical results?

'So that it may not seem that I am reasoning in an unwonted manner, I suggest that the reader consider the example of light—a most convincing example: light, too, is contained in the body of the Sun, whence, in company with the motive force, it is propagated throughout the whole Universe. Who, then, I ask will dare say that light is a material substance? In its behaviour, active and passive, such as reflection and refraction, light obeys the law of position and even that of quantity: to such a degree that it may be dense or rarefied, and can be regarded as a surface wherever it is received by something capable of being illuminated. Nevertheless, as is said in the *Astronomiae Pars Optica*, light, similarly to the motive force, *is not* present in the intermediate space between the source and the illuminated object, even though it passes across the space; yet, in a way, it *has been there*.[14] If light be propagated without [requiring] time, whereas a [motive] force moves in time, it will be seen that the reason for [these] two [different behaviours] is the same, provided that the matter be properly considered. Light as such acts in an instant; but when material substance intervenes, then it, too, acts during time. It illuminates surfaces in an instant, for in that case there is no question of an affection of material substance, because all illumination occurs as a result of surfaces or *quasi*-surfaces, and not as a result of corporeality *qua* corporeality. On the other hand, light bleaches colours with time, because it acts on matter *qua* matter, and it warms matter by expelling the contrary coldness, which is fixed within the body of the matter and not on its surface.

'Similarly with regard to the motive force, which comes from the Sun, perpetually and instantaneously, to the place where there is a suitable moving body; but it moves in time because the moving body is material. If preferred, this comparison may be expressed as follows: as light is to illumination, so is the [motive] force to motion. Light does everything possible for its brightness to be a maximum; colour, however, does not reach the highest degree of brightness, for colour combines its particular species with the illumination from light to produce something of a third nature. Similarly, the [motive] force does not cease until the planet moves with the same velocity as itself. Yet the velocity of the planet is never as great as that of the *species*, because of the resistence of the surrounding medium, *i.e.*, the matter of the aetheric aura[15] [which fills all space], or because of the tendency of the moving body itself to come to rest (others would say, because of the weight; for my part, I do not identify them *simpliciter*, not even in the case of the Earth). The periodic time of the planet results from a combination of all these factors with a decrease in the motive force.'

In fact, as we have just seen, Kepler—and, to some extent, Copernicus before him—no longer regarded gravity (heaviness) as one of the constituent and fundamental attributes of bodies.[16] Bodies are small, or large, dense, or rarefied, and consequently possess a certain *moles*, or size; but they are no longer heavy, or light, in themselves. Therefore, they no longer possess a tendency to move with natural motion upwards, or downwards, as do bodies according to Aristotelian physics; they have only the inability to move of their own accord, without a driving force; they possess a natural *inertia*, a tendency to come to rest, a resistance to motion.[17] Properly speaking, gravity is only the resultant of mutual attraction between cognate bodies—stones, or the Moon and the Earth. As for the planets, they do not attract each other; nor are they attracted by the Earth; nor by the Sun. Therefore, they are not 'heavy'; yet, they possess in proportion to their *moles* an inertia by means of which they resist being driven by the *species motrix* emanating from the Sun; as the *moles* are different, so are the resistances. From this arises the differences in velocity among the planetary motions. The planets move not only quicker, or slower, according as the motive force acting on them is greater, or less; but also they move more, or less, slowly according as the resistance, *i.e.*, the *moles* (size, or density), afforded by their bodies is greater, or smaller.[18]

The similarity between motive force and light, developed at length by Kepler, does not amount to identity. The Sun, however, is not only a source of light: we have to ask ourselves whether among its other properties there is not one which will provide a deeper understanding of the nature, as well as the action, of the motive force or of its *species*.[19]

> 'We have dealt with the subject of this [motive] force which embraces and carries the bodies of the planets along with it; its relationship with light; in what sense it is cognate with the latter; and the nature of its metaphysical being.
>
> 'Starting with an analysis of the effluent *species*, it now remains to examine the nature of its source considered in the more fundamental aspect of its arrangement. It would seem that something divine and comparable with our own soul[20] resides in the body of the Sun, whence this *species* would emanate to drive the planets in a circle, in the same way that the *species* of motion enters the stones from the soul of him who throws them, this *species* being that by which the stones are moved, even when the hand that throws them is withdrawn.[21] However, rather different conditions need attention by those who consider things seriously. In fact, this force, which extends from the Sun to the planets, moves them in a circle round the body of the Sun, which remains in the same place. Now, this cannot happen, nor be understood by the mind, except in one way: the force itself must travel the same path on which it carries the planets, as occurs in ballistae and all violent motions.[22]
>
> 'Seeing that this *species* is immaterial, that [it extends to] a boundless distance without any lapse of time, and is similar to light in all other respects, it is not only necessary, because of the nature of the *species*, but even probable, because of its relationship with light, that it is divided according to the division of the body or source [from which it emanates], and that ever since the beginning of creation each particle of this immaterial *species* tends to that part of the Universe [direction] which corresponds to a fixed part of its body [source]. If it were not so, it would not be a *species*, and would not emanate from the body [source] in straight lines, but in curves. Consequently, the *species* being moved in a circle, in order that by this motion it should confer circular motion on the planets, it is necessary for the body of the Sun or its source to move with it[23]; not however [by a motion of translation] from one place in the Universe to [another] place, for I place the body of the Sun at the centre

of the Universe, as did Copernicus. Yet, it is necessary for this body to rotate about its centre or stationary axis, whilst remaining in the same place.'[24]

It is not easy to understand Kepler's view of the relationship between the *species* and the body from which it emanates.[25] He definitely says that it is emitted in straight lines from each [physical] point of the source-body; but we do not know if each of these points corresponds to one single line, or to several as in the case of light. On the other hand, it is clear, that the directions in which these straight lines extend are fixed with respect to the source-body, and not with respect to the Universe; it is equally clear, that if the source-body be set in motion, then the whole array of straight lines will move with it, and especially, if the source-body be given a rotatory motion, then the whole array of straight lines will rotate as well. In short, everything happens as though we were in the realm of pure geometry, or, better still, in the realm of radiative geometry in which straight lines drawn from a body would form, literally, *one body* with it.

There is no doubt as to the origin of this conception. Geometrical optics provided Kepler with his model; thus[26]:

'So that the comparison may render the force of my argument all the more obvious, I wish to remind the reader that in the *Astronomiae Pars Optica* I have shown that vision results from reception by the eye of luminous rays emitted by the surface of the thing which is being viewed. Imagine an orator in the midst of a large crowd of people and turning about so as to make a complete rotation. The listeners who are directly in front of him see his eyes, but those who are behind him are deprived of the sight of his eyes. Whilst he performs a turning movement, he directs his eyes towards all his listeners; consequently, all of them in succession are able to have sight of his eyes during a very short time. They are able to do so, because the ray or *species* of the colour of the eyes of the orator penetrates the eyes of the spectator. Thus, by turning his eyes in the narrow space occupied by his head, he turns at the same time the rays of this *species*, which trace out the much larger sphere round about him wherein the eyes of the spectators are situated.

'Indeed, if this *species* of colour did not move in a circle together with the rays [emitted from his eyes], the spectators would not have sight of his eyes. It is clear, therefore, that the immaterial *species* of light either moves in a circle or is motionless together with the thing (which turns or is motionless), of which it is the *species*.'[27]

So, they are not *new rays*, which the eyes of the orator emit successively in different directions to which they are directed; they are the same rays, rigidly fixed to his eyes, and, so to speak, emitted once and for all; they are rays which turn with him, and pass from one listener to another, without being destroyed, or absorbed, as a result of reception in, or by, the eyes of the latter. This would, undoubtedly, be inadmissible and even inconceivable were we dealing with a material entity, though this is not the case—at least, Kepler seemed to think so—when dealing with an immaterial entity. So, he continued[28]:

> 'As, therefore, the [motive] *species* from the source, or the motive power which moves the planets, moves in a circle about the centre of the Universe, I conclude therefrom, without fear of being absurd, and in agreement with the example quoted, that the object itself, namely, the Sun, of which it is the *species*, turns also.
>
> 'It may be proved by the following argument also: motion which is local and independent on time should not be ascribed to a pure immaterial [motive] *species* . . . simply . . . because this [motive] force is deprived of matter. However, it has been shown that the force turns in a circle; that an infinite velocity cannot be ascribed to it (apparently it would in fact then confer a like infinite velocity upon bodies themselves); and, consequently that it completes its circuit in a certain time. It follows, that the force does not perform this motion by its own nature, but moves in this manner only because the body on which it depends moves [with a rotatory motion]. It even seems that this argument allows us to conclude that within the confines of the solar body there is not something immaterial [endowed with rotatory motion] by the rotation of which the motive *species* emanating from this immaterial [entity] would make itself manifest.[29] For, once again, local motion in time cannot be legitimately ascribed to something which is immaterial. Therefore, the body of the Sun itself turns in the abovesaid manner. The poles of the Zodiac are fixed by the poles of its axis of rotation (the line drawn from the centre of the body through its poles to the fixed stars); and the ecliptic is fixed by the greatest circle of the body.
>
> 'Furthermore, we see that the planets are not carried along all with the same velocity, but that Saturn takes thirty years, Jupiter twelve years, Mars twenty-three months, the Earth twelve, Venus eight and one-half, Mercury three months [to complete their circuit]; and that the sphere of the motive power emanating from the Sun (as well in the lowest position of Mercury as in the highest position of Saturn) nevertheless turns completely with the same velocity as the body of the Sun and in the same period of time. (There is nothing absurd in this, seeing

that the motive force which emanates is immaterial and, by its nature, could move with infinite velocity—provided that it were possible to confer this motion upon it by some source or other:—in fact, it could not be prevented from moving in this manner either by the weight of which it is bereft, or by the resistance of the corporeal medium.) Consequently, it is clear that the planets are incapable of partaking of the speed of the motive force. Saturn is less apt [to do so] than Jupiter, seeing that it returns later to its original position, even though the sphere of the motive force near the path of Saturn turns just as rapidly[30] as does the sphere of this motive force near the path of Jupiter; and so on up to Mercury, which like the superior planets is undeniably slower than the force by which it is carried along.

'It follows, therefore, that the planetary globes have a material quality which, by its inherent nature from the beginning of things, inclines it to a state of rest or loss of motion.[31] Now, in the conflicts that arise from the opposing effects of these things [the action of the motive force and the tendency to a state of rest], the planet which is subjected to the action of a weaker force[32] carries off the greater victory; it moves more slowly; and the one which is nearer to the Sun has less success [in its opposition].

'To continue, reasoning by analogy shows that all the planets, even the most inferior one, Mercury, possess in their own right a material force [which allows them] to free themselves to some extent from the action of the sphere of solar motive force.[33] In consequence, the rotation of the solar body takes place in a much shorter time than the periodic time of any of the planets. For this reason the Sun turns where it is situated in a period of time which is necessarily less than three months.'

From analogical reasoning Kepler even estimated that the period of the Sun's rotation on its axis should not exceed three days.[34]

'If you prefer to lay down a diurnal period for the Sun['s rotation] in order that the Earth's diurnal rotation, through some magnetic force, may be produced through a daily rotation of the body of the Sun, I shall not reject it out of hand.'[35]

Indeed, this rapid rotation did not seem to be inappropriate to the nature of the body in which the initial act of all motion is to be found.

'Now, this opinion [concerning the rotation of the Sun's body as being the cause of motion in the other planets] is very nicely confirmed by the example of the Earth and the Moon.[36] This happens because the principal [and monthly] motion of the Moon is derived entirely from the Earth, which is its source (for the Earth has the same relation to the Moon, as the Sun has to the other planets). Consider, then, in what manner our Earth confers motion on the Moon: in effect, whilst the Earth together with its immaterial *species* turns on its axis twenty-nine and one-half times, the action of the said *species* on the Moon is such that, during this time, it conveys the Moon once only on its circle. Nevertheless, it is to be noted that in any given time the centre of the Moon covers a path round the Earth's centre which is twice as long as that covered by any point immediately below it on the great circle of the equator on the Earth's surface. In fact, if equal distances were covered in the same time, the Moon should return [to its original position] only on the sixtieth day, seeing that the amplitude [radius] of its circle is sixty times the amplitude of the Earth's globe.[37]

'Hence, it is obvious, that the power of the Earth's immaterial *species* is very great, and undoubtedly the body of the Moon is highly rarefied, too, and its resistance [to motion] is very weak. . . . In fact, if the Moon by the force of its material substance offered no resistance to the motion conferred [on it] externally by the Earth, it would be carried along with the same velocity as the Earth's immaterial *species* . . . and would complete its revolution [round the Earth] in the same space of twenty-four hours during which the Earth rotates; for, though the tenuity of the Earth's motive *species* at a distance of sixty semi-diameters is very considerable, nevertheless the ratio of unity to nothing is the same as that of sixty to nothing. Consequently, if the Moon offered no resistance, the Earth's immaterial *species* would prevail completely.[38]

'If anyone were to ask my opinion of the body of the Sun from which emanates the motive *species* [which makes the planets rotate], I should tell him to pursue the analogical argument further, and to examine more closely the example of the magnet.[39]

'The strength of a magnet resides throughout the whole of its body, because it increases with the size (*moles*), and decreases as the size becomes smaller. In the case of the Sun, the motive force seems to be so much stronger [than in other celestial bodies] that it is probable that its body is the densest of all bodies in the whole Universe.

'In the same way that the force with which a magnet attracts iron spreads out orbicularly so as to form a certain sphere within which the iron is attracted [by the magnet], and all the more strongly, the further it penetrates the sphere[40]; in the same manner the force which moves the planets certainly spreads outwards in a sphere from the Sun and becomes weaker as the distance increases. However, a magnet does not attract [iron] by all its parts, but possesses filaments or rectilinear fibres (the seat of the motive force), extended in length in such a way that a strip of iron placed at the mid-point between the heads [poles] of the magnet is not attracted by it, but only brought parallel to the fibres.[41] So, it is conceivable that there is no force in the Sun attracting the planets [towards it], as in the magnet (otherwise they would progressively draw nearer to the Sun until they completely joined up with it), but only a controlling force[42]; and consequently it possesses circular fibres extending round about it in the plane which is defined by the circle of the Zodiac.[43] Therefore, as the Sun turns eternally [on its axis], the sphere of the motive force or the *defluxus* (effluvium) of the *species*, which is spread throughout the whole of planetary space by the magnetic fibres of the Sun, turns also with the Sun: and [completes its revolution] in the same time as the Sun. We have, therefore, the same state of affairs with the motive force of the Sun as with the magnetic force of a magnet which follows the movement of the latter and draws along with it the piece of iron placed within its sphere of action.

'The example provided by the magnet is certainly a very good one and agrees so well with reality that it could very nearly be regarded as representing the truth. Why, then, do I make use only of the magnet as an example? The Earth, as has been proved by William Gilbert, an Englishman, is a large magnet, which, according to the same author (a supporter of Copernicus[44]), completes its diurnal revolutions in the same manner as I have conjectured for the Sun: and it does so because it possesses magnetic fibres which cut its line of motion at right angles. These fibres form circles round the poles of the Earth parallel to its motion. I have, for very good reasons, already said that the Moon is carried along by the rotation of the Earth and the translatory motion of the magnetic force, though with one-thirtieth of the velocity.[45]

'I am well aware that the Earth's filaments and their motion fix the plane of the equator, and that the Moon follows in preference the path of the Zodiac. Apart from this one exception, everything is in perfect agreement. The Earth is entirely within the lunar orbit, in the same way that the Sun is within the orbits of the other planets: and as the planets are eccentric with respect to the Sun, so is the Moon with respect to the Earth. . . .

'The conclusion is therefore plausible: because the Earth moves the Moon by its *species*, and is a magnetic body; and because the Sun moves the planets in a similar manner by the *species* which it emits, therefore, *the Sun, too, is a magnetic body.*'

The conclusion is, indeed, plausible; and what is more, the Sun is, unquestionably, a magnetic body. However, we might ask if its action on the planets does reduce to a magnetic action on this account; in other words, if the *species motrix* should be identified with, or merely likened to, a magnetic force. I myself believe that, as in the case of light, there can be only a comparison, or at most, identification of the action of the motive force with the directional, non-attracting action of the magnet.[46]

My interpretation seems to be confirmed by the very fact that, immediately after having declared the Sun to be a magnetic body, Kepler reverted to a consideration of the similarity between the motive force and light, and once more raised the problem of the possible influence of occultation (conjunction) of one planet by another on their motion. Ought not such an occultation to bring about an arrest of motion?

No doubt such an occultation is a rare event, and we may assume that God has disposed the planetary orbits in such a manner as to avoid it as far as possible, for example, by giving them different inclinations. Still, it is a fact that occasionally the Sun and two planets are in the same straight line. Hence, if the *species motrix* were identical with light, or if the latter served as a vehicle for the former, the result would be not a stoppage but a slowing down of their motion —the planets are too small in ratio to the Sun, and too far away from each other for a true occultation to take place. We could suppose, for example, that the very slow displacement of the apsides and nodes results from this cause.[47]

Kepler, however, thought quite differently, and as before pointed out that analogy is not the same as identity[48]; immaterial properties, or species, though similar in some respects, are nevertheless different in others.[49] For example, light is generally stopped by a screen, but magnetic force is never, or hardly ever, stopped. Now, the analogy between motive force and magnetic force is even more valid than the analogy with light, but it still remains an analogy. Kepler therefore wrote: To the question

'*Whether or not there is an absence of motion as well as an absence of light in consequence of an occultation of planets*,[50] one should reply in the first place that the analogy between light and motive force should not be falsified by a rash confusion of their properties. Light is stopped by anything opaque; it is not stopped by the body as such,[51] simply because it is light, and does not act on the body itself but on its surface, or as if were one. The [motive] force acts on a body regardless of its opacity: also, as it is not resisted by the opacity, it is not stopped by the opacity. Consequently, if I had not been able to find in nature examples where rays of light, even when intercepted, nevertheless retain their efficacy in places where they cannot reach, I should have kept light entirely separate from the motive force.

'However, let us consider the example of the magnet so as to remove the objection to interference [by occultation]. Its force is in no way prevented from acting by the interposition of a substance (undoubtedly

> because it is immaterial), but passes through strips of silver, copper, glass, bone, or wood, and attracts a piece of iron placed behind them as if nothing had been interposed [between the magnet and the iron]. However, the action is stopped by interposing a magnetic plate. The reason for this is immediately obvious: the plate behaves in the same way as the magnet itself. Consequently, its force exceeds that [of the magnet] which is farther off, and situated behind it. If the magnetic force suffer interference by interposing a strip of iron, it is because the iron itself has a magnetic nature, and immediately impregnates itself with the magnetic force and makes use of it as though it were its own.' Now, as 'the motion of the planets is not stopped by the conjunction of two [planets], it follows that there is a greater difference between the nature of the Sun and that of the other stars [planets] than there is between the nature of the magnet and that of iron: whereas the iron immediately absorbs the force of the magnet, the planets do not do so with respect to that of the Sun.'[52]

Having successfully removed the obstacle which the absence of any effect from the occultation of one planet by another seemed to raise against the similarity of the *species motrix* to light and magnetic force —the objection is only in respect of identity—it now remains to be seen whether or not this similarity involves consequences which are incompatible with the very data that Kepler's theory was intended to explain. He has indeed said that, whilst light is propagated 'orbicularly', the *species motrix* is propagated 'circularly'; but had he any right to say so? To the contrary, are not we obliged to assume that the *species motrix* spreads throughout all space in exactly the same way as light and magnetic force, and consequently, that the attenuation is not a function simply of the distance but rather of its square or even its cube?[53] In fact:

> 'It was shown in chapter XXXII that the motion of the planets becomes stronger or weaker in simple proportion to the distance; but it appears that the force emanating from the Sun should increase or decrease in the duplicate or triplicate proportion of the distance. Consequently, the increase and decrease [in speed] of planetary motion will not be [a function] of the attenuation of the force emitted by the Sun.'

The objection which we have touched on above[54] is certainly a very serious one. Fifty years later, Ismaël Boulliau regarded it as quite conclusive.[55] It must be admitted that Kepler's answer to it is rather weak, to say the least.

Any notion of a dependence on a triplicate proportion can easily be set aside, for it would mean that the motive force fills all space as it spreads outward; but this is precisely what does not happen, seeing that it is immaterial. However, it is more difficult to explain why the motive force does not obey the same law as light during its propagation and attenuation, namely, that of duplicate proportion. In fact, Kepler himself admitted that 'if the magnetic filaments from the Sun arranged themselves according to the longitude of the Zodiac', that is say, formed circles parallel to the equator in, or on, the body of the Sun, then it was nonetheless true—as in the case of light—that the individual points (real, not mathematical) of these filaments emit rays in all directions, not only towards the Zodiac, but also towards poles. Still, Kepler regarded it as true 'that a certain amount of the *species* will be present in the regions of space situated above the poles of the Sun', though the amount would be very small, seeing that 'the filaments from the body of the Sun do not extend towards the poles', and that these filaments are not the seat of the motive force except in so far as they turn with the body of the Sun. Now, the Sun rotates in a fixed direction 'from right to left'; and it is a motion in this direction which the *species* communicates to the planets,[56] and not a motion towards the poles, in which direction the Sun itself executes no movement. It must not be forgotten that the *species motrix* transfers to the objects on which it acts something of the motion of the source from which it emanates: it is not motive, so to speak, in an absolute manner and in so far as it emanates from the Sun, but only in so far as it emanates from a Sun in rotatory motion. Kepler's explanation was as follows[57]:

'However, the objection can be used in the opposite sense, thus: if as much light be dispersed in the larger sphere as is condensed in the smaller sphere, there will not be, on the other hand, so much [motive] force in the two [spheres], because the force is not regarded [as being propagated] orbicularly in a sphere, as in the case of light, but in a [plane] circle in which the planet is situated. Indeed, it has been assumed earlier on that the magnetic filaments from the Sun extend only in length, and not towards the poles or in other directions.

'It will be objected that the consideration of light and of force comes to the same thing, and that the argument is faulty. For, as in the case of light, the rays do not emanate from single points and circles on the body corresponding to points and circles on the sphere—thus, at γ they do not come from the single [point] α (otherwise it would be impossible to ascribe any density to light, because it would have no quantity in its origin, seeing that it emanated from a point); on the contrary, individual points on the imaginary spherical surface receive rays from the whole hemisphere of the body which provides the illumination; so, the ray δ as well as ray ε arrives at γ:—force will function in the same manner. If the magnetic filaments of the solar body arrange themselves according to the longitude of the Zodiac, and even if one single great circle subjacent to the Zodiac or to the ecliptic should be the nearest to the planet's orbit; and, furthermore, if the other smaller circles (which finally reduce to a point at the poles) adjust themselves to the corresponding circles on the sphere of the planet; nevertheless, [motive] rays emanate from all the filaments of the solar body, and these rays converge on the individual points of the path of the given planet as well as the poles themselves which are above the poles of the body of the Sun; consequently, the body of the planet is carried along according to the degree of density of this *species* composed of all the abovesaid filaments.

'However, it does not follow—as one might fear—that the planet is moved indiscriminately in any direction, in the way in which the Sun illuminates equally in all directions.

'In fact, the Sun's magnetic filaments do not of themselves produce motion; they do so only because the Sun, by rotating very rapidly where it is placed, carries them round and causes them to turn together with the motive *species* which emanates from them. Consequently, the planet will not go to the left, because the Sun turns perpetually to the right. Nor will the planet go towards the poles (even though a certain amount of the *species* from the Sun's body is present also at these points); for the filaments from the Sun's body do not extend towards the poles, nor does the Sun itself turn in that direction, but only in the direction in which it is urged by the filaments.

'This being granted, it follows, no matter how little the planet be carried towards the poles, that the Zodiacal region, intermediate between the poles, should be on the whole the only one in which all the planets ought to be situated, if, without any deviation in longitude, they but refrained from their own proper motion (this is treated later in chapter XXXVIII).

'For [that portion of] the *species* from the Sun's hemisphere which reaches some point in the Zodiac (let us say ζ in the accompanying diagram) [emanates] entirely from semicircular filaments all tending in the same [direction], as from θ towards λ, from λ towards μ, etc. On moving towards the poles of the Universe, for example to η, the [*species*] will then include also components emanating from the filaments direct from the pole of the Sun's body ν, as well as those from the whole circle $\lambda\mu$, which surrounds the pole ν; therefore, the *species* will contain filaments tending in opposite directions: for the opposite portions λ and μ of the circle go in opposite directions. Consequently, the *species* $\theta\eta\mu$

extending towards the poles is less apt to produce planetary motion.' (Fig. 9.)

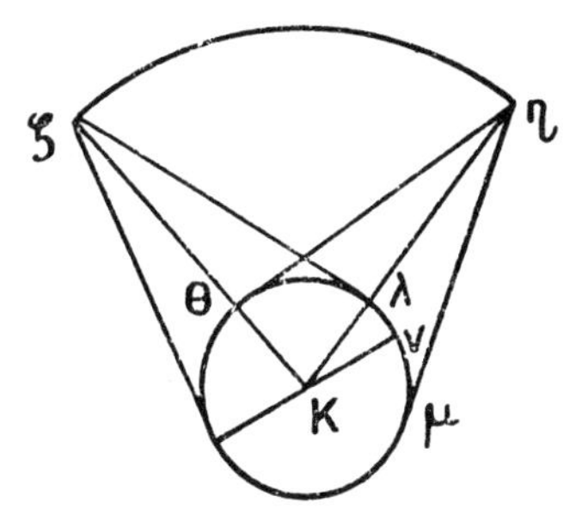

Figure 9.

I have already said that Kepler's argument is not very sound[58]; but that should not be held against him: the nature of the motive force was obscure and little understood. On the other hand, it was definite that the planets obeyed a very exact physical law in their motion round the Sun—velocity inversely proportional to the distance. It was equally definite, in consequence of the most fundamental laws of dynamics—velocity proportional to the motive force—that this law of velocity implies a corresponding law of force which explains the former. It is clear, therefore, that the 'analogical' objections were unavailing, and we must conclude—or assume—that[59]:

> 'the amount of force, or of light, propagated in the larger sphere γ is exactly the same as in the smaller one β: in fact, nothing is lost on the way and the motive *species* reaches in its entirety as far as we wish, being attenuated only as a result of extension of the spheres, so that at given points on the spheres, for example at α and β, it will be not so dense here and denser there in the inverse proportion of the distances $\alpha\beta$ and $\alpha\gamma$. This is the only reason for the lessening [of its power].'

The analogy between *species motrix* and light emerges triumphant from criticism. Kepler, when summarizing the dynamic structure of the solar system in the preface of his *Astronomia Nova* brought the whole weight of his argument to bear in saying[60]:

> 'the Sun, remaining in its place, turns as though upon a lathe, and sends into the vastness of the Universe an immaterial *species* from its body, similar to the immaterial *species* of its light: which *species*, in consequence of the rotation of the solar body, rotates in the manner of a very

> rapid whirlpool which extends throughout the whole extent of the Universe, and carries the planets along with it, bearing them in a circle with a stronger, or weaker, *raptus* according as the density of this *species* is less, or greater, in conformity with the law of its *effluxus*',

that is to say, according as the planets are farther away from, or nearer to, the Sun.

VI

The Individual Motive Forces

The mechanism of planetary motion seemed therefore to be definitely established, but a fresh difficulty immediately arose. In fact, the planets being subjected to the *raptus* of the solar whirlwind—and to it alone—ought to describe concentric orbits about the central body, and their motion in these orbits ought to be perfectly uniform. The 'force' emanating from the Sun (or from the Earth), whether it be considered after the manner of light or of magnetic force, is a purely motive one; it produces forward movement; it does not attract, nor does it repel. Consequently, the planets in their motion, should have no reason to approach the Sun (which does not attract them, as the Earth does the Moon); nor should they have reason to move away from it—in Kepler's universe the circular motion of celestial bodies does not develop centrifugal force[1]; so, they would revolve eternally at the same distance from their motive source, which acting in a uniform and constant manner—the supreme rule in celestial mechanics was that the revolution of the Sun, and hence its *species*, takes place in a uniform and constant manner[2]—would confer on them likewise a uniform and constant motion. Now, we know that this is not the case. Therefore, in order to explain their true motions, it becomes necessary, in addition to having the common solar motive force, to endow each planet with its own individual motive force whose action accounts for the eccentricity of the orbits, and, at the same time, the exact non-uniformity of the motions round the Sun.

However, we must not anticipate, but must follow the progression of Kepler's thought:

'*Apart from the common motive force, the planets are endowed with their own individual force; and the motion of each of them is compounded of two causes.*'[3]

'Up to now, the motive force of the Sun has been [regarded as] uniform, and only having varying degrees of strength according to the amplitude of the different circles. Consequently, the planet, if it remained at the same distance from the Sun, would revolve round it in a very uniform [manner], and would not experience any increase, or decrease, from the motion of the Sun. If, however, a certain inequality be observed in the operation of this force, it results from the fact that the planet is moved from a given distance from the Sun to some other; consequently, it is exposed to the action of different degrees of strength of the force [emanating] from the Sun. We then ask ourselves, what is the reason that the planet moves nearer to, or away from, the Sun? seeing that the solid spheres have no existence, as was shown by Tycho Brahe.[4] Is this result also produced by the Sun? I say that the Sun is in some measure responsible; but there are other causes, too.

'The examples provided by natural things and the relationship between celestial and terrestrial things demonstrate that the workings of simple bodies are all the simpler the commoner they are; but that variations, if there be any (as in the case of planetary motion, where there is a change in the distance from the Sun), result from concurrent external causes.

'Thus, in the case of a river, the ordinary property of the water is to fall towards the centre of the Earth; but, as there is no direct way thither, it goes downwards wherever it finds a depression; it stagnates where it falls on level ground; it is carried off with impetuous force where it is urged by the greatest slopes; there are [places] where it becomes a whirlpool if, as a result of a rapid fall, it drops on rocks.

'When water, by its own force, does nothing more than fall towards the centre of the Earth, [we observe] the single effect of a single property; but stagnation, deviation, cataracts and whirlpools and all changes result from external, adventitious causes.

'Other pleasing examples, more relevant to our investigation, are provided by the propulsion of ships.

'If a line or cord be thrown across and over a river from one bank to the other, and if a pulley running on the cord and attached to another cord be tied to a boat floating on the river, then the ferryman, by suitably setting the rudder or oar fixed at the back of boat, and keeping it still, can make the boat cross [the river] from one bank to the other simply by the force of the flowing water, whilst the pulley runs on the cord above. On very wide rivers, ferrymen make their boats come and go, or turn round . . . without touching the bottom, or the banks [with a boat-hook], but merely by the action of the oar [rudder], thereby converting the fall of the river by itself to their own use.

'Now, the force sent out from the Sun into the Universe by the *species* is a rapid torrent, which carries all the planets and possibly all the aetheric aura[5] from West to East; but in itself, it is unable to bring the bodies [of the planets] towards the Sun, nor to repel them; . . . It follows, therefore, that the planets themselves, like the boats . . . have motive forces . . . by means of which they accomplish not only their movement towards, or away from, the Sun, but also (as is possible according to our reasoning) the declinations and latitudes . . .' at the same time making use of the force of the cosmic torrent.

The 'example' of the ferryman, who performs complicated movements by turning the rudder of his boat in a certain way and so setting it in a suitable position (direction) with respect to the river current, is only an 'example'; he assures us that it is possible to utilize the Sun's action, in conjunction with that of the planet itself, to produce the eccentric motion of the latter. However, he definitely does not provide us with a proper explanation, as will appear from a thorough discussion of this concept.[6] Furthermore, this 'example' presupposes the existence of a 'ferryman', *i.e.*, a planetary intelligence, capable of calculating the appropriate angle that must be given to the 'rudder' in order to set the planet at the required distance. This is certainly not an overwhelming objection: indeed, there is nothing absurd in ascribing souls, or even intelligence, to the planets. In the present case, however, and particularly if it be a question of making the planet describe a rigorously circular orbit, the action of this intelligence will encounter difficulties, which if not totally insurmountable, are certainly not negligible.[7]

It becomes necessary to consider without bias in what manner and by what means the individual motive force of the planets can bring the latter to describe an eccentric orbit. Fortified by the knowledge of our earlier discoveries, we should expect the orbit to be perfectly circular. In pursuing this investigation, which will likewise bear on the nature of the planetary motive forces, Kepler states that we ought to take the following 'most true' axioms as the basis of our proofs[8]:

'1. The body of a planet naturally tends to a state of rest in any place where it is put by itself.
'2. By means of the force [which emanates] from the Sun, it is carried from place to place to its appropriate Zodiacal longitude.
'3. If the distance of the planet from the Sun were not modified, the result of this movement would be a circular orbit.
'4. For a given planet and two possible circular orbits, the periodic times will be in the duplicate proportion of the distances or the amplitudes of the orbits.[9]
'5. The bare, isolated force residing in the planet is not sufficient to carry

> its body from place to place, for it [the planet] is without wings or feet wherewith it could press on the aetheric aura.
> '6. Nevertheless, the decrease, or increase, in the distance of the planet from the Sun results from the [action of the] force which is peculiar to the planet itself. These axioms are conformable to nature, agree with each other, and have been proved in what has gone before.'

Unfortunately, these 'most true axioms' do not lead directly to the desired solution. Two or three years later Kepler wrote[10]:

> 'It is quite unbelievable how much trouble these motive forces have given me . . . they ought to have enabled me to determine the [variation] in distance of the planets with respect to the Sun and to calculate the equations of motion . . . but, in fact, they only gave wrong results which did not agree with observations. Now, this did not result from the fact that I made a mistake in introducing them, but under the spell of vulgar opinion I had attached them, so to speak, to the mouths of the orbits. Bound by such chains, they were unable to perform the service I expected of them.'

Kepler is here referring to the difficulties which are set out at length in chapter XXXIX of the *Astronomia Nova*—and, in fact, the twenty succeeding chapters also—which difficulties were caused not only by his desire to 'save' the circular motion of the planets, but also by his stubborn adherence to the thousand-year old dogma according to which planetary motions could only be 'saved' by reducing them to circular motions.[11] His desire to 'save' was quite understandable: after all, one does not lightly abandon something in which one could see a very great discovery, and even the discovery which 'realized the permanent ideal of astronomy: and his adherence to the age-old dogma makes us aware of one of the most important, and most reasonable, authorities, namely his technical authority—the impossibility of treating planetary motions, and computing orbits and positions otherwise than on the basis of the geometrical methods developed by the Greek astronomers (Fig. 10a).

Consequently, it is most interesting, especially for us who are not familiar with these methods, to find Kepler explaining how 'with the aid of geometric diagrams' we can represent the laws of planetary motion[12]; and at the same time (we have already touched on this problem[13]) he shows us to what extent the introduction of dynamical considerations into astronomy modifies the conditions under which geometrical reasoning can be applied by imposing conditions to which it must conform.

'Let us assume the orbit of the planet to be a circle as hitherto believed [when he wrote chapter XXXIX of the *Astronomia Nova*, Kepler no longer believed it, but, although he asks us to engage in this detailed, patient analysis of possible, or rather impossible, conditions for circular motion in order to make it easier for the "philosopher" to admit that the orbit in question is not a circle,[14] there is no doubt that he is describing the progress of his own thought[15]] and that [the circle] is eccentric with respect to the Sun, the source of all power. Let this circle be *CD*, described about the centre *B* with the distance [radius] *BD*; let *BC* be the line of apsides; let the Sun be [at] *A*, and let the excentricity be *BA*.

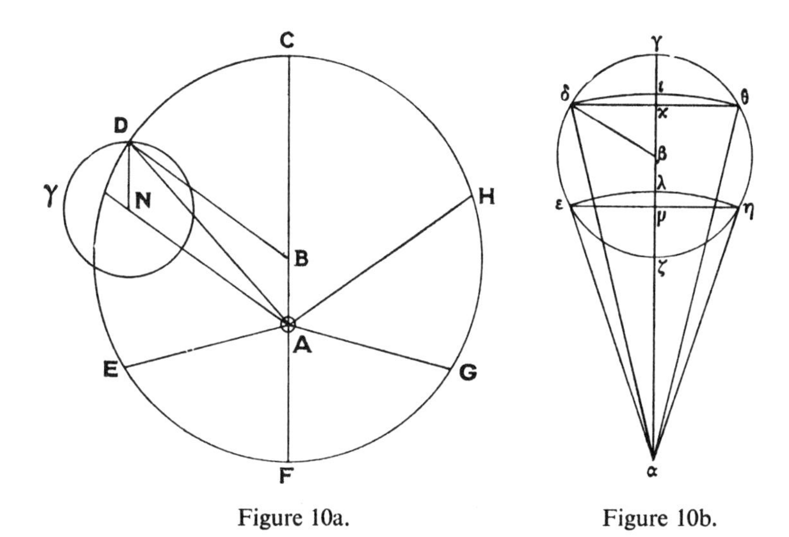

Figure 10a. Figure 10b.

Divide the eccentric [*CDEF*] into any number of equal parts, starting at the point *C* of the line of apsides, and join the dividing points to *A*. [The straight lines] *CA*, *DA*, *EA*, *FA*, *GA*, *HA* will give the distances of the terminal points of the said equal divisions with respect to the source of power. With centre β and distance [radius] $\beta\gamma$ equal to *AB*, draw the epicycle $\gamma\delta$ and divide it into the same number of equal parts as the eccentric, starting with the point γ; [the points of division will then be γ, θ, η, ζ, ε, δ]. Produce the line $\gamma\beta$ [to α] and make $\beta\alpha$ equal to *BC*; join [the point] α to the terminal points of the equal divisions on the epicycle by the lines $\gamma\alpha$, $\delta\alpha$, $\varepsilon\alpha$, $\zeta\alpha$, $\eta\alpha$, $\theta\alpha$. These lines will be equal respectively to the distances traced out on the eccentric starting from *A*. . . . Then, with centre α and distance [radius] $\delta\alpha$ draw the arc $\delta\theta$ to cut the diameter $\gamma\zeta$ in ι; and with the same centre α and the distance [radius] $\alpha\varepsilon$ draw the arc $\varepsilon\lambda\eta$ to cut the diameter $\gamma\zeta$ in λ; finally, join the terminal points of the divisions equidistant from the aphelion γ of the epicycle by the lines $\delta\theta$ and $\varepsilon\eta$, which will cut the diameter in [the points] χ and μ; then $\alpha\delta$, or $\alpha\iota$, will be longer than $\alpha\chi$, and $\alpha\varepsilon$, or $\alpha\lambda$, will be longer than $\alpha\mu$.' (Fig. 10b.)

A contemporary of Kepler, on reading this passage and its continuation, would say to himself: but that is obvious; everyone, since the time of Ptolemy, and even since Apollonius, knows that motion on an eccentric can be reproduced by an epicyclic motion; the two systems are equivalent, as Kepler himself had demonstrated at length at the beginning of the *Astronomia Nova*.[16] This was immediately confirmed by Kepler.

'Now, if it were possible for the planet by its own force (*vi insita*) to describe a perfect epicycle, whilst its orbit [deferent] would be a perfect circle, then it would be necessary to assume that similar arcs are described in the same time on both the epicycle and the eccentric.[17] Therefore, it would become immediately apparent by what means and in what measure the distance $\alpha\iota$ becomes equal to AD; for, $\alpha\iota$ and $\alpha\theta$ being equal, the planet in moving from γ to θ makes the distance $\alpha\theta$ necessarily equal to AD.'[18]

A contemporary of Kepler would be somewhat surprised by the expression, 'If it were possible . . .'. He would be even more so by the continuation of the passage just quoted[19]:

'Apart from the fact that, if we say that the planet moves by its own power from one place to another, from γ to θ, we seem to be in conflict with our axiom, this notion implies many absurdities. Particularly that which makes it necessary to ascribe the same velocity to the centre N of the epicycle in its motion round the Sun as well as to the planet D [in its motion] round the centre B of the eccentric, with the result that these motions would accelerate and retard simultaneously. Seeing that these accelerations and retardations arise from the greater, or less, distance between the body of the planet and the Sun, it would be necessary to assume that the centre of the epicycle, which is always at the same distance, moves with greater, or less, speed according as the planet is farther away from, or nearer to, the Sun.'

Kepler's contemporary, whether Ptolemaean or Copernican, would no doubt reply that there is no absurdity in an epicycle and its centre moving with equal (angular) velocities: the Copernican would add—provided they are constant. The Ptolemaean would have no difficulty in assuming non-uniform motion of the centre of the epicycle, and if he made any objection against non-uniform motion of the epicycle, it would be because he considered it unnecessary; and not impossible, or absurd. Quite to the contrary: it was Kepler, with his insistence on wanting to apply considerations of moving forces to things to which they cannot and ought not to be applied, that he would regard as being absurd. In other words, he would say (if he were familiar with the terms we are going to use), that it was the substitution of dynamics for kinetics—in itself an absurd substitution—which was the source of the absurdities allegedly discovered by Kepler in the classical system of epicyclic motion.

From his point of view, the Ptolemaean is perfectly right; but it was precisely this point of view which Kepler found to be absurd. For the latter, whose deep and most fertile intuition lay in a negation of the division of the Universe into two parts, and whose mind with all its meandering tirelessly pursued the aim of unifying celestial physics and terrestrial physics, the moving celestial bodies were no longer entities of an entirely different nature from that of things in our sublunary world. They were bodies comparable with those round about us; bodies like the Earth; bodies which, though animated, were material in the most emphatic and strictest meaning of the term; that is to say, gross and dense (more or less) and consequently even *inert* (more or less, too), and therefore in need of a motive force capable of overcoming their *inertia*.[20] In Kepler's view, the application of a dynamic outlook to astronomy was the first condition for his success. Now, from the dynamic point of view it was obviously absurd that the motive force should act on the centre *N* of the epicycle, seeing that this is a purely imaginary point without any physical reality; and it was even more absurd that the force should cause it to move (assuming that the force were able to act on it) with a varying velocity, and not at constant velocity, as one would expect. The height of absurdity, finally, would be to suppose that the motive force *acting on the body of the planet*—the only tangible thing capable of coming under the influence of its action—could give rise to a steady motion of the point *N*. Kepler continued[21]:

> 'Furthermore, although the force which carries the planets round is much more rapid than any of the planets, as was shown in chapter xxxiv, we ought to consider in our imagination one single ray of this force [radius vector] *AN* emanating from the Sun, in other words a line on which the centre *N* of the epicycle is perpetually situated,[22] and which together with the centre *N* will be sometimes slow, sometimes rapid. Once again, this is contrary to what was previously said, namely, that

the [motive] force at a given distance always produces the same velocity. As for the planet, we should assume that it detaches itself from this imaginary ray [whilst moving] in the opposite direction, and that it does so irregularly in equal times, according as the ray itself is rapid or slow. In this way, we should, doubtless, come close to the geometric concepts of the Ancients, but we should stand aside from physical speculations. . . . Therefore, it is impossible for me to conceive how all these things could be brought about naturally.'

The mechanism of planetary motion would be simplified (or would seem to be so as far as we are concerned) if, instead of directing our attention to the epicycle $D\gamma$, we limited our considerations to its diameter [radius] *ND*. In fact, in this case, the planetary intelligence would only have to keep this diameter, or radius, in a fixed direction so that it remained always parallel to itself,[23] and kept the planet at a fixed distance from the point *B* (centre of the eccentric) and, of course, from the point *N*.

Unfortunately, the simplification is illusory. In fact, it is impossible for any intelligence whatsoever to estimate the distance between itself and points such as *B* and *N* which are not defined in space by some system of reference.

'I do not deny that a centre can be imagined together with a circle about it. I merely say, that if this centre exist only in the mind, and be not [defined] in a distinctive manner, then it is impossible for a real body to execute a perfectly circular motion about it.'[24]

The terrestrial geometer, who draws circles on paper, and restricts himself to their evaluation, is on that account in a much better position than the planetary geometer. The former fixes the position of the point *B* with respect to the (real) body of the planet; the latter would have to fix the position of the planet with respect to the (imaginary) point *B*.

The conclusion to which Kepler was led by these, and other, considerations provides a very significant and curious statement. It may be stated thus: *The epicyclic motion required to realize an eccentric and perfectly circular orbit cannot be produced by purely natural means.* It cannot be realized without the intervention of a planetary intelligence (in addition to a motive soul) capable of making calculations, and causing the planet to perform motions, variable both in speed and direction on its imaginary epicycle, which would make a perfect eccentric circle of its true orbit. Kepler arrived at his conclusion much more slowly than would appear from chapter xxxix of the *Astronomia Nova,* which was written, or rather re-written, *after* and not *before* his abandonment of circularity, and radical rejection of planetary intelligences in the mechanism of planetary motion.

This conclusion is not readily acceptable, for neither an intelligence, nor a soul, can impart translatory motion to a body. Furthermore, as Kepler showed by subsequent reasoning,[35] the planetary intelligence in question would not be capable of performing the necessary calculations, seeing that the basic data for them would not be available, namely, (a) the distance of the planet from the points N and B, which distances could not be apprehended because these points are not located in any manner in the homogeneous expanse of the aetheric aura; and (b) the angles DNA and DBA, which would be just as difficult to ascertain, because they are purely imaginary.

Kepler thought that the situation would be improved (for the intellective soul), and that the problems (of computation) to be considered would be simpler to solve, if the device of the epicycle were abandoned and the planet were assumed to have only a motion of libration on the diameter $\alpha\gamma$ (or $A\gamma$), *i.e.*, on the radius vector of the planet,[26] in present-day terminology. In order to determine the distances to be covered ($\gamma\iota$, $\iota\lambda$, $\lambda\zeta$) on this radius vector, it would be a necessary condition for the planetary intelligence to base the calculations, not on the (imaginary) arcs $\gamma\delta$, $\delta\varepsilon$, $\varepsilon\zeta$, etc., with all the attendent insuperable difficulties, but on a true phenomenon, really observable by the senses, namely, the change in apparent diameter of the solar disc as a function of the distance from the planet.[27]

In fact, Kepler considered that there was no doubt (a) that these changes were discernible, for the dimensions of the solar disc exceed zero for all the planets—even at the distance of Saturn the value is 3′; (b) that these changes enabled the distance of the planet from the Sun to be evaluated—in the case of Earth, the distance of 229 solar diameters corresponds to a value of 30′ for this diameter, and 31′ corresponds to 222 diameters[28]; (c) that these changes were in reality, or at least most probably, apprehended by the planets.[29] It should be noted, however, that the planetary intelligence would not be able to calculate the distances directly from the apparent changes in the size of the solar disc, seeing that the distances $\gamma\iota$, $\iota\lambda$, $\lambda\zeta$ are not a uniform function of the increase and decrease of the size in question (in other words, the distance of the planet from the Sun): $\iota\lambda$ is larger than $\gamma\iota$, and $\lambda\zeta$ is smaller than $\iota\lambda$.[30] The planetary intelligence would therefore need to have memorized a table of equivalents similar to the Alfonsine, or Prutenic, Tables; this would be very improbable. Furthermore, the hypothesis in question, namely, that dealing with libration, as well as that dealing with motion on an epicycle, founders on a general objection: the inherent animal force (*vis insita*) of the planet cannot give rise to a motion of translation, and hence neither to motion of libration. Seeing that the latter cannot be explained by the action of the Sun alone—in any case, the Sun could act only in one direction[31]—nothing remains but to find a means of co-operation between the force of the planet and the force of the Sun, as was attempted above. However, have not we already shown that it is impossible by natural means to make a planet execute the motions required for a perfectly circular orbit? Undoubtedly; but the correct conclusion is that this orbit is not a circle. Let us abandon, therefore, this assumption which we have adopted for the sake of conforming to tradition, and which is contradicted by observational data; we shall then see that most of the difficulties we have been discussing will disappear; even the difficulty of being obliged to endow the planetary soul with fore-knowledge of a table of equivalence between the angle (*sine* of the radius vector) under which the diameter of the solar disc is seen from the planet and the distance (*arc sine* of the radius of the epicycle) to be covered on the radius vector which links it to the Sun.

We can say at once that shedding the belief in circularity of the planetary orbit in favour of an oval orbit did not involve abandonment, in Kepler's view, of the use of 'circles' for establishing the path described by a planet in the aetheric aura. On the contrary, the 'circles' seemed to him to be better adapted to their new task than to the old one. Paradoxically, whilst developing a new computational method (that of areas) for determining the equations of celestial motion, Kepler refused for many years to follow his own inspiration, and it was only right at the end, when he had replaced the *oval* by the *ellipse*, that he then decided to dispense with their use, and, in the words of Borelli, to banish them from the heavens.[32]

VII

From the Circle to the Oval

I have already quoted[1] the passage from the preface of the *Astronomia Nova* where Kepler bemoaned the unbelievable trouble that his attempt 'to harness the planetary driving forces to the mouths of the orbits' had caused him. It is now time to quote the sequel where Kepler, with legitimate pride, summarized the great discoveries of the *Astronomia Nova* (in inverse chronological order[2]): ellipticity of planetary orbits, which meant a radical break with the millenary tradition of astronomy; the law of areas, which completed the overthrow of circularity by depriving it of its technical bases, seeing that the law furnished a new means of establishing and computing the equations of planetary motion; the *physical* explanation of celestial motions, which hallowed the victory of the unitary concept of the Universe and science[3]:

> 'My wearying work came to an end only when [I had established], by means of extremely laborious proofs and numerous observations, that the path of a planet in the sky is not a circle, but is a *perfectly elliptical oval path.*
>
> 'Geometry showed that such a path would be described if the individual motive force of the planet were assigned the task of carrying out libration of the planetary body on a straight line extending to the Sun [the radius vector]. Such a libration would give rise not only to this path, but also to the equations for the eccentricity—correct and in agreement with observation.
>
> 'Lastly, the roof was put on this structure, and it was proved geometrically that this libration must be produced by a corporeal, magnetic faculty. Consequently, the individual motive forces of the planets prove to be, most probably, nothing more than affections of the planetary bodies themselves, such as is the case with a magnet, which tends towards the pole and attracts iron. Therefore, the whole system of celestial motions is administered by faculties which are merely corporeal, *i.e.*, magnetic, with the sole exception of the rotation of the body of the

> Sun in its fixed position in space, for which there seems to be need of a force derived from a soul.'

The 'extremely laborious proofs', to which Kepler refers, were concerned with his repeated efforts to base the fundamental principle of his dynamics—force, and therefore velocity, inversely proportional to the distance—on a physical (concrete) hypothesis, as in the case of the innumerable calculations which he was obliged to carry out in order to determine accurately the true path of Mars and to find the mathematical form of the equations of its motion. The task was all the more arduous, firstly, on account of the setback with the *hypothesis vicaria*, which had shown him the impossibility of ascribing more than a calculative and approximate value[4] to the classical methods (uniform angular velocity with respect to the *punctum aequans*) in which he had initially placed his confidence; and secondly, the few calculations of the position of Mars, which he had made in 1602 using the method described above, had shown him, contrary to what he had so confidently asserted to Herwart von Hohenberg and to Maestlin, that the path of Mars *is not a perfect circle, but is rather an oval.*[5]

Let us be quite clear on the matter. There was nothing surprising in the fact that the orbit of Mars (or any other planet) was not a perfect circle—I mean, from the point of view of traditional astronomy: it was accepted as a matter of course. In fact, these orbits (except in the case of the Sun) were not circular for Ptolemy (his orbits, Kepler said later, were spiral), nor for Copernicus, for whom they had 'bulges'. The true orbits were extremely complicated lines, resulting from the fact that the planets were considered to move on sufficiently large epicycles in order to account for the inequalities in their motions: only the centre of these epicycles described circles —moreover it was not true for all the planets.[6] Kepler, who thought he had gained his first victory in the conflict with Mars by showing that its orbit was *circular, like that of all the other planets*, experienced an almost complete setback.[7] As I have already said, it was probably this setback, as much as weariness, that turned him aside from the pursuit of Mars and prompted him to engage in other interests, and all the more so, seeing that the oval path of his elusive enemy proved to be extremely difficult of computation, not only by the old methods but also by the new method that he had developed meanwhile. In passing, it may be noted that it was rather remarkable and very characteristic of Kepler's spirit—and of his confidence in Tycho

Brahe's observations, too—that he accepted the oval form of the orbit of Mars with good grace, and did not hesitate to acknowledge that his great discovery—the perfect circularity of all orbits—was only a mistake. Nevertheless, it is clear that the verdict of empirical data, to which Kepler himself ascribed crucial importance, fell on well-prepared ground, and that Kepler's belief in the circularity of orbits had been already greatly shaken, first of all by the setback with the *hypothesis vicaria*, and subsequently, possibly even more so, by the *physical* impossibility (as we have just explained) of making a planet describe a perfect eccentric circle by natural means. It is equally certain that acceptance of the oval was rendered all the easier for him by the fact that it appeared accompanied, so to speak, by the mechanism by which it is produced. It was a very simple mechanism, most 'classical', at least in appearance[8]; the planet moves by *uniform motion* on a circle (epicycle) whose centre, in its turn, moves with a similar, but *non-uniform*, cyclic motion (eccentric deferent) round the Sun—a motion whose velocity (linear as well as angular) depends on the distance of the planet from the Sun.

A letter to Fabricius, dated 1 October 1602,[9] details this mechanism and at the same time explains a new method of calculating the planet's equation of motion. The method was based on physical reality and not on abstract, unreal geometrical constructions, and was particularly adapted to the 'non-classical' reality of this motion which embraced the fundamental law: velocity inversely proportional to distance, or time of travel directly proportional to the latter. (Fig. 11.)

> 'If the Sun did not drive Mars along, then Mars, between *E* and *F*, in passing through *I*, *H*, *G*, would describe a circle [about *B*], whose eccentricity would be 9165 [parts] of 100 000 contained in *AB*.
>
> 'Nevertheless, a circle is described with respect to the line *AE* and centre *A*, though the planet moves on this circuit with non-uniform motion. The motion of Mars comprises dependence on the motive force from the Sun and a reaction against it. At the beginning of the period of revolution, or anomaly, it exerts its effort in the opposite direction [to the motion of the solar motive force]; at *H*, *i.e.*, a little after quadrature, it is pulled directly towards the Sun; at the mid-point of its path, *F*, it exerts its effort in the same direction [as the motive force]. It exerts this effort only with respect to the motive force of the Sun; and this effort is equal [constant], for in equal times, it describes equal arcs about the centre *B*, which moves non-uniformly. Now, when I say that the planet moves by exerting its effort having regard for the motive force

of the Sun, I am explaining the physical cause of the planet's motion; and when I say that it is related to the centre *B*, I am explaining the manner in which we imagine the motion to take place.'

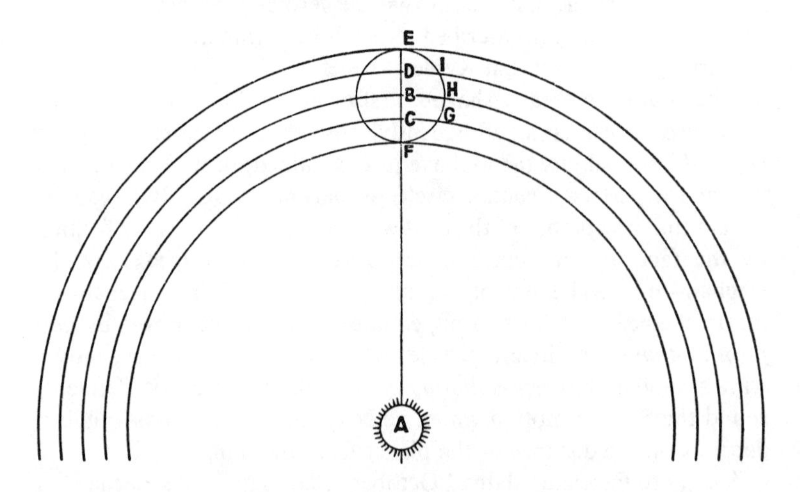

Figure 11.

A —Sun.
AB—Radius of the orbit of Mars.
BC—True eccentricity.

It is obvious, Kepler continued, that the planet is not referred to an imaginary centre, as we make it to be, when we have a sheet of paper in our hands. By describing the small circle, the planet is well able to regulate itself by the Sun, whose apparent dimensions alter with the distance . . .

'I shall now explain the laws of motion of the point *B* [centre of the epicycle]. It must be supposed that, whatever be [the ratio] of the [distance] *AF* to the [distances] *AE, AD, AB, AC*, such is [the ratio] of the time of travel of Mars round the Sun on one minute of arc (1′) when it is at *F* to the time of travel on the same path when it is at *G, H, I, E*; for the motive force of the Sun is circular, and if the planet did not fall from *E*, it would be carried in a perfect circle starting at *E* [with radius *AE*]. Similarly, if it remained at *I*, it would be carried along the circle passing through *DI*, but it would complete its course in less time, because the motive force in the smaller circles is more closely gathered together, and hence denser as well as of greater effect, than in large [circles] further away [from the Sun]. Therefore the motion of

Mars about the Sun is slow at *E*, moderate at *H*, and rapid at *F*; but the motion about the centre *B* remains constant. This is how the *oval figure* [of the orbit], to which I have referred in my earlier letters,[10] is produced; for it [the planet] passes more rapidly [from *E*] to *H* at the distance *AH* [from the Sun], than *B* can travel one quadrant of the circle round the Sun. Consequently, the distance [of the planet from the Sun] will be shorter towards the sides, *i.e.*, in the median longitudes, than is required by the ratio of a perfect eccentric circle; and the difference at perigee [perihelion] will be greater than at apogee [aphelion]; which is the definition of an oval. . . .

'If you wish to know the degree of speed in which Mars happens to be, seek its distance from the Sun by means of the simple anomaly; and if you wish to know the place [where it occurs], you must know how much motion was produced by all the distances accrued from apogee [aphelion] in the time corresponding to each of them. Now, as these distances are infinite in number because the points *F*, *G*, *H*, *I*, *E* are infinite in number, and as they [the distances] are not equally [uniformly] distributed along the path round the Sun, that is to say, those which are closer together are nearer, for Mars is slower [in that part of the circuit corresponding thereto], and those which are farther apart are shorter; you realize how difficult it is to compute the eccentric positions starting from the true or physical hypothesis.[11]

'The way to do it is as follows: divide the circuit into 360° and imagine that the planet moves uniformly so long as it is in one [of these degrees]. The time of travel for all the degrees calculated in this manner one after the other will give the anomaly of each. The same applies to the Earth. This is the whole hypothesis of Mars: the ratio of the orbit of the Earth to the orbit of Mars is equal to 100 000:152 518, the semi-diameter of the small circle of the Earth is 180.'

The method advocated by Kepler, and which in fact he used to compute with infinite patience the series of distances and lengths of path, is decidedly clumsy, inelegant, and hardly 'geometric' at all. Its great importance, as far as we are concerned, lies in the fact that it represents the first stage in his thought, which culminated two years later in the calculation of surfaces and in the law of areas. Kepler's text, as well as the parallel texts to be quoted shortly, furnish good examples of the difficulties—which we have forgotten—that were initially encountered by the infinitesimal calculus, a method of calculation indispensable for a science wanting to base its equations on 'true, physical and real hypotheses'.

The number of points on the path is infinite, and to each point there corresponds a different distance, and hence a different velocity.

Now, the infinite cannot be known. What, then, can be done, except to substitute a finite unity? and to be satisfied with an approximation where it is not possible to secure absolute accuracy?[12] Certainly, a closer approximation could be made by calculating the distances, not degree by degree, but minute by minute, or even second by second, as Kepler, with his tongue in his cheek, suggested to Fabricius (4 July 1603): but that would not solve the problem. The situation would be vastly different if one knew the law governing the generation of the oval curve; unfortunately, it was not known. Had the orbit been an ellipse, there would have been no problem; it would have sufficed to refer to Archimedes and Apollonius. Unfortunately, it was not an ellispse, but an ovoid curve.[13]

In order to avoid the difficulties I have just mentioned, Kepler tried to tackle the problem in another way; starting from the idea that all the distances of the planet from the Sun are contained within the plane of the curve (orbit), he substituted the sum to infinity of the distances in question by the surface, or segment of surface, in which they are contained. Here again, mathematical difficulties were encountered: it was not known how to evaluate ovoid surfaces: 'Inform me, (he wrote to Fabricius), of a geometrical construction for the squaring and division of ellipsoids in a given ratio, and I shall straightway tell you how to compute [orbits] according to the true hypothesis.' Therefore, he asked him—as he had formerly asked Herwart von Hohenberg—to enlist the help of great geometers, his friends: 'Therefore, arouse your Netherlandian minds that they may help me here.'[14] A geometrical solution of this problem was needed in order to evaluate the surfaces in question, to determine the totality ('collection') of motive forces acting on the planet, and hence the 'accumulation' of times and path elements; as well as to evaluate the difference between the surfaces corresponding to an oval orbit and those corresponding to a circular orbit. In fact, as we do not possess one, all that can be done is to divide the surface in question into a certain number—360—small 'triangles', calculate their area one after the other, and then add them all together. In short, do for areas what had been done for distances, and once more be satisfied with an approximation; even an approximation of the second degree.

The stages in thought which led Kepler to develop the calculation of surfaces, and which led finally to the discovery of the law of areas, are described much more fully in the *Astronomia Nova* than in the letters to Fabricius quoted above. That account reveals far better

Kepler's deep inspiration; his desire to constitute a new method for establishing the equations of planetary motion—a method that would be copied from the dynamic reality of these motions, and would give, so to speak, a true representation of them.

At the beginning of chapter XL (the last chapter of Book III), which is devoted to an exposition of *An imperfect yet adequate method of calculating the equations* [of planetary motions] *based on the physical hypothesis,*[15] Kepler says that 'in order to prepare the way for the natural form of the equations', 'extensive investigations were needed' into the dynamic structure of planetary motions, which investigations, as we have seen, ended in recognition of the impossibility of a natural production of a perfectly circular orbit. No doubt, this is because the form of the equations can appear 'natural' only to someone who is convinced of the necessity for introducing dynamic considerations into astronomy, and because the new method shows itself to be faulty for a circular path, but successful for an oval one. Kepler went on to say[16]:

> 'My first mistake was in having assumed that the orbit on which planets move is a circle. This mistake showed itself to be all the more baneful[17] in that it had been supported by the authority of all the philosophers, and especially as it was quite acceptable metaphysically.'

The mistake—the pursuit of the chimera of circularity—was certainly a serious one; and Kepler was right in saying that it cost him much effort and time to discard it. However, we, who know the sequel, must admit that the time was not entirely wasted, and that the 'error of circularity'—to say nothing of the fact that it was really imposed as a result of studying the Earth's orbit—was an inevitable and indispensable stage in the development of his thought. How would he have managed to come to the ellipse, had he not started with the circle? Kepler himself was really aware of this, for in his account he makes us retrace the steps in the development of his thought by starting with the circle.[18]

> 'Let us assume, therefore, that the planet's orbit be a perfect eccentric; for in the case of the Sun's orbit, the space which the oval takes away from the eccentric is extemely small. The consequences that necessarily follow from this difference in the case of the other planets will be discussed later in chapters LIX and LX.
>
> 'As the times required by the planet to cover equal portions of the eccentric depend on the distances of these portions [from the Sun], and

as the distances of the individual points [on the path] change throughout the whole length of the semicircle, I had great difficulty in finding out how the sum of the individual distances [from the related points on the segments] could be obtained. If we do not obtain the sum of all these distances, which are infinite in number, we shall not be able to find the time required [to traverse] one of these segments; and, consequently, it would not be possible to establish an equation; for the sum total of the distances to the time of duration of one complete revolution is the same as the ratio of any part of the sum of the distances to the corresponding time.'

Kepler's reasoning is both subtle and false. To assume that the ratio of 'the sum of the distances' on the whole path to 'the sum of the distances' on part of the path is equal to the respective times required to traverse them, is to make a mistake, as it so happens, in the *circularity hypothesis*. Furthermore, taken literally, the expression 'sum of all the distances' presupposes a knowledge of an infinity of different distances—an impossibility according to Kepler himself, who in a similar case went to the extent of invoking the authority of Aristotle[19]; finally, no matter what the lengths of the segments in question, these 'sums' will necessarily be infinite 'sums' between which no ratios can be established. Of course, Kepler was not aware of the fact. Hence, his solution, in the description of which he has undoubtedly faithfully followed the steps in the development of his thought, is extremely interesting. His solution admits that it is not possible to 'sum up' an infinity of distances; for there cannot be a ratio between infinite quantities. However, as he told Fabricius, these infinite sums total, which are beyond computation and incapable of being performed, can be replaced by finite sums of a finite number of distances, which will be equivalent to them, at least approximately; and ratios can be found between these finite sums which will be approximately equivalent to the ratios between the infinite sums total.[20] Kepler continued:

'For this reason, I started by dividing the eccentric into 360 parts, as though they were the smallest parts, and I assumed that the distance did not change along one such part.[21] I then determined the distances from the beginning of the said parts or degrees, and I added them together.[22] Then, for the period of revolution, although it amounts to 365 days and 6 hours, I assigned another, round value, and assumed that it corresponded to 360 degrees, or the whole circle that astronomers call the mean anomaly.[23] Finally, I put the ratio of any distance to its time as equal to that of the sum of the distances to the sum of the times.[24]

Finally, I added together all these numbers for the individual degrees. By comparing these times or degrees of the mean anomaly with the degrees of the eccentric, or with the number of parts up to the point whose distance is required, we obtain the physical equation,[25] to which, in order to obtain the complete equation, must be added the optical equation obtained at the same time as the distances.

'Seeing that this procedure is mechanical and laborious, and that the equation it provides for any particular degree cannot be computed without taking into account all the other [degrees], I tried to find a different method.[26]

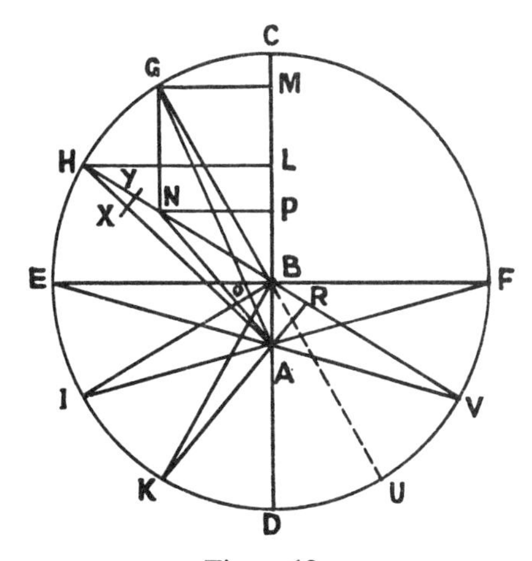

Figure 12.

'When I took into account the fact that there is an infinity of points, and therefore an infinity of distances, on the eccentric, it occurred to me that the surface of the eccentric comprised them all; for I remembered that Archimedes, of old, when he sought the ratio of the circumference to the diameter, divided the circle into an infinity of triangles. There we have the hidden power of his proof through the impossible. Therefore, instead of dividing the circumference into 360 parts, as previously, I cut the surface [of the circle][27] into as many parts by drawing lines [radii vectores] starting at the point from which the eccentricity is measured [point *A* on diagram *B*].

'Let *AB* be the line of apsides; *A* the Sun (or Earth in the case of Ptolemy); *B* the centre of the eccentric *CD*, whose semicircle *CD* is divided into a certain number of equal parts *CG*, *GH*, *HE*, *EI*, *IK*, *KD*; and let the points *A* and *B* be joined to the points of section. *AC* will then be the greatest distance, *AD* the shortest, and the others in order [of decreasing length] *AG*, *AH*, *AE*, *AI*, *AK*. (Fig. 12.) Now, as [the areas of] triangles of the same height are proportional to their bases, and as the sectors, or triangles *CBG*, *GBH* and the others, having for base the smallest parts (minima) of the circumference, which do not differ from straight lines, have all the same height equal to the radii *BC*, *BG*, *BH*, it follows that all these triangles will be equal. All the triangles are contained in the area *CDE*, and all the arcs or bases in the semicircle *CED*. Hence, by compounding, the ratio of the surface *CDE* to the arc *CED* is [equal to that of] the surface *CBG* to the *CG*; and, by permuting [the terms of the proportion], the ratio of the arc *CED* to [the arc] *CG*, *CH* and so on is the same as [that of] the surface *CDE* to the surface *CBG*, *CBH* and so on. No fault is committed in this way by considering surfaces instead of arcs, and by taking the surfaces *CBG*, *CBH* instead of the angles of the anomaly of the eccentric *CBG*.

'Furthermore, in the same way that the straight lines drawn from *B* to the infinite parts of the circumference are all contained in the surface of the semicircle *CDE*, and the straight lines drawn from *B* to the infinite parts of the arc *CH* are all contained in the area *CBH*, so are the straight lines drawn from *A* to the infinite parts of the circumference or of the arc. Now, seeing that these two [assemblies of straight lines], those [starting] from *B* and those [starting] from *A*, fill one and the same semicircle *CDE*, and as those which start from *A* are the very distances whose sum is required, it seemed to me that we could conclude that we should have the sum of the infinite distances [from *A*] up to *CH* or *CE*, if we calculated the surface *CAH* or *CAE*: not because the infinite could be reached, but because I considered that this surface contained the measure of the effect of the whole collection of these distances over the accumulated times,[28] and, consequently, that a knowledge of this surface gave this measure, without the necessity of reckoning up even smaller parts.

'Whence, the ratio of the surface *CAE* is to one half the time of restitution [period of revolution], which we designate by 180°, as the areas *CAG*, *CAH* are to the times of travel over *CG* and *CH*. Thus, the arc *CGA* becomes a measure of the time[29] or the mean anomaly, which corresponds to the arc *CG* of the eccentric, seeing that the mean anomaly measures the time.

'However, as we have just explained, the portion *CGB* of this surface *CAG* measures the eccentric anomaly, whose optical equation is the angle *BGA*. Consequently, the residual surface, namely the triangle *BGA*, is the excess (at this place) of the mean anomaly over the eccentric anomaly; and the angle of this triangle *BGA* is the excess of the eccentric anomaly *CBG* over the mean anomaly *CAG*. Hence, a knowledge of one and the same triangle provides the two parts of the equation which corresponds to the mean anomaly *GAC*.'

The new method seemed, therefore, to provide the desired solution.

Unfortunately, there was a difficulty; for, as Kepler explains[30]:

> 'My reasoning contains a paralogism, though it is of no great importance.[31] When Archimedes divided the circle into an infinity of triangles, these triangles were right angled with respect to the circumference, because their apices [were] at the centre *B* of the circle; but the triangles whose apex is at *A* do not have the same relationship to the circumference, because the straight lines drawn from *A* to the circumference cut it obliquely in all points, except *C* and *D*.
>
> 'The error could also be found by trial, as I have done, when I took the distances *AC*, *AG*, *AH* for each whole degree of the angles *CBG*, *GBH*, and then added them together . . . The result was a sum greater than 36 000 000, although [the sum] of the 360 distances starting from *B* comes to 36 000 000 exactly.[32] If these two sums were a measure of the same surface, they ought to have been equal.
>
> 'The error may be explained as follows. Draw any straight line (other than *CD*) through *B* to cut the circumference: let the line be *EF*, and join the points of section *E* and *F* to *A*. As the point *A* is not contained in the line *EF*, *EAF* will form a triangle; therefore, *EA* plus *AF* is greater than *EF*. However, the surface of the circle contains the sum of all the [lines] *EF*, namely, a sum which is smaller than that of all the [lines] *AE* and *AF*, because such a triangle is formed between *A* and all the opposite points of the eccentric, except between *A* and the points *C* and *D*, where we have a straight line instead of a triangle.
>
> 'It may be mentioned in passing, that we can show in the same way that the distances starting from *A*, which correspond to each of the 360 whole degrees of the angle having its apex at *A*, when taken together are less than 36 000 000. In fact, draw any straight line (other than *CD*) through *A*: let the line be *EV*, and join *E* and *V* to *B*. Then, in the triangle *EBV*, the straight lines *EB* and *BV* taken together exceed the sum of *EA* and *AV*, the two opposite distances. All the 360 *EB*, *BV* taken together make 36 000 000; consequently, all the *EA* and *AV* taken together make less than 36 000 000.'

Hence, the sum of the radii vectores originating from *A* is greater, or less, than that of the radii vectores originating from *B*, depending on the way the calculation is made, or rather—seeing that we must use the right expression, and that it is quite obvious that Kepler made some saving in his 'mechanical and laborious' method of calculation in the second case—depending on the way the points of division are taken on the circle; that is to say, either with 360 angles of 1° at the apex *B*, or with the same angles at the apex *A*. It is precisely this aspect to which Kepler seems not to have directed his attention; he was more preoccupied with the argument presented in the form of a paradox against the reliability of the method of exhaustion inspired by Archimedes.

Now the points of the two methods of division do not all coincide, and the sums of the radii vectores originating from *A* are therefore different; at least there is nothing surprising in the fact that they are.[33] The geometrical reasoning transferred by Kepler from the first to the second case is perfectly correct, and there is really no paradox. All the same, there remains a difficulty, which is connected with the nature of the problem to be solved, namely, what is the right method of dividing the circle in order to obtain an infinite number of radii vectores? However, Kepler, who was not conscious of the difficulties in going from the finite to the infinite, nor of the real paradoxes which gave so much trouble to a Cavalieri or a Torricelli, was not greatly embarrassed.

Did he not say that the 'paralogism' was 'of no great importance'? Was it because the differences between the results of the summations in question are small and, in short, that they can be assimilated without too much mistrust to the surfaces, which can be used for computation? Possibly; but there is another reason; almost by a miracle, there were compensating errors.

In Kepler's words[34]:

> 'To repeat, then, what has already been said: this method of [calculating] the equation is by far the most rapid, and it is based on the causes of natural motions, as explained above; furthermore, in the theory of the Sun, or of the Earth, it agrees perfectly with observation; nevertheless, it is false in two respects: *primo*, it supposes that the planetary orbit is a perfect circle, which is not true, as will be shown later in chapter XLIV (however, if we suppose the orbit to be elliptical,[35] this method has no defects; this should be noted): *secundo*, it makes use of a surface which does not exactly measure the distances from the Sun to all the points [of the path]. Now, these two causes of error, as though by a miracle, cancel each other in the most exact manner, as will be shown in chapter XLIX.'

Chapter XLIX is devoted to proving the perfect ellipticity of the orbit of Mars (and, hence, of the other planets, too); but towards the end of 1604, when he was writing chapter XL of the *Astronomia Nova*, Kepler had not reached that point. The reference to chapter XLIX, as well as the note in parenthesis on ellipticity, are certainly very much later additions made by Kepler during the final revision of his work.

When he was writing chapter XL, he believed that the marvellous compensation of the double error present in his computation of surfaces related to the case of the oval, and not the ellipse.

Proof of this is contained in a letter which Kepler wrote to his teacher Maestlin[36] (15 March 1605). This most interesting letter, to which I shall return, reveals the last—or last but one—stage in the development of Kepler's thought in regard to calculating the equations of motion by means of surfaces. In this letter, Kepler dropped the passage at the end, borrowed from Archimedes, which he had used in chapter XL, and jumped boldly to the concept of an assembly of distances on the surface containing them all.

After having described the physical structure of the solar system to Maestlin, Kepler wrote[37]:

> 'It was not enough to form a true hypothesis; it was necessary, in addition, to subject it to [the test of] calculation. A tremendous work, and infinitely complicated! Nevertheless, with God's help I accomplished it, and I believe that you will agree that it is sufficient [for my purpose] if one of the three anomalies *DCE*, *DBE*, *DAE* be given, then the other two can be calculated. . . . Let us assume that the anomaly of the eccentric *DBE* be given . . . and that the mean anomaly *DCE* has to be calculated. Now, the mean anomaly measures the time taken by the planet to [cover] the arc *DE*; but the times are proportional to the distances. Now, as all the distances are contained in the sector *DAE* (as I explain in detail in my book on Mars), the surface *DAE* can be determined.[38] The result is easily obtained, for if it be assumed that the path is a perfect circle, then the sector *DBE* is given. It then remains only to determine the surface of the triangle *DAB*. The base *BA* and the height of this triangle are given, and hence the *sine* of angle *DBE* (Fig. 13). As [the areas] of triangles of the same height are proportional to their bases, and [the areas] of triangles with equal bases are proportional to their heights, it follows that we know the ratio of any surface *DAB* to the surface of a triangle *DAB* whose angle at the apex *B* [*DBA*] is a right angle. Now, when we know the value of the maximum surface *DBA*, we know the other surfaces also (I reason thus: if the surface of a circle, according to Adrianus Romanus, or anyone else, be 360°, or a corresponding number of minutes, or seconds, how many will there be in the surface of the triangle?), and these other surfaces betoken the physical, second part of the equation in such a way that we obtain the surface *DAE*, *i.e.*, the mean anomaly; or its Ptolemaic measure, the angle *DCE*.'

The key to Kepler's reasoning, (which is rather elliptical in the way it has been set forth), seems to me to lie in the fact, that, *contrary to his predecessors*, he pictured the planet as linked to the (imaginary) radius vector, which moves or propels the planet round the Sun.[39] He was also the first to treat the *distance* of the planet from the Sun (centre of the system) as an integral part of the scheme, or, if preferred, of its equation of motion: Copernicus did not introduce distance into the calculations.[40] Consequently, the *assembly of distances*, represented by the surface of the triangle having a circular

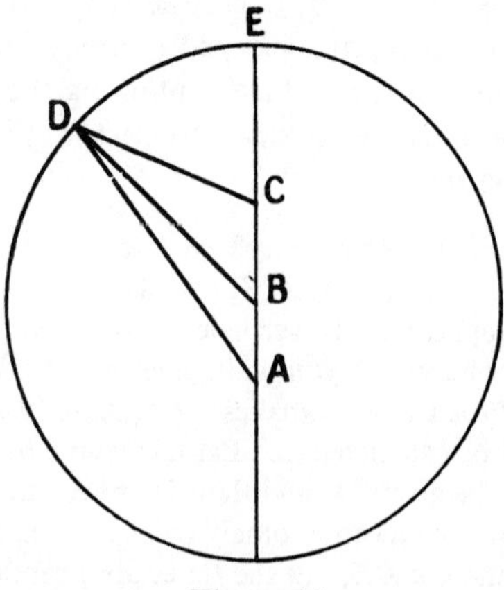

Figure 13.

base and limited in the plane of the orbit by the radii vectores between the start and finish of the motion in question, became in his view a determinative element in the mechanism of planetary motion. We should be wrong, however, in regarding this conception to be anything more than it really is, namely, a consequence, pushed to the limit, of the dynamic point of view applied to planetary motion—a point of view in which the *velocity* of the planet at each point of its path is a function of its *distance* from the source of motion. We should be wrong, particularly when expressing Kepler's thought, in using the terminology found in modern text-books and speaking of the surface in question as the surface 'swept out' by the radius vector of the planet. It was something quite different for Kepler: it is the 'place' where the assembly of distances is to be found, this assembly or 'sum' determines the assembly or 'sum' of the 'times' taken by the planet to cover a corresponding element of its orbit; or, conversely, the assembly of velocities and elements of the path whose 'sum' determines, or constitutes, the length of the path which has been traversed. The surface in question contains all the distances—the term used by Kepler is most exact and very characteristic: he said that the distances *insunt planitiae*—and that is why the surface can represent their 'sum', although in itself it is by no means this 'sum', and is not made up of the 'distances'.

This concept of surface was not a primary concept, supplied intuitively; as we shall see later, it was a derived concept, even an approximate one. Kepler's primary concepts were those of distance, and motive force.

In fact, if, as did Kepler, we consider the planet's motion on its orbit as being bound up with that of its radius vector, we cannot fail to notice, in the case of uniform motion in a perfectly circular orbit about the centre *B*, that the distances covered by the planet in unit time will be equal, and, hence, the surfaces determined by the radius vector (the circular sectors) will be equal also; they will contain as many 'degrees', 'minutes', or 'seconds', *i.e.*, sectors, or 'triangles' having one degree, one minute, or one second of the orbit for base. It is quite clear that we must go further and regard these sectors, or triangles, as being made up of the 'sums' of radii vectores, or 'distances' from the centre. It is equally clear, that in this case, the surfaces corresponding to the different times will be proportional to these times, or what is the same thing, to the distances covered by the planet on its path.

However, if the motion be not uniform, and is referred to a point *A* which does not coincide with the centre of the circular orbit, as in the case of the planets; if, furthermore, the velocity of the said planet's motion be inversely proportional to its distance from *A*, the radius vector linked with the planet, which from now onwards covers unequal portions of its orbital path in equal times, will be determined by 'triangles' which are no longer identical, but quite different and even dissimilar. This notwithstanding, they may be regarded as being formed, or constituted by, the assemblies or sums of different distances (to the point *A*) from points on the path described by the planet.[41] In this manner we arrive at the law of areas, which Kepler formulated in a rather unusual way, when he told Maestlin that the *surface DAE* is nothing but the mean anomaly, and that it measures the time of travel through the segment *ED* of the orbit in exactly the same way that Ptolemy treated his angle *DCE*. Unfortunately, in the present case, which deals with a perfectly circular orbit, this beautiful, elegant law does not apply. It is demonstrably wrong—as Kepler well knew.[42] Fortunately, it is wrong only for a circular path; now the planets do not move in circles, but in oval paths: thus Kepler wrote:

> 'You will raise a double objection[43]: first, if we assume that the path is a perfect circle, then the surface of the circle does not measure, or does not contain, the distances starting from *A*; and this is true, as I have shown. Secondly, we assume—which is not the case—that the path is a circle: in fact it is an oval. I say that one of these objections removes the other; in other words, the two errors compensate each other.'

VIII

From the Oval to the Ellipse

When Kepler seriously resumed the study of Mars in 1604, he was convinced, and had been for a long while, that the orbit described by his enemy in the sky was not a circle, but an oval.

Naturally, this was not enough; it was not sufficient to know that the orbit was approximately an oval; it was necessary to define its nature[1] more closely. Furthermore, it was not enough to have a general idea of the geometrical structure and physical mechanism responsible for the orbit in question; it was necessary to be able to calculate the essential elements, namely, the distances and dimensions of the radii.

Now, the very reasons which had led Kepler to reject the circle in favour of the oval, namely, the curious results which led to the *hypothesis vicaria*, and the non-agreement of calculations made in accordance with the 'physical hypothesis' (calculation of surfaces, or 'calculation of triangles', as Kepler called it) with observational data, forced him not to build upon results secured by these two methods.

Consequently, there was nothing more to be done except to make a fresh start, commencing with a redetermination of the constants of the orbit of Mars (line of apsides, eccentricity, etc.) as accurately as possible; and to determine empirically as many successive positions as possible of Mars in its orbit, and in this way, to define it more accurately.

All this is unfolded by Kepler at the beginning of Book IV of the *Astronomia Nova*, where he explains also how and why he was obliged to renounce circularity.[2]

In chapter XLI he reminds the reader how, *by assuming that the orbit of Mars is a circle*, it is possible from three observations of the planet to determine its exact position on the eccentric (η, κ, θ), the eccentricity ($\alpha\gamma$) and the position of the line of apsides ($\varepsilon\delta$).

Using as basis, observations from the years 1595 (25 October) and 1590 (31 October and 31 December), the value 9768 (the radius $\gamma\eta$, or $\gamma\kappa$, being taken as equal to 100 000) is obtained for the eccentricity ($\alpha\gamma$). As for the line of apsides, it is found that it should occupy the position 27° 8′ 36″ (Fig. 14).

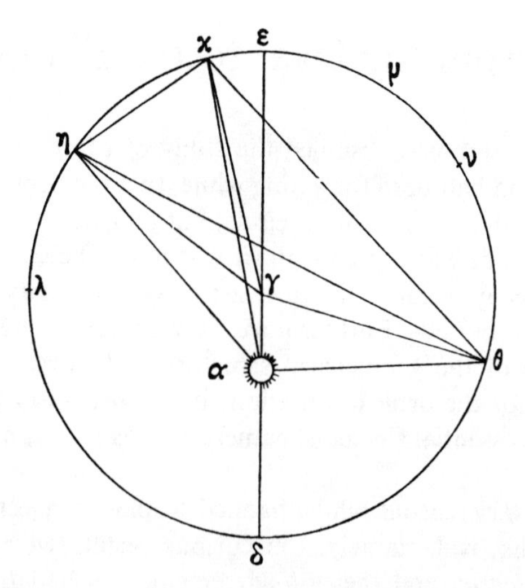

Figure 14.

Unfortunately, if, instead of the three observations of 1590, we take others as a basis of our calculations, we shall obtain different results; for example, instead of the positions η, κ, θ, occupied by Mars in its orbit, we take λ, μ, ν. Furthermore, all these results, differing among themselves, would differ also from those obtained by using more precise methods; in particular, the eccentricity comes to 9264 and not 9768.[3] The inevitable, or at least most probable, conclusion is that the points of the orbit do not fall on a circle; the orbit of Mars is not circular.

Let us be quite clear on the matter: this is the conclusion to which *we* are forced. It was forced on Kepler when he rewrote chapter XLII of the *Astronomia Nova*, but not two years before. The discovery that the results obtained from different observations did not agree dealt another blow to his belief in the circularity of planetary orbits;

his belief was shaken, but not destroyed. Kepler says as much at the beginning of chapter XLII[4]:

> 'You have realized, dear reader, that we must make a fresh start; for, as you see, three positions on the eccentric of Mars and as many distances [from the Sun] do not agree with the determination of aphelion when referred to a circle. This caused me to suspect that the planet's orbit is not a circle; and for that reason, it is not possible to determine other distances starting from three [known] distances. Furthermore, the distance of each position must be calculated from the observations relating thereto; in the first place those at aphelion and perihelion.'

Chapter XLII is devoted to finding these two fixed points on the orbit using nine observations made when Mars was close to one or other of them, and at intervals separated by periods corresponding approximately to its period of revolution. The chapter ends—we must not forget that Kepler is describing one stage in the development of his thought where he had not yet abandoned his belief in circular orbits—by confirming his thesis of the uniformity of planetary orbits with respect to the true Sun; and hence, the necessity of bisecting the orbit of Mars, as well as that of the Earth and all the others.[5] Having definitely established this, Kepler then proceeded to apply his new 'physical' method of framing and calculating the equations of motion to the case of Mars, always assuming perfect circularity for its orbit.[6] Though, in the case of the mean longitudes, the results from the 'physical' calculation agreed rather well with those given by the *hypothesis vicaria*—for the eccentric anomaly equal to 90° the difference was only 24 seconds—the agreement was not so good in other parts of the path. For eccentric anomalies of 45° and 135°, the difference amounted to 8 minutes.[7] An error of this size could not be attributed to the *hypothesis vicaria*; so, Kepler was obliged to conclude that there was some defect in the 'physical' calculation. It was clear, that this defect—surfaces not corresponding exactly to the 'sums' of the 'distances' which 'regulate the speeds and times'—was not able of itself to account for the error in question. Indeed, the difference between the sum of the distances and the surface was very small, whilst that revealed between the results of the physical calculation and the *hypothesis vicaria* was considerable. Furthermore, the physical calculation would have given times which are too short. In fact, the error makes them too long, particularly in mean longitudes.

As the times are directly proportional to the distances, it means that the physical calculation made use of *distances which were too large*. In other words, by assuming the true orbit of Mars to be a circle, it has been placed *too far away*.

Kepler continued by explaining that the systematic differences between the results of physical calculation and those supplied by the *hypothesis vicaria* could not result from[8]:

> 'having abandoned the double epicycle of Copernicus and of Tycho Brahe which makes the planet's orbit oval, and having adopted the perfect circle of Ptolemy. . . . In fact . . . the Copernican orbit is not curved towards the centre, as we require here, but, on the contrary, is farther away [from the circle] by 246 parts [of the radius], which would have increased the error still further in this instance, where we assume that the times are proportional to the distances.'

In order to get rid of the error, it is clear that the circle must be abandoned; but this does not mean a return to Copernicus, or to Tycho Brahe. Quite the contrary; it must be assumed that the orbit is a curve which is contained entirely within the circle originally adopted.

This is precisely what Kepler declared when he resumed, once again, a comparison of the results of calculation with those of observation: the observed distances are shorter than the calculated ones. The differences, though small, could not be ignored. For[9]:

> 'If anyone should wish to ascribe these differences to errors of observation, he would betray the fact that he had not taken account of the force of our proofs; and would have to charge me with gross falsification of Tycho Brahe's observations.'

Account must be taken also of the evidence:

> 'The planet's orbit is not a circle; but [starting at aphelion] it curves inwards little by little and then [returns] to the amplitude of the circle at perigee. An orbit (line) of this kind is called an oval.'

We may add, and this is of great importance, that the evaluation of the surfaces, as we have already said (p. 240), agrees much better (and even perfectly) with an oval orbit than with a circular one. Now, as this evaluation was based on dynamic concepts, it follows that the 'ovalness' of planetary orbits—in the circumstances, that of Mars—is found to be conformable with the new celestial dynamics, and hence to be confirmed by it.

Kepler then continued[10]:

> 'It was assumed in chapter XLIII that the surface of a perfectly (circular) eccentric was very nearly equivalent to the [assembly of] distances of

any number of equal parts of the circumference of this eccentric [reckoned] from the source of motive force; and, consequently, that the portions of the surface were measures of the times during which the planet remained in the corresponding parts of the circumference of the eccentric. However, if the surface, whose boundary is traced by the planet's orbit, be not a perfect circle, but in its width is of smaller dimensions than the length of the line of apsides; and if, notwithstanding, this surface, circumscribed by an irregular orbit, give a measure of the times that the planet takes to [accomplish] its full circuit and equal portions of it, then it follows that the decreased surface measures a time equal to that [which was measured] by the surface not so decreased. Consequently, those portions of the decreased surface [which are] nearer aphelion and perihelion will measure a longer time, for in their case, the decrease is very small; on the other hand, those portions [which are] in mean longitudes will measure a shorter time than in the former instance, for that is where most of the decrease in the total surface occurs. If, therefore, we use this decreased surface to frame the equations [it will follow] that the planet, when near aphelion and perihelion, will be slower than indicated by the first—and correct—form of the equations; and in the mean longitudes it will be faster, because the distances there will be less. Hence, the times which are shortened here, will be increased by way of compensation above and below at aphelion and perihelion, in the same way that a sausage, when squeezed in the middle, swells at the ends as a result of the compression.

'Now, if contraries be cured by contraries, it is obvious that we are in possession of the best remedy for expurgating the faults from which our hypothesis suffers, as we have seen in chapter XLIII. In fact, if the planet moved slower near aphelion than it should do according to the hypothesis of circularity, this retardation would, (or, at least, would be able to) make the excess of 8′ 21″ in the first octant disappear; and its acceleration in the mean longitudes would play the same part for the deficiency in the third octant.[11] Here we have a fresh argument which proves that the planet's orbit does in fact deviate from the circle assumed [by us] and curves laterally inwards towards the centre of the eccentric.'

Unfortunately the remedy in question—to use Kepler's imagery—works only when it is suitably prescribed; that is to say, if the distances of the planet from the Sun be decreased in the required proportion as a result of the 'ovalness' of the path; in other words, if the figure and dimensions of the oval be correctly determined. Now, that is an extremely difficult thing to do. Kepler confessed that the argument he had just put forward was not decisive; it had not sufficient force to make him renounce circularity. Much to the contrary.[12]

'For, after having long toiled in trying to reconcile this form of the equations [with the circular orbit] I was so struck by the absurdity of the measure [of times by the surfaces] that I abandoned the matter completely; and it was only when [the calculations made] by the methods described in chapter XLI revealed to me the deviation of the distances, that I reverted to this manner [of calculating] the equations.'

The order in which the arguments were adopted is of little consequence. The result was: *the orbit of the planet* [*Mars*] *is not a circle, but an oval.*[13]

I have recounted above the stages in this discovery which overturned the traditional principles of astronomy, and which, as far as Kepler was concerned, presented afresh, but with increased acuteness, the problem of the mechanism by which planetary motion is produced—a problem which he had already propounded, and the solution of which he had sought in vain for circular motion. Here is what Kepler himself said[14]:

'As soon as I knew from Tycho Brahe's very accurate observations that the planet's orbit is not exactly circular, but deviates therefrom at the sides, I thought I then knew the source and natural cause of this deviation. I was, in fact, greatly concerned with the methods in chapter XXXIX, and I recommend the reader to read this chapter again very carefully before going any further. In this chapter I ascribed the cause of the eccentricity [of the orbit] to a certain force located in the body of the planet; it follows that I ascribed the origin of the deviation to that force also . . . but, as the proverb says, *canem festinum coecos parere catulos.* [In my haste I came to premature conclusions.] In chapter XXXIX I found myself in very great difficulty because I could not show sufficient probable cause why the planet's orbit should be a perfect circle (there were always some absurdities ascribed to this force which had its seat in the body of the planet); but now, when I found from observations that the planet's orbit was not perfectly circular, the strong impression this discovery made on me immediately led me to believe that the [causes] which, in chapter XXXIX, seemed absurd and unsuitable for the production of a circle, could take on a more probable appearance, and by their action provide the planet with an orbit exactly in agreement with observation[15]. If I had pursued this course rather more wisely, I should have immediately come to the truth; but I was blinded by the desire [to be finished with it] and did not pay attention to all the details of chapter XXXIX, with the result that I held to the first idea that came to my mind—an idea that seemed to me probable because it imposed uniform motion of the epicycle. Consequently, I became involved in further mazes. . . .'

One of the ideas described in chapter XXXIX—the least suitable according to Kepler—explained the production of the eccentric, yet perfectly circular orbit by a combination of non-uniform planetary motion on its epicycle with an equally non-uniform motion of the centre N of this epicycle about the point A.[16] 'This, as Kepler recalled, seemed to me absurd.' On the other hand, it seemed to him that the absurdity would be avoided, if, instead of a non-uniform motion, a *uniform* epicycle motion were ascribed to the planet. In this case, especially by assuming that the planet moves with an angular

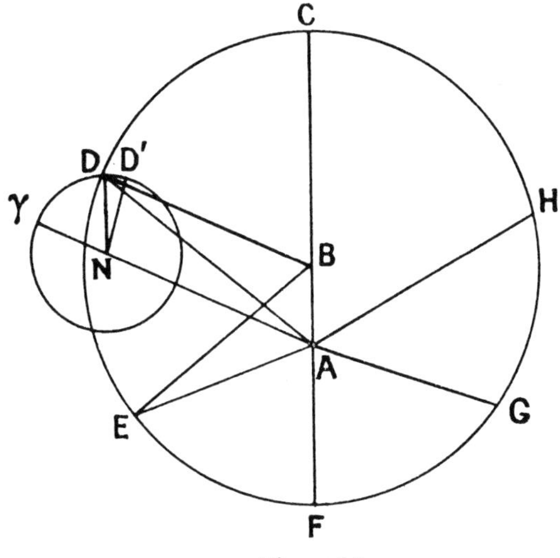

Figure 15.

velocity equal to the mean velocity of the centre of the epicycle about the point A, this velocity at the start of the path (near aphelion C) will be greater than that of the centre N; angle $\gamma ND'$ will therefore be greater than angle NAC; the radius ND' will not be parallel to AC, but will be inclined to this line, and the planet will appear, and will be, to the right of point D at D'. The situation will be reversed at perihelion: the angular velocity of the planet on its epicycle will be less than that of the point N; it will therefore be retarded with respect to the latter and will be nearer to AF (from F) than if it moved with an angular velocity equal to that of point N (Fig. 15).[17]

'This was exactly what was required by observation—and it was also evinced by the physical equations formed by summing the distances

> *AC, AD*—namely, that the planet moves more rapidly at the sides of the eccentric, and consequently that the distance from the Sun becomes smaller. Now, seeing that this agreement was of a most persuasive kind, I immediately concluded that the inflection of the planet towards the interior resulted from the fact that the force moving the planet and fixing its distance [from the Sun] in accordance with the law of circularity takes precedence of the Sun's force in its action; thus, the former force causes the planet to progress by equal amounts in equal periods of time, and causes it to make uniform approach to the Sun in agreement with the law governing the epicycle; whilst the latter [the Sun's force] controls the planet with varying degrees of force, and causes it to advance in a non-uniform manner, the more slowly the farther away it is. Consequently, the distance of equal arcs on the epicycle increases towards C (aphelion) and F (perihelion), and decreases in the mean longitudes . . .'[18]

Kepler concluded sadly,

> 'This was how it came about that my mind was confirmed in the mistake which, sometime before, I had started to put behind me, namely, the notion that it was the function of the planet's own force to guide it on an epicyclic path.'

There is nothing more significant than Kepler's words. Nothing reveals the almost insurmountable power of the tradition of, or obsession with, circularity more than this unexpected return of Kepler to concepts which he himself had submitted to thorough criticism by proving the impossibility of epicyclic motion in the absence of solid spheres, and by trying to replace this mechanism, which cannot be realized in nature, by a libration of the planet on its radius vector.[19] We have the impression that Kepler only accepted his own criticisms grudgingly; and that he put them joyfully aside as soon as a deceptive mirage inspired him with the hope of getting out of the difficulty by proven means.[20]

Kepler's deception lasted a rather long time. The whole of 1604 and the beginning of 1605 were devoted to working out the theory of the 'oval', and comparing it with observational data.[21] We might be tempted to call it a year wasted . . . but we are well aware, and Kepler reminds us of it later, that the human mind advances towards truth only by eliminating errors.

The first question that presented itself to Kepler as a result of adopting the 'oval' orbit was concerned with determining the surface which it 'circumscribed'. To do this, it was necessary to ascertain the surface area of 'the lune between the oval and the circle whose axis is the diameter'. Unfortunately, the 'lune' in question has an irregular shape, because the oval is wider at aphelion and narrower at perihelion: it is wider at the top, and narrower at the bottom. 'If only the orbit were an ellipse . . .',[22] wrote Kepler to Fabricius. The *Astronomia Nova* abounds with the same sentiment[23]: 'If our figure were a perfect ellipse, the problem would have been solved already by Archimedes. . . .' Fortunately, the difference between our oval and the ellipse is very small. In fact, that portion of the lune which exceeds the ellipse at the bottom is very nearly compensated by the portion which is lacking at the top. We can assume therefore, without any great error, that our oval is a perfect ellipse, and make our calculations accordingly.

Now, the surface of the lune between the ellipse and the circle was stated by Kepler to be almost exactly equal to that of the small circle whose diameter is the eccentricity. Taking the radius as 100 000, the eccentricity is 9264. If we take the surface area of the circle as 31 415 900 000, that of the lune will be 269 500 000 and of the oval 31 146 400 000. At its greatest width the lune will measure 0·008 58 of the radius.

Furthermore, if we assume for the oval surface, as in the case of the circle, that the surface represents the 'sum' of the distances and therefore the 'sum' of the times taken to travel the various paths, then we have a way of applying the evaluation of surfaces (the 'physical' calculation) to the new conditions. The error is not very great and would seem to be negligible. Unfortunately, the results—I shall spare my reader details of them—did not confirm Kepler's hope: calculations did not agree with observations; the distances in the first quadrant were too short. Therefore, the oval was somewhere between the auxiliary ellipse and the circle.

Kepler then said to himself—and we now witness an exact repetition of what happened two years previously when he still believed in circularity of the orbit—after all, the circle is not the 'sum' of the distances, and neither is the oval. So, it would be better to abandon the geometrical procedure in favour of the arithmetical, that is to say, calculate one by one the distances of small portions (1 degree) of the orbit and find their sum. That was what he did: and three times over, assuming different values for the eccentricity.[25] The result was disastrous.[26] Kepler informed his reader:

'You could say that we had reached the worst state of affairs . . . though,

> my friend, if such a result had been capable of perturbing me, I should have been able to save myself all this labour and remain satisfied with the *hypothesis vicaria.* Know then that errors show us the way to truth.'

Further attempts; further frustrations[27]; yet some progress. He gradually confirmed that the width of the lune assumed as a basis of calculation—0·008 58—was much too large, and ought to be approximately halved.[28] A fresh series of calculations of the positions of Mars starting with 22 different observations, and made according to the new method, confirmed the result: the width of the lune, *i.e.*, the shortening of the radius of the eccentric circle was only 0·004 32.[29] Fabricius had noted the same fact about the same time when he wrote to Kepler, 27 October 1604, saying, 'Your oval is too narrow.'[30] In fact, the calculated distances differ more from the observed distances as we go away from the apsides (aphelion)—this is the opposite of what ought to have been found according to chapter LIV. There, the distances calculated according to the law of circularity were too great; here, the hypothesis of the oval makes them too small. 'It is therefore clear', wrote Kepler,[31] that the orbit is not a circle, and that it does not differ from a circle as much as it ought according to the theory of the oval given in chapters XLV and XLVI. D. Fabricius has found the same. He very nearly forestalled me in discovering the truth. For, seeing that the perfect circle deviates [from reality] in the opposite direction, we both concluded that the truth was in between.' Kepler, in fact, exaggerated the merits of Fabricius, who came to the conclusion that the whole Keplerian enterprise was thereby condemned, and that it was necessary therefore to revert to the traditional circles. Kepler, of course, did not take it that way. What was condemned, was the mechanism by which it was produced, and in which he had believed with such unfortunate obstinacy. It was that which went up in smoke.[32] As for the planetary orbit, it was indeed oval; but slightly larger. It was almost a perfect ellipse.[33]

The mechanism of epicyclic motion having disappeared in smoke, Kepler found himself driven back to the idea which he had already outlined two years before when he still believed in the perfect circularity of planetary orbits; this idea involved motion of libration (oscillation) of the planet on the radius vector linking it with the Sun. It is really astonishing to see how Kepler, in his theory of the 'oval', retraced exactly the same path that he had taken in his theory of the 'circle'. Precisely as in the earlier instance he did not place the planet simply on the radius vector, but on the diameter of an (imaginary) epicycle: and, as before, the epicycle, although it had lost all real function, nevertheless retained that of providing a measure of the oscillatory motion.[34] It is difficult to part with old friends; and it was difficult, even for Kepler, to free himself from the mechanics of circular motions.[35]

It was at the beginning of 1605 that Kepler realized that the epicyclic motion to which he was so strongly attached was not leading him to the desired conclusion, and that it was necessary to replace it by libration. His letter, dated 5 March 1605, to Maestlin[36] clearly reveals the stages in the development of his ideas:

> 'With regard to the motion of Mars, I shall explain myself more clearly. . . . The distances from the Sun do not fall on a perfect circle, but on an oval, the shape of which I have found after infinite labour by replacing the eccentric [circle] by a concentro-epicycle in the following

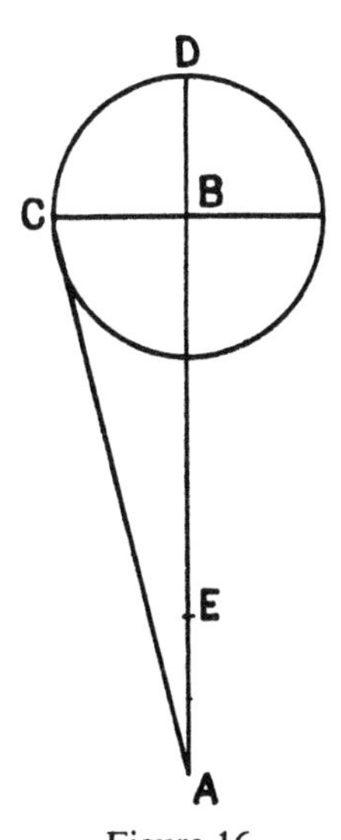

Figure 16.

> manner: Let *A* be the Sun, *AB* the radius; make *BD* equal to 9300, and *AE* similarly; let *E* be the point of equality [equant point] of *B* about *A*, and [also that of the motion] of *D* about *B* (Fig. 16). (I am adapting myself here to the hypotheses of the Ancients to the detriment of truth.) This being granted, let the distance *BA* be taken for the distance *CA*. Then, the orbit of the planet in mean longitudes will deviate from

circularity according to the amount by which *BA* is shorter than *CA*. Thereby, it is finally demonstrated that it is false that the planet moves on an epicycle; this is what I have most strenuously maintained for a long while, but which has always been contradicted by observations. For the planet does not revolve on an epicycle, but oscillates on the diameter of this epicycle: if this be not assumed, then we have differences of 15′ and more between parallaxes [calculated] of the annual orbit and the observed [parallaxes]. Furthermore the point *E* is not truly an equant point. . . . Consequently, we know that we are not dealing with a natural hypothesis. . . .

'That which cost me the greatest amount of work and trouble was to weld my two false hypotheses into a single true one.[37] I made a thousand attempts, which I have no doubt told you about in my earlier letters.[38] The solution was possible only by investigating the [real] causes in nature, namely: the body of the Sun is circularly magnetic and it rotates in its place, and thereby causes the sphere of its force to rotate with it; this sphere of force does not attract, but has the power of promoting motion. On the other hand, the bodies of the planets [are not in themselves endowed with motivity, but] are inclined to remain at rest in whatever part of the Universe they are placed. Consequently, in order that they should be moved by the Sun a constraining force is needed, whence it follows that those which are more remote from the Sun are pushed more slowly, and those which are nearer are pushed more rapidly, that is to say, the eccentric moves uniformly with respect to the equant point.[39] On the other hand, every planetary body must be regarded as being magnetic, or *quasi*-magnetic; in fact, I suggest a similarity, and do not declare an identity. It must be assumed also that the line [axis] of this force [*quasi*-magnetic for the planets] is a straight line having two poles, one retreating from the Sun, the other pursuing it.[40] This axis, through an animal force,[41] is [constantly] directed approximately towards the same parts of the Universe. As a result, the planet, carried along by the Sun, turns towards the Sun, first its retreating [repelling] pole, then its pursuing [attracting] pole. As a consequence we have the increase and decrease in libration.[42] I cannot conceive any other means [of producing it]. For both in retreating from, and approaching, [the Sun, the planet] does so according to the measure of the angle which the line [drawn] from the Sun to the centre of the body [of the planet][43] makes with the axis [of the planet], and this *ceteris paribus*. This is what I have previously said in the geometrical hypothesis[44]: it is attested by observation, that the planet performs librations, and particularly that during libration it moves slowly in the vicinity of the apsides of the epicycle, and more quickly in the mean positions; whereas in its *raptus* round the Sun, it moves slowest at aphelion, and quickest at perihelion.[45] Furthermore, the superior semi-diameter of the libration is traversed in a longer time than the equal inferior semi-diameter; for the magnetic force of the planet itself acts also less strongly when the planet is remote from the Sun; this is exactly what happens in the case of magnets.'[46]

In the *Astronomia Nova*, chapter LVI, Kepler tells us that it was chance that enabled him to find a way out of the maze. It was one of those accidents that happen only to those who deserve them. In fact, having been forced to admit that the width of the lune between the planet's orbit and the eccentric circle of his theories was only 0·004 29 and not 0·008 58 of the radius, he never ceased to meditate on the reason for it, and the way in which such a lune could be produced.'[47]

'Thinking on this matter, I fell quite by chance on the *secant* of the angle 5° 18′, which is the measure of the largest optical equation [corresponding to an eccentric anomaly of 90°].[48] Now, when I noticed that the [*secant*] was equal to 100·429, it was as though I had awakened from a dream and saw a fresh light. I started to reason as follows: in the mean longitudes, the optical part of the equation is the greatest. In mean longitudes, the lune [is greatest], that is to say, the decrease in the distances is greatest, and this decrease is just as large as the excess of the *secant* of the largest optical equation—100·429—over the radius—100·000. Therefore, if in the mean longitudes we take [the length] of the radius instead of [that of] the *secant*, we obtain what is suggested by observation. I shall draw, therefore, the general conclusion: if in the scheme outlined in chapter XL we take *HR* instead of *HA*, *VR* instead of *VA*, *EB* instead of *EA*, and so on, we shall have the same result for all other places on the eccentric as that which has been obtained here for the mean longitudes.' (Fig. 17.)

Let us explain. Kepler found that in order to obtain the correct distances of the planet (Mars) from the Sun it was necessary to substitute *EB* for *EA*, and *HR* for *HA*. Now, *EA* is equal to *EB* sec *BEA* (*EB* sec φ), or,

$$EB = \frac{EA}{\sec \varphi} = EA \cos \varphi.$$

As for *HR*, it equals $HB + BR$, or, $HB + AB \cos RBA$. But *EB* and *HB* are radii of the eccentric circle, *AB* is the eccentricity, and as for the angle *RBA*, it is equal to angle *HBC*, *i.e.*, the eccentric anomaly. Let us take the radius of the eccentric circle as equal to unity, designate the eccentricity by *e*. We shall obtain for the length of the radius vector: $RV = 1 + e \cos \beta$. In the case of *EB*, *i.e.*, when the eccentric anomaly β is equal to 90°, $e \cos \beta = 0$, and $RV = 1$.[49]

The solution to the problem of the motion of Mars—and by the same token, of all the other planets, too—was accordingly established.

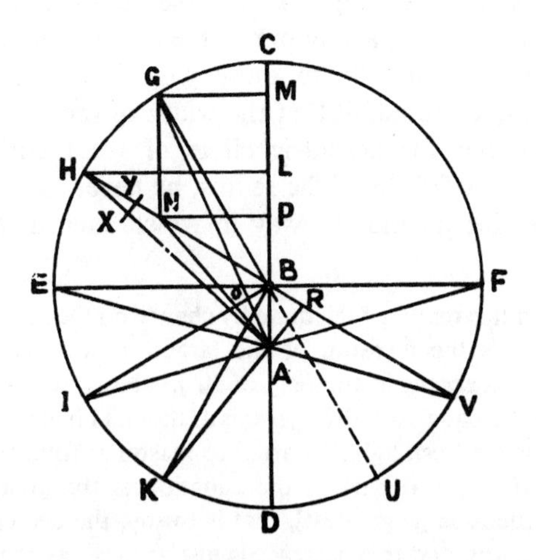

Figure 17.

However, Kepler, who was unable, as I have said, to free himself from the system of circularity, and who was haunted by the ghost of the epicycle, failed to notice it. So, instead of accepting purely and simply the solution as discovered, he continued by saying[50] (Fig. 18):

'As a result of the equivalence [of this system with that in the diagram on p. 219] we must substitute $\alpha\chi$ for $\alpha\delta$, or $\alpha\iota$ and $\alpha\mu$ for $\alpha\varepsilon$ and $\alpha\lambda$ in the small diagram of chapter XXXIX.[51]

'The reader will do well to refer once more to chapter XXXIX. He will find that I have already envisaged there, by a consideration of physical causes, all that is confirmed here by observations; namely, that it was conformable to reason that the planet should perform a certain libration on the diameter (perpetually pointed towards the Sun) of a kind of epicycle.

'In that place he will find that there was no other objection to this notion than the fact that, in order to obtain a perfect circle, we were obliged to make the upper and lower parts $\gamma\iota$ and $\gamma\zeta$ of the libration corresponding to equal portions of the arc of the eccentric) unequal; that is to say, it was necessary to shorten the upper parts, and lengthen the lower parts. Seeing that we have now rejected the circular orbit for the planet, and as we have substituted $\kappa\alpha$ [and] $\mu\alpha$ for $\delta\alpha$ [and] $\varepsilon\alpha$, *i.e.*, for $\iota\alpha$ [and] $\lambda\alpha$, it follows that the said parts of the libration, namely, $\gamma\kappa$ and $\mu\zeta$ are equal. Thus, that which hindered us in chapter XXXIX, has been transformed into an argument in behalf of the discovery of truth.

'If the median parts $\kappa\mu$ be larger than the exterior ones $\gamma\kappa$, $\mu\zeta$, this, as will appear in the next chapter (LVII) is conformable to nature; whereas in chapter XXXIX we were obliged to consider it as being contrary to its laws.

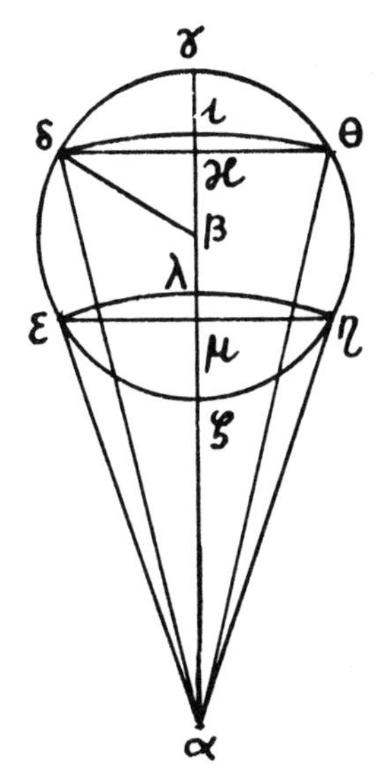

Figure 18.

'Similarly with regard to the difficulty of accepting that the change in the Sun's apparent diameter can serve as an indication by which the planet regulates its approach and recession.[52]

'It has been proved[53] by very accurate observations that the planetary orbit in the aetheric aura is not a circle, but is oval in shape, and that the planet oscillates on the diameter of a small circle in such a manner that, whilst traversing [equal] arcs on the eccentric (Figs. 17 and 18), it proceeds over the diametric distances $\gamma\alpha$, $\kappa\alpha$, $\mu\alpha$, $\zeta\alpha$, and not over the circumferential distances $\gamma\alpha$, $\delta\alpha$, $\varepsilon\alpha$, $\zeta\alpha$, *i.e.*, $\gamma\alpha$, $\varepsilon\alpha$, $\lambda\alpha$, $\zeta\alpha$, which would produce a perfect circle. Whence it appears that a lune will be cut off from one-half of the eccentric circle; and the width of this lune at any place is [equal] to the difference of the distances in question, *e.g.*, $\iota\kappa$, $\lambda\mu$. This being admitted, not because of *a priori* reasons, but because of observations (as I have already said), it follows that our physical speculations will proceed more justly than hitherto. Indeed, the agreement of this libration with the space covered on the eccentric does not arise from any mental or rational process—the reasoning faculty of the planet co-ordinating equal parts of the libration $\gamma\kappa$, $\kappa\mu$, $\mu\zeta$ (which are not even equal) with equal arcs *CD*, *DE*, *EF* on the imperfect eccentric; but it does arise in a natural manner, which depends, not on the equality of the angles *DBC*, *EBD*, *FBE*, but on the extent of the angle *DBC*, *EBC*, *FBC*, which perpetually increases; this angular extent is approximately proportional to the *sine* in geometry where a rise (increase), through being continuously lowered, is gradually transformed into a fall (decrease). [This] is more probable than if we were to say that the planet sharply turns its rudder; which, as we have already said in chapter XXXIX, is clearly contrary to observation.

'Now, seeing that the measure of this libration points its finger to Nature, the cause of it will be a natural one; consequently, it will not be the planet's intelligence, but a natural faculty, or possibly even a corporeal one.'

So, in the last analysis, the replacement of the circular orbit by an oval one proved to be something much more important than a mere change from one orbit or 'equation' to another: in fact, whereas the circular orbit—whether produced by epicyclic motion, or by oscillatory motion—had shown itself to be incapable of realization by purely natural means, or forces, and necessarily implied the intervention of a planetary soul, or intelligence,[54] the oval orbit, which results from oscillation of the planet on its radius vector in accordance with a *sine* law,[55] gives us hope of a natural explanation—a hope which we shall find was realized later. Kepler had always sought a physical explanation of planetary motions[56]; that is to say, an explanation dependent only on natural forces—or, at most, on 'animal' forces—and to the exclusion, as far as possible, of purely spiritual factors.[57] It is understandable that he should then turn with fresh enthusiasm to an investigation of the possible and impossible, probable and improbable, causes of these motions—causes to which he had already given considerable attention, though without any positive result, in the celebrated chapter XXXIX of *Astronomia Nova*.[58]

It will be recalled that in chapter XXXIX Kepler had suggested the possibility of explaining the motion of planets towards and away from the Sun by the whirlpool action of the motive *species* rotating with the Sun (Fig. 19); (he used the analogy of the coming and going of a ferry-boat urged on by the river current). Through a careful analysis of this 'example' we are soon made to realize that the 'example' in question is not adequate. In fact, if we assume that the 'rudder' of the planet, and therefore the planet itself, rotate continuously, this rotation should take twice the time taken by the planet to go round the Sun; the planet ought to present continually changing faces towards the Sun and the Earth. Now, the Moon, whose motion cannot be explained differently from that of other planets, does nothing of the kind, but always offers us the same view. Furthermore,

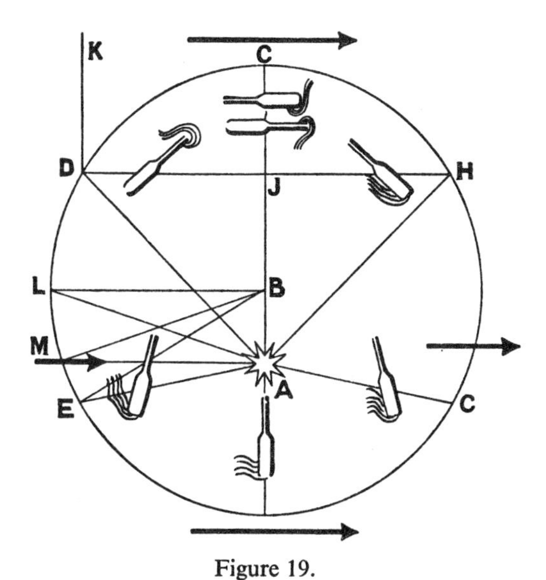

Figure 19.

the 'example' calls for the action of a material reality—water—whereas the motive *species* of the Sun is nothing of the kind.[59] To reject the materialistic analogy is not a set-back; on the contrary, it enables us to make a step forward. The immaterial *species* of the Sun has magnetic properties. Why, then, should not the 'rudder' of the planet have the same? In other words, why not assume, as Kepler had explained to Maestlin, that the planets are immense spherical magnets? This is all the more probable seeing that the Earth is definitely one, as Gilbert had proved. The planet, like all magnets, would then have two poles; one of which would draw it towards the Sun, and the other would repel it. Nor would there be any need, as in the previous 'example', for the planet to rotate; it would suffice for its axis to maintain a constant direction, as is the case with the Earth's axis. This maintenance of the direction of the planet's magnetic axis could be ascribed just as well to a natural power of the magnet; we know that there are two forces in a magnet—one 'directing' (or repelling), the other 'attracting'. It would suffice, therefore, to suppose that the 'directing' forces of the planets are much more powerful than their 'attracting' forces so that the latter are unable to modify the position of the axes to any appreciable extent. This task could be entrusted also to 'animal' forces.[60] Under these conditions, the planet during its passage round the Sun would present first one pole, and then the other, to the Sun, and would approach, as well as move away from it. The objection that the Sun, being a simple body, ought

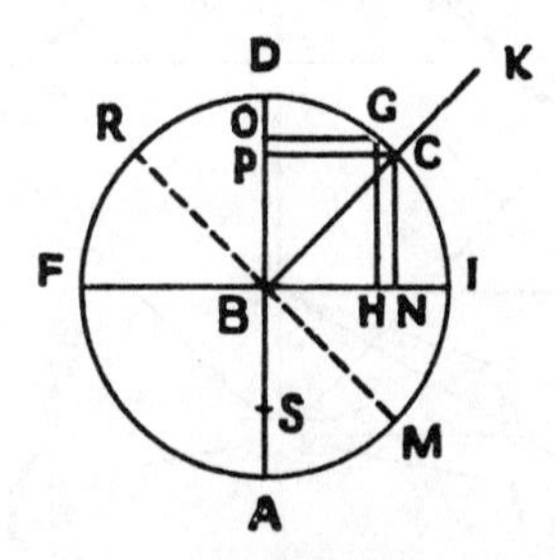

Figure 20.

to function in one way only, could be overcome by supposing it to be 'neutral' like a piece of unmagnetized iron, and by ascribing the difference in behaviour to the dual nature of the poles of the planet.

Kepler fully realized that everything could just as well be accounted for by the action of an intelligence; indeed, the latter would be able to control its action on the changes in the apparent dimensions of the disc or solar diameter, or even the true anomaly,[61] which is something it could not do in the case of the circular orbit. However, this hypothesis is by no means necessary; we can even regard it as superfluous, seeing that the intelligence in question, in order to carry out these motions, should be endowed with an animal force also, and even a natural force, which by themselves, and without the help of the intelligence, are quite sufficient to produce the required motions. In fact, a more detailed study of the behaviour of a magnet placed in the same position as the planets with respect to the Sun shows that it is subjected to an attractive force which is measured by the *sine* of the true anomaly, and consequently, that it performs oscillations which are measured by the *versed sine* of the eccentric anomaly.[62]

Observations show that the planets make these oscillations uniformly and that their amplitude is measured by the *versed sine* of the eccentric anomaly.[63] The necessary conclusion is that the planets are magnets, and that the natural—or animal—causes provide an explanation of the behaviour.

This is asserted even more clearly in the preface to the *Astronomia Nova* quoted above[64] than in the text.

The matter could have stayed there. Kepler could have rested on his laurels, claiming that he had reached 'the goal which was the end purpose of all this work', that is to say, he 'had found a physical hypothesis which not only provides distances in agreement with observation, but also accurate equations', and thereby freeing us from the necessity of borrowing them from the *hypothesis vicaria*. He might have continued in not understanding the true significance and deep import of his discovery. Fortunately—or was this once more an act of Providence?—in computing the distances and apparent and true positions of Mars by his new method, he made a mistake, as a result of which the curve was distorted and the positions of Mars in the upper part of the orbit deviated by $5\frac{1}{2}'$, and in the lower part deviated by $4'$.[65]

Some months later, when the difficulty had been overcome, Kepler gave Fabricius the following account[66]:

> 'I must now tell you what progress I have made with Mars. When, as a result of assuming a perfectly circular orbit, I saw that the distances [Sun–Mars] were too large, and by the same amount were for my [auxiliary] ellipse, which was very nearly oval, they were too small, I could, quite properly, have drawn the following conclusion: the circle and the ellipse are geometric figures of the same kind; both deviate from the truth, but in opposite directions; therefore, the truth must be in between; and between two elliptical figures there cannot be anything else except an ellipse.
>
> 'For this reason the orbit of Mars is an ellipse in which the lune needed to make it a circle has one half the width of that belonging to the previous [auxiliary] ellipse. The width of this lune was 858/100 000; it should now be 429/100 000. This is the fraction by which the distances computed for a perfectly circular orbit must be shortened in the mean longitudes. That is the truth. See, however, how, in the meantime, I

have once more indulged in delusions, and given myself fresh worry. See, how—as the saying goes, "He who never doubts, is never sure of anything"—I tremble before the revealed truth. My old ellipse with a shortening of 0·008 58, had a natural cause . . . as a matter of fact, not very conformable to nature; but if the ellipse had a shortening of 429, any natural cause was missing. . . .'

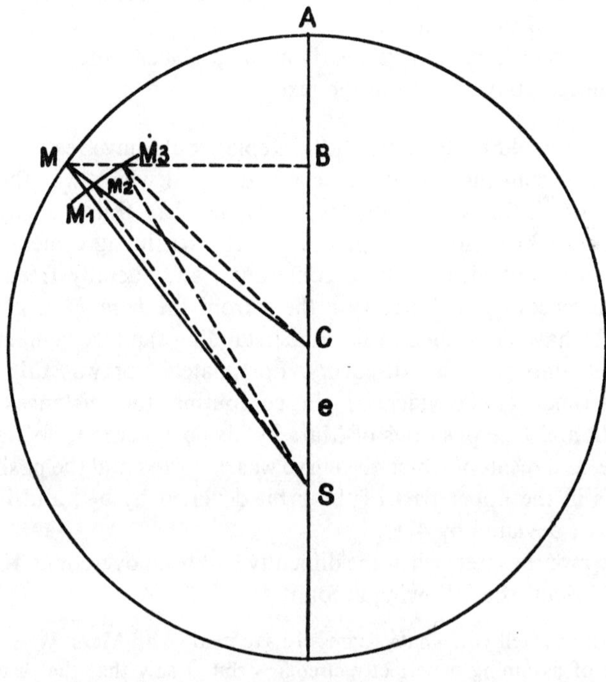

Figure 21.

Kepler's reasoning, which led him to make the mistake, is rather involved; but its basis is simple. Reduced to essentials, it can be set forth more or less as follows[67]:

To find the position of Mars with respect to the Sun Kepler started by drawing the line of apsides and the eccentric circle which Mars would have traversed if its orbit had been circular (but which in fact it does not traverse). Mars being at the point M at a given moment, its distance from the Sun would be MS. Now, we know that this is not the case, and that this distance should be shortened and diminished by the amount of oscillation which the planet makes on the diameter of its (imaginary) epicycle, whose radius is equal to e (the eccentricity); the amount is e.versin ACM.[68] It will then be at the distance SM_1 (and not SM) from the Sun. However, where will it be located? Kepler considered that it ought to remain on the radius (CM) of the eccentric circle, and therefore at the point M_2, such that $SM_1 = SM_2$.

Observation failed to confirm his reasoning: Mars was found at the point M_3 to the right of the calculated position. His reasoning was quite unfounded: there was, in fact, no reason why the planet should be on the (imaginary) radius of the eccentric circle which it does not traverse.[69]

Kepler continued the account to Fabricius of the development of his ideas—the discovery of the oval in 1602, the labours of 1603 and 1604—up to the substitution of oscillation on the diameter for motion on an epicycle.

> 'Ah, poor me! It was only at Easter that I noticed that an orbit formed in that manner is not elliptic—as given by my proof—(if I had been more circumspect, I should have remembered that I had already proved it in my *Commentaria de Motibus Stellae Martis*), but that in the octant positions it approaches closer [too much] to the circle and so becomes enlarged.
>
> 'My argument as follows was faulty: libration along the diameter of the epicycle is equivalent to an ellipse in the mean longitudes and apsides; therefore it is equivalent everywhere. That is wrong; consequently, as in the old, false hypothesis, neither the distances nor the equations for the eccentric fulfil their purpose. Oh! the fertile harmony of the two things, which never ceased to guide me in such perplexities.
>
> 'But I have the answer, my dear Fabricius: the orbit of the planet is a perfect ellipse (Dürer often calls it the oval), or deviates therefrom by no more than an imperceptible amount.'

Fabricius did not reply to Kepler's triumphant announcement. At least, not immediately; but eighteen months later, after he had finished working out his own theory of the motion of Mars, he sent Kepler a letter in which, basing himself on the views of Tycho Brahe and Copernicus, he formulated a general criticism of the new concepts introduced by Kepler into astronomy, namely, abandonment of the principle of circularity in celestial motions. The objections raised by Fabricius were certainly not original; but that is precisely why they are of importance to us: they reveal, in conjunction with Kepler's reply, the relative merits of the supporters of tradition, and the exact nature of Kepler's innovation. By your oval or your ellipse, wrote Fabricius,[70] (and it is interesting to note that in his condemnation he made no distinction between one or the other) you destroy circularity and uniformity of motion, which seems to me quite absurd. The sky is spherical, and consequently, there are circular motions, regular and uniform with respect to its own centre. The celestial bodies, too, are perfectly spherical. Furthermore, it is certain that all their motions, uniform with respect to their own centres, are carried out on circles and not on ellipses. In fact, as the centre of an ellipse is not everywhere equidistant from the circumference, a motion with uniform velocity on an ellipse cannot be uniform with respect to its own centre.

We have seen that non-uniform motions with respect to *their own centres* were fully admissible in Ptolemaic astronomy, which was satisfied with their uniformity with respect to some point or other (the equant point). However, that was what Copernicus protested against; and Fabricius, though not by any means a Copernican, accepted—as did Tycho Brahe—the criticism Copernicus made about Ptolemy: celestial motions must be uniform with respect to their own centres; uniform motions with respect to any other point are excluded, even if circular; *a fortiori*, those which are not circular are excluded. So, it is not the path of the effective orbit described by the heavenly body with which we are concerned; it is its structure, or infra-structure. Fabricius continued by telling Kepler that if (whilst retaining the principle of circularity) he could explain the ellipticity of an orbit by means of a combination of circular motions—naturally, uniform—the situation would be quite different, for then, his hypotheses would not transgress the fundamental principles of natural philosophy.

Kepler's reply to the objections raised by his friend is equally revealing: he fully recognized the justness of his insistence on uniformity—today, we would say, invariance; but he disputed that the only way of ensuring it was by means of circularity. He reminded Fabricius that the principle of uniform circularity is not by any means followed by the *true* motions of the planets[71]:

'You say, that through my oval, uniform motion has been suppressed. Well then! it is suppressed also by Ptolemy's spirals and equant circle; and if Copernicus proposes to re-establish uniform motion, he will not succeed as regards the composite orbit of the planet; because the planet moves non-uniformly on its orbit, and in addition, this orbit deviates from a circle, as Copernicus himself admits.'

Then, no doubt remembering that Fabricius had not disputed this, he continued:

'However, you say that the fundamental elements by which this motion is realized, namely circles, are confined to uniform motion. I agree; but these motions do not agree with the phenomena. Furthermore, according to my theory, the fundamental elements which produce the planetary motion are similarly constant. The difference is to be found only in the fact that you use circles, and I use corporeal forces. Howbeit, the perfectly regular rotation of the Sun is constant as far as I am concerned; the rotation of the immaterial, magnetic *species* of the Sun is constant; the action of this *species* or motive force on the planet at a fixed distance is constant; the undoubtedly very slow rotation of the axis of the planetary body which causes precession of the apogees is constant and perfectly circular; the magnetic force which causes the bodies of the Sun and the planet to approach, or move away from, each other is constant, depending on the angle which the planet's axis makes with the direction [radius vector] towards the Sun. . . .

'Yet, when you say that there is no doubt but that all motions take place on a perfect circle, [I reply] that it is not so for compound motions, *i.e.*, true motions. In fact, according to Copernicus they take place, as I have already said, on an orbit which is enlarged at the sides, and according to Ptolemy and Tycho Brahe even on spirals. When you speak of the components of motions, you speak of something which is only imagination, and which does not exist in reality; for nothing performs the circuits in the sky except the body of the planet itself; [there is] no sphere, no epicycle; you who have been initiated into Tychonian astronomy should know that. Now, if we adhere to the fundamental claim that nothing moves except the body of the planet, the question then arises: what is the path traced out by the circumvolution of this body? To that, I reply, not conjecturally, but on the basis of knowledge founded on geometric proofs, that the orbit of the body will be oval, very much as given by Copernicus, who, however, gave motion also to epicycles and circles [spheres] besides the body of the planet. The simplicity of Nature must not be judged by our imagination.'

We now realize what it is that distinguishes Kepler from his predecessors and contemporaries. It is not the fact that he regarded the *true* orbit of the planet to be an oval and not a circle; it is the primordial importance which he ascribed to this fact; and he did so because astronomy for him was a science of reality, and should reveal to us that which truly takes place in the Universe. Secondly—we might even reverse the order of the factors—*real forces* and, consequently, a true mechanism, namely, that of oscillation, must be invoked by way of explanation; there cannot be recourse to the imaginary elements of a purely artificial structure.

Let us now return to the *Astronomia Nova*. In chapter LVIII, which was written a little before the letter just quoted—the date is probably October 1605—Kepler set forth the story of the great discovery in a slightly different manner; it is more concise, and more dramatic.[72]

According to the *Astronomia Nova*, Kepler, having verified the lack of agreement between calculation and observation, and being tired out, discarded oscillation on the diameter:

> 'I returned to the ellipses with the intention of applying an hypothesis quite different from that of oscillation.[73] The circle is too wide; the ellipse [auxiliary, in chapter XLV] is too narrow. The deviation in each case is of the same order; but only another ellipse can be placed mid-way between the circle and the ellipse. Consequently, the orbit of the planet is an ellipse, and the [maximum] width of the lune between it and the circle is one-half that of the former case namely, 000 429.'

In the circumstances, Kepler was immediately convinced. On the other hand, what he could not understand, in spite of all his efforts which nearly drove him to distraction, was the reason why the planet should prefer to describe an elliptical path instead of confining itself to oscillation.[74]

> 'Poor fool!' Kepler apostrophized himself; 'as if oscillation on the diameter could not lead us unequivocally to the ellipse . . .',[75]

as we shall see in chapter LIX of the *Astronomia Nova*, where it will be proved, not only that the elliptical orbit is the only one that can be adopted by the planet, but also that the area of the eccentric circle is a measure of the sum of the distances of the planet from the Sun.[76]

IX

Astronomy with the Ellipse

Chapter LIX, in which the first two laws of Kepler—those concerned with the ellipticity of planetary orbits and the areas described—are proved and related to each other, is undoubtedly the crown of the *Astronomia Nova.* Nevertheless, it must be admitted that it is more involved and confused than any other. At heart, Kepler agrees. Although he declares that the confusion of which he may be accused does not really exist; nevertheless, he finds it necessary to justify his having used 'non-geometrical' arguments besides other extremely complicated and subtle ones: he is dealing with difficult problems, and it is impossible to treat them in a simple manner; let him who will try to simplify Apollonius. . . .[1]

Kepler was certainly right to mention the difficulty of the problems he was trying to solve, and we can appreciate it even better than he, for we know full well that they could not be treated 'geometrically' with the means at his disposal. To do so, it needed the development of the infinitesimal calculus combined with the incomparable virtuosity of Newton.[2] The ponderous, awkward, involved character of his proofs cannot be held against him; nor the absurdity of some of his arguments: in fact, he ought to be admired for them. It is nonetheless true that Kepler achieved a connection between the two laws only at the price of terrible confusion, as it seems; and he even finished with a contradiction which, fortunately, he did not notice. The fact is that the real problem confronting him was insoluble as far as he was concerned, seeing that it was a question not only of finding a connection between the law of areas and the elliptical orbit, but also of deducing both of them from the dynamic law of distances[3]; that is to say, to deduce true propositions from false premises (which, after all, is possible, and Kepler did it frequently), but also by inference to link together propositions which are strictly speaking incompatible.

At the beginning of chapter LIX Kepler brings together a series of propositions relating to the ellipse which enable him to state positively that the corresponding elements of the ellipse and the circle are in the same ratio (Fig. 22, p. 275)—this was an important matter for him, because it was to be the middle term of his argument.

The first of these propositions, which Kepler calls 'pro-theorems', was borrowed from Apollonius *via* Commandinus, and states[4]:

> 'I. If an ellipse be inscribed within a circle so that its longer axis touch the circle at opposite points, and a diameter be drawn through the centre and the points of contact, and perpendiculars be dropped on to the said diameter from other points on the circumference; then all these perpendiculars will be divided by the circumference and the ellipse in the same ratio.'

The second proposition adds[5]:

> 'II. The surface of the ellipse thus inscribed in the circle bears the same ratio to that of the circle as the ratio in which the abovesaid lines are divided.'

The third proposition is particularly important, for it is the one which authorizes Kepler to replace the surface of a circle by the surface of an ellipse, and in this roundabout way to revert to the ellipse:

> 'III. If lines be drawn from a certain point on the diameter to the points of section of a given perpendicular with the circumferences of the circle and the ellipse, the spaces cut off by these lines [in the circle and the ellipse] will bear the same ratio to each other as the sections on the perpendiculars.'

The fourth proposition prepares the way from geometry to dynamics, and allows Kepler to 'explain' the retardation and acceleration (in respect of circular motion) of planetary motion on its elliptical orbit[6]:

> 'IV. If, by means of the abovesaid perpendiculars, the circle be divided into any number of equal arcs, the ellipse is [thereby] divided into unequal arcs; the ratio between those arcs near the ends of the longer axis is greater than that between those arcs in the median positions.'[7]

'Pro-theorem' VII is rather curious. Basically, it defines the ellipse, or rather, explains (with supporting proof) how to find the foci of a given ellipse. Naturally, Kepler did not yet make use of this term —he did not do so until the *Epitome Astronomiae Copernicanae*[8]— but limited himself to saying:

'If from the end-point of the shorter semi-diameter of the ellipse a line be drawn equal to the longer semi-diameter in such a way that its end-point be on the longer semi-diameter, then the square [raised] on the distance between this point and the centre will be equal to the gnomon which is formed between the square [raised] on the longer semi-diameter and [that which is raised] on the shorter semi-diameter.'

With 'pro-theorem' VIII we reach the crucial part of the exposition: the justification for using the computation of areas (surfaces) in the case of the elliptical orbit. This was a problem that Kepler had already touched upon many, many times in connection with the oval, when he asserted that the error involved in this computation in the case of the circle (an error resulting from the non-equivalence of the whole assembly of the eccentric distances and of the surface of the circle) disappeared in the case of the oval.[9] Only at this stage does he provide us with a 'geometrical' proof.

'Pro-theorem' VIII resumes, with further explanation, the statement in chapter XL: it states[10]:

'VIII. If a circle be divided into any number, or an infinity,[11] of parts, and if the points of division be joined with some point other than the centre within the circle, and if they be joined also with the centre; then the sum of the lines starting from the centre will be less than the sum [of those starting] from the other point. In fact, two lines drawn from an eccentric point to the opposite points on the circumference will be almost equal to the sum of the two [lines] drawn from the centre to the said opposite points when they are close to the line of apsides; on the other hand, two lines [drawn from the eccentric point to two opposite points] in the median positions will be [when taken together] much greater than [the sum] of those drawn from the centre to the same points.'

It follows that the surface of the circle cannot serve as a measure of the 'sum' of the distances; and that, we must not forget, means that it does not represent dynamic reality (determined by the 'distances') of planetary motions.

On the other hand, if we replace the distances between the eccentric point and the diametrically opposite points on the circumference, term by term, by their projection on the diameter joining them in the 'sum' in question, or, in Keplerian terminology, if we replace 'circumferential' distances by 'diametrical' distances, then the assembly of these lines will be equivalent to the surface of the circle. Or, to quote Kepler[12]:

'IX. But if, instead of the lines which [start] from the eccentric point we take those which are determined by the perpendiculars [drawn] from this point on to those [lines] passing through the centre [diameters joining the opposite points in question]—that is to say, if we take the diametrical distances instead of the circumferential ones, as they have been called in chapters XXXIX and LX—then their sum is equal to the sum of the lines which are drawn from the centre.'

'Pro-theorem' X[13] informs us that the surface of the ellipse is no better adapted to serve as a measure of the sums of the eccentric distances than the circle. If the ellipse be divided into any number of equal arcs, and if straight lines be drawn from some eccentric point to diametrically opposite points on the periphery of the ellipse, then their sum, exactly as in the case of the circle, will be greater than that of the lines (diameters) passing through the centre.'

With 'pro-theorems' XI and XII we arrive finally at the heart of the problem. In chapter LVI,[14] Kepler had shown (a) that if the distances of Mars from the Sun, computed for an hypothetical circular orbit, were shortened in the proportion of the *secant* of the radius of the optical equation, then the true distances corresponding to an elliptical orbit were obtained; and (b) that these distances were equal to the projections of the 'circumferential' distances on to the corresponding diameter ('diametrical' distances). The argument now proceeds in the opposite sense, and Kepler shows that by replacing the 'circumferential' distances by 'diametrical' distances, that the ellipse is changed into a circle, seeing that the 'circumferential distances' of the latter are exactly equal to the 'diametrical distances' of the former.[15]

'XI. I can now pass on to the proof. If, in an ellipse divided by perpendiculars dropped [on to the diameter] from equal arcs of the circle (as in pro-theorem IV), the points of division on the circle and on the ellipse be joined with the fixed point of pro-theorem VII,[16] then, I say, the lines which are drawn to the circumference of the circle are circumferential; but those which [are drawn] to the periphery of the ellipse are diameters which correspond to an equal number of degrees [reckoned] from the apside of the epicycle. This means, that in Fig. 22 the lines *MN* and *NY* are equal to the lines *KT* and *TJ* respectively.'

Whence it follows—and this is of far-reaching consequences—that the assembly ('sum') of the 'distances' from the focus of the ellipse to its 'circumference' is now exactly measured, whereas, up to now we knew that it was only 'larger' than its surface: it is strictly equal to that of the circle having the major axis as diameter. Seeing that (according to 'pro-theorem' III) the ratio of the corresponding parts of the ellipse and the circle is always equal to the ratio of the former to the latter[17]:

> 'It follows [from "pro-theorem XIII"] that the surface of the circle, taken as a whole, as well as in its individual parts. is a true measure of the sum of the distances from the Sun of the arcs on the elliptical orbit of the planet.'

Hence, the computation of 'surfaces' is established as being perfectly accurate for a planet moving in an elliptical orbit, subject to the requirement that the surface being measured is not an ellipse, but a circle[18]; and also by the requirement that the arcs of the circle forming the bases of the surfaces in question shall be determined, not by producing the radius vector drawn from the Sun to the planet until it intersects the circle, nor by an intersection with the circle of lines drawn from the centre of the ellipse to the planet (*i.e.*, radii of the circle), but by the intersection with the circle of perpendiculars raised on the major axis of the ellipse and passing through the points occupied by the planet in its orbit.

In other words, it is the sector *AKN* which will measure the 'sum of the distances' from the Sun to the planet moving along the arc *AM* of the ellipse. Now, the times being proportional to the distances, the surface *AKN* will be a measure of the *time* taken by the planet to move from *A* to *M*; which means that the ratio of this time to the total duration of the planet's revolution round the Sun will be the same as that of segment *AKN* to the total surface of the circle *AKECJ*. Seeing that the corresponding parts of the circle and the ellipse bear a constant ratio to each other, it follows that the ratio of the segment *AMN* of the ellipse to the surface of the ellipse is the same as that of the segment *AKN* of the circle to the surface of the circle; and, therefore, the surface of the segment *AMN* of the ellipse is an exact measure of the time taken by the planet to traverse the arc *AM*. This is undoubtedly a highly satisfactory result, seeing that it finally gives a 'geometrical' proof of the law of areas. It was nonetheless rather surprising to Kepler—how is it possible to accept, in effect, that two different surfaces, one of which encloses the other, can both of them serve as a 'measure' of one and the same phenomenon? Or, more precisely, how is it possible to accept that each of these surfaces can correspond to the same 'sum' of 'distances'?

Consequently, he felt obliged to discuss and justify the matter in a long and curious 'dubitation'.

Max Caspar in his valuable commentary to the *Astronomia Nova* considers that Kepler had not noticed that the law of areas was an immediate consequence of his geometrical reasoning, and as a result created unnecessary difficulties for himself.[19] All that he had to do, was to repudiate the law of distances, which was obviously incompatible with the law of areas. Had not he himself, sometime before, ascertained that these two laws are not identical? The law of areas was confirmed by observation. Therefore, it was necessary to abandon the other, and not to try and make them agree.[20]

No doubt! but, as Max Caspar himself moreover admitted, Kepler could not do so, seeing that the law of distances formed the very basis of his dynamics. I would add, that he was even less able to do so, because it was by using this law as a starting point that he had deduced the law of areas. Therefore, he was obliged to demonstrate their equivalence; that is to say, show that the surfaces *AKN* and *AMN* are, both of them, equivalent measures of the time of travel; or, what is the same thing, according to the time measured by the surface *AKN*, the planet will reach the point *M* (on the line *KL*), and not a point situated further to the left on the line *KH*, as it seems natural to believe,[21] and as he himself believed.

Basically, Kepler's remarks are only repetitions of what he had said, more briefly, previously. If the periphery of the ellipse be divided into equal parts, or if equal parts on the circumference of the circle be co-ordinated with those parts on the periphery of the ellipse fixed by the radii of the circle, then the assembly of distances from the focus of the ellipse to the terminal points of these small arcs will not correspond to the surface of the sector *AKN* of the circle; the surface *AMN* will not be in the required ratio to it; the law of areas will not be applicable. This appears from the following consideration[22]:

> 'Let us assume the orbit *ABC* of the planet to be divided into equal parts. The planet will remain that much longer on the arc near to *A* than on the arc near to *C* in the proportion that *NA* is greater than *NC*. Now, *NA* and *NC* together form the major axis of the ellipse, whereas *HB* is its semi-axis minor. That being so, the sum of the times taken by the planet to traverse the arcs at *B* and at the diametrically opposite point will be shorter than the sum of the times required to traverse equal arcs at *A* and at *C*. In order for the times of travel on arcs near to *A* and *C* (taken

> together) to be shorter, and those on the arcs near to *B* and its opposite point to be longer, it is necessary for the arcs near to *A* and *C* to be shorter and [for the arcs] near to *B* and its opposite point (taken together) to be longer. This is precisely what is achieved by the perpendiculars *KML*',

that is to say, by the perpendiculars dropped on to the major axis of the ellipse from the end-points of the equal arcs into which the circumference of the circle *AKC* was initially divided; also, if the end-points of all the unequal arcs resulting from this operation be joined to the foci of the ellipse, then the surface of the ellipse *AMN* can be substituted for the assembly of lines drawn from all the points such as *M*.

Taken literally, this 'reasoning', in which lines and surfaces are inextricably mixed and confused, is quite simply false, or even worse than that. Yet, in its very confusion, it is characteristic and revealing, and really seems to confirm the impression produced on us by similar kinds of reasoning on the part of Kepler which we have already encountered many times.[23] It seems to me, that the matter could be summarized by saying that there is a deep-seated duality in the Keplerian mind: the geometer does not think quite like a physicist.

When dealing with pure geometry, Kepler naturally thinks like a geometer: points are then points; straight lines and curves are 'real' straight lines and curves. However, the situation changes when he is dealing with physics, or with astronomy: points now become 'physical' points; straight lines become lines of force, which like rays of light, are not one-dimensional entities. When we came to motion, the situation changes even more, for motion implies displacement and time. Now, Kepler thinks of motion as a function of time far more than as a function of space, as we have already frequently noticed; and it is for this reason that of the two expressions for the fundamental law of celestial dynamics (law of distances), expressions which are formally equivalent, namely: (a) *velocity* inversely proportional to distance; and (b) *time* directly proportional to distance, Kepler definitely preferred the second. Thus, at least in the *Astronomia Nova*, he never thinks of *instantaneous velocity*—a paradoxical notion which implies negation as well as displacement of time—but always of the velocity of a body on an element of its orbit, or, better still, the *time* in which a body traverses this element. He always speaks of the time taken by a body to traverse a given arc—small or large—or of the time during which the body 'rests' in a position situated at a given distance from the eccentric point of the circle, or of the ellipse; a time which, strictly speaking, is equal to zero. For us, Yes! but for Kepler, No! For him, an interval in time will never be nothing, any more than an interval in space will be nothing; because the 'point' in question is not a point: it is an element of duration, or of interval. It has dimension. It is an 'indivisible'[24]; or, to use non-Keplerian terminology, it is a differential.

It is undoubtedly for this reason that Keplerian 'distances' are not geometrical lines joining the point of origin to points on the orbit (circular or elliptical); in the first place, they are the 'distances' between the Sun and the planets; and, in the second place, they are the 'distances' between the point of origin and the terminal points of a path, or of an element of the path. It is for this reason, also, that they nearly always occur in pairs: for even when Kepler changes the conditions back again by starting from the orbit instead of the 'distances', it is from the end-points of the equal, or unequal, arcs, into which he has divided the curves to be compared, that he draws his 'distances'. These portions of arc may be *finite*, or *infinite*, in number, as he has just said: but that does not alter the matter. It is obvious that these portions of arcs still remain arcs, even if they be infinite in number. No doubt, they are infinitely small, but they are still arcs. They do not become points; and so will be capable of having equal and unequal magnitudes.

These are the first mumblings of the infinitesimal calculus: the first indication of the concepts and methods—so bold and fruitful, yet logically so confused—of the *Nova Stereometria Doliorum Vinariorum*.[25]

However, let us return to Kepler's 'proof'. 'Pro-theorem' XIV recalls[26] that if the ellipse *AMC* be divided into any number of equal arcs, if the distance from *N* be assigned to each of them, and if the surfaces *AMN*, *ABN*, *ABCNA* be taken as the sum of the distances of *N* from the points of the arcs *AM*, *AB*, *ABC*, then the same mistake will be made as in chapter XL, when trying to apply this method of computation to the perfect circle: the result obtained will be too large.[27] On the other hand, if the ellipse be divided into the same number (equal to the number of parts in the previous operation) of unequal parts by first of all dividing the circle into equal arcs (contrary to the method in 'Pro-theorem' X), and then, if perpendiculars *KL* to the major axis *AC* be drawn from the end-points of these arcs so that the ellipse is divided into the same number of unequal arcs, and if, finally, the surfaces of the corresponding sectors of the ellipse be taken as the sum of the distances *N* from the points of these arcs, then the error in question will be completely eliminated and fully compensated. The next two pages are devoted to the proof, by 'subtle' (yet confused) considerations, to show that this compensation is equally valid at the apsides, where, as a result of the greater 'contraction' of the arcs of the ellipse in these regions, the 'distances' which link them to the focus are 'denser' than in the median parts of the orbit, where they are 'rarer'.[28]

This being accepted, Kepler then proceeds to the final proof ('Pro-theorem' XV) of the fact that[29]:

> 'The arc of the ellipse, whose time [for the distance travelled] is measured by the surface *AKN*, must terminate on [the perpendicular] *LK*, and therefore must be *AM*',

by assuming as the major premise of his reasoning the validity of the law of areas for an ellipse, the validity of which he had just established correctly. The minor premise was given once more by 'Pro-theorem' III, where it is proved that:

> 'The surface *AKC* is related to the surface *AMC* in the manner as the surface *AKN* is related to the surface *AMN*. As the ratio of equimultiples is the same, it follows that the surface of the circle *AKN* is the measure of the sum of as many diametrical distances (such as *KT*, *TJ*), or distances on the ellipse from *AM*, as there are parts in *AK*. Whence it appears reasonable that a larger quantity (density) of distances can be ascribed to those parts of the ellipse near to *A* and *C*, namely, as many as there are sections formed on the perimeter of the ellipse by the perpendiculars *KL* drawn from the equal arcs of circle *AK*.'

Kepler seems to be conscious that all his 'subtle' considerations did not add much to his original argument. He continued as follows[30]:

> 'However, in order that anyone who feels perplexed by the subtlety and complexity of [my] arguments shall not be in doubt as to the truth of this proposition [he must know] that it was first revealed to me by experience, and in the following manner. I had calculated the particular degrees of the eccentric anomaly by taking the diametrical lines *KT*, *TJ* instead of the [circumferential] distances from *N*; then, I added them together, taking them in order, and adding each degree to the sum of the preceding ones. This sum total was 36 000 000, as it should be. Now, when I compared the partial sums with the total sum in such a way that (according to the law of proportions) the sum of 36 000 000 had to 360° (the

arbitrary designation of the total time of revolution) the same ratio as each of the partial sums to its corresponding lapse of time: I obtained exactly the same result—even to seconds—as if I had multiplied one half of the eccentricity by the *sine* of the eccentric anomaly, and compared [the product] with the surface of the circle, which, similarly, would have the value 360° (the arbitrary designation of the total time of revolution). However, as I believed that the true distance *MN* should have its end on the line *KH*, that is to say, that it was *ZN*, and that I had consequently determined the true anomaly corresponding to the mean anomaly *AKN* as being *ZNA*, the equations deviated very decidedly from the data of my *hypothesis vicaria* . . . in the region of 45° the difference between the true anomaly and truth was over 5½′; and in the region of 135° it was about minus 4′. Yet, when I applied *AM* in such a way that it terminated on *KL*, and that the true anomaly, corresponding to the mean anomaly *AKN*, was *MNA*, my calculation agreed precisely with the *hypothesis vicaria, i.e.*, with observation. As I was then certain of the fact itself, I was constrained to seek a causal explanation of it starting from the [dynamic] principles admitted at the beginning [of this work], an explanation which I have set down for the reader as clearly and skilfully as possible. Still, if anyone believe that the obscurity of this analysis derives from the confusion of my mind, I, for my part, shall admit having committed the fault of not having wished to leave the questions [dealing with the causes of celestial motions] without investigation, even though they be most obscure and not really necessary for the exercise of astrology in which many [people] see the sole purpose of celestial philosophy. Howbeit, with regard to the content [of my investigations], I should like to ask my critics to read the *Conics* of Apollonius. They will then realize that there are subjects which no favourable disposition of the intellect can make so clear that they can be treated in such a way as to be understood by casual reading.' Kepler then concludes triumphantly by saying: 'If the physical causes, which I had assumed from the start to be explanatory principles, were not true, they would never have been confirmed throughout an investigation so acute as this.'

Chapter LX, the last of Book IV of the *Astronomia Nova*, sets forth the method of finding the two parts (physical and optical) of the equation, as well as the true distances of the planet from the Sun, 'starting from the physical hypothesis (of the ellipse) of planetary motion, which is most true and conformable to reality. It was not possible to do this, heretofore, together and at the same time by using the *hypothesis vicaria*.' Thus, the truth of the new hypothesis and the falsity of the *hypothesis vicaria* were confirmed.[31]

The contents of this chapter LX are too technical to be studied in detail; but Kepler's technique in itself is so interesting and full of meaning that it seems really necessary at least to give an example of it.[32]

'We have assumed (in chapters LVI, LVIII and LIX) that the planet [oscillates] on the diameter [radius vector] directed towards the Sun, and it is as a result of approaching, and moving away from, the Sun that the elliptical orbit is realized, and that the planet remains at a particular point in

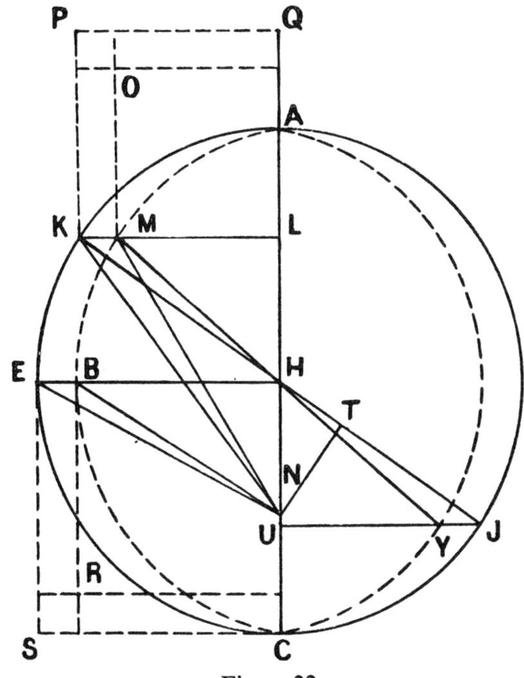

Figure 22.

its orbit for a time that corresponds to the distance of the point in question from the Sun; furthermore it is very convenient [to use] the procedure that I have developed in chapter LIX to make a rapid summation of any number of the times of travel. For it has been proved, that if a perpendicular be dropped from a point on the circle on to the major axis of the ellipse inscribed within the circle (*e.g.*, in the accompanying diagram the line *KL* is drawn to *AC*) so as to cut the ellipse in *M*, and if the Sun be placed at *N*, then the sum of all the distances from the Sun *N* to the points of the arc *AM* will be contained in the area *AKN*. (Fig. 22.)

'If, then, we take the arc *AM* of the ellipse, which is determined by the [given] arc *AK* of the circle, the area *AHK*, which is the sector of arc *AK* is [similarly] given, and this arc determines its measure with respect to the whole circle taken as equal to 360°.

'Now, if the arc *AK* be given, the *sine KL* is given also; but the area *KHN* bears the same ratio to the surface *HEN* (as was proved in chapter XL) as *KL* to the *sinus totus* (radius *EH*): and, as the eccentricity *HN* is given, one half of it multiplied by *HE* will determine the area *HEN*, whose size is [must be] calculated at the start, once and for all, for the purpose of knowing the value (size) of this small surface [*HEN*] if the area of the whole circle equals (corresponds to) 360°.

'Once the surface *HEN* is known, it is quite easy, by the rule of proportion, to find the area *HKN*. In fact, the surface *NEH* bears the same ratio to the surface *NKH*, or to its value in degrees, minutes and seconds, as does *EH* to *KL*. If we add it to the value of *KHA*, we obtain *KNA* as a measure of the time that the planet spends on *AM*. This constitutes one part of the equation which I call (a) *physical*, namely, the area *AKN*; however, I construct tables in such a way that this equation has no need to be mentioned, and there is no separate column giving (b) the *optical* part of the equation, namely, the angle *NKH*. Moreover, the terms, mean anomaly, eccentric anomaly and true anomaly are more familiar to me. (c) *The mean anomaly* is the [mean] time arbitrarily determined, and its measure is the area *AKN*. (d) *The eccentric anomaly* is the path (orbit) of the planet from apogee, namely, the arc *AM* of the ellipse, or [the arc of the circle] which determines it, namely, *AK*. (e) *The true anomaly* is the apparent size of the arc *AK* [seen] from *N*, namely, the angle *ANK*.

'The angle of the true anomaly is obtained as follows: if the arc *AK* be given, its *sinus complementi* [*cosine*] *LH* is given also. The ratio of the *sinus totus* [radius] to *LH* is equal to that of the total eccentricity to the part [fraction] which must be added to 100 000 (or subtracted, if less than 90°) in order to obtain the true distance of Mars from the Sun, *i.e.*, *NM*. Now in triangle *MLN* the angle whose apex is *L* is a right angle, and [therefore] *MN* is given, as well as *LN*. In fact, *LN* is composed of *LH*, the *sinus complementi* of the distance [arc] *AK* starting from apogee, *i.e.*, of the eccentric anomaly, and of the eccentricity *HN*. Under 90°, the difference between *LH* and *HN* must be taken instead of their sum, and instead of the complement of the eccentric anomaly, its excess must be taken; nor will the angle *LNM* of the true anomaly remain unknown. It is easy to see what change [in this procedure] must be made for the other half of the circle.

'*Vice versa*, if the eccentricity and the true anomaly be given, it is possible to determine the value of the eccentric anomaly. However, the procedure—whether we follow the "demonstrative" [geometrical], or the analytical, method—is rather more complicated and laborious.[33] On the other hand, if the mean anomaly be given, there is no geometrical method for deriving therefrom the true anomaly, or the eccentric anomaly; for the mean anomaly consists of two parts [surfaces], one of which is a sector and the other is a triangle. Now, if the former [the sector] be measured by the arc [of the eccentric circle] we obtain the second [the triangle] by multiplying the *sine* of this arc by the value of the larger triangle, and neglecting the last [figures]. However, the ratios between arcs and their *sines* are infinite in number. Therefore, if their sum be given, it is impossible to say what is the value of the arc and what is the value of the *sine* corresponding to this sum, unless we first of all ascertain which surface corresponds to a given arc; that is to say, we must construct tables and then use them *a posteriori*.'

That is my opinion, Kepler continued, adding that if this semi-empirical method of computing be singularly lacking in geometrical elegance, then this defect prompts him all the more to exhort mathematicians to provide the solution to the following problem[34]:

'*The area of part of a semi-circle being given, together with a point on the diameter, to find the arc and the angle corresponding to this point, taken as apex, whose sides and the arc contain the area in question.*' Or, otherwise expressed: '*To divide the area of a semi-circle from a given point in a given ratio.*' Kepler concluded by saying, 'It is enough for me to believe that [this problem] cannot be solved *a priori*, on account of the heterogeneity of the arc and the *sine*; and if someone would point out my mistake and show me the way [to do it], he would be a great Apollonius in my opinion.'

However, we cannot go any further until this new Apollonius appears on the scene. Therefore, Kepler considered that he had done everything that it was humanly possible to do; and that was more—much more—than anyone before him had done. In fact, he had succeeded in observing the unobservable planet, and to build a theory of its motion. So, he addressed a veritable victory bulletin to the Emperor Rudolph II; together with an appeal for financial aid to continue the war against the rest of the celestial army.[35]

'I bring to Your Majesty a noble prisoner whom I have captured in the difficult and wearisome war entered upon under Your auspices. I am not afraid of his refusing the name of captive, nor of his being indignant thereat, for this is not the first time that he has borne it. The terrible god of war, on a former occasion when he bitterly put aside his shield and arms, allowed himself to be captured in the net of Vulcan. . . .

'Hitherto, no-one had more completely got the better of human inventions; in vain did astronomers prepare everything for the battle; in vain did they draw upon all their resources and put all their troops in the field. Mars, making game of their efforts, destroyed their machines and ruined their experiments; unperturbed, he took refuge in the impenetrable secrecy of his empire, and concealed his masterly progress from the pursuits of the enemy. The Ancients experienced this many times; and Pliny, the indefatigable explorer of the mysteries of Nature, said: *Mars is an unobservable heavenly body*. . . .[36]

'For my part, I must, above all, praise the activity and devotion of the valiant captain Tycho Brahe, who, under the auspices of Frederick and Christian, sovereigns of Denmark, and then under the auspices of Your Majesty, every night throughout twenty successive years studied almost without respite all the habits of the enemy, exposing the plans of his campaign and discovering the mysteries of his progress. The observations, which he bequeathed to me, have greatly helped to banish the vague and indefinite fear that one experiences when first confronted by an unknown enemy. . . .

'During the uncertainties of the combat, what disasters and what scourges have there not been to desolate our camp. The loss of an illustrious chief, sedition of the troops, contagious diseases, all these have contributed to augment our distress. Domestic joys and misfortunes took their toll of time from these concerns; a fresh enemy, as I report in my book *De Stella Nova*, fell on the rear of our army.

'The soldiers, deprived of everything, deserted in mass; the fresh recruits were not familiar with the manœuvres, and as a crowning misfortune, provisions were lacking.[37]

'Finally, the enemy resigned himself to peace, and through the medium of his mother, Nature, he sent me his admission of defeat, and surrendered himself as a prisoner on parole; Arithmetic and Geometry brought him without resistance to our camp.

'Since then, he has shown that we can trust his word; he requests only one favour of Your Majesty: all members of his family are in the heavens; Jupiter is his father, Saturn his grandsire, Mercury his brother and Venus his love and sister. Being accustomed to their august society, he misses enjoying with him, as he does today, the hospitality of Your Majesty. To do so, the war must be pressed with vigour; there are no further risks now that Mars is in our power.

'Yet, I beseech Your Majesty to remember that money is the sinew of war, and to have the bounty to order Your treasurer to deliver up to Your general the sums necessary for the raising of fresh troops.'

Kepler's eloquent appeal went unheeded. In the affairs of this world, armies carry greater weight than the heavenly hosts. The Emperor's treasurer had difficulty in paying the former; he could not undertake a planetary war. He was even unable, at least immediately, to supply the sum needed to hallow the victory over Mars, that is to say, to pay for the printing of the *Astronomia Nova*. It was completed in 1606, but did not see the light of day till three years later, in 1609.[38]

III. FROM CELESTIAL PHYSICS TO COSMIC HARMONY

I

The *Epitome Astronomiae Copernicanae*

In his letter of dedication to Emperor Rudolph II, Kepler claimed for himself no more than the credit of having gained the victory in the war against Mars. He certainly deserved very great credit for this victory, which, from Kepler's point of view, was of secondary importance. As we have seen, a study of the motions of Mars was only a means of arriving at something much more important, namely, the development of a new astronomy, that is to say, of a celestial physics whose laws, revealed by a study of the motions of Mars, would nevertheless be of general utility and be capable of application to all the planets without exception.[1] This is the celestial dynamics which we find set forth in the *Astronomia Nova* in an almost definitive manner; his later works, the *Harmonice Mundi* and the *Epitome Astronomiae Copernicanae*, include an addition of prime importance to Kepler's astronomical work, namely, the discovery of his 'third law' and his archetypical and physical explanation, though they change little in the dynamical sub-structure. The *Epitome* stresses some aspects of Celestial Physics; introduces more accurate information on some features[2]; completes others; systematizes and simplifies the presentation and proof of various theorems[3]; but does not make any essential modifications to the concepts of the *Astronomia Nova*, except in one particular, which will be dealt with later. Consequently, we can spare ourselves and the reader, too—at least to some extent[4]—all the repetition which a detailed study of this work would entail.

Nevertheless, the differences between the celestial physics of the *Astronomia Nova* and the *Epitome* have a certain interest for us, who are concerned with the development of Kepler's thought. The *Epitome Astronomiae Copernicanae*[5] is the last of Kepler's great works; it is the most mature,[6] the most systematic,[7] the most carefully elaborated.[8] Furthermore, he sets forth his discoveries within the framework of his general conception of the Cosmos, and not in the relative isolation of a technical work on astronomy, as he did in the *Astronomia Nova.* Consequently, all the archetypical and other arguments which played so great a part in the *Mysterium Cosmographicum*, and to which Kepler made only passing reference in the *Astronomia Nova*, now reappear; and with even greater force. That which strikes the reader of the *Epitome* most of all, is Kepler's faithfulness to Christian Pythagorism (or Neo-Platonism). Indeed, like a fire that has remained hidden in the cinders without having lost its heat, it bursts forth again in the *Harmonice Mundi* as well as in the *Epitome*, where the two fundamental theses of the *Mysterium Cosmographicum*—the concept of the Universe as an expression of the Holy Trinity, and the concept of the 'archetypical' character of the five regular solids—are found once more. We might go even further and say that the expression 'faithfulness' is inadequate, and that Kepler, in his later works, shows himself to be more Pythagorean than ever. Having realized the impossibility of securing exact agreement between the scheme of the five regular bodies and observational data, he subsequently treated the scheme as a mere approximation, though he did not abandon it at all: he completed it by, and subordinated it to, a higher, and more Pythagorean, archetypical structure, namely, that of numbers and harmony.[9] On the other hand, at the beginning of Book IV of the *Epitome*, which book is devoted to an exposition of the 'celestial physics', Kepler says[10]:

> 'The philosophy of Copernicus allots the principal parts of the Universe to regions distinct from the latter. As in the case of the sphere, which is the image of God, the Creator and Archetype of the Universe (as was proved in the first book[11]), there are three regions, which are symbolic of the three persons of the Holy Trinity—the centre symbolizes the Father; the surface, the Son; and the intervening space (*medium*), the Holy Ghost—in the same way there are three principal parts of the Universe, each being in a particular region of the (total) sphere: the Sun is at the centre; the sphere of the fixed stars is on the surface; and, finally, the planetary system is in the intermediate region between the Sun and the fixed stars.'

This is what Kepler had already discovered twenty-five years before. As will be seen, he had made some progress in the meanwhile: so he explains[12]:

> 'The perfection of the Universe consists in light, heat, the motion and harmony of its motions, which are similar to the faculties of the soul:

light resembles the sensitive, heat the vital and natural, motion the animal, harmony the rational. Most certainly, the beauty of the Universe is found in light, in heat, in life and vegetation; in motion a kind of *quasi*-power [is found]; and in harmony, contemplation wherein Aristotle found bliss. Now, as in any affection three things are necessarily united, the cause "by what means" (*a qua*), the subject "in what" (*in quo*) and the form "under which" (*sub qua*): therefore, the Sun, with respect to all the affections of the Universe plays the part of the efficient [cause]; the region of the fixed stars [that] which gives form, has capacity, and sets a boundary; the *intermedium* plays the part of the subject. . . . For all these reasons the Sun is certainly the chief body in the whole Universe. Indeed, as regards light, the Sun itself is, on the other hand, the most resplendent [body] and, as it were, the eye of the Universe; on the other hand, as the source of light, or its focus, it illuminates, colours and embellishes the other bodies of the Universe. The *intermedium* in itself is not luminous, but is pellucid and transparent; it is the path by which light is distributed from the source; it contains the globes and all the creatures which enjoy the light poured on them from the Sun. The sphere of fixed stars serves as the bed in which this river of light flows; it resembles an opaque, illuminated wall which reflects and enhances the light from the Sun. It could be very well likened to a lantern which excludes draughts. The brain, which is the seat of perception in animals, communicates to the whole animal all its senses and by the action of a collective sense ensures that they are all ready to hand, by stimulating them in some way and ordering them to keep watch. In this comparison [of the Universe with an animal] the Sun behaves according to its own collective sense; the globes in the intervening space according to the perceptive organs; and the fixed stars according to the objects perceived.

'As regards heat, the Sun is the hearth of the Universe at which the globes in the intervening space warm themselves; the sphere of the fixed stars retains the heat so that it shall not be dissipated—with respect to the Universe it plays the part of a protective wall, of a fur, or of a coat, if I may use the imagery of the Psalms of David. The Sun is a fire, as was claimed by Pythagoreans, or otherwise an incandescent stone or mass, as Democritos [claimed]; in comparison therewith, the sphere of fixed stars is composed of ice, or otherwise is a crystalline sphere. If some vegetative faculty belong to the whole aetheric aura throughout the full extent of the Universe, as well as to terrestrial creatures—a conjecture which is suggested as much by the manifest power of the Sun to inflame as by physical considerations concerning the origin of comets —we ought then to believe that this faculty is rooted in the Sun as the heart of the Universe, and that it is propagated (being carried by light together with heat) throughout the immensity of the Universe; in like manner the seat of heat and of the vital faculty of animals is found in the heart, and the seat of the vegetative [faculty] is found in the stomach; whence, by the aid of spirits, these faculties extend to, and penetrate into, the other members of the body; the region of the fixed stars assists this growth by concentrating the heat like a skin covering the Universe. . . .

'As regards motion, the Sun is the prime cause of planetary motions and is the prime motive force of the Universe on account of its bodily nature. The moving [bodies], *i.e.*, the planetary globes, are placed in the intermediate space; the region of the fixed stars provides, as it were the moving bodies with the place and base on which they rest; the [base] itself is motionless and motion is reckoned by comparison with it.[13] In the case of animals, the brain is the seat of the motive faculty; the body and its members are given motion. The Earth provides the base for the animal's body; the body for the arm or the head; the arm for the fingers; these are the bases, in themselves motionless, on which depends the motion of each part.

'Finally, as regards the harmony of the motions, the Sun occupies the one position from which the planetary motions give the appearance of harmoniously regulated masses.[14] The planets, which move in the intermediate space, form the objects or terms between which harmonic relationships exist. The sphere of the fixed stars or the circle of the Zodiac provides a measure of the amount of apparent motion. Similarly, in the case of man, it is the intellect that abstracts the Universals and forms numbers and proportions which have no existence outside the intellect.[15] However, individuals endowed with senses are the foundation of the Universals; the individual, discrete units [are that] of numbers; the real terms [are that] of proportions; finally, memory, which is (as it were) the place of quantity and time, has a certain resemblance to the sphere of fixed stars, as the seat and repository of sensations. Furthermore, sensations never give rise to a judgement that is not in the mind; the *affectus* of joy (pleasure) never engenders a sensual perception that is not in the heart.

'Therefore then, the vegetative [power of the aether] corresponds to the nutrition of animals and plants; heating corresponds to the vital faculty; motion to the animal faculty; light to the sensitive faculty; harmony to the rational faculty. It is on this account that the Sun has the best claim to be taken as the heart of the Universe, the seat of life and reason, and the chief of the first three members of the Universe; and, in the philosophical sense, there is truth in the panegyrics by which the poets extol the Sun calling it the king of the heavenly bodies; the Sidonians and Chaldeans, the queen of the skies (a turn of speech also found in German)[16]; and the Platonists, the king of intellectual fire.'

Having fully demonstrated the eminent nobility of the Sun, Kepler continued by explaining that its central position results, more or less, automatically, even without invoking Trinitarian symbolism. Indeed, as the Ancient Pythagoreans recognized and as Copernicus did, too, in his turn, the central position, notwithstanding everything said about it by Aristotle,[17] is definitely the most honourable one and is accorded to the Sun in virtue of its perfection: being the source of life, heat, light, it could not be placed anywhere else except at the centre of the Universe.

The same argument applies in respect of the part it plays as the source of motion; a part which necessarily implies a central position, because the Sun by rotating on itself causes the planets to move—this rotation is no longer deduced, as in the *Astronomia Nova*, from planetary motions, but is, on the contrary, a basic fact, which can be verified by telescopic observation of sun-spots.[18]

That the Sun is the source of planetary motions is proved *a contrario*, by the impossibility of finding another cause; particularly, of explaining their motions by the agency of souls or intelligences belonging to the planets—the absence of solid spheres excludes translatory motion by the souls. In fact[19]:

> 'Aristotle himself readily admits that a [celestial] body could not be transported from place to place by a soul, if the latter were deprived of the contrivance of the sphere occupying the whole circuit, and if there were not in addition some motionless body on which the sphere rests.'

As for the intelligences:

> 'It is not possible for the planetary globe to be moved in a circle by an intelligence alone; for the spirit, lacking sufficient animal power to produce motion, does not have the motive force on which to call, and cannot be understood and perceived by the inanimate globe, which furthermore does not possess the faculty of obeying orders from intelligence, or of moving of its own accord.'

If it be objected that the absence of spheres guiding the planets makes it all the more necessary to have the controlling action of planetary intelligences,[20] then we shall be completely deceived. Such action would be both impossible and pointless: for[21] 'even if it were assumed that it sufficed for the intelligence to wish to move [the body] in such and such direction in order to produce motion, the production of orbital motion would be nevertheless impossible.' In the first place, if the motion of the planets were the result of the operation of intelligences, they would move on circles, and indeed on concentric circles,[22] and not on ellipses as they in fact do. Furthermore, they would be incapable of tracing an eccentric orbit (not even a circular one) in the skies; the possibility of an elliptical orbit would be even less.[23] On what would they be able to regulate their motions?

> 'On the contrary, the elliptical shape of the planetary orbits and the laws of motion by which such a figure is traced, reveal more of the nature of balance and material necessity than of the concept and determination by a mind.'[24]

We repeat. In order to move material bodies—especially when it is a question of translatory motion—corporeal forces must be available to act upon them. Heavenly bodies are no exception to this rule. We have seen that that was Kepler's first conviction, the one that guided him to the establishment of celestial physics, and that enabled him to explain the existence of differences in the periodic times of planets,[25] by introducing the relationship between motive force and resistance (inertia). Starting from this relationship, the *Epitome* succeeds in deducing the law, discovered in the meanwhile, which relates the periodic times to the distances of the planets from the Sun. Kepler's persistence in striving to prove his law of inertia is also understandable; it is the law which he applied subsequently, not only to translatory motion, but also to rotation in position, which was something he had rather neglected to take into consideration in the *Astronomia Nova*[26]:

> '*Why do you say that a celestial body, composed of matter, cannot be moved circularly merely at the order* [*of a mind*]*? In fact, celestial* [*bodies*] *are neither heavy nor light, but particularly fitted for circular motion, and consequently do not resist the motive mind.* Even if a celestial globe be not [heavy] in the manner that a stone is said to be heavy on this Earth, nor light like fire, nevertheless, on account of its material nature, it has a natural inability to move itself from place to place; it has inertia or natural immobility in virtue of which it remains at rest in any place where it is set by itself. Consequently, in order for it to be able to abandon its position and rest, a certain force is required; and this force must be greater than [that of] its substance and bare body, and be able to overcome its natural inertia. Now, such a faculty is beyond [the power] of an intelligence; [it is] an indication of life or the companion of substance.[27]
>
> '*How do you prove that the substance of celestial bodies resists their motive forces and is overcome by them, as in the case of a balance the weights* [*are overcome*] *by the motive faculty*?—It is proved, *primo*: by

the periodic times of rotation of the individual globes on their axes[28]; thus, the Earth rotates daily, the Sun rotates in about 25 days. If the inertia of the matter composing a celestial globe, which is equivalent to a certain weight, were nothing, then there would be no need of any force to move the globe: and if the smallest motive power were ascribed to it, there would be no reason why the globe should not rotate immediately. However, seeing that the changes undergone by the globes are accomplished within fixed times, which are longer for one planet, and shorter for others, it is obvious that the inertia of matter is not related to the motive force as nothing to something. *Secundo*: it is proved by the circumvolution of the globes as a whole round the Sun; for the Sun is a motive force, unmatched in its excellence, and by its unique rotation moves the six globes, as will be explained later. Now, if the globes had no natural resistance [to motion], and this were not in a fixed proportion, then there would be no reason (cause) for them not to follow the rotation of their motive force very exactly and not to rotate with it in the same period of time. In fact, they all move in the same direction as that in which the motive force rotates; however, not one of them fully attains the speed of its driving force, and they succeed one another more and more slowly. Thus, they combine the swiftness of the motive force with the inertia of their own substance.'

Moreover, this same inertia of matter explains the regularity of planetary motions by replacing the unavailing, impossible spheres: for if there were no spheres[29]:

'*By what means do you suspend the globes, composed of material substance as you admit, and especially the Earth, so that each of them remains within the limits of its own region, even though deprived of bonds with the solid spheres?*—Seeing that the solid spheres certainly do not exist, it is necessary to have recourse to the inertia of matter, which causes each of the globes to remain at rest in whatever part of the Universe it be placed (motive forces being neglected); this results from the fact that matter *qua* matter has no faculty for transferring its body from place to place.'

Then:

'*What is it that makes the planets go round the Sun, each within the limits of its own region, if there be no solid spheres, if the globes themselves can do nothing but remain stationary, and if they cannot be moved from place to place by any soul in the absence of solid spheres?*—Although things far removed from our understanding and without positive example[30] are difficult to explain (and it would be absurd for us to discuss them, as Ptolemy has rightly reminded us), nevertheless it is true that if we give heed to probability, taking care not to assert anything contrary to our reason, then it is fairly clear that we should not have recourse either to a mind which would make the globes rotate by the dictates of reason, or as it were by order; or to a soul which would preside over this motion of revolution, and which would give the impression [of motion] on the globes[31] through a uniform thrust of forces, as occurs in rotation about the axis. [The origin of] the motion of the primary planets round the Sun is to be attributed solely and uniquely to the body of the Sun placed in the middle of the whole Universe.'

Moreover, confirmation of this is provided by the following considerations[32]:

'1. It is definite, that a planet follows the Sun more slowly the farther away it is [from the Sun], and in such a manner that [the times] of the periodic motions [of the planets] are in the sesquialteral [$\frac{3}{2}$] proportion[33] of their distances from the Sun. Therefore, we deduce that the Sun is the source of motion. 2. We draw the same conclusion [in respect of the motion] of individual planets, namely (as we shall see later), from the fact that the nearer any planet approaches the Sun, the more rapidly does it move, and in a proportion which is exactly double.[34] 3. The nobility and power of the body of the Sun, which, with its absolute fullness of form, is the grandest and largest [of the heavenly bodies] and which is the source of life and heat whence comes all life in plants, are not contrary to that opinion. In fact, heat and light[35] can be regarded as *quasi*-contrivances, peculiar to the Sun, for conferring motion on the planets. 4. The probability of these considerations implies the rotation of the Sun in space on its stationary axis in the same direction in which all the other planets move, the period being shorter than that of Mercury, which is the nearest and quickest of them all.'

Thereafter, a fresh question arose. It was a question that Kepler had not raised in the *Astronomia Nova*; or, at least, he had made only passing reference to it. What makes the Sun turn on its axis? and not only the Sun, but also the Earth and the other planets? In fact, in the *Astronomia Nova*, as I have pointed out, Kepler had not yet expressly extended to rotatory motion the fundamental law of his dynamics which makes the velocity of motion depend on the relationship between the motive force of the prime mover and the inertia of the moving body; consequently, he contented himself with vague answers, borrowed either from traditional sources or from Gilbert.[36] It was not so in the *Epitome*. Here, the problem is discussed at length; and, morever, he came to adopt in a more general way the solution put forward in the *Astronomia Nova*: souls cause the planets, the Sun and the Earth to rotate—at least, they ensure the continuity and uniformity of their motion.

Naturally, there is no lack of reasons for endowing the Earth with a soul, and Kepler willingly enumerates them.[37] They are: (1) the perpetual, sensible subterranean heat; (2) the generation of metals, minerals and fossils; (3) the fire and light producing properties—cognate with the soul—possessed by certain products of Earth, such as sulphur, marcasites, etc.; (4 and 5) the formative power (of Earth) which produces animated beings such as flies, insects, marine monsters, besides geometrical shapes such as snow-flakes and crystals; (6) the perception by the Earth of celestial phenomena (aspects and relationships between angles) in accordance with which meteors (atmospheric phenomena), volcanic eruptions and so on are produced.

Yet, these general, physical reasons only confirm others which are, properly speaking, dynamical.[38]

In fact, '*if the Earth, in consequence of its material substance resisted circular motion* [rotation], *its diurnal motion would be violent, and, therefore, could not be perpetual*'; an inevitable conclusion from Aristotle's dynamics, as well as from the dynamics of *impetus* which was favoured by Kepler for terrestrial things,[39] but which he had not, so far, extended to the heavens. He does so, subsequently, at least in respect of *rotation*—but, curiously enough, not for orbital motion—and it was the inadequacy of the hypothesis of an *impetus* (impressed on the Earth at the very moment that it was created) to explain its rotation, and particularly its *constancy*, that justified, in the last analysis, the attribution of a soul to the Earth, as well as to the Sun.

Indeed, although it is incontrovertible that the material inertia of the Earth's body renders it very suitable for receiving the *impetus* of rotatory motion, which is impressed on it in the same way as is done by setting a wheel violently in rotation (that is to say, the force required is greater when the material substance is heavier [denser]), nevertheless, this motion remains, properly speaking, violent. For this reason, it persists much longer; whereas feathers and similar bodies do not easily accept the *impetus*, and consequently are unsuitable for throwing from slings and engines of war. On the other hand, a motion that is impressed on a body by the [substantial] form of its material substance, or by its soul, should not be regarded as of that kind, seeing that nothing is more natural to material substance than its form, and nothing is more natural to the body than its faculty or its soul. Consequently, in virtue of its material nature a magnet tends to move downwards, but the inherent force of its corporeal nature confers on it upward motion towards another magnet, which motion is contrary to the former, yet is by no means violent. Similarly, the movements made by animals against the natural weight of their bodies are not deemed to be violent.

All this being taken for granted, what is the result as regards the Earth?[40]

'*What is it that confers on the Earth a circular motion about its axis? Is it an extrinsic, or an intrinsic, cause, a natural power, or a soul?*—I say that not only is each of these [hypotheses] probable, but even that all three can be consistent with each other, and, no doubt, are so.

'*In the first place explain the external motive cause.*—If children can make a top spin in one direction or the other[41] with a motion which is all the more uniform and constant when the *impetus* is applied with greater skill, and if they can make the top complete a large number of gyrations once it has been set in motion by the *impetus* it has received, so that it will continue to spin until it falls in consequence of its motion being reduced through unevenness of the plane of the table [friction], encounter with the air, and its weight, [if, I say, children can do this], why should God not be able in the beginning to give the Earth, as from the outside, an impulse of this kind; which impulse, remaining constant in power, would have produced all the subsequent rotations, even those of the present day, although there could already have been twenty times one hundred thousand of them.[42]

'The following circumstance confirms this hypothesis: in the same way that a top continues to spin in the direction in which it was started, so does the Earth rotate always in the same direction, for there cannot be any reason why the Earth should rotate in one direction more than in the opposite one, were it not that the direction of rotation had been ordained by the Creator in the very beginning [at the creation]. As for [the hypothesis of a corporeal faculty]: 1. It is probable that the first *species* of rotatory motion remaining in the Earth was transformed or condensed (coagulated) into a corporeal faculty and that it was implanted also in the fibres of the Earth arranged according to the path of the motion.... For if the *species* of motion can be brought in from outside, can separate itself from the motive cause, can impress itself on the moving body of the top by force of impulse or drive, and can persist therein for some time ... as if it possessed an existence of its own ... although this *species* is there only as a guest, there is nothing to prevent the *species*, (by means of which the Creator in the beginning stimulated the Earth's globe to motion), from penetrating more thoroughly and lastingly into the body of the Earth and into the circularity of its fibres; there is nothing to prevent the *species* from settling itself there as a special form of its [the Earth's] body; there is nothing to prevent the *species* [from ceasing] thereafter to be merely the guest of the Earth, (like the *impetus* in the top), and becoming a lawful occupant, victorious and triumphant over its material substance. 2. We can appeal also to the following argument: the vigour of the [Earth's] rotation does not diminish, but has still, to-day, the same speed as formerly . . . a fact which agrees more with an intrinsic cause than an extrinsic one. 3. If it be rather probable that the cause [of the constancy] of the direction of the axis is to be found in the corporeal form determined by the straight fibres parallel to the axis, then it seems . . . even more probable that it will be found in another condition of the axis, determined by the circular fibres, and in the motive faculty which depends on it; in fact, this globe is mobile with respect to the circular fibres.[43] 4. It would be equally appropriate to ascribe the difference in directions of the motion to this corporeal form; there are, in fact, many examples (instances) where the form of a body causes motion in a particular direction. Thus, a magnet attracts iron towards one of its sides, but repels it from the other; similarly, the globe on account of its circular state which has its origin in the motion impressed upon it in the beginning, rotates in the direction towards which its circular fibres are directed.'

The scheme outlined by Kepler (in which that of Gilbert can be seen) seems rather to assume an impossibility, namely, motion of a purely material being by its own accord. To this Kepler replies[44]:

'Even if this corporeal form of the fibres were put forward as the one and only cause of motion, nevertheless, that which causes motion and that which is moved would not be identical. That which is true for a falling stone is just as true here for the circular fibres of the globe. It would be a different matter as regards the circular fibres themselves to the extent that they are arranged in a circle, and as regards the form of the body to the extent that it is circular, and as regards the motive faculty of this body. To summarize and bring together all that has been so far noted: the same globe, on account of its straight fibres, will remain at rest and resist motion, and on account of its circular fibres, will move; on account of its material inertia . . . it will gather *impetus*, and yet on account of the form, extended by its circular fibres, it will move.'

At first sight, the dynamics of the *impetus*, strengthened by the theory of the fibrous structure of the planets, is quite capable of explaining their motion of rotation; but it is a deception, and Kepler continues[45]:

'*By what arguments do you superpose a soul on this prime motion, which soul is situated in the body of the Earth?*—They are numerous, being based partly on the motion itself, partly on other tokens of the soul, and partly on examples of bodies in the Universe.'

'*What are the arguments based on the motion itself?—Primo*, even if the Earth were apprised by these circular features (fibres) and therefore seemed capable of giving motion [to itself], these features would seem to be, however, the instruments of the motive cause rather than the motive cause itself. For example, in the human body, the nerves, muscles, ligaments, articulations, bones, although they are perfectly adapted to motion, they are not, however, the primary cause of motion, but merely the instruments of the soul for the purpose of moving the body.

'*Secundo*, the constant vigour of this rotation, or its uniform velocity, finds more certain support in the soul than in the corporeal faculty, for the measure of this velocity is determined by the ratio between the forces of the driving power and the inertia or resistance of the material substance, namely, by the excess or the victory of the former over the latter. Similarly, as we have already said, the internal form of the body tires less than the *species* of the motion conferred on it from without, because the former united to a natural object is directly and continuously conjointed with it; whereas the latter having migrated into a foreign body becomes weaker the longer it remains there and, in a manner, withdraws itself longer from the source: similarly, the force of the soul is more certain and more constant than the corporeal form; in fact, the soul carries out its unparalleled actions without detriment to their source, seeing that it is an entelechy which continually restores itself; but the corporeal form is subject to time and does not persist without detriment and without becoming weaker, albeit insensibly, with the passage of time, for it is said: *death cometh even to stones and marble.—Tertio*, nature, the very origin of this motive power as recorded above, comprises something more noble and more exalted than the corporeal form; for if it be a *species* emanant or derived from the prime motive cause, which cause is nothing but God himself who conferred [on Earth] the beginning of its motion, what else could it be except a prime mover, itself non-mobile, *i.e.*, a soul? In fact, God himself is super-essentially the prime, creative motive cause, in such a manner as to be, transcendentally, essential motion; that is to say, eternal generation of which all souls and [all] minds are shadows bearing its image as shadows [bear the image] of bodies.'

Let us now deal with the Sun. Kepler was never in any doubt as to its 'animation'. Some of the arguments he puts forward are rather surprising. Let us start with the dynamical argument: as in the case of the animated nature of the Earth, the Sun's rotation could very well be the result of an *impetus* imparted by God to its body at the very moment of creation. Nevertheless,[46] 'the constancy and eternity of this motion, in which all life in the Universe has its origin, is more conveniently explained by the action of a soul.'

Kepler was right: for if the *impetus* of the Sun's rotatory motion, in view of its enormous mass, were infinitely greater than that of the Earth, nevertheless all the *impetus* is consumed in producing motion. As for the other reasons for assuming a solar soul, they provide a curious combination of the perfection of its nature with the fact that it is subjected to changes[47]:

> '*Have you other arguments, besides that of motion, which make it probable that a soul resides in the body of the Sun?*—1. A very strong proof is provided by the matter of the solar body and its luminosity; it seems to be a quality of the body of the Sun which derives from its state of being informed by a very powerful soul; especially as it is definite that its matter is the densest of all bodies in the Universe. It is, therefore, reasonable to believe that the greatest forces belong to that soul which dominates and sets such resistant matter ablaze. 2. I think that we must assume a soul rather than some inanimate form, because sunspots and the uneven illumination of its various parts at different times show that the Sun is not the seat of one single, continuous and perpetually uniform energy throughout all parts of its body, but that it is susceptible of motion, change and vicissitudes similar to those which are produced, *mutatis mutandis*, in the body of the Earth. Consequently, from its innermost parts [*lit*: from its viscera] something like clouds (possibly consisting of soot) loom here and there, form spots, and disappear after their matter has been consumed, thereby restoring brightness to those parts which they covered. Now, seeing that these changes are perpetual, they indicate the action of a soul rather than that of a simple form. 3. Furthermore, light in itself, no less than heat, is something related to a soul . . . [On Earth], nothing in fact becomes inflamed, that is to say, becomes luminous, unless heat be engendered in the body by a soul. . . . Now, it is clear that light is something related to our flames because, when condensed by concave mirrors or convex lenses, it burns like flames and live coals. Consequently, it is fitting that the body of the Sun, which is the source where light resides, should be endowed with a soul that is the author, guardian and continuator of this fiery illumination. 4. The function of the Sun in the Universe seems not to have been conceived except on the assumption that, illuminating everything it contains on this score light within its body; similarly, heating everything [it contains] heat; invigorating everything [it has] corporeal life; moving everything [it has] the principle of motion and motion itself, and therefore possesses a soul.'

Naturally, 'animation' of the Sun, any more than that of the Earth, does not imply the attribution of anything other than dynamical, or at most, sensory, powers to their 'soul'; for, though, in another context,[48] Kepler asserts the existence of a solar intelligence, or at least the existence of intelligences inhabiting the Sun, these intelligences are purely meditative ones and contribute nothing to its motion: therefore, to the question[49]:

> '*To the soul of the Sun do you join a mind or intelligence which would regulate the motion about its axis?*' he replies: 'No mind is needed for implementing the motion; for the direction in which the Sun rotates was fixed from the very beginning of things; and the constancy of the revolution as well as of the periodic times depends . . . on the constant ratio of the motive power to the resistance of the material substance. Now, maintenance of the axis of the Sun's body in the same direction is a state of rest rather than the work of a mind, because, since the first beginning of things no impression of this (rotatory) motion has been received by the axis. In fact [the position] of the mean circle [the equator] between the extremities of the axis [the poles] necessarily follows the direction of the axis; and the axis remaining [stationary], it will remain perpetually subordinated to the same fixed stars. Finally, the very gripping of the bodies of the planets, which the Sun moves in a circle by turning on its axis, is a corporeal property, and not an animal or mental one.'

The dynamic rôle of the Sun and the mechanism of its action on the moving celestial bodies is explained in the *Epitome*, in much the same way as in the *Astronomia Nova*, by the action on the planets of a *species motrix* emanating from the Sun. However, as will be seen later, the action of this *species* is defined in a very interesting manner, and in particular the action of the 'planetary motive forces' is henceforth supressed[50]:

> '*Is it then the Sun which makes the planets move round it by the gyration of its body? And how does it do so, seeing that the Sun lacks hands by which it can lay hold of a planet removed to so great a distance, and by turning on itself, how can it carry* [*the planet*] *along?*—By way of hands, it has the property of its body by which it emits straight lines throughout the whole extent of the Universe; which property, from the fact that it is a *species* of the [solar] body, rotates together with this body after the manner of a very rapid whirlpool, and passes over the whole extent of the circuit at whatever distance it reaches just as rapidly as the Sun rotates about its centre in its narrow space.
>
> '*Can you illustrate that by an example?*—We can certainly avail ourselves [of the example] of the sympathy of the magnet for the bar of

iron impregnated by the magnet whose strength it has imbibed by rubbing. Turn the magnet near the bar, and the bar will turn immediately. If the manner of taking hold be of a different kind, nevertheless you notice that in this case also there is no contact between the bodies.

'*The example is undoubtedly convincing, but it is obscure: explain this property and to what kind of things* [*it belongs*].—In the same way that there are two bodies, the one that causes motion and the one that is moved, so there are two forces by which the motion is administered: the one is passive, and is related more to the material substance, namely the similarity of the body of the planet with the body of the Sun in respect of corporeal form, and [the fact that] one part of the planetary body is friendly towards the Sun, and the opposite part is hostile; the other force is active and comparable rather to the form, namely, the fact that the body of the Sun has the strength to attract the planet by its friendly part and to repel it by its hostile part, and also to keep hold of it if it be so placed as not to turn either its friendly or hostile parts towards the Sun.'

Now we understand: henceforth, it is not the planets which draw themselves towards, or thrust themselves away from, the Sun; it is the Sun which acts in both senses. Here is a concept which is not without difficulties. Kepler continues the discussion as follows.[51]

'*How does it come about that the whole body of the planet is similar or related to the body of the Sun, and nevertheless that one part of the planet is friendly towards the Sun, and the* [*other*] *part is hostile?*—In the same manner, assuredly, as in the case of attraction of a magnet by a magnet: the bodies are related, and nevertheless attraction occurs at one part, and repulsion at the other part. Furthermore, this "friendship" and "hostility" are so called after the effect—approach or flight—and not after any dissimilarity in the bodies. . . . In the case of magnets, the diversity [of the opposite parts] derives from the position of the parts in the whole. . . . In the sky it is rather different, for the Sun possesses the active and energetic faculty to attract, or repel, or hold a planet, and not on one side only, as in the case of the magnet, but by all parts of its body. Consequently, it is credible that the centre of the body of the Sun corresponds to one extremity or to one side of the magnet, and the entire surface to the other side of the magnet. In the planetary bodies, that part or extremity which faced the Sun, at the very beginning of things and at its first establishment, is related to the centre of the Sun and is attracted by the Sun; but that part which was [turned away] from the Sun and turned towards the fixed stars is similar to the nature of the surface, so when it turns towards the Sun, the Sun repels the planet.'

The structure ascribed by Kepler to the solar magnet is an impossible one. As far as we are concerned, this is of little importance. What does matter, is Kepler's decision to give it this structure in order to get rid of the planetary motive forces and to concentrate all 'action' in the body of the Sun. However, before coming to the profound reasons which gave rise to this decision, we must consider an obvious difficulty in the Keplerian concept, one which is common, moreover, to the schemes of the *Epitome* and of the *Astronomia Nova*, where, however, it was not discussed. The difficulty is that, as a result of the action of 'attraction' or of 'repulsion' by the Sun, ought not some of the planets to approach and make contact with the Sun, and others to recede towards the fixed stars? Kepler replied that such would indeed be the case, if the Sun did not rotate on its axis. He continues[52]:

> '*So that I may have a better understanding of the action of the Sun's rotation, tell me what, in your opinion, would happen if the Sun did not rotate?*—The magnet does not cease to attract a magnet which presents its friendly part until it has made contact and joined up; if it faces the hostile part, either the latter is converted, after which it is attracted as before; or, if it cannot be converted, then it is repelled and removed from the sphere of influence, unless it be prevented [from doing so]. We must imagine the same state of affairs with the Sun, namely, if it did not rotate on its axis, then none of the primary planets would accomplish their revolution round the Sun, but some of them would continuously approach the Sun until they joined it through contact; those planets which presented the other side to the Sun would then thrust towards the fixed stars; and those which presented their side [in between] would remain quite stationary as a result of the Sun's attractive property being balanced by the repulsive [property in them].'

It is easy to see that Kepler's argument is both faulty and ill-considered, or at least, is incomplete and specious. It implies, in effect, the total inability of the Sun to 'convert' the planetary *quasi*-magnets; or, what comes to the same thing, he ascribes to the planets an invincible force of resistance to such 'conversion'. If it were not so, no planet would remain stationary in the sky, for they are precisely those which would present 'the side in between' to the Sun, and being subjected simultaneously to attraction on their 'friendly part' and to repulsion on their 'hostile part' would inevitably be 'converted' by the Sun, and would go to join up with it.[53] However, we shall not dwell on the matter. We are going to examine Kepler's concept in greater detail, from the viewpoint of the true conditions, that is to say, those of a Sun which is not stationary, but which rotates on its axis[54]:

> '*What is it therefore that takes place now that the Sun rotates on its axis?* —To be sure, when the body of the Sun rotates (this force rotates also in the same way as when a magnet rotates), the tractive force of each of its parts is directed towards and to other directions of the Universe. Seeing that the Sun seizes the planet by this property of its body, either by drawing [attracting] it, or by repelling it, or by [being] undecided between the two, it carries the planet around, and possibly it carries round all the intervening aetheric aura also.[55] Thus, in attracting and repelling it confines, and in confining it conveys in a circle.'

It will be recalled that Kepler devoted much discussion to elaborating the scheme for the action of the *species motrix* in his *Astronomia Nova*: on the other hand, he had not considered in detail its application to the various planets, or to the problem of the periodic times. It is expressly referred to in the *Epitome*[56]:

> '*If that were so, then all the planets would complete their course in the same time as the Sun.*—Undoubtedly, if there were nothing more; but it has already been pointed out that, in addition to the rotating force of

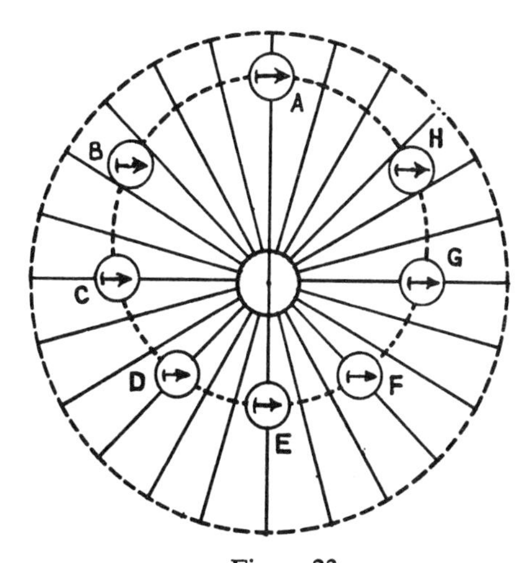

Figure 23.

> the Sun, there is in the planets themselves a natural inertia [opposing] this motion, through which they are constrained, by reason of their material substance, to remain in their place. Consequently, the vectorial power of the Sun, and the weakness or material inertia of the planet vie with each other. Each shares the victory; the former displaces the planet

from its position; the latter releases part of its body, *i.e.*, of the planet, from the chains which bound it to the Sun, in such a way that it is apprehended by ever changing parts of this circular force which surrounds the Sun, namely, by that force which immediately succeeds the one from which the planet had freed itself. In diagram 23, the *species* of the rotating solar body is represented by the circle drawn in broken lines, and it is understood that this circle is traced by any one of the planetary positions *A, B, C, D, E, F, G* or *H*. Let the Sun with its *species* turn from right to left; let the planet *A* be seized first of all by that part of the Sun's *species* which is designated by the radius *A*; let the radius *A* move during some space of time to the position *D*, and let it carry the planet along, though resisting and shedding its bonds, in such a way that in the same lapse of time [in which the radius arrives at *D*] it (the planet) is thrust only from *A* to *B*. The first radius, therefore, leaves the planet at *B*, which is retarded by the space *BD*; but the planet does not stay there, for, at *B*, this first radius is replaced by the radius *H*, which in its turn seizes the planet; in fact, the distance between *A* and *D* is the same as that between *H* and *B*.'

The constancy of the rotation of the Sun's body, which is assured by the action of its motive soul, implies that of the whirlpool of its force or *species*; from which, in its turn, we have constancy in the periodic times of the planets.[57] As for their differences, they are easily explained by the differences in the ratios between the motive force of the radii of the *species* and the resistance which the planet offers to it. In fact[58]:

'*Why do the planets disengage themselves from this* raptus *to different extents, with the result that, according to Copernicus, Saturn covers only 240 miles whilst Mercury covers 1200?*—1. Because this force, issuing from the body of the Sun becomes weaker in the various intervals in the same proportion as they increase, or in proportion to the extent of the paths described by these intervals; this is the main cause. 2. To this must be added the effect of inertia or the greater or less resistance offered by the planetary globes, as a result of which the [abovesaid] proportion corresponds only to one half; but more concerning this later.[59]

'*How does it come about that the force emanating from the body of the Sun is weaker in a larger interval at* A, *than near to the Sun at* E*? What reduces it or makes it weaker?*—The fact that this force is corporeal and partakes of quantity: that is why it can disperse itself and become weaker. Seeing that there is as much force diffused in the more spacious sphere of Saturn as is condensed in the smaller sphere of Mercury, it follows that it is very tenuous in all parts of the sphere of Saturn, and consequently is very weak; but it is very dense in that of Mercury, and for that reason is very strong.'

The weakening of the 'motive force' and hence of the planet's speed with, and in proportion to, the distance is an essential proposition in Kepler's celestial dynamics. The *Epitome* maintains it with the addition of an important correction, which I have already mentioned and shall refer to again later. In contrast with the *Astronomia Nova*, the *Epitome* is satisfied with putting forward the 'example' of the balance, and does not discuss the problem of attenuation of the motive force. Perhaps, Kepler realized the fallacious character of the arguments in the *Astronomia Nova*, and assumed that the attenuation of the *species*, like that of light, took place in the double proportion of the distance, whilst maintaining that its dynamic (motive) action decreased only in direct proportion to the distance, seeing that it functioned only in a plane.[60]

Now, if the *species*, or its power, decrease with the distance, and the planets in their courses round the Sun approach, or move away from, the Sun, will not their motion be thereby affected?[61]

'*The body of the planet is always the same, but it is repelled, as you maintain, by the Sun and attracted towards it; it traverses, therefore, different stages of the vectorial force; consequently, the proportion of the force at the body of the planet does not remain constant.*—Certainly not, if we consider the parts of one revolution, and it is for that reason that the same planet is faster in one part of its revolution at *E* than at another at *A*, as will be explained later. Notwithstanding this, the total vectorial force . . . the sum [of all the forces] present at all stages through which the planet passes in one revolution has always the same abundance in each circuit.'

Let us now consider in greater detail the mechanism of planetary motions as put forward in the *Epitome*.[62] We have already seen that, contrary to what was said in the *Astronomia Nova*, which explained the orbital motion of planets by a conjoint action of the Sun and the planets themselves, the *Epitome* ascribes the whole action to the Sun. It is the Sun, which, by means of the motive radii of its *species*, leads the planets round about, and also repels and attracts them according as they turn a 'friendly' or 'hostile' side towards it. The planets take no part in this process. They have no δύναμις, but only ἀδυναμιαί, resistances which they offer to being 'moved' or 'converted', the former in virtue of their material substance, the latter in virtue of their structure. Now, it is a curious thing, that if the structure of the planetary bodies be on the whole more complex in the *Epitome* than in the *Astronomia Nova*, it is no less modelled on the 'example' of the magnet.

In fact, by referring to 'sides' of planetary bodies friendly or hostile to the Sun, we have considerably simplified, or even falsified, reality; for it is not as bodies that the planets possess friendly or hostile sides; they have them only because they are magnetic bodies. Furthermore, they are friendly or hostile to the Sun not on account of parts of their bodies, but on account of parts of the magnetic fibres which pass through them.[63] It is precisely because these fibres, or filaments, rigidly maintain their direction in space (or nearly do so), that the planets resist the converting action of the Sun.

We can, therefore, put the question, which Kepler put forward so effectively, as to the significance of suppressing the action of the planetary motive forces[64]:

> '*If the planetary globes have an internal magnetic structure, why do not you ascribe to them* [*the action*] *of moving away from the Sun, or of approaching the Sun, according to the difference between the sides of their body, as you did in the* Commentaria de Motibus Stellae Martis?— 1. Because Astronomy brings evidence that this moving away from, and approach towards, the Sun occur on the line which is so to speak stretched towards the Sun, and that the revolution [round the Sun] which is combined with [the libration] does not change the direction; now, the magnetic fibres are only rarely directed towards the Sun. 2. Because two very different [actions] would be ascribed to these magnetic fibres. First of all, they would direct themselves towards the same sides of the Universe, which is a condition similar to a state of rest; then they would move the body [of the planet] sometimes towards the Sun, sometimes in the opposite direction. Now, this process [performed] as a result of attraction and repulsion agrees better with the gripping of, and carrying round, of the bodies, which actions are performed by the Sun. 3. Because it is more likely that the force of the Sun's active *species* extends to the planets, than that theirs extends to the Sun, so that the planets can thrust themselves away from, or draw themselves towards, the Sun. The Sun, indeed, is an immense body, and the planets are very small [bodies]; light and heat manifestly reach us from the Sun; the Sun conveys the planets. As regards other solar forces, we know [that they extend their action to the planets], but we have no such clear examples of the extension [of the action] of planetary force to the Sun. 4. It will be shown later that the action of the Sun communicates a certain inclination to the fibres of the [planetary] body: it is, therefore, probable that the libration of the entire [planetary] body derives from the Sun, rather than

being innate in the planet, that is to say, as far as the planet is concerned, it is a state which depends on another body, and is not the result of an action or a motion of its own.'

Here we have very respectable reasons for transferring all action to the Sun; but there is still another, even more important one. Kepler considered that it was only in this way that the central position of the Sun in the Universe could be maintained. So, to the question[65]:

'*Do you assert that this force is common to the Sun and the planets, and communicates reciprocally forces of repulsion and attraction, in the same way that the force is common to two magnets?* Kepler replies: 'This is precisely the fifth reason why this force of repulsion and attraction is not accorded to the planets themselves, namely, because it is not reciprocal . . .; in fact, if the force of the planet extended to the Sun, the latter should be displaced from its position at the centre of the Universe (by an amount inversely proportional to their bodies); or at least the Sun should oscillate, being pulled sometimes to one side and sometimes to the other, when several planets being located all on one side act on the Sun by the similar faculty possessed by each of them.

'*You will note that in this way you do not avoid an unfortunate consequence. The Sun pushing the planets by the active* species *of its body, as by a pole, pushes itself away in proportion, and drawing the planet as by a hook, draws itself somewhat towards the planet.*—If we deny mutual attraction and repulsion, we avoid [this consequence] entirely. For, *primo*, if this planetary force be not extended to the Sun, neither the form nor the disposition of the bodies will be affected by it; furthermore, the consequence in question will not result from material necessity alone, independently, as it were, of the will of the Creator; for, such is the volume, the density of the Sun's material substance, and its power of attraction and repulsion, that the Sun does not move from its position. Thus, when a ship is fast on the sands, it can still be pulled free, moved and shifted from its position by two hundred horses; but one hundred horses by themselves, although providing one-half of the required force, would not, however, move the ship by one-half [the distance: they would not move it at all], because, between moved and not moved, there is no mean term, seeing that they are contradictory terms.'

Yet once again, we find Aristotelian dynamics exerting its hold on Kepler's mind, and rushing to assist his cosmology. *Mutual* forces are one thing, *unilateral* action is another; and if it be a question of an exterior force acting on a body to set it in motion, then the force must be greater than that of the opposing resistance,[66] whether it be internal or external, of the body concerned. Therefore, the Sun is able to move the planets in the same way that it illuminates them, heats them, attracts and repels them without being under the necessity, for that purpose, of stirring from its place at the centre of the Cosmos to which it has been assigned by God.

Let us now return to the planets[67]:

'Let us start at the moment when the magnetic fibres present their side to the Sun in such a way that their two ends are equidistant from the Sun; this takes place, according to the diagram [Fig. 23, p. 299] when [the planet] is at its greatest distance from the Sun. The Sun, then, neither repels nor attracts the planet, but hesitating as it were between the two, nevertheless takes hold, and through the rotation of its body and the *species*, which it sends forth, causes the planet to move from *A* to *B* by overcoming its resistance, and in its turn [being] overcome by the resistance; consequently, it allows the planet to escape, so to speak, out of its hands, that is to say, from the previous radii[-vectores], *A*, of the *species* and takes hold of it again by those that follow, *H*, and this occurs in proportion to the force of the *species* in this interval. In this way, therefore, the planet is pushed forward whilst the magnetic fibres, in consequence of their directional power, always tend towards the same region of the Universe.[68] It follows that the "friendly" side of the Sun is gradually turned towards the planet,[69] and the "hostile" side is turned away from it. Consequently, the [planetary] globe starts to be very slightly attracted by the Sun; the attraction is quite small, as the distances from the two extremities of the Sun differ but slightly. As a result of this attraction, the planet deviates somewhat from the large circle it started [to describe] at *A* and moves inwards nearer to the Sun, adopting as it were a more confined circuit [in the radius of action] of a stronger force, because the force is denser. Consequently, the planet frees itself therefrom to a less [degree] and is moved more rapidly.

'This attraction is very slow [weak] at the start near *A*; but it is very rapid [strong] in the position where the "friendly" hemisphere of the planet's body[70] faces the Sun whilst the "hostile" [hemisphere] is completely hidden behind the former, that is to say, in the position where the magnetic fibres are directed towards the Sun, as happens near *C*, which corresponds to one quarter of the entire circular path. Then, near *D*, the Sun's attraction becomes weak again, but the speed of circular motion continues to increase, because the interval [distance] between the Sun and the planet continues to decrease (as a result of the attraction). Now, the falling off in attraction at the start near *C* is almost nothing [is very weak], but near *D* it becomes more and more apparent as the "hostile" part of the planet reappears and begins to face the Sun. When one-half of the path has been completed at *E*, the two hemispheres of the globe moving round [the Sun] are once again turned equally towards

> the latter. At this moment, all attraction ceases; the planet is then at its *minimum* distance from the Sun, and therefore, its motion is quickest, for it is contending with the densest, and therefore with the greatest, force to which it is subjected, and frees itself but very little from this ambient force.
>
> 'However, as soon as the globe has been carried from this position *E* on its orbit towards *F*, the part which is "hostile" to the Sun comes nearer [to the Sun] than the other "friendly" part, and turns more and more towards the Sun; consequently, the planet starts to be repelled by the Sun, and moves from a more confined and denser sphere of the *species* into a more spacious, less dense and weaker [sphere]. In consequence, its motion decreases, and in the reverse order [to the preceding]; it decreases slowly at first from *E* to *F*, but when all the "hostile" hemisphere or its fibres are pointed directly towards the Sun, and the "friendly" side is hidden, then the planet is very rapidly [strongly] repelled, and its [orbital] motion slows down anew and tends to the mean [velocity]. This occurs at *G*, the other quarter of the circular path. When the planet has been carried further towards *H*, its repulsion is still further diminished, until it disappears completely at *A*, where the planet returns to its original position, being thrust to its greatest distance from the Sun.'

In fact, the mechanism of planetary motion is rather more complicated, because the direction of the magnetic fibres does not remain entirely constant, but is affected to some extent by the 'converting' action of the Sun. Indeed, it would not make sense to withhold from the Sun the power 'of doing less', *i.e.*, of converting (turning) the 'friendly' fibres towards itself, seeing that the Sun is credited with the power 'of doing more', *i.e.*, of urging the planet on its orbit and of attracting (or repelling) it. It must be remembered, however, that the power of attracting or repelling, inherent in the Sun's *species*, is relatively much weaker than its power 'of urging',[71] and that the magnetic fibres themselves offer a considerable resistance to any action tending to change their direction. It follows that the 'conversion' takes place much more slowly than the progression of the planet on its orbit: consequently, the magnetic fibres are, generally speaking, directed towards the Sun (except in the quadrant positions) because the force of attraction succeeds in turning the magnetic fibres only towards *H* or *Q* during the time that the propelling force of the *species* carries the planet from *P* to *I*, or to *N*. Similarly, during the time that the planet moves to *R*, the attracting force succeeds only in re-aligning the fibres and setting them in the direction which they occupied at the start. If the attracting force be greater in the second quadrant than in the first, because the planet is nearer to the Sun, then the propelling force is also greater for the same reason, and the planet consequently moves faster[72] (Figs. 24 and 25).

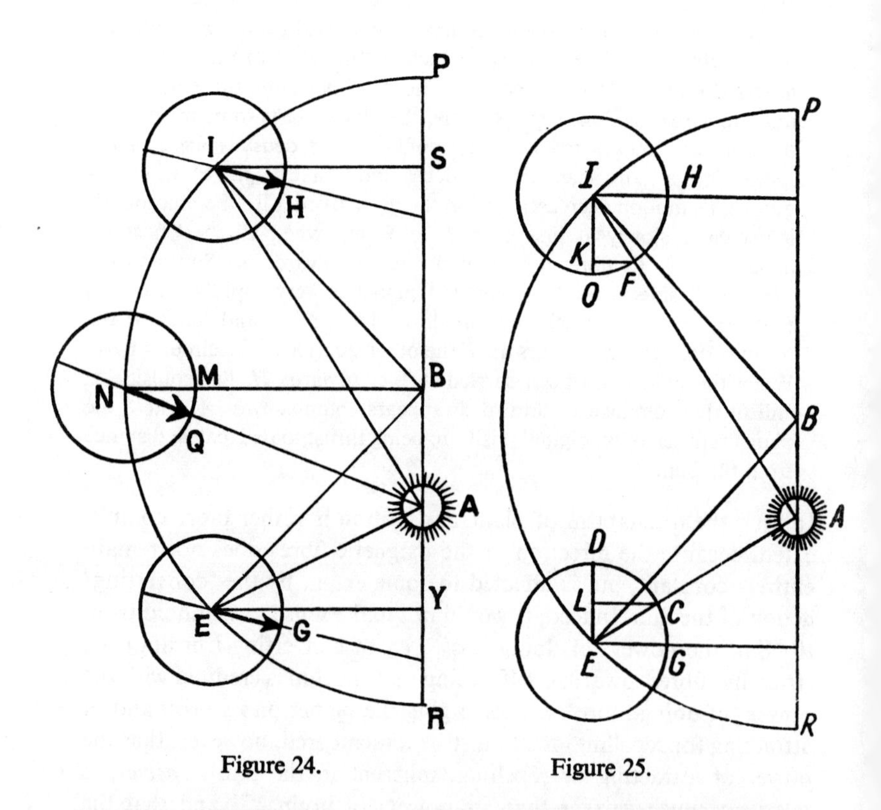

Figure 24. Figure 25.

That is precisely the reason why the planets do not fall into the Sun: they have no time to do so, as it were, for, in view of the disproportion between the attracting and the propelling forces, their motion towards the Sun is so much slower than their motion *round* it, that, in the time required for half the circuit (from *A* to *E*, or from *P* to *R*), they succeed in getting closer to the Sun by a very small distance only, namely, by an amount equal to twice their orbital eccentricity.[73]

We must now consider the way in which Kepler, starting from his modified physical concepts, proved that the resulting planetary orbit is an ellipse, and that the motion on this orbit conforms to the law of areas. The reader will not fail to notice the progress made by Kepler since the *Astronomia Nova*.

In order to give an account of the mechanism of the orbital motion, it is necessary to be able to determine the variations in the propelling and attracting forces throughout the whole path; in other words[74]:

> 'The angles which the radii [vectores] of the Sun make with the magnetic fibres of the planetary globe must be taken into account. The *sines* of the complements of these angles [the *cosines*] are a measure of those portions of the forces which act[75] [on the planet]. This results from the fact that the efficient causes of the libration are the [motive] rays of the Sun and the magnetic fibres of the body of the planet, [that is to say], two physical lines; therefore it is right to let the measure of the force of libration depend on the angle between these two lines, and in fact on the *sine* of the angle[76] (Fig. 25).
>
> 'Let *A* be the Sun, and *E* the centre of the planetary body; *RP* the line drawn through *A*, the Sun, and the centre of the orbit, *B*; let *EG* and *IH* be magnetic fibres almost perpendicular to *RP*... *G* and *H* being the Sun-seeking ["friendly"] ends of the fibres. In Lib. IV, fol. 583 [*G.W.*, vol. VII, p. 334; *O.O.*, vol. VI, p. 375] it is assumed that the fibres remain nearly parallel to each other during revolution of the body [of the planet round the Sun]; consequently, at *P* and *R* there is no reason for attraction or repulsion seeing that [at *P* and *R*] where the ends, both Sun-fugitive and Sun-seeking,[77] are equally distant from the Sun; whereas, at intermediate positions, where the Sun-seeking or Sun-fugitive ends are pointed directly towards the Sun,[78] the force of libration is *maximum*. Let *AE* and *AI* be rays from the Sun. Draw the lines *ED* and *OI* parallel to *RP*, and from the points *F* and *C*, where the rays from the Sun cut the mean globes of the planet, [draw lines] perpendicular to them, namely *CL* and *FK*. Then, the angles between the [motive] rays of the Sun and the fibres will be *AEG* and *AIH*, the complements of these angles, [will be the angles] *CED* and *FIO*, or the arcs *CD* and *FO* and their *sines* *CL* and *FK*; *IH* and *EG* being the *sinus totus*.[79] It follows, therefore, that the ratio (modulus) of the Sun's total forces, effective at *I* or *E*, to that part of the forces which act on the planet when the fibres are in the positions *EG* and *IH*, is the same as that of *EG* or *IH* to *LC* or *KF*.'

More simply expressed, this means that the forces of attraction (or repulsion) acting on the planet will be proportional to the *sinus complementi* (*i.e.*, *cosine*) of the angle *AIH* between the fibres and the radius vector from the Sun. Or, that this angle, being equal to angle *PAI*, which in astronomical terminology is called the true anomaly (*coaequata*), the force causing libration, and therefore the magnitude of the libration, will be proportional to the *sinus complementi* of the true anomaly.

The *sine* (or *cosine*) law already put forward by Kepler in the *Astronomia Nova*, which law, according to him, governed the relationships between natural forces (*e.g.*, between the degree of illumination and heating of a surface exposed obliquely to luminous rays) does not owe its prominent position to the fact that it is merely the expression (or consequence) of a still more fundamental law; for example, we are already familiar with the part played by the lever and the balance in Kepler's mind.[80] In fact, in the balance, as he explained at length, the weights hanging from the ends of the beam act only in proportion to the *sine* of the angle of displacement of the beam. Similarly in the case of the magnetic fibres, which could, in fact, be likened to the beam of a balance with the [motive] ray of the Sun as its supporting pillar; the magnetic fibres are affected by the attracting forces of the Sun only in proportion to the *sine* [*cosine*] of the angle of inclination to their length, or strictly speaking, to half their length. . . .[81]

A measure of the librating force (attracting or repelling) having been established, as well as the law governing its variation during the course of the planet round the Sun, Kepler then asked himself if his reasoning were not invalidated by the fact that the direction of the magnetic fibres is not absolutely constant, but is also affected by the Sun. Naturally, he said, it was not so: and, as we shall see, it was even a favourable circumstance[82]:

'In the first place, remember that when the planet is at the apsides, that is to say, at the beginning of its orbit, the angle between the radius [vector] from the Sun and the fibre is a right [angle]. On the other hand, it has been proved in Lib. IV, fol. 593 [*G.W.*, vol. VII, p. 339; *O.O.*, vol. VI, p. 379] that the fibre *NQ* in the diagram [Fig. 24] tends (is pointed) towards the Sun, in other words, when it joins with the radius [vector] *NA* from the Sun, . . . and coincides with it when quadrant *PN* of the orbit, reckoned from the apse *P*, is reached [by the planet], and it was proved that the arc of the orbit starting from the apse then measures the complement of the angle in question.[83] It is then only necessary to prove that the angles, such as *HIA* [eccentric anomaly], which the fibres between zero and the right angle make with [the radius from] the Sun, are completed by the mean arcs of the orbit, such as *PI*, so that together they make [an angle of] 90°.

'The proof is as follows: on p. 596 [*G.W.*, vol. VII, p. 341; *O.O.*, vol. VI, p. 379] it was said that *IS* is to *NB* approximately as the *sine* of the angle *HIS* is to the *sine* of the angle *QNB*.[84] This is the reason why the ratio of *IS* to *NB* has been used, although from physical considerations this is true rather in respect of the *sines* of the angles *IAP* and *NAP*. In fact, the *sine* of angle *AIB* is to the *sine* of angle *ANB* as the *sine* of angle *IAP* is to the *sine* of angle *NAP*. (In fact, as *BI* is to *BA*, so is the *sine* [of] *BAI* to the *sine* [of] *BIA*; and as the same [radius] *BI* or *BN* is to *BA*, so is the *sine* [of] *BAN* to the *sine* [of] *BNA*; and as the *sine* [of] *BAI* or [of] *IAP* is to the *sine* [of] *BAN* or [of] *NAP*, so is the *sine* [of] *AIB* to the *sine* [of] *ANB*.) Consequently, on comparing these various ratios, we find that the angle *HIS* is equal to the angle *AIB* and [that the angle] *QNB* is equal to the angle *ANB*, and after subtracting equal angles we find [that the angle] *SIB* is equal to the angle *HIA* (similarly, angle *BNB* is equal to angle *ANA*).

'However, the measure of [angle] *SIB* is [the arc] *IN*, and the measure of [angle] *SBI* is [the arc] *PI*. Hence, the measure of [angle] *HIA* will be [the arc] *IN*, the complement of arc *PI*. Therefore, the arc of the orbit *PI* being given, *SI*, the *sine* of this arc, is given also, *i.e.*, the measure of the increase in libration.'[85]

This does not mean, as we might believe, that the *sine* [*cosine*] of the orbital angle measures the acceleration of the planet's centripetal motion; it measures the velocity, proportional to the force of attraction, or of repulsion, which is exerted on the planet; on the other hand, the *ratios* of these *sines* measure the acceleration.

Having obtained a measure of 'the increase in libration', or of its force, at any moment during motion and therefore at any point in the planet's orbit, we must then enquire what is the sum total of these librations, in other words[86] . . . '*what is the measure of that part of the libration performed from the start* [of the course, *i.e.*, from the apse] *up to a given moment?*' It is determined by the *versed sine* of the arc traversed on the orbit. In fact, the ratio of the major axis of the ellipse to the total libration, or, what comes to the same thing, that of the semi-diameter of the [circular] orbit to the eccentricity, is equal to the ratio of the *versed sine* of each of the arcs of this orbit (starting at the apse) to that part of the libration performed whilst the planet traverses the said arc.

The proof is simple enough, but interesting, because, contrary to the methods used in the *Astronomia Nova*, Kepler, in the *Epitome*, employs infinitesimals in a very daring way (Fig. 26).

In the accompanying diagram, which represents a perfectly circular orbit with centre B, PR is the line of apsides, A is the Sun, AB is the eccentricity, and $AF = 2AB$ is the total libration of the planet. Divide the circle into equal small parts, starting at P.

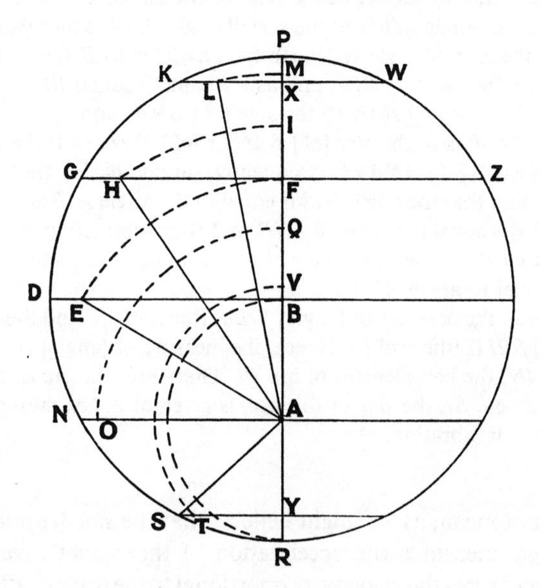

Figure 26.

By the previously proved propositions, the increases in libration (which will be designated by PM, MI, IF, FQ, QV, and VB) will be in the same ratios as the *sines* KX, GF, DB, NA, SY, RR of the corresponding arcs PK, PG, etc. As pointed out by Kepler, this is only true as the degree of division tends towards infinity, and when the *sine* at R is regarded as being equal to the [zero] arc RR. This being assumed, the argument is as follows:

'The points P, M, I, F, Q, V, B, represent the increases in libration; transpose them, by relating them to the distances of the planet from the Sun A, and then, with A as centre, describe the arcs ML, IH, FE, QO, and VT with the intervals [radii] AM, AI, AF, AQ and AV. The elliptical orbit of the planet will then be represented by a line falling from P to R and passing through L, H, E, O, and T; and the distances of the planet from the Sun will be AP, AL, AH, AE, AO, AT and AR, whilst the *versed sines* of the corresponding arcs *PK*, *PG*, etc., will be *PX*, *PF*, *PB*, *PA*, *PY* and *PR*. I say, therefore, that the whole diameter *PR* [considered] as the *sagitta* of the *arcus PDR* will bear the same ratio to the total libration *PB* [*FA*] as the *sagittae* of the individual *arcus*[87] to the increments of the individual librations, namely, *PX* to *PM*, will be as *PF* to *PI*, *PB* to *PF*, *PA* to *PQ*, *PY* to *PV*.

'Now, it has been stated that the portions *PM*, *PI*, etc., of the libration[88] are [amongst themselves] in the ratio of the *sines KX*, *GF*, etc. Therefore, the portions *PX* and *PF*, etc., of the whole *sagitta PR* will be, amongst themselves, in the same ratio as the *sines KX*, *GF*, etc., provided that, as before, the division be infinite, and that the point *R* subtend, so to speak, a line *RR*. Consequently, by permutation [we find that] the portions of the libration are proportional to the parts on the *sagitta*, and therefore that each whole portion of the libration, reckoned from *P*, corresponds in the same proportion to its whole *sagitta*.

'*However, how do we know that the parts* PX *and* XF *of the diameter* PR, *regarded as* sagittae, *are in the ratio of the* sines KX *and* GF *by which they are specified?*'[89]

We might even ask if we really know this, *i.e.*, if this proposition be entirely correct. In the *Astronomia Nova*, Kepler had tried to prove it by starting from 'the anatomy of the circle' . . . and did not succeed. Furthermore, he had given it no more than an approximate value.[90] At that time, he had not yet read Pappus, and, moreover (we might even say, as a result), he had hesitated to use infinitesimals. Everything had now changed: thanks to Pappus and a daring use of the method of infinitesimals, the proof is done in no time[91]:

'Pappus has proved in his *Mathematical Collections*, Lib. V, Prop. XXXVI, that if a sphere (such as *PGZ*) be cut by parallel planes (such as *KW*, *GZ*, etc.) the ratio between the parts of the surface of the sphere [cut off] and that of the corresponding parts on the axis (*PR*) was the same. For example, the ratio of the spherical surface *KPW* to the part *PX* of the arc is the same as that of the surface *KWZG* to the part *XF* and so on.

'However, if the spherical surface be considered to be divided into an infinite number of zones of equal width, then each of these zones, such as *KW* or *GZ* would be like a circle without height. Now, the circles *KXW* and *GFZ* are to each other . . . as their semi-diameters *KX* and *GF*, etc. For this reason, those parts, such as *PX* and *XF*, on the axis *PR*, which correspond to them, will be the ratio of the *sines* by which they are specified.'

We could then ask, by what right does Kepler now substitute lines for surfaces, radii of circles for heights of spherical calottes or rings;

we might even deny that he had any right, but then we have logical niceties of which Kepler was unaware. As far as he was concerned, the passage is quite sound, and its correctness is confirmed by, or depends on, the fact that it leads to a correct conclusion, namely, ellipticity of the planetary orbit.

The ellipticity of planetary orbits, which had been asserted, or at least mentioned many, many times by Kepler, was in fact a presupposition in the physical theory of the motion of celestial bodies in the *Epitome*. Was not it precisely for the purpose of letting them describe ellipses, that he gave them a fibrous structure and endowed the Sun's *species* with the ability of 'converting' their magnetic axes to some extent and of inclining them towards itself by the exact amount needed to substitute a libration proportional to the eccentric anomaly (from which it is a necessary consequence) for a libration proportional to the *cosine* of the true anomaly (from which the ellipse is not deducible)? It was only after he had developed the mechanism of libration and proved that it is measured by the *versed sine* of the arc traversed by the planet, or, more exactly, by the arc of the circle circumscribed about its elliptical orbit, that Kepler proceeded to a formal proof of the statement[92] '*from this libration there results* [for the planet] *an elliptical orbit, as has been confirmed by observation.*'

The line of argument in the *Epitome* is in the reverse order of that in the *Astronomia Nova*; and it is much simpler. Kepler starts by reminding us, as we already know from Apollonius, that the ellipse has the property of cutting all perpendiculars dropped on to its major axis from points on the circle circumscribed about this axis; that the ellipse has, in addition, two *quasi*-centres located on the major axis at equal distances from the centre of the circumscribed circle (point of intersection of the major and minor axes), which points it is his custom to call 'foci'[93]; and the sum of the distances from any point on the ellipse to these foci is always equal to the major axis. It follows that the distance from the points of the ellipse located on the minor axis to the foci is equal to the radius of the circumscribed circle (or to one-half of the major axis); in the accompanying diagram, $EA = DB$ or PB or BR (Fig. 27).

Observations reveal that the planets, starting from the apse at P and having traversed one quarter of their orbit, are distant from the Sun by an amount which is exactly equal to the radius of the circumscribed circle, that is to say, they are at E. This result implies an elliptical orbit with the Sun at one of the foci. According to the law of planetary motion which has been assumed (libration proportional to the *versed sine* of the eccentric arc), the ratio of the libration corresponding to arc *PG* with respect to the libration corresponding to arc *PD* (one-quarter of the path), will be equal to the ratio of the *versed sine* of arc *PG*, *i.e.*, *PF*, to the *versed sine* of arc *PD*; but the libration corresponding to arc *PD* is equal to the eccentricity (*BA*) and the *versed sine* of arc *PD* is equal to the radius (*PB*). Therefore,

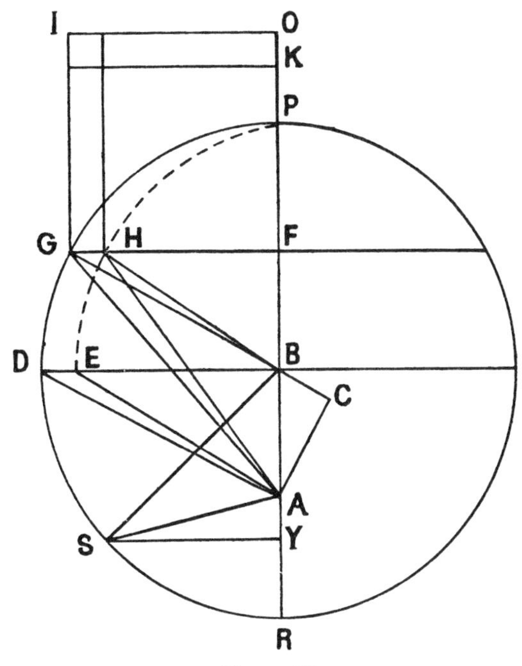

Figure 27.

the libration corresponding to arc *PG* is to *BA* as *PF* is to the radius *PB*. Subtract this from the *maximum* distance *AP*, and find a point *H* on *GF* whose distance from *A* is equal to the remainder (*AP* minus the libration).

Then, the ratio of *GF* to *GH* will be equal to that of *DB* to *EB*, which is precisely the characteristic of an ellipse.[94] Kepler, still making use of infinitesimals, adds that the ratio of the length of the circumscribed circle to the length of the ellipse is approximately equal to that of *DB* (the radius) to the mean proportional between

DB and *BE* (the radius and one-half of the minor axis).[95] In fact, if we divide the circle and the ellipse by an infinity of ordinates (such as *GF* and *DB*), then the small arcs adjoining *P* will be to each other in the ratio of the ordinates (*GF* and *HF*), whilst those near *D* and *E* will be equal. 'Thus the ratio of *DB* to *EB* [that of the ordinates] is gradually removed and disappears in the simple ratio of equality. However, the complete arcs, starting from *P*, have a ratio to each other which is composed of all the ratios of all these particular tiny [arcs], and consequently never reach the ratio of *DB* to *EB*.'

As for the ratio of the area of the ellipse to the area of the circle, it is, like that of the segment of the circle to the corresponding segments of the ellipse, equal to the ratio of the major to the minor axis of the ellipse, as was proved by Apollonius.

Having established the ellipticity of the orbit, Kepler then passed to a proof of his second law, which relates the time taken to travel a certain distance on the orbit with the area swept out by the radius vector[96]; or, in Kepler's words, the law relating the time during which a planet remains on a segment of its orbit[97] to the surface of the elliptical sector determined by this segment[98]:

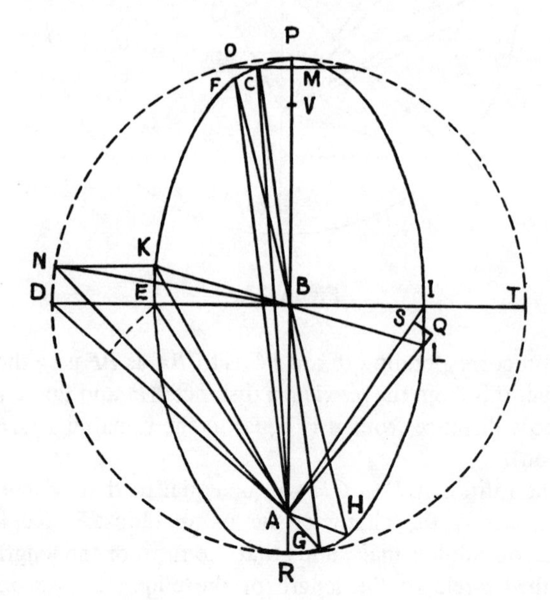

Figure 28.

'*In what manner is the surface of the ellipse suitable for measuring the time during which a planet remains on a segment of this arc?*—In exactly the same manner that, by dividing the circle into equal parts, the ellipse is divided into unequal arcs, which are smaller at the apsides and larger in the mean longitudes[99]; and this is done as follows:

'With centre *B* and distance (radius) *BP* draw a circle whose diameter is *PBR*; and in this circle draw the line *PR* as [representing] that of the apsides; on this line let *A* be [the place of] the Sun, the Source of motion towards *R*. *AB* will be the eccentricity and *BV* will be equal to *AB*. Now draw an ellipse with its foci at *A* and *V*, and tangent to the circle at *P* and *R*. This ellipse *PERI* will represent the orbit of the planet; its minor axis will be *EI*, and the diameter of the circle, perpendicular to *PR*, will be *DT*.

'Divide the semi-circle *PDR* into very small, equal parts; *PONDRT* will be points resulting from this division, and from these points draw lines perpendicular to the line of apsides *PR*, such as *OM*, *NK*, which will cut the ellipse in the points *C* and *K*. If we join these points of section *CKEI* to *A*, the Sun, I say that the duration [of travel] of the planet on the arc *PC* is measured by the area *PCA*; similarly the duration [of travel] on the arc *PCK* is measured by the area *PCKA*; and that [of travel] on *PER*, one-half of the orbit from apse *P* to apse *R*, is measured by the area *PERP*, which is one-half of the area of the entire ellipse *PERIP*';

this is a direct conclusion from the theorem of Apollonius which Kepler has already quoted.

As a result of this division, the middle parts of the orbit are larger than those near the apsides in the ratio of the semi-axis major to the minor axis[100]; the middle portions will be larger in the inverse ratio of their distances from the centre of the ellipse. Whence[101]:

'It follows that if we take together [two] small arcs of the orbit close to the two apsides, and two larger arcs near the mean longitudes, and we then measure the time taken by the planet to travel on these arcs [by the areas of the corresponding sectors of the ellipse], we shall have equal areas as a measure of these times, because the first two arcs taken together are at the same distance from the Sun as the latter two, taken together.'

The formulation of the law of areas in this passage from the *Epitome* is identical with that which we have already found in the *Astronomia Nova*. On the other hand, the proof is different; it is much more direct, more elegant, more geometrical, for Kepler has abandoned the roundabout method with the circle: and he has abandoned the summation of velocities and times corresponding to the 'lines of force' on the path of travel. The proof still depends on the same fundamental principle of his dynamics—times proportional to distances (velocities, therefore, inversely proportional to distances)—and employs, as before, the 'trick' of 'taking together' the two opposite arcs on the path of travel, by which means he avoids considering the intermediate portion (he comes to it later), but applies the principle of equality to 'pairs' on the path of travel.[102]

'In fact, if *PC* and *RG* be equal, then the areas *PCB* and *RGB* are equal also. Similarly, if *KL* and *LI*, larger than the previous ones, as has been shown, be equal, then the areas *KEB* and *LIB* will be equal also.

'Now, it has already been proved that [the ratio] of *PB* to *BE* is the same as that of *KE* to *PC*. The triangles *PBC* and *BEK* (rectilinear, or nearly so) are therefore ἀντιπεπονθότα, because the height *BP* of the one is to the height *BE* of the other as the base *KE* of the first is to the base *PC* of the second. Therefore the areas *BEK* and *PBC* are equal. It follows that the areas *BEK* and *BIL* taken together are equal to the areas *BPC* and *BRG* taken together: but *BPC* and *BRG* taken together are equal to the [areas] *APC* and *ARG* taken together, because the equal heights *BP* and *BR* taken together are equal to the heights *AP* and *AR* taken together. Also, the areas *BEK* and *BIL* taken together are equal to *AEK* and *AIL* taken together, because . . . the triangles *BEK* and *AEK*, as well as [the triangles] *BIL* and *AIL*, raised on the bases *EK* and *IL* (or near thereto, at *E* and *I*), have the same heights *BE* and *BI* and the same bases, namely, *EK* and *IL* respectively. Therefore the areas *EAK* and *IAL* are here ascribed to the longer arcs *KE* and *LI* (taken together), and the areas *APC* and *ARG*, equal to the first, are ascribed to the arcs *PC* and *RG* taken together. For, as it has been proved, the distances of the arcs *KE* and *PI* from the Sun, namely *EA* and *AI*, taken together, are equal to the distances of the arcs *PC* and *RG* from the Sun, namely *PA* and *AR*, taken together, as has already been proved.'

Let us pause awhile, for we have, indeed, come to a point of great interest. We have already seen how Kepler, in the *Astronomia Nova*, by a confusion of thought and through incorrect reasoning, had tried to bring his dynamical law (velocity inversely proportional to distance) into harmony with the law of areas, and even to deduce the second law from the first. It was a fruitless attempt, seeing that the two laws are strictly incompatible, a fact which Kepler, with admirable insight,[103] had realized in the meantime.[104] Consequently, he introduced an extremely important correction in the *Epitome*, or rather, he introduced explanatory information on the method of applying this law to orbital motions, for he fully upheld its absolute validity.

In fact, we have come to a glaring contradiction, for[105]:

> '*If equal areas be assigned* [*to pairs of arcs*] *which are unequal and are the same distance from the Sun; and if the times, or the duration of travel, over the unequal* [*arcs*], *which are the same distance from the Sun, must also be unequal in accordance with the axiom which we have previously used: how, then, can equal areas correspond to* (*measure*) *unequal times?*— If the pairs of arcs [taken together] be indeed unequal amongst themselves, they are, nevertheless, equivalent to equal arcs in respect of those portions of the periodic time associated with them.
>
> 'It has already been said, that, the orbit being divided into very small equal parts, the time [taken] by the planets to traverse them will increase in proportion to the distances between them and the Sun. This should not be understood as being applicable indiscriminately to all of these [very small] parts, but only to those which are at right angles to the Sun, such as *PC* and *RG*, where the angles *APC* and *ARG* are right angles; with regard to the others, which are at an oblique angle [to the Sun], this relation between velocity and distance must be understood as being applicable only to that which (in each of these small parts) relates to the motion of the planet round the Sun. In fact, seeing that the planet's orbit is eccentric, it follows that two elements of motion combine to define it (as was proved above); one of them is the motion produced by a [motive] force of the Sun, which carries the planet round the Sun; and the other is that of libration towards the Sun [produced] by another property of the Sun distinct from the former.'

As a result of the conjoint action of these two forces, the one propelling, the other attracting, and the compounding of the resulting motions, the planet describes an oblique arc *KI*, whose end points *L* and *I* are not equi-distant from the Sun. So, in order to separate in one's mind the effects of the two causes in question, and to isolate the component elements of the orbit, we proceed as follows[106] (see Fig. 28, p. 314); we:

> 'produce *AL* to *Q* so as to make *AQ* a mean between *AL* and *AI*; with centre *A* and radius *AQ* draw the arc *QS* to cut the longer [radius] *AI* in *S*. The arc *QS* will then represent the first element of the compound motion, and the difference between *AL* and *AI*, namely, the parts *LQ* and *SI* joined together, will represent the second element of the motion which must, on the score of reason, be separated from the former. In fact, it has nothing to do with [the question of] periodic time, seeing

that it is subject to different laws, and that we have already treated [its motion] when dealing with libration.'

Taken literally, Kepler's statement is faulty, because it is precisely on account of its libration that a planet describes an elliptical orbit; and the law of areas, as he conceived it, is valid only for the elliptical orbit: in fact, if the planet described an eccentric circle, the law of areas would not apply.[107] However, Kepler was right in reminding us that the motion of libration obeys a law of its own, and that the principle of proportionality between times and distance is applicable *only to motion of translation.* He went on to say[108]:

'Now, this second element of motion can only be separated from [the first] by dividing the orbit into unequal parts, as we have just described. For [the amount] by which [the arcs] *KE* and *LI* taken together exceed [the arcs] *PC* and *RG* taken together derives from the second element of motion; and, this excess having been eliminated, there remains from the first element [of motion] something which is equal to *PC* and *RG* taken together.'

Kepler proves this as follows.[109] In the first place, it will be recalled that the distances *AE* and *AI* are equal to the radii *BP* and *BR.* With centre *A* draw arcs passing through *E* and *I.* One of these arcs 'will cut and throw towards *K*' a part of the surface *AEK* equal to that which the other arc adds to the surface *ALI.* Therefore, triangles (sectors) having a straight line for their base 'are born' from the oblique triangles (sectors) *AEK* and *ALI,* and the sum of their surfaces is equal to that of those of the oblique sectors. It was proved above that the areas *AEK* and *ALI* taken together are equal to the areas *BPC* and *BRG* taken together. Because *AE* and *AI* are equal to *BP* and *BR,* it follows that the base of the right angled triangles (which replace the oblique triangles) are equal to the bases of triangles *BPC* and *BRG.*

'Therefore, that part of the oblique bases *KE* and *LI* which relates to the motion round the Sun is equal to the arcs *PG* and *RG* taken together; the libration of the Sun hardly affects the path of these arcs, seeing that *AP* and *AC* (and similarly *AR* and *AG*) differ only by a negligible amount.'

It is obvious: Kepler realized that he had made a mistake in the *Astronomia Nova* by being too hasty in extending to the whole orbit a relationship which is valid only at the apsides, without appreciating the true significance of the fact that a planet is in a very different condition at the apsides (the direction of its orbital motion is perpendicular to that of the radius vector), and, consequently, it was a mistake to assume that the orbital velocity, instead of the velocity perpendicular to the radius, (they coincide at the apsides), is inversely proportional to the distance from the source. The principle of continuity played him a bad trick, and caused him to make a mistake which he could have avoided if he had adhered more firmly to the fundamental law of all mechanical action, namely, that of the lever (or of the balance). However, if the first mistake be corrected, and the velocity perpendicular to the radius be substituted for the orbital velocity (which *is not* inversely proportional to the distance), then the error disappears, as was proved for the orbital elements near the apsides.[110]

'The same may be proved for other small parts of the orbit. Consider, for example, [the arc] *CF*; produce the [straight lines] *CB* and *FB* to *G* and *H*; add [the arc *CF* to the] corresponding [arc] *GH*; join the four points [*C*, *F*, *G*, *H*] to *A*, the source of motion. It has been proved above that *CA* and *AG* taken together, as well as *FA* and *AH* (taken together), are equal to *PA* and *AR* (taken together), that is to say, to the major axis [of the ellipse]; for this reason also, as before, the areas *ACF* and *AGH*, taken together will be equal to the [areas] *APC* and *ARG*, taken together although [the arc] *CF* is a trifle longer than [the arc] *PC*, and [the arc] *GH* is a trifle longer than [the arc] *RG*, in consequence of dividing [the ellipse into unequal parts]. But the new arcs drawn about centre *A* with distances (radii) *AC* and *AG*,[111] and cutting the [straight lines] *AF* and *AH* [produced], when taken together, will be equal to the arcs *PC* and *RG* [taken together], for the radius of one of them (*DC*) is greater than that of the other (*AG*) to the extent that the angle which subtends (measures) the first, *i.e.*, *CAF*, is smaller than the angle which subtends (measures) the second, *i.e.*, *GAH*. Therefore, the angles *CAF* and *GAH*, taken together, will always be equal to the angles *PAC* and *RAG*, taken together.

'Seeing that the equality (uniformity) of the first element of the planet's motion, *i.e.*, its progress round the Sun, depends on the equality of the angles about *A*, the Sun, particularly of the two opposite [angles] taken together, it [follows] that the area of the ellipse will be distributed equally between the arcs which subtend these angles, that is to say, the two [opposite] areas, [taken together], will be always equal to the other two [taken together]. Therefore, it is quite justifiable (but only when we consider pairs of angles) to take the area as a measure of the time: in fact, the times taken [by the planet to traverse] equal arcs—not any arcs,

> but those [which correspond] in their progress round the Sun—must be equal.'

Consequently, everything is in order and Kepler concludes: '*Recte igitur area pro mensura temporis constituitur.*'

Is the correction introduced by Kepler into his law of velocities sufficient? In other words, is it equivalent to the true law, the discovery of which is usually associated with Newton, and according to which the (orbital) velocity of a planet at a given point on its orbit is proportional to the perpendicular drawn to the tangent at that point? Max Caspar thought so. I confess that I should have liked to believe it myself. Unfortunately, Caspar's arguments have not

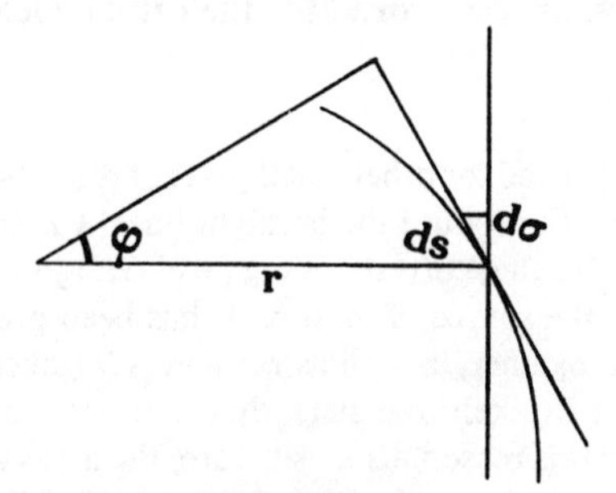

Figure 29.

convinced me. They are, nevertheless, extremely interesting and reveal, as it seems to me, a curious and instructive situation. Therefore, we shall have to examine them very carefully (Fig. 29).

'The disagreement between the law of distances and the law of areas', wrote Caspar,[112] 'is completely removed from now onwards. For, if Kepler had assumed, heretofore, that the (orbital) velocity were inversely proportional to the distance r, he now says that the distance r is inversely proportional to the component of the velocity perpendicular to the radius r.

'Let ds be an element of the curved orbit, and $d\sigma$ (Kepler particularly refers to *particulae minutissimae*) the component perpendicular to the radius r; then,

$$\frac{d\sigma}{dt} = \frac{c}{r}.$$

Because $d\sigma = \cos\varphi \, . ds$, therefore

$$\frac{ds}{dt} = \frac{c}{r . \cos\varphi},$$

that is to say, the velocity is inversely proportional to the perpendicular to the tangent.'

There is nothing wrong with Caspar's proof, and his conclusion must be accepted: Kepler's formula is equivalent to Newton's. This does not detract from Newton's glory: the replacement of a formula, or of a law, by another, which is objectively equivalent, very often means considerable progress.

However, it is certain that Kepler did not notice the equivalence in question, and did not substitute Newton's formula for his own.[113] I feel that we can, and should, go even further, and ask ourselves if he could do so; in other words, were the two formulae, which we regard as equivalent, the same for him, too—or, rather, would they be the same for him.

The question may seem strange: I shall be told, that a mathematical proof is not an argument *ad hominem*; if it be true for us, then it must be true for Kepler also, as well as for anyone else. This is perfectly true, when dealing with pure mathematics,[114] but we are not dealing with pure mathematics; we are dealing with physics, a 'mixed' science in which our mathematical formulae express and translate into symbols the fundamental concepts—the axioms—relating to the structure of real phenomena. In the present instance, we are concerned with motion. Now, the concept of motion underlying Caspar's proof is quite different from that of Kepler; it implies the principle of inertia and the preponderance of the straight line over the circle. Consequently, the elements of motion—lateral and centripetal—which comprise the orbital motion are straight lines; and the elements of this latter motion are (infinitesimal) straight lines whose direction is that of the tangent to that point on the (curved) orbit occupied (for a moment) by the moving body. However, these elements are not all straight lines in Kepler's view; those connected with libration undoubtedly are; but those connected with lateral motion are (infinitesimal) circular arcs. Even the infinitesimally short tangents play no part in his arguments. It would, therefore, be impossible for him to recognize, that is to say, accept, the validity of Caspar's (Newtonian) argument, even though, like Newton, he had assumed that the straight line and the curve merge together when they are infinitely small.[115] He would have explained the agreement of results by the two lines of argument in that manner. Having said this, it is still nonetheless true that, by an argument which *we* should not accept because it is based on concepts very different from ours—

and, what is more, false—Kepler succeeded in reconciling his dynamic law with the (true) law of areas. We shall revert to this.

Up to now, as Kepler emphasized, we have been dealing with the areas of pairs of opposite, corresponding triangles (sectors): we have not considered the part played by the areas of the individual triangles as a measure of time. However, we come immediately to the last stage in the proof the law of areas[116]:

> '*Now therefore, if the area of the ellipse be justifiably distributed between the pairs of opposite arcs, prove that the individual (single) triangles give exact measures of the times required to traverse the individual arcs.*—According to our axiom, the time required by the planet [to traverse] the arc *PC* bears the same relation to the time required to traverse the equal arc *RG* as the distance *AP* of the former from the source of motion [*A*] bears to the distance *AR* of the latter. Now, the area of triangle *PCA* is to the area of triangle *RGA* (whose base *RG* is equal to the base *PC* of the former) as the height *PA* of the former is to the height *RA* of the latter. Therefore, the time required by the planet to traverse the arc *PC* is to the time required by the planet to traverse the equal arc *RG* as the area of the triangle *PCA* is to the area of triangle *RGA*.
>
> 'Similarly, it may be proved that the time required for the planet to traverse [the arc] *CF*, of equal effect to [the arc] *CP*,[117] bears the same relation to the time required by the planet to traverse [the arc] *GH* as the area *ACF* bears to the area *AGH*, the sum of the [last] two areas being equal to the sum of the former, and so on. Consequently, the total area of the ellipse, starting at *A*, being divided into triangles [having the arcs *PC*, *CF* . . . *RG*, *GH*, etc., for their bases], is distributed amongst them in the same proportion in which the whole periodic time is distributed amongst [the times required to traverse] these arcs. Therefore, the [areas of] the individual triangles give a very true measure of the corresponding times required to traverse their arcs [which are the bases of the triangles].'

The proof is completed. The law of areas, henceforth freed from the 'obscurity' by which it was surrounded in the *Astronomia Nova*,[118] reconciled with the dynamic law of planetary motion which determines the ellipticity of their orbits, and thereby certain of being founded on physical reality and of expressing the true structure of the solar system, consequently replaced the artificial inventions of ancient astronomy—for example, the equant—and completed the defeat of circularity.

But wait: is not it 'unreasonable' and 'unusual', if not 'unduly bold', to resort to calculations with the ellipse rather than with the circle? No! says Kepler, who undoubtedly often heard this reproach; calculation with the ellipse, whilst being strictly equivalent to that with the circle,[119] is much more convenient. However, its decisive superiority over calculation with the circle does not lie in its convenience, but in its truth, that is to say, in its agreement with the reality of things. This is precisely the reason, namely, because it is physically impossible, much more than because it is not entirely true, that the Ptolemaic concept of the equant must be rejected. So, too, must be all the others, which, like the theory of planetary motions put forward by Fabricius who refused to examine causes and made shift with purely mathematical constructions, falsify the very meaning of astronomy, the knowledge of reality.[120]

We can now sum up. Kepler's celestial physics, as put forward in the *Astronomia Nova*, explains the motions of celestial bodies, or at least, their motion of translation—the rotation of the Sun, the Earth and the other planets on their axes demands, in effect, the intervention of a 'soul', at least motive—through the action of magnetic and *quasi*-magnetic forces. The Sun, the planets, the Earth and the Moon, in fact, all the celestial bodies are magnets. However, they are not only magnets; they are also *material bodies*, even before they became magnets, and *as such* (*i.e.*, material bodies) they are sources of *quasi*-magnetic forces whose action, combined with that of the true magnetic forces, regulates the motions of the wanderers in the sky.

In concreto, the Sun, rotating on its axis, sends out into space (in the plane of the ecliptic) a motive whirlpool which carries the planets round and impresses on them a circular motion round the Sun; at the same time the planetary magnets, in accordance with a mechanism which has been fully described above, causes the planets to approach and recede from the Sun. As a result of being subjected to this twofold influence, the planets do not describe circles in the sky, but describe ellipses having the Sun at one of their foci.

As a slight anachronism, or anticipation, we might say that, in Kepler's view, the celestial motions are explained by the action and superposition of two fields: a dynamic field caused by the rotation of the Sun (and of the Earth) which produces the orbital motion of the planets (and of the Moon) about the central body; and a magnetic field, which by causing them to approach and recede from the central body periodically (there is a like phenomenon in the case of the Earth and the Moon), determines the elliptic form and the kinetic law of their motion.

In the case of the Earth–Moon system, a gravific field (mutual attraction[121]) is added to the two fields already mentioned, but it is curious that this field seems to play no part in astronomy as such: it comes into the theory of tides, but not into the theory of lunar motion.[122]

The *Epitome* retains this scheme, but replaces the 'magnetic field', that is to say, the action of the true magnetic forces acting *between* the planets and the Sun, by the action of a *quasi*-magnetic force, which, like the *species motrix*, emanates from the Sun, and which attracts or repels the planet, according as the planet presents its 'friendly' or 'hostile' side to the Sun; the Sun for its part is not attracted or repelled through this action.[123] This substitution of unilateral action for interaction (magnets *mutually* attract, or repel), was necessary so that the Sun should not relinquish its state of immobility at the centre of the Universe.

This was, undoubtedly, an important modification, and it does credit to Kepler's sagacity, for it shows that, in his opinion, heliocentrism meant immobility of the Sun as well.[124] However, *in concreto*, celestial mechanics was only slightly affected. Especially as the attribution of the *quasi*-magnetic force of attraction and repulsion to the *species*, henceforth responsible both for propelling the planets and imparting to them a libratory motion, made the structure almost entirely similar to that of the true magnetic force to which, in the *Astronomia Nova*, Kepler assigned a 'directing' function besides an 'attracting' (and 'repelling') function. On the other hand, attraction, pure and simple, plays no part in the system of the *Epitome*, any more than it does in the *Astronomia Nova*.

Reverting to the image of 'the field', we could say that the modification introduced into the *Epitome* (compared with the *Astronomia Nova*) consists essentially in a limitation of the size of the planetary 'fields' with respect to those of the Sun: whereas the latter extend to the furthermost limits of the moving Universe, the former consist only of 'streams' of varying size, which in any case are sufficiently small not to engulf the Sun, nor to encroach on one another.[126]

The result produced by the action of one or other of these mechanisms is exactly the same: the planets describe elliptical orbits in the sky, and their motions are governed by the following two laws: (a) the velocity of the planet is not constant,[127] but at any point of its orbit is inversely proportional to its distance from the Sun; and (b) in spite of, or because of, the perpetual change in the velocity of its orbital motion, the area swept out[128] by the radius vector drawn from the Sun to the celestial body in question is strictly proportional to the duration of the motion; in other words: *in equal times, the areas swept out by the radius-vector are the same, although the planet traverses different distances.*

II

The *Harmonice Mundi*

Kepler's 'celestial physics' had given a partial answer to the question concerning the mathematical laws according to which God had regulated the planetary motions, and the physical means He had used to do so. His 'celestial physics' explained why the planets move faster when they are close to the Sun, and slower when they are farther away. It had fixed the shape of their orbits; and revealed the law underlying their motions. However, it had done so only for each planet taken in isolation; whereas the main problem, namely, the relationships between the individual orbits, and of the distances (Sun–planets) to the periodic times, had not advanced one step nearer to its solution. The solution put forward in the *Mysterium Cosmographicum* definitely did not agree with reality, at least not completely. The distances resulting from the concept of the five regular solids gave no better agreement than those obtained from the 'exact observations' of Tycho Brahe, or those given by Copernicus. As for the problem of the relationship between the periodic times and the distances, the *Astronomia Nova* had certainly pointed out a way, which finally (in the *Epitome Astronomiae Copernicanae*) enabled the required solution to be found, but he had not pursued the matter. Hence, the question remained open, and it was not till twelve years later, in the *Harmonice Mundi*,[1] that Kepler gave an answer to it.

The partial set-back to polyhedral cosmology, together with its partial success, could obviously only be satisfactorily explained by assuming that God, when he created the Universe, did not adhere *only* to the relationship between the five regular solids.[2] This is rather understandable: Kepler's first solution was too simple, or even over-simple. The existence of five, and only five, regular solids satisfactorily explained the number of planets, as well as their general arrangement; but it was unable to explain details. The five regular solids formed, as it were, the boundaries of the Universe; but a determination of what is contained within them demanded the application of other principles. Besides the regular solids, there were others, such as the star-pointed polyhedra,[3] which the Divine Architect was able to, and did, take into account. In fact, however, it was the purely geometrical concept of the structure of the Universe which showed itself to be inadequate, and not only the concept which took the five regular solids as model for the structure. It was too static. It would probably have suited a Universe at rest, but not one in motion; and especially not a Universe in which the planets do not move in concentric circles with constant velocities at fixed distances from the Sun, but describe eccentric orbits, periodically approaching, or moving away from, the Sun, and changing their velocity at every moment. Now, he who speaks of *velocity*, undoubtedly conveys the notion of distance; also, and perhaps more especially he conveys the notion of *time*. It was because time was ignored, that the *Mysterium Cosmographicum* failed to reveal the true structure of the Cosmos. Purely geometrical relationships were unable to give expression to it. Harmonic relationships were needed; for God, whilst being a geometer, was not solely an architect—a fact which the ancient Pythagoreans had certainly apprehended. He was also, even primarily, a musician.[4]

Vice versa, a God, who was merely a geometer, would no doubt have been satisfied with a Universe constructed on the basis of the sphere and the five regular solids, a Universe in which the planets would revolve eternally on concentric circles, that is to say, without ever changing their distances from the Sun, nor their speeds, nor the order of their progression. However, for a God, who is a musician, such a Universe in which the planets would, each of them, eternally give out the same 'note', even if the ensemble of these notes produced a concord, would be unacceptable. He who speaks of music, has in mind variety, not monotony. Therefore, ought not a God, who is a musician, to assign to each of the planets its own proper musical phrase instead of one single 'note'? and from these individual phrases to develop in the course of time a polyphonic and contrapuntal harmony? To achieve this, what matter if the strict bounds of geometrical ratios be transgressed? what matter if the result of natural necessities be subjected to the physical mechanism of eccentric motions?[5]

No doubt, at the time of the *Mysterium Cosmographicum*, Kepler had taken harmonic relationships into account, and he had favoured

Copernicanism for its possession of an orderly principle in which time clearly intervened, namely, the principle of agreement between temporal and spatial relationships, that is to say, the fact that the periodic times of the planets are, *grosso modo*, consistent with the distances of the planets from the Sun. However, as he admits in the *Harmonice Mundi*, he had not understood the prime importance of it at that time. Moreover, if he had understood it, it would have availed him little; for then, and even five years later, in 1599, when he had resumed his study of the harmonic relationships, and had decided to write a book on the harmonies of the Universe, he did not have the essential elements of the solution, nor even of the correct circumstances of the problem. In particular, he lacked as much a true understanding of stellar reality as of the nature of harmony, or, more precisely, the nature of the elements between which it was required to find the harmonic relationships.[6] These matters are explained in the wonderful pages of the *Harmonice Mundi*, which is a gripping account of his astronomical system and, at the same time, an enthusiastic introduction to the delights of 'harmonic' contemplation, the supreme joy of the soul conceiving the music of the spheres, which the divine Pythagoras believed he had heard[7]:

> 'At the outset, my readers should understand that the old astronomical hypotheses of Ptolemy, as set forth in the *Theoricae* of Peuerbach and the writings of other abridgers, must be kept removed from our present investigations and completely banished from the mind, for they are unable to give us a true description either of the arrangement of the heavenly bodies, or of the laws governing their motions.
>
> 'Furthermore, I cannot do otherwise than replace them simply by the Copernican theory of the Universe; and (if it be possible) convince all men of its truth[8]; but seeing that the majority of students are always but slightly familiar with truth, and that the theory according to which the Earth is one of the planets and moves amongst the stars round a motionless Sun always seems absurd to most of them, therefore let them, who are disturbed by the strangeness of this doctrine, know that my harmonic speculations have their place even in Tycho Brahe's hypothesis.
>
> 'In fact, whereas this author agrees with Copernicus in everything concerning the arrangement of the heavenly bodies and the laws governing their motions, it is only the annual motion of the Earth, as understood by Copernicus, that he removes from the whole system of planetary orbits and from the Sun, which both authors agree is the centre of the system. Now, the [relative] motion which results from this transfer is exactly the same; so, the Earth, according to Tycho Brahe, occupies at any moment

the same position as that given by Copernicus in the planetary system of the Universe, even if not in the vast immense space of the sphere of the fixed stars.

'The same circle is drawn on paper either by moving the tip of the compass round the other fixed point, or by fixing the paper to a turntable which moves under a steady compass. So it is in the present instance. According to Copernicus, the Earth describes its orbit between the outer circle of Mars and the inner circle of Venus by the true motion of its own body; whereas, according to Tycho Brahe, the whole planetary system (in which, among the other orbits, are also those of Venus and Mars) rotates [about the Earth] like a board fixed to a wheel, so that the space contained between the orbits of Mars and Venus appears to the stationary Earth like the wheel to a turner's stylus: as a result of this motion of the system, the Earth, which remains stationary, describes the the same circuit round the Sun in the space between Mars and Venus, as it describes (according to Copernicus) by the true motion of its own body [round the stationary Sun.]

'Now, the harmonic speculations envisage eccentric planetary motions as seen from the Sun; but it is easy to understand that, if the observer were on the Sun [he would believe himself to be stationary] however great its motion might be, [and] the Earth, although at rest (we accept Tycho Brahe's view for the moment) would nevertheless appear to him to describe its annual circuit amongst the planets in a period of time intermediate between the periods of the other planets [Venus and Mars]. Hence, even those who are weak in the faith,[9] and are consequently incapable of imagining the motion of the Earth amongst the stars, can nevertheless find pleasure in the exalted contemplation of this most divine mechanism; it is sufficient if they interpret what is said [here] concerning the Earth's motions on its eccentric [as referring] to the appearance of these motions when viewed from the Sun, the Earth remaining still, in the manner set forth by Tycho Brahe.

'It is clear, however, that the true disciples of Samian philosophy should have no reason to be envious . . . for the pleasure that they will derive from these delectable speculations will, in many respects, be all the more exquisite, seeing that they, too, accept mobility of the Earth and immobility of the Sun: in fact, the pleasure will come from the highest perfection of contemplation.

'*In the first place*, then, my readers must realize that, nowadays, all astronomers accept that all the planets, with the exception of the Moon which is the only one to have the Earth for its centre, move round the Sun[10]; it must be mentioned that the Moon's orbit is not large enough to be drawn on the accompanying diagram[11] [with dimensions] in proportion to those of the other planets. To the five planets shown on this diagram, a sixth is added; and this planet, too, describes its orbit round the Sun, either by its own proper motion (the Sun remaining stationary), or [by remaining] stationary itself whilst the whole planetary system rotates about it.

'*Secondly*, the following fact is equally well established: all the planets move on eccentric orbits; that is to say, they change their distances from the Sun in such a way that in one part of the orbit they are very far from the Sun, whilst in another part they are very close.[12] In the accompanying diagram, three circles have been drawn for each planet, not one of which represents the true eccentric orbit of the planet; however, [the radius of] the innermost one, for example *BE* in the case of that for [the circle of] Mars, is equal to the longest radius of the eccentric orbit; the orbit itself, represented by *AD*, touches *AF*, the outermost of the three circles, at *A*, and touches *CD*, the innermost circle, at another point *D* [the apsides] (Fig. 30).

'The orbit *GH*, represented by the dotted line drawn through the centre of the Sun, indicates the path of the Sun according to Tycho Brahe. If the Sun move along this path, then each point of the planetary system describes a similar path . . . and if one of these points, namely the centre of the Sun, be in a certain part of its orbit—here, in its inferior position—then all parts of the system will be found, each of them, in the inferior positions of their own proper orbits. Because of lack of space, the three circles of Venus have merged into one, contrary to my intentions.

'*Thirdly*, the reader should know that in my *Mysterium Cosmographicum*, published twenty-two years ago, [I showed] that the number of planets, or spheres, surrounding the Sun had been fixed by the All-wise Creator as a function of the five regular solids[13] on which Euclid wrote a book several centuries ago, and called it the *Elements*, because it consists of a series of propositions. Now, it has been proved in the second book of the present work, that there cannot be a greater number of regular solids, that is to say, the five regular plane figures cannot be put together to form a solid in more than five ways.

'*Fourthly*, with regard to the relationships between the planetary orbits, the connection between two neighbouring orbits is always such that they are approximately[14] proportional to the spheres of the five solid bodies; that is to say, to the ratio between the spheres circumscribed about, and inscribed within [this body]. In fact, when I had completed my determination of the [planetary] distances, using the observations made by Tycho Brahe, I discovered that if the angles of the cube be set on the innermost sphere of Saturn, then the centres of the sides [of the cube] almost touch the median sphere of Jupiter; and if the angles of the tetrahedron rest on the innermost sphere of Jupiter, then the centres of the sides [of the tetrahedron] very nearly touch the outermost sphere of Mars; also, if the angles of the octahedron reach to any

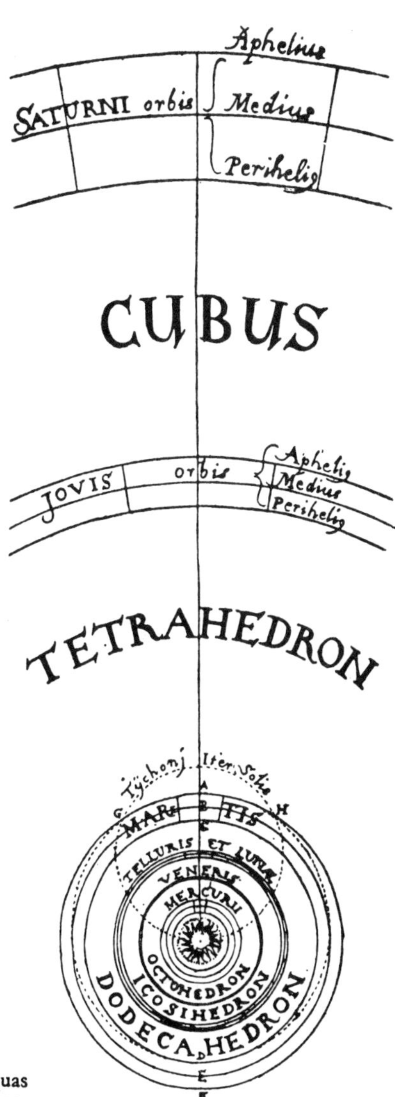

III. Numerus corporum Harmonias facientium quam causam habeat.

IV. Magnitudo Sphaerarum, quas permeant illa corpora, quam causam habeat?

Figure 30.

one of the spheres of Venus (for all three of them are contained within a very small space), then the centres of the octahedron come lower than the outermost sphere of Mercury [and cut it]; finally, coming to the relationships between the spheres of the dodecahedron and the icosahedron—they are the same—we find that the closest [relationships] to

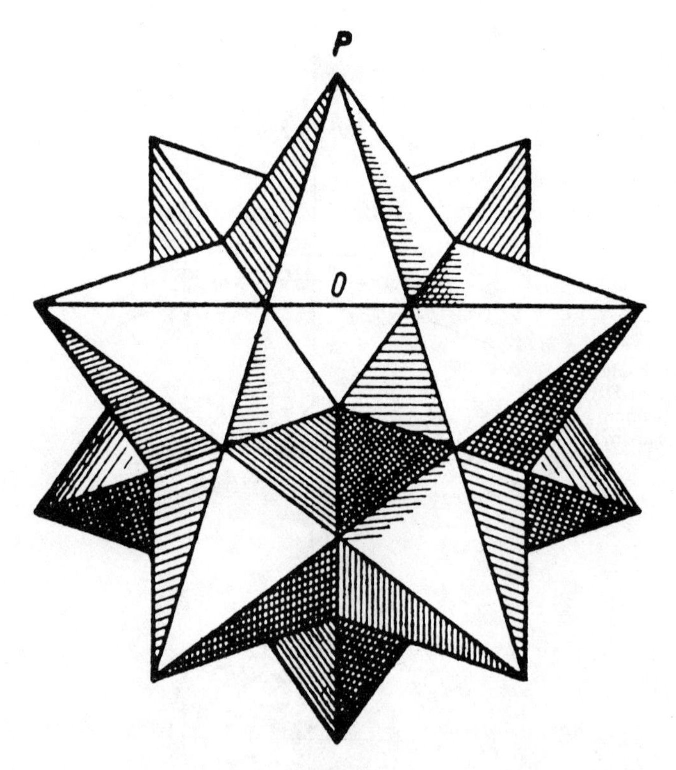

Figure 31.

them are given by the distances, or the spheres, of Mars and Earth, and of Earth and Venus. If we take as a basis of calculation [the distances] from the innermost sphere of Mars to the median sphere of Earth, and from the median sphere of Earth to the median sphere of Venus, we find that the ratios between them are the same; for the mean distance of Earth is the mean proportional between the least distance of Mars and the mean distance of Venus; but the ratios between the [true] spheres of the planets are always greater than those of these two groups of orbits shown in the diagram; furthermore, the centres of the sides of the dodecahedron do not touch the outermost sphere of the Earth, nor do the sides of the icosahedron touch the outermost sphere of Venus; and this *hiatus* is not filled by the radius of the Moon's orbit, added to the greatest distance from the Earth and subtracted from the least distance. However, there is another relationship connected with a certain [semi-regular] figure. If a dodecahedron—to which I have given the name *echinus* (hedgehog), because it is formed from twelve stars of five [projecting] sides, and consequently very close to the five regular solids—be placed with its twelve points on the innermost sphere of Mars, then the sides of the pentagons, which are the bases of the rays or points of the hedgehog, will touch the median sphere of Venus (Fig. 31).

'In short: the cube and the octahedron merge fairly well with their conjugate planetary orbits, the dodecahedron and the icosahedron do not extend to them, but the tetrahedron is exactly in contact with the two orbits in question.

'In the first instance there is a deficiency, in the second there is an excess, in the third there is equality with respect to the distances of the planets.

'From these considerations it appears that the precise relationships between the planetary distances were not derived solely from the regular solids; for the Creator, the very source of geometry and who, as Plato says, geometrizes eternally, does not depart from his model. Certainly, the same conclusion could have been deduced from the fact that all the planets change their distances periodically, and in such a way that each has two particularly important distances, namely the *maximum* and the *minimum*, with respect to the Sun; consequently, the distances of each planet can be compared in a fourfold way, namely, we can compare their *maximum* distances, or their *minimum* distances, or their relative mutually opposed distances, *i.e.*, the greatest as well as the least.[15] Thus, the possible ratios of the distances between two neighbouring planets are twenty in number, whereas the number of solid figures is only five. However, it is reasonable to believe that the Creator, if he paid attention to the relationship between the spheres from a general point of view, also took into consideration the particular relationships between the various distances in the individual orbits: His Providence must be just as perfect in the one case as in the other, and the two aspects must be linked to each other.

'Now, if we consider all this with due attention, we shall certainly come to the conclusion that there is need of some supplementary principles beyond that of the five regular solids[16] in order to establish the diameters and eccentricities of the orbits.

'*Fifthly*, in order to arrive at the motions between which the harmonic relationships have been established, I must, for the benefit of my readers, stress the fact that, starting with the extremely accurate observations of Tycho Brahe, I showed in my *Commentaria de Motibus Stellae Martis* that equal diurnal arcs are not covered with equal velocities on the same eccentric, but that: (a) *in equal parts of the eccentric, the* [*different*] *times bear the same ratio to each other as the distances from the Sun*, the source of motion; and, on the other hand, assuming the times to be equal, for example, one natural day in each case, then: (b) *the true diurnal arcs corresponding to them on one single eccentric orbit are inversely proportional to the distances from the Sun*[17]*; I showed also that:* (c) *the orbit of the planet is elliptical, and that the Sun, the source of motion, is at one of the foci of this ellipse;* it follows (d) *that the planet, when it has covered of its entire circuit, starting at aphelion, is* (e) *distant from the Sun by an amount which is exactly* [equal] *to the* [arithmetical] *mean of its maximum distance at aphelion and its minimum distance at perihelon.* From these two axioms it follows (f) that *the mean diurnal motion of the planet on its eccentric is equal to the true diurnal arc of this eccentric at the moments when the planet has completed one quarter of the eccentric reckoned from aphelion.* Furthermore, it follows (g) *that any two* [true] *diurnal arcs of the eccentric situated at equal true distances, the one at aphelion and the other at perihelion, when taken together, are equal to two mean diurnal arcs*, and consequently, seeing that circles bear the same ratio to each other as their diameters, it follows (h) *that the ratio of one mean diurnal* [arc] *to the totality of the mean* [arcs], equal amongst themselves throughout the entire circuit, is the same as *that of the mean diurnal* [arc] *to the totality of true arcs* of the eccentric, equal in number but unequal amongst themselves.'[18]

Kepler then points out that the possession of a true astronomical system, whilst indispensable, is not enough. At least, in so far as it remains theoretical and abstract, and does not include exact knowledge of planetary distances and velocities. Especially the latter, for velocities, and not distances,[19] provide the immediate basis of celestial acoustics[20]; (the old Pythagoreans were mistaken on that score). It was the need to acquire this knowledge, namely, to work out the data for all the planets from Tycho Brahe's observations, that explains, much more than other hindrances, Kepler's delay in carrying out his project of developing the harmony of the Universe.

We come now to celestial acoustics. According to Kepler, the 'note' corresponding to each planet is determined by the velocity (angular, not linear) of its daily motion. More precisely, the frequency of this 'note' corresponds to this angular velocity measured in seconds. Seeing that the velocity in question is not constant, but changes during the course of the planet's circuit round the Sun, this 'note', too, does not remain constant, but extends over a musical interval whose range is a function of the eccentricity of the orbit of the particular planet. If the eccentricity be known, it is quite easy to calculate the musical interval. For the purpose of this calculation, the least velocity—and hence the frequency—of the planet at aphelion is taken as fixing the fundamental note of the planet. The number of vibrations given out by this note in unit time, namely, one second, corresponds to this velocity.

We shall now make these calculations and compare the slowest motions at aphelion (the lowest note) with the fastest motions at perihelion (the highest note). The accompanying Table A shows that

TABLE A. ANGULAR VELOCITIES OF THE PLANETS AT APHELION AND PERIHELION

Divergent interval	Convergent interval				
		Saturn	$\frac{1' 46'' \, a}{2' 15'' \, b}$	$\frac{1' 48''}{2' 15''} = \frac{4}{5}$	Major third
$\frac{a}{d} = \frac{1}{3}$	$\frac{b}{c} = \frac{1}{2}$				
		Jupiter	$\frac{4' 30'' \, c}{5' 30'' \, d}$	$\frac{4' 35''}{5' 30''} = \frac{5}{6}$	Minor third
$\frac{c}{f} = \frac{1}{8}$	$\frac{d}{e} = \frac{5}{24}$				
		Mars	$\frac{26' 14'' \, e}{38' \, 1'' \, f}$	$\frac{25' 21''}{28' 1''} = \frac{2}{3}$	Fifth
$\frac{e}{h} = \frac{5}{12}$	$\frac{f}{g} = \frac{2}{3}$				
		Earth	$\frac{57' 3'' \, g}{61' 18'' \, h}$	$\frac{57' 28''}{61' 18''} = \frac{15}{16}$	Major semitone
$\frac{g}{k} = \frac{3}{5}$	$\frac{h}{i} = \frac{5}{8}$				
		Venus	$\frac{94' 50'' \, i}{97' 37'' \, k}$	$\frac{94' 50''}{98' 47''} = \frac{24}{35}$	Minor semitone
$\frac{i}{m} = \frac{1}{4}$	$\frac{k}{l} = \frac{3}{5}$				
		Mercury	$\frac{164' 0'' \, l}{384' 0'' \, m}$	$\frac{164' 0''}{394' 0''} = \frac{5}{12}$	Octave + minor third

the ratios, or intervals, are almost entirely consonant: in fact, the divergence is generally less than a semitone, except in the case of the Earth and Venus whose orbital eccentricities are too small.[21]

These are the results obtained for each planet by itself. As for the relationship between the 'notes' given out by the whole assembly of planets, or, in the first instance, the relationship between the 'notes' given out by neighbouring planets, the matter may be considered in two different ways. On the one hand, we can compare the velocities of the planets at their greatest separation; and, on the other hand, their velocities when the planets are closest together; that is to say, the least velocity (the lowest note) of the superior planet at aphelion with the greatest velocity (the highest note) of the next following planet at perihelion (this constitutes the divergent interval): or, the highest note (the fastest motion at perihelion) of the superior planet with the lowest note (the slowest motion at aphelion) of the inferior planet (this constitutes the convergent interval).

On referring once more to the table we observe that there is almost perfect consonance in both cases, except for Mars and Jupiter, where the ratios of distances agree however with those required by the theory of regular solids. We shall see, shortly, by considering the planets all together, that the ratios of their motions, or of the 'notes' which they give out, produce a most harmonious concert, and not confusion or cacophony; but we must not apply our time-scale to the celestial concert. Though the harmony (consonance) between two planets is relatively frequent, it becomes more and more difficult to realize this harmony when the number of planets is increased. If all of them be considered together, then the time required for such a consonance to be produced (supposing, as is reasonable, that it was realized at the very beginning of creation) is so long, that it possibly measures the duration of the Universe from the Creation to the Day of Judgement.

Let us now continue our analysis of astral music. By relating sounds to motions (frequencies to angular velocities), Kepler was able to associate a note, or rather, a musical interval with each of the planets. It is quite obvious that these notes, or these intervals, do not all fall in the same octave. In order to decide to which octave the highest and lowest notes of the planets belong, Kepler took the octave containing the notes of Saturn as his basis, and divided the numbers expressing the greatest and least velocities of each planet by 2 as many times as was necessary to give a final ratio between them smaller than $\frac{1}{2}$; in other words, he divided them by a power of 2 that would give a quotient less than double the figures expressing the velocities of Saturn. The exponent of 2 then indicates the octave to which the note belongs, and the quotient indicates the note in the fundamental octave of which the note given out by the planet is the exact 'harmonic'.[22]

TABLE B

Velocity of Saturn at		aphelion	divided by	$2^0 = 1' 46''$
,,	,,	perihelion	,,	$2^0 = 2' 15''$
,,	Jupiter	aphelion	,,	$2^1 = 2' 15''$
,,	,,	perihelion	,,	$2^1 = 2' 45''$
,,	Mars	aphelion	,,	$2^3 = 3' 17''$
,,	,,	perihelion	,,	$2^4 = 2'23''$
,,	Earth	aphelion	,,	$2^5 = 1' 47''$
,,	,,	perihelion	,,	$2^5 = 1' 55''$
,,	Venus	aphelion	,,	$2^5 = 2'58''$
,,	,,	perihelion	,,	$2^5 = 3' 3''$
,,	Mercury	aphelion	,,	$2^6 = 2' 34''$
,,	,,	perihelion	,,	$2^7 = 3' 0''$

If we assume that the note of Saturn at aphelion is G_1, then the lowest note of Earth will be G also, because the two notes are represented by practically identical numbers, namely $1' 46''$ and $1' 47''$ [$1' 47''.2^5$], but this G will be pitched five octaves higher. The number corresponding to the highest note of Mercury is $3' 0''$ [$3' 0''.2^7$], which is very close to $\frac{5}{3}$ times $1' 47''$; this note is therefore E, and it is pitched seven octaves plus a major sixth above the lowest note of Saturn; and so on.

The tunes (in modern notation) played by the individual planets will be as follows[23] (Fig. 32):

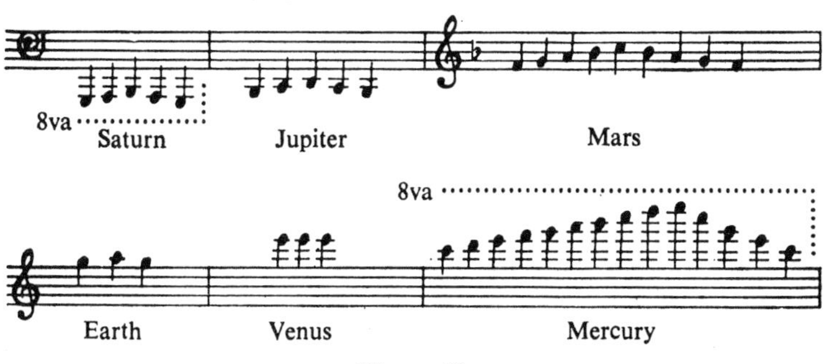

Figure 32.

The differences between the eccentricities of the planets is immediately obvious; the almost circular orbit of Venus stands in contrast with the highly eccentric orbit of Mercury. The gap between Mars and Jupiter is also most obvious.[24]

A glance at the tables reveals a close relationship between the position of the octaves, or more correctly, the 'notes' ascribed to the planets and the places they occupy in the solar system; the position of the octaves, which corresponds to the pitch of the notes or the frequency, gets higher as we approach the Sun. There is nothing startling in this, seeing that the pitch of the notes has been linked to the velocities of the planets, and we all know that the planets move quicker the nearer they are to the Sun, and move slower the farther away they are. What is not known, is the exact relationship between distance and velocity, or the periodic time. If this were known, it would be possible to compute the distances from the times, that is to say, from the harmony of the Universe . . . for it is obvious that there must be an exact mathematical relationship between the two; and it is equally obvious that the Divine Musician took this relationship as the basis for ordering the motion of the planets and their distribution in the heavens.

We are now faced, even more urgently, by the problem that Kepler had raised in the *Mysterium Cosmographicum*, and which he had not been able to solve at the time. Now he has found the answer.[25]

> 'At the time of the *Mysterium Cosmographicum*, I did not clearly see my way . . . but after I had determined the true orbital distances, as a result of incessant labour over a long period of time and by making use of Tycho Brahe's observations, finally, the true relationship between the periodic times of the orbits occurred to me, and if you ask me when exactly I made this discovery, it is as follows: The idea came to me on 8 March of this year 1618, but as the calculation was not successful, I rejected it [the answer] as being false. However, the idea came to mind again on 15 May, and through a fresh assault overcame the darkness of my reason so completely and in such harmony with my seventeen years of labour on Tycho Brahe's observations as well as my present studies, that I thought at first that I was dreaming and that I was assuming as an accepted principle something which was still the subject of investigation. Nevertheless, the principle is indisputably true and quite correct: *the periodic times of any two planets are in the sesquialteral ratio to their mean distances,*[26] *i.e.*, of their orbits.'

How did Kepler arrive at this ratio? Unfortunately, we do not know. Kepler generally keeps us well informed about the successive stages in the development of his thought, but this time he is silent. Did he arrive at it *by trial and error*, as R. Small and J. B. Delambre surmise,[27] that is to say, by trying various combinations of powers? It is not impossible, seeing that he need not have made a large number of trials. If the law $(T^2/R^3) = C$, which governs the relationship between periodic time and planetary distance cannot be deduced directly from his fundamental dynamic law, velocity is inversely proportional to distance (when each planet is considered individually this leads to the relationship: time is proportional to distance; and with different planets, it leads to the relationship: time is proportional to the square of the distance[28]), then, because both of these relationships are wrong (one being too large, and the other too small), it is rather obvious to try the sesquialteral ($\frac{3}{2}$) ratio which lies between the two.[29] From the dynamic point of view, it seems also rather natural to explain the matter by a corresponding relationship between the *moles* of the planetary bodies.[30]

Certainly, these are quite natural steps in the development of thought. I must confess, that they seem a little too much so. We are not accustomed to such simple developments from Kepler.... Furthermore, when he says that he first had the idea on 8 March 1618, and abandoned it because it was not confirmed by his calculations—which could mean that the calculations he had made (and in which he had doubtless made a mistake) were not confirmed by the sesquialteral ratio—but that he came back to it on 15 May of the same year, when he found it to agree perfectly not only with empirical facts but also *with his present studies* (*of harmony*), he seems to imply that these later considerations were responsible for setting him on the way towards discovery of his law.

We are also tempted to believe that he already had other ideas, both daring and profound, at the time of the *Harmonice Mundi*. (In particular, they included a perfectly clear formulation of the concept of *mass*, which replaced the confused notion of *moles*.[31]) Starting from these ideas, the sesquialteral ratio was worked out *a priori* in the *Epitome*. Unfortunately, yet again... ! In the *Harmonice Mundi*, it was not (as in the *Epitome*) this ratio which was derived from that of the dimensions of the celestial bodies; on the contrary, it was these latter dimensions which were adapted to the ratio by assuming[32] that 'the volume (*moles*)[33] of the bodies is proportional to the periodic times'; furthermore, the proportionality is given in a way which is expressly rejected in the *Epitome*, where it is not the volume (*moles*), but the mass (quantity of matter), that is to say, the volume (*moles*) multiplied by the density of the planet, which is co-ordinated with the ratio of the periodic times as cause and not as the effect.[34]

Once in possession of the law of periodic times, Kepler set to work again, and finally arrived at the following ratios of the maximum and minimum planetary velocities,[35] not without an occasional 'adjustment' so as to obtain a more perfect harmony, as he himself admits:

Planet	*Ratio of velocities*
Saturn	81 : 64
Jupiter	8000 : 6561
Mars	36 : 25
Earth	3125 : 2916
Venus	250 : 243
Mercury	12 : 5

Using these ratios, he recalculated the eccentricities, the mean motions and the mean distances. The results are summarized in the following tables:

Eccentricities (the radius of the planetary orbit being taken = 100 000):

Saturn	5 882
Jupiter	4 954
Mars	9 091
Earth	1 730
Venus	710
Mercury	21 551

MEAN DISTANCES
(the radius of the Earth's orbit being taken = 1000)

	Calculated from harmony at		From the observations of Tycho Brahe at	
	Aphelion	Perihelion	Aphelion	Perihelion
Saturn	10 118	8994	10 052	8968
Jupiter	5464	4948	5451	4949
Mars	1661	1384	1665	1382
Earth	1017	983	1018	982
Venus	726	716	729	719
Mercury	476	308	570	307

If we now assume the *minimum* distance of Saturn from the Sun (at perihelion) to be 8994 semi-diameters of the Earth's orbit, we can calculate the distances according to the scheme of the five regular solids, and can then compare them with the results obtained from harmony. The results are as follows:

If the semi-diameter of the sphere of Saturn circumscribed about the cube be equal to 8994 semi-diameters of the Earth's orbit, then:

The semi-diameter of the sphere inscribed		Distances calculated from harmony	
In the cube (Jupiter)	= 5194	Mean	= 5206
In the tetrahedron (Mars)	= 1649	Aphelion	= 1661
In the dodechedron (Earth)	= 1100	Aphelion	= 1018
In the icosahedron (Venus)	= 781	Aphelion	= 726
In the octahedron (Mercury)	= 413	Mean	= 392

If instead of the tetrahedron we put an echinus in the sphere of Mars, we obtain a value of 726 for the distance of Venus at aphelion, which agrees with harmony and with observation; and if we relate our calculation to the square on the base of the octahedron placed in the sphere of Venus, we obtain a value of 406 for the distance at aphelion, and 376 at perihelion. The difference is less, but remains; and we know that it must remain.

The harmonic theory thus emerges victorious after being confronted by facts. Furthermore, it explains the partial set-back in the concept of the *Mysterium Cosmographicum*. Looking back, Kepler said[36]:

'All that has been said so far will become clearer by giving an account of my discoveries. Twenty-four years ago, when I engaged in these speculations, I first of all asked myself if these planetary spheres were not equi-distant from each other (in fact, according to Copernicus, the spheres are separated from each other, but do not touch): for I could imagine nothing finer than a relationship of equality. However, this relationship has neither head nor tail: for this material equality did not yield a fixed number of moving bodies, nor a definite size for the intervals. I then thought about similarity, that is to say of the same ratio between intervals and between spheres; but the result was just as defective. To be sure, it followed that there were different intervals between the spheres; on the other hand, [they were] not unequal in an irregular manner, as required by Copernicus; and neither the value of the ratio, nor the number of the spheres could be inferred. I then passed to a consideration of the regular plane figures: the intervals were obtained from the inscribed circles: but not as a fixed number. I came to the five [regular] solids: in this instance the number of bodies and the size of the intervals were almost correct. With regard to the apparent deviations I had recourse to the perfection of astronomy [through Tycho Brahe]. Now, astronomy was perfect during those last twenty years; and yet the intervals deviated from the solid figures; and the reason for the very uneven distribution of the eccentricities amongst the planets appeared not at all. It was obvious: in the structure of the Universe, I had been looking only for stones, [and] for the most elegant form suitable to stones; unaware [as I was] of the fact that the Architect had made them in the articulate image of an animated body. Thus, gradually, especially during the last three years, I came to Harmonies, and abandoned the scheme of regular solid bodies as far as details were concerned; in fact, they [the Harmonies] held themselves apart from the form which, in the last analysis, is imposed by the hand; and the forms from the material substance which is [represented] in the Universe by the number of bodies and the entire amplitude of the gaps; furthermore, they [the Harmonies] provided the eccentricities which the latter [the regular solids] were unable to give: they gave, so to speak, nose, eyes and other members to the statue, whereas the latter [the regular solids] prescribed only the rough external quantity of its mass.

'For this reason, in the same way that it is not usual for the bodies of animated beings, or the volumes [blocks] of stones to be shaped in conformity with the absolute standard of some geometric form, but that they should deviate from the external spherical shape, however elegant it might be (nevertheless, retaining the exact measure of its volume), in order that the body may acquire the organs necessary to life and that the stone may receive the image of the animated being; so, the proportions which should be prescribed for the planetary spheres by the solid figures, were less [in value], and as they affected only the body and its material substance, they ought to yield to the Harmonies as much as is necessary in order to make them come closer and [contribute] to the beauty of the motions of the spheres. . . .'

Once again, the individual harmonies (consonances) must bow before the Harmony of the Universe.

Perfect agreement has now been secured. The secret of cosmic order has been discovered. We can easily understand that Kepler was surprised to receive this signal favour from the Lord, and that it fell to him to have the glory of revealing to the world the mysteries of Creation. Still, the ways of Providence are imperscrutable, and

Kepler could do no more than give thanks to God[37] . . . *Gratias ago Tibi Creator Domine*. On the other hand, he was aware that his ideas were too modern, too extraordinary, too conflicting with the tradition of a thousand years for him to have any hope of their being immediately accepted. No matter! 'The die is cast. My book is written; it will be read by the present age, or by posterity. . . . It could wait a hundred years to find a reader. Did not God wait six thousand years for one to contemplate His works?'[38]

III

Harmony of the Universe in the *Epitome*

With the *Harmonice Mundi*, Kepler had realized and even surpassed, the purpose which he had formerly set himself: the structure of the Universe, and even the profound justification of this structure, had been revealed to his astonished eyes. For the future, Kepler could therefore peacefully enjoy the happiness, or bliss, that devolved upon the faithful disciples of the Sage of Samos.

Unfortunately, this happy state did not last long. Kepler very soon realized that the solution was incomplete; the apprehension of ideas was not perfect. The problem of the dimensions of the Universe had not even been touched. Furthermore, and this was something even more serious, the law relating planetary distances with the periodic times had been accepted as a fact: it had not been *deduced* from considerations of harmony; nor had it been linked with the general constitution of the Cosmos.

It was for the purpose of supplying this omission that the later chapters of the *Epitome Astronomicae Copernicanae* were devoted to the *Harmonice Mundi*. We shall now deal with the deduction of the 'sesquialteral' ratio from 'harmonic' and 'archetypal' considerations as given there by Kepler. However, before doing so, we ought to stay and consider a most extraordinary undertaking of this extraordinary mind, namely, an overall estimate of the structure and size of the Universe; in other words, a disclosure of the mind of the Divine Mathematician before the creation of the Universe, which Kepler, emboldened by his recent successes,[1] felt more and more certain of being able to achieve through his own mind. After all, why not? In fact, did not God create man in His own image, a 'creature of meditation' capable of understanding; and, conversely, did not He create the Universe as a function of this 'creature of meditation'? Hence, there was an obligation to put him in a position which would enable him to achieve his supreme purpose, the intellectual contemplation of God's handiwork. Consequently, it is perfectly clear, that the Earth, the dwelling place of this creature of meditation, acquires a primordial rôle, a key-rôle, in building the structure of the Universe. This is what he says, without equivocation, in the *Epitome*[2]: Kepler's imaginary questioner asks[3]:

> '*In your opinion, with what must one start in an investigation of the proportions between* [*celestial*] *bodies?*—With the Earth, because it is (1) the dwelling place of the creature of meditation; (2) who is assuredly the image of God the Creator; (3) [with the Earth] because we read in the divine Moses that 'In the beginning God created the heaven and the earth'; (4) because the Earth's orbit is the common measure [of the spheres of the planets] and even forms a mean proportional between the limits of the superior planets and of the inferior [planets]; and, finally, (5) because every arrangement of these proportions loudly proclaims that, by making the [dimensions] of the bodies and the intervals between them suited to the body of the Sun as the primary measure, God the Creator started with the Earth.'

This being accepted, it is nonetheless true that the Sun is the central and principal body in the Universe, being the visible image of God the Invisible,[4] and that its dimensions could not, or at least, ought not to depend on those of the Earth. No doubt; but the effective dimensions of the body of the Sun are arbitrary, precisely because it is the first; everything else is in relation thereto[5]:

> '*What, in your opinion, is the reason for the dimensions of the body of the Sun?*—The following reasons persuade us that the globe of the Sun is the first of all bodies in the Universe in the archetypical order of creation, as well as in its temporal order: (1) According to Moses, the work of the first day was [the creation of] light, by which we may understand the body of the Sun; (2) The body of the Sun received priority amongst natural things, not only as regards its bulk, but also as regards the point of time in which it was created. Now, therefore, the first body, from the very fact that it is the first, bears no proportion to those that follow; but those that follow [are determined] with respect to the first. For this reason there is no archetypal reason for the dimensions of the Sun, and a globe twice the size would not have been different from the one that now exists: for, at the same time, the remainder of the Universe, and man with it, would similarly have been twice their present size.'

A question now arises: What is the basis of the relationship which the Creator established between the body of the Sun and that of the

Earth? We shall see very shortly, that this basis is not purely architectonic, but essentially harmonic and epistemological; and the first element to be fixed was not that of their dimensions, but of their relative distances[6]:

> '*In what way, therefore, were the dimensions of the Earth adapted to those of the Sun?*—By taking into account the conditions of vision of the Sun. Earth was the future dwelling place of the creature of meditation, as a function of whom the whole Universe was created. Now, contemplation (or meditation) has its origin in the vision of stars, seeing that the number of things to contemplate has its origin in the number of things to be seen. Therefore, the first visible [thing] was the light of the Sun, because: (1) it was the work of the first day; and (2) of all visible things it is the most excellent, the foremost, the original and of necessity had to be the cause of visibility of all other things. It follows, therefore, that the vision [apparent dimension] of the Sun from the Earth was taken as a basis in accordance with which the bodies of the Universe have been proportioned; similarly with regard to the superior celestial bodies, the dimensions of the regions of the Universe corresponding to them were determined by a proportional dividing of the Earth's orbit.'

We now come to details[7]:

> '*What is the appearance* [*apparent size*] *of the Sun's diameter?*—According to the ancient observations of Aristarchos and the most recent observations of our times it is constant and is such that if the Earth were placed at its greatest distance from the Sun, and a circle were described from the centre of vision, the diameter of the Sun would occupy exactly one seven hundred and twentieth part, *i.e.*, one half degree, of this circle; or, in other words, the angle between the lines touching the two sides of the Sun's boundaries would be one seven hundred and twentieth part of four right angles.
>
> '*What, in your opinion, is the reason for this figure?*—The archetypal cause of the first thing must be sought also among primary things; but there is no geometrical reason for dividing the circle into 720 parts starting from a figure with the same number of sides, for this figure is derived by bisection of a 45-sided figure which cannot be constructed geometrically, as is proved in Book I of the *Harmonice Mundi.*[8] It follows that section of the circle in this manner is justified on grounds of harmony. Now, it appears to be necessary for the circle of the Zodiac, (in which the true motions of all the planets, as well as the apparent motion of the Sun must be performed), to be divided into parts pertaining to harmonic numbers based on the appearance [apparent size] of the primary body. The smallest number with which we can fix all the

divisions on a monochord, required to establish a system covering the range of two scales (major and minor), is 720, as was proved in the *Harmonice Mundi*, Book III, chapter VI.[9] Seeing that the motions of all the planets, as I have proved in Book V of the *Harmonice Mundi*, have been accommodated to these two scales, it was appropriate that the primary body itself, which is the leader of this chorus, should, by the appearance of its diameter as seen from the Earth, divide this circle of the Zodiac—the indicator and measure of harmonious motions—according to the division of the monochord, that is to say into 720 parts, for the benefit of terrestrial inhabitants, namely, creatures of meditation. Now, 720 is twice 360, three times 240, five times 144, six times 120, eight times 80, ten times 72, twelve times 60, fifteen times 48, sixteen times 45, eighteen times 40, twenty times 36, twenty-four times 30—which gives the greatest number of divisions in aliquot parts. '*What is the result of assuming this hypothesis with regard to the interval [distance] between the Sun and the Earth; and what is this ten foot pole we use to measure planetary orbits?*—If the diameter of the Sun as seen from the Earth must occupy one half-degree, then the view-point, or, rather, the centre of the Earth, must be distant from the Sun by an amount which is equal to 229 semi-diameters of the globe of the Sun's body, or a trifle more, as is taught in geometry.'

The distance between the Earth and the Sun, and also the dimensions of the Sun, specified in this manner, nevertheless leave the dimensions of the Earth quite undetermined. Kepler continues[10]:

'*I have the distance; tell me also the size of the Earth's globe as determined by its causes.*—The previous considerations are not yet sufficient for a determination of the dimensions of the Earth; for this purpose we require a supplementary axiom. It is clear, seeing that the Earth was destined to be the future dwelling place of a creature capable of measuring, that it should, through its own body, provide a measure of the bodies of the Universe, and its semi-diameter—seeing that it is a line—[should be the measure] of the lines, that is to say, of the distances. As the measure of bodies is different from the measure of lines, and as the first proportions established by the Creator were the ratio between the body of the Earth and that of the Sun, as well as [the ratio] between the diameter of the Earth and the distance (interval) from the Earth to the Sun, nothing is more suitable to a well-ordered and harmonious construction than to suppose that these two ratios are equal. We must then assume that the number of times the body of the Earth is contained in the body of the Sun is equal to the number of times the semi-diameter of the Earth is contained in the distance between the Sun and the Earth, that is to say, [we must assume] that the ratio of the body of the Earth to the body of the Sun is the same as that of its semi-diameter to the distance between their centres.

'*How do you deduce the quantity* [*dimension*] *of the Earth's semi-diameter from these two axioms?*—Let us suppose the Sun's semi-diameter to be 100 000 parts; the distance between the centres of the Sun and the Earth will be [equal to] 22 918 116 such parts; the cube of 100 000, *i.e.*, 1 000 000 000 000 000, is to be divided by the distance 22 918 116; the [cube] root of the quotient (which is the *sine* of 0° 15′ 00″) must be found; it is 6606. This is the Earth's semi-diameter; for the number of times 6606, the Earth's semi-diameter, is contained in 22 918 116, the distance between the Sun and the Earth, is the same as the number of times the cube of 6606 is contained in the cube of 100 000, the Sun's semi-diameter, that is to say 3469 and one-third times. . . . The Sun's semi-diameter will contain the Earth's semi-diameter slightly more than 15 times, but the body of the Sun will contain that of the Earth rather more than 3469 times.[11]

'*You almost treble the quantity* [*value*] *which the Ancients assigned to the greatest distance between the Sun and the Earth, for they estimated it to be 1200 semi-diameters of the Earth; and as for the proportion of the bodies you increase it 20 times, for they* [*the Ancients*] *made the Sun only 166 times larger than the Earth: have you no fear from astronomical observations?*[12]—Not at all; for the Ancients made the Sun so close that they were obliged to assign to it a parallax of three minutes. Consequently, Tycho Brahe concluded that the parallax of Mars—seeing that it is nearer to the Earth than to the Sun—must be greater than three minutes.[13] However, I observed that the parallax of Mars is never capable of being perceived. The distance from Mars [to the Earth], even when it is very near, and *a fortiori* that of the Sun, is greater than 1200 semi-diameters. The diameters of Mars and Venus can be observed with the old instruments and also with the new Netherlandian telescope, and are found to be no more than several minutes. Therefore, if the Sun were as close as the Ancients claimed, these planets, each in its proportion, would be very close also; this was deduced by Tycho Brahe from [the views of] Copernicus. If Mars were so close [to the Earth], it ought also to be smaller, for its diameter to be such as we see it. Mars would then be smaller than the Earth, that is to say, the superior would be smaller than the inferior; in which case, there would be no analogy between the size of the bodies and their order, which is not compatible with the beauty of the Universe.'[14]

Thus, astronomical observations do not invalidate, but on the contrary confirm the increase in distance between the Sun and the Earth, as postulated by Kepler, and even confirm the analogy between sizes and distances. . . . Having obtained the dimensions of the Earth, we can now go further:

'*Which is the body the determination of whose dimensions immediately follows the determination of those of the body of the Earth?*—That of the Moon, a secondary planet: (1) because this celestial body is particularly affected by the Earth,[15] in that it helps the growth of terrestrial creatures, is observed by the creature of meditation who inhabits the Earth, and because it is through [observing] the Moon that observation of the stars begins; (2) because the ratios of the constituent proportions are almost the same.

'*State the foundations of the relationships between the Moon and the Earth both as regards the* [*dimensions of the*] *bodies and their distance.*—(1) The Moon at its greatest separation from the Earth should occupy one 720th part of a circle by its visible diameter, as much in virtue of this number itself (whose perfection has been set forth above), as on account of eclipses of the Sun, a spectacle ordained by the Creator in order to give the creature of meditation the opportunity to learn the nature of the motion of celestial bodies. Now, in order to produce these eclipses in a perfect manner, it is necessary for the semi-diameters of the Sun and the Moon, at their greatest distance from the Earth, to appear equal; the Moon will then be able to cover the Sun exactly. . . . (2) It is also appropriate for the ratio [between the dimensions] of the bodies of the Earth and of the Moon to be to the ratio between the distance of the Moon and the semi-diameter of the Earth as the ratio between the bodies of the Sun and the Earth is to the ratio between the distance of the Sun and the semi-diameter of the Earth[16]: that is to say, each of these ratios should be similar to the others. In fact, the Moon, being a secondary and terrestrial planet and made to obscure the Sun, should follow the example (model) of the proportions between the orb of the Sun and that of the Earth.[17]

'*What is the consequence?*—From the two axioms as laid down there are two consequences, each of which in itself is endowed with great probability: furthermore, although they do not follow from the preceding [axioms], they may be used as axioms, for in themselves they are most worthy of confidence. This first is this: seeing that the ratio of the proportions relating to the Sun [and to the Earth] is one of equality, in other words, seeing that the body of the Earth is contained in the larger body of the Sun as many times as the Earth's semi-diameter is contained in the distance or semi-diameter of the sphere of the Earth or of the Sun . . . it will be the same in respect of the Earth and the Moon: the body of the Earth will contain the smaller and more confined body of the Moon as many times as the semi-diameter of the Earth is contained in the distance or semi-diameter of the sphere of the Moon. . . . This proposition acquires its value as an axiom from the fact that the Earth is the habitation of a creature capable of measuring: for this reason, [the Earth], through its body, provides a measure of the smaller body of the Moon, in the same way that it provided a measure of the body larger than itself, namely, the Sun; similarly, and in the same way, it provides a measure of the semi-diameter of the sphere of the Moon through its [the Earth's] semi-diameter, because the sphere of the Moon alone is placed about the Earth as the sphere of the Earth is placed about the Sun: for this reason a determination of the dimensions of the sphere and the body of the Moon, much more than that of the bodies of the other planets, is apt for the Earth; certainly as much as was the determination of the dimensions of the sphere and of the body of the Sun. So, it is appropriate, if nothing prevent, to put the ratio of equality here, seeing that it is the first and foremost [of the ratios]. Furthermore, we can deduce from these two premises, (the proof is very long and may be found in my *Hipparchos*[18]), that the semi-diameter of the Moon's orbit, or its distance from the Earth, is a mean proportional between the distance or the semi-diameter of the Earth's sphere and the semi-diameter of the body of the Moon; consequently, the ratio of the Earth's semi-diameter to the semi-diameter of the sphere of the Moon is equal to the ratio of the latter to the semi-diameter of the sphere of the Earth or of the Sun. A certain equality between the two proportions is found there, too, and is very probable, for the sphere of the Earth with respect to the Sun is as the sphere of the Moon with respect to the Earth.'

All these arguments are very probable and plausible; but[19]:

'*Are observations in good agreement with such a distance between the Moon and the Earth?*—Perfectly. In fact, Tycho Brahe found [that] the distance between the Moon and Earth at perigee was a little less than 54 terrestrial semi-diameters . . .; at apogee, [this distance] is greater than 59 and less than 60 semi-diameters, whereas, according to the principles developed above, the former is equal to 54 and the latter is equal to 59 [semi-diameters].

'*What, then, is the ratio between the diameters of the Sun and of the Moon?*—The same as that between the orbit of the Sun and the orbit of the Moon, or of the latter to the body of the Earth, namely, that which is between the numbers 59± and 1.'

Having fixed the dimensions of the two 'luminaries' in this manner, Kepler then passed on to a study of the structure of the Universe, *i.e.*, the solar system as a whole; the laws which govern it are not the same as those which had determined the relationships between the dimensions and the spheres of the Sun, the Earth and the Moon. In particular, epistemological considerations no longer intervene: henceforth, considerations of architectonic and harmonic expediency are uppermost[20]:

'*What is the ratio between* [*the dimensions*] *of the planetary globes?*— Nothing is more conformable to nature than to make the order of magnitude of the globes identical with that of the spheres. Consequently, of the six planets, Mercury has the smallest body, seeing that it is the innermost and has the most confined orbit; Venus is a trifle larger, but yet smaller than the Earth, because it revolves in a smaller orbit than does the latter, though the orbit is larger than that of Mercury; Mars follows [and is a little larger than] the Earth in order of magnitude, for its orbit is outside [that of the Earth] and larger [than that of the latter], but it is nevertheless the innermost of the superior [planets]; next comes the globe of Jupiter, which is the middle one of the superior [celestial bodies]; finally, the largest of the planetary globes is that of Saturn, because it is the highest [most distant]. Now, as bodies have three dimensions, either according to their diameters, or according to their surfaces, or according to the spaces contained within their surfaces, that is to say, according to their corpulences (volumes), and as the ratio between the diameters is double for that of the surfaces and treble for that of the bodies[21]; it is agreeable to reason that the ratio between the intervals (distances) should be accommodated to one of these three [ratios] between the globes. For example, as Saturn is nearly ten times higher [more distant] than the Earth from the Sun, either the diameter of Saturn will be ten times larger than the diameter of the Earth, the surface one hundred times that of the Earth, and the body one thousand times that of the Earth; or the surface of Saturn will be ten times that of the Earth in order that the ratio between the bodies [volumes] shall be sesquialteral [$\frac{3}{2}$] with respect to the ratios between the intervals, and that Saturn shall be thirty times larger than the Earth, and its speed one-thirtieth[22]; but the ratio between the diameters will be only one half of the ratio between the intervals, namely, a little more than treble. Or, again, the bodies themselves will have a ratio between the intervals such that Saturn, being ten times higher [more distant], will be only ten times larger than the Earth; on the other hand, with respect to the surfaces, it would have only two-thirds of the ratio between the intervals, and with respect to the diameters, one-third; thus the diameter of Saturn would be only a little more than twice the diameter of the Earth.

'*Of these three ways, the first does not agree with archetypal reasons, nor with observations.*'[23]—In fact, in that case the planets should have been much larger than they really appear. As for the archetypal reasons, Kepler explains that: "in the same way that we have already assumed that the ratios between the bodies of the Sun and the Earth, of the Earth and the Moon were the same as [the ratios between] the semi-diameter of the Earth and the semi-diameters of the spheres [orbits], so, now, we must assume that the ratios between the planetary bodies follow those of the semi-diameters of the spheres. Thus, Saturn by the magnitude [volume, *moles*] of its body will be a little more than ten times larger than the Earth, Jupiter more than five times, and Mars one and one-half times; but [the *moles* of] Venus will be a little smaller than three-quarters of that of the Earth, that of Mercury a little larger than one-third.'

Kepler ceased not to be preoccupied with modification of the orderly principle to which I have alluded. Furthermore, he considered that he ought to justify it[24]:

'*Ought not the bodies of all other planets to be determined in ratio to the body of the Sun, in the same way that* [the moles of] *the Earth* [*was determined in ratio to that of the Sun*], *and in virtue of the same laws as* [the body of] *the Earth?*—Not at all; for the result would be that the dimensions of the bodies of the planets would follow an order the reverse of that of the spheres: Mercury would be the largest, and Saturn the smallest with a diameter less than one-third of that of the Earth,'

which would be absurd and, what is more, contrary to fact, that is to say, observable reality.

Up to now, in his analysis of the structure of the Universe, Kepler had only introduced the ratios between the geometrical dimensions of the celestial bodies—and 'spheres'; but we know quite well that celestial objects are not pure geometrical objects: they are material bodies, analogous, though doubtless dissimilar, or at least not entirely similar, to the material of the Earth.[25] In any case, the analogy is sufficiently close, simply on account of materiality, to enable us (and we must) to consider them all under a common aspect, and to introduce a new dimension, namely *density*,[26] into the analysis of their structure.

It is this density which, combined with the volume (*moles*) of the body, will determine the abundance (*copia*) of matter contained in it; and it is this *copia* or quantity of matter—a term which we could translate by *mass*—which in its turn will determine the *inertia*, that is to say, the resistance of the body in question to motion, and hence its velocity, according to the fundamental principle of Aristotelian dynamics: velocity is proportional to the motive force and inversely proportional to the resistance to motion.[27] As a result, the time required for each planet to complete its circuit will be found to be fixed; and as the density, as much as the volume, in the case of the planets is related to their distance from the Sun, their periodic times will be determined finally as a function of this distance. We now ask[28]:

'*What can we deduce from the rarity and the density of the six globes?*—

'*First*, it is not fitting for the density of the matter to be the same in all; for, where a certain multiplicity of bodies [seems] necessary, a variety of properties must also be assumed, in order that [the bodies] be truly multiple. Now, the chief property (condition) of bodies *qua* bodies is the internal arrangement of their parts; for inequality in volumes (*molium*) does not happen to bodies themselves except through the surfaces which, in a way, delimit their volumes; furthermore, one internal part of a body does not differ from one part of another by virtue of this limitation of volume [but only by its internal structure]. However, the main argument in favour of the dissimilarity between the matter [of celestial bodies] is deduced from the difference in periodic times, which would not occur if the density of the globes was the same, as we shall see later. In fact, in this case, the variation between masses would be parallel to that of the motive forces.[29]

'*Secondly*, it is appropriate for a body nearer to the Sun to be denser. In fact, the Sun is the densest body in the whole Universe, evidence whereof is its immense and manifold force, which could not but have a proportionate cause; furthermore, those places near the centre involve a certain notion of confinement which [expresses itself] by condensation of matter in a confined place.

'*Thirdly*, the rarity, in any case, ought not to be ascribed to bodies in proportion to their largeness, nor denseness in proportion to their smallness. For example, the distance and size of the globe of Saturn (in accordance with what has been developed above) are to the distance and size of the globe of Jupiter almost as 10 is to 5. I say, then, that the density of the matter in the globe of Saturn to its density in the globe of Jupiter should not be put down as having the [inverse] proportion, namely, that of 5 to 10. If we were to do so, we should infringe another law of variety by ascribing the same, and not a different mass (abundance of matter, *copia*), to all the planets. In fact, if we multiply the volume of Saturn, 10, by its density, 5, we obtain a quantity (abundance, *copia*) of matter equal to 50, that is to say, as much as if we multiplied the volume, 5, of Jupiter by its density, 10. Now, it seems preferable and better that neither the volumes of the globes of different densities should be equal amongst themselves, nor the density in globes of unequal volumes, similarly; nor that the abundance (*copia*) of matter should be distributed in equal portions amongst all the globes differing in volume and density;

but rather, in order that everything should be diversified, [it is preferable] that the order (I mean order, and not proportion) in which the mobile globes succeed each other, starting from the centre, should be co-ordinated not only with the volumes of the bodies and their rarity, but also with the very abundance of the matter they contain. . . . In other words [it is fitting] that, if Saturn has an abundance of matter [mass] of 50, then Jupiter has less than 50, in any case, more than half [of 50], 25 (let us say, 36): therefore the bodies will be as 50 to 25, the abundance of matter (mass) as 50 to 36, the rarity as 50 to 36, or as 36 to 25, and the density, inversely, as 25 to 36, or as 36 to 50.'

Such an arrangement is obviously much more suitable than one that would ascribe the same mass to all the planets, for that would produce unbearable monotony throughout the Universe. Furthermore, if we assume identity of mass, as Kepler admits having done in 1618, we come to conclusions which do not agree either with reason or with experience. Kepler continues[30]:

'. . . some time ago, when I tried to pursue the idea of equality in the abundance of matter [in planetary bodies], I was obliged to assign the ratio of the periodic times themselves to the size of the bodies, namely, [to assume] that, as [the period of] Saturn is 30 years, and that of Jupiter is 12, similarly, the size of the globe of Saturn to that of Jupiter would be as 30 is to 12. Now, this proportion is too large, and has been disproved by my own observations as well as by those of Remus.[31]

'*Fourthly*, the following reasons persuade us that the proportion of the abundance of matter must be taken as being exactly one half of the proportion between the volumes or the size (and thus be one and one-half that of the diameters of the globes and three-quarters [of the proportion] of the surfaces). First of all, the proportion of the abundance of matter, as well as that of density, become [in this instance], both of them, equal to one-half of the ratio between the distances from the Sun; and thus they are mutually compared as equal parts of this proportion: where the abundance of matter is greater, the density is less in the larger body in question, which is the best mediation of all. For example, if Saturn were twice as high [distant] as Jupiter, it would be one and one-half times heavier[32] and one and one-half times rarer; conversely, if Jupiter were one and one-half times denser, Saturn, in its turn, would be twice as high [distant] rather than heavier, and twice as large rather than rarer.

Furthermore, by dividing the proportion between the intervals into two, a geometrical [proportion] follows: in the same way that, between two distances (intervals) from the planets to the Sun (let us assume that they are 1 to 64), two mean proportionals, 4 and 16, must be introduced in order to fix the two remaining dimensions of the bodies so that the bodies of the mobile globes shall be in the ratio of 1 to 64, the surfaces as 1 to 16 or 4 to 64, and the diameters as 1 to 4 or 4 to 16 or 16 to 64; so also, between the two distances (intervals) of these same two planets and the Sun, 1 to 64, a single mean proportional of 8 is put in order to fix physically, in the interior of the bodies, the structure of their material substance, which is something unique[33]; so that, once again, the space of these globes (their volumes) will be as 1 to 64, the abundance of matter, as well as its rarity in the smallest compared with the rarity in the largest, will be as 1 to 8 or 8 to 64, and conversely, the density as 8 to 1 or 64 to 8. . . .[34]

If, starting from these principles, we now calculate the densities of the planetary bodies, always seeking the mean proportional between the distances of two of these bodies from the Sun, or more exactly, between the diameters of the two spheres or orbits, then relating and finally reducing all the results to some common round number, we shall obtain the figures given in the table below, to which figures I have added [the names] of terrestrial substances which closely correspond to them, as may be seen in my book *Auszug aus der Messekunst Archimedis*, written in 1606[35]:

Saturn	324	The hardest gem stones.
Jupiter	438	Lodestone.
Mars	810	Iron.
Earth	1000	Silver.
Venus	1175	Lead.
Mercury	1600	Mercury.

'As for gold whose density, calculated on the same basis, is 1800 or1900, we reserve it for the Sun.'

This is a curious outcome of the doctrine of harmony, and it very nearly re-establishes the alchemical doctrine, which finds a connection between celestial bodies and metals.

However, that is not the most important consequence of the doctrine, which is the derivation of the law governing the periodic times of the planets on their paths. Kepler started by announcing this law as a fact at the beginning of the *Epitome*, Lib. IV, pars II,[36] and it was not till fifteen pages later that he came to a detailed discussion of it[37]:

' *You said, at the beginning of this speculation on motion, that the periodic times of the planets are exactly in the sesquialteral proportion of their orbits or circles; I ask, what is the cause of this?*—There are four causes which combine to determine the duration of the periodic time. The first is the length of the path; the second is the weight or abundance of matter to be transported; the third is the strength of the motive force; the fourth is the volume or space occupied by the matter to be carried along. It resembles a mill, the wheel of which is turned by the *impetus* of the stream: according as larger and longer vanes are attached to the wheel, so a larger force is imparted to the machine by the impetuous flow of the stream. . . . Similarly, with the celestial whirlpool of the Sun's *species* . . . which causes the motion [of the planets], the more spacious the body, the more broadly and deeply is it bathed in the actuating force (cf. Fig. 33, *A.D*): and, *caeteris paribus*, the more rapidly is it transported, and

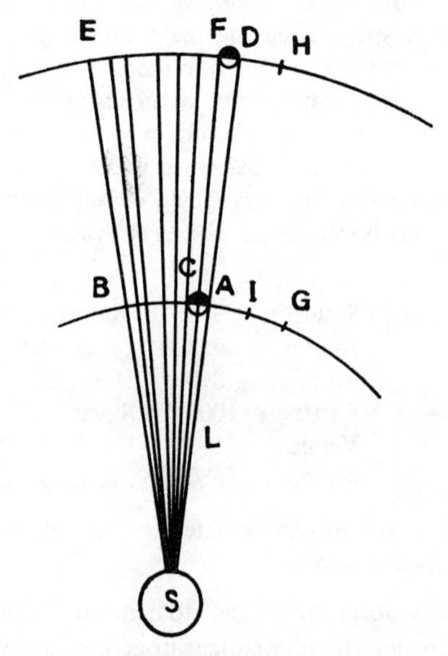

Figure 33.

the more rapidly does it complete its circuit. The circular orbits of the planets bear a simple proportion to the intervals: as *SA* is to *SD*, so is the whole circle *BA* to the whole circle *ED*; but the weight or abundance (*copia*) of the matter in the various planets is proportional to the intervals divided into two parts [$\sqrt{2}$], as was shown above,[38] in such a manner that the higher [more distant] one always has more matter; [for which reason it] is moved more slowly and acquires more time in its circuit, having need [already] of more time in addition on account of [the much greater length of] its path. For, if we take the mean proportional *SK* between *SA* and *SD*, the distances of the two planets, then the ratio of

SA, the smaller, to *SD*, the greater, is the same as the ratio of the abundance of matter in planet *A* to that of planet *D*. On the other hand, [the action of] the third and fourth causes are mutually compensating. . . . Therefore the simple proportion between the circuits and the halved proportion between the intervals [corresponding to the abundances of matter] taken together form a sesquialteral proportion: for this reason the periodic times bear a sesquialteral proportion to the intervals: thus, if *SD*, *SK*, *SA*, *SL* form a continued proportion, then the periodic time of planet *A* to that of another planet *D*, will be as *SL* to *SD*.'

There is some doubt, however, concerning the compensation between the third and fourth causes[39]:

'*In comparing two planets, prove that the weakening of the actuating force is exactly compensated by the space occupied by the mobile bodies of the planets.*—The volumes or space occupied by the bodies are in simple and direct proportion to the distances (intervals), as was shown above. The ratio of *SA* to *SD* is the same as that of the volume of a planet at *A* to the volume of another planet at *D*. Now, the density and the strength of the motive force are similarly proportional to these same simple intervals, but inversely; hence, the ratio of the interval *SA* to [the interval] *SD* is the same as that of the strength of the *species CA* to the strength of the *species FD*. Consequently, the more this force is weakened, the more it is utilized [by the planet]; suppose that Saturn be moved by a force which is one tenth of that which moves the Earth, but that, on the other hand, in virtue of its body it utilizes in its region a force which is ten times greater than that utlized by the body of the Earth in its own region; now, if the total force utilized by Saturn in virtue of its volume be divided into ten portions, which would be equal in spatial extension to [the spatial extension of] the entire force utilized by the Earth, then each of these portions or spaces of the force [utilized by] Saturn would have only one-tenth part of the force possessed by the single one utilized by the Earth; consequently, these ten [portions], the individual tenth-parts of the force being united together, have only as much power as the single part [of force] which carries the Earth along. If, therefore, in the bulk of Saturn's globe, which is rarer, there were no more matter than is present in the smaller body of the Earth, which is denser, then the globe of Saturn would be moved in its orbit by a distance equal to the length of the total terrestrial orbit, and would cover its own orbit in ten years. Now, it has nearly three times as much matter and weight as the Earth; therefore, it needs three times as long, namely, thirty years.'

It seems unnecessary to comment on these passages. They are classic in their blinding clarity; unsurpassable in the easy confidence with which the Imperial Mathematician develops the mind of the Divine Musician and corrects the mistake that he had made in the *Harmonice Mundi*, a mistake which consisted in not having understood that the principle of variation, laid down by the Architect of the Universe as the guiding conception in its construction, could not involve exceptions; and, therefore, that not only the distances and the motions, but also the material structure and the dimensions of the bodies in the Cosmos ought to adapt themselves to this requirement and embody the simple proportions which form the very basis of harmony.

However, we have not yet completed our study of the Cosmos; we have determined the dimensions and the structure of the mobile Universe, but the latter is only part of the Whole. As regards this part, what is its relationship, or preferably, what are its relationships, to the Whole?

Kepler, therefore, pursues his quest[40]:

'*Finally, what, in your opinion, is the ratio of size between these three principal regions of the Universe, the space in which the Sun* [*is contained*], *the space or region of the moving bodies, and the space of the whole Universe, namely, the region limited by the sphere of the fixed stars?*—'If the rules of Copernicus[41] do not extend to a determination of the height of the fixed stars, and if this height seems to be like to infinity, because, in comparison with it, the whole interval between the Sun and the Earth (which interval, in the opinion of the Ancients, contains 1200 terrestrial semi-diameters, and according to us, 3469) is insignificant; nevertheless, reason, aided by what it has already discovered, opens the way to achieve precisely that.

'However, to begin with, we must glance at the example of the Earth and the spheres of the Moon and the Sun, because the proportions of the whole Universe are derived from the proportions of the Earth itself; and the region, limited by these three bodies and their motions, is, as it were, a small universe. For the Earth—at least in appearance, and in reality according to Tycho Brahe—plays the same part in the sphere or region of the Sun, as does the Sun, according to Copernicus, in the region of the fixed stars. As the Sun is [in reality] at the centre [of the sphere] of the fixed stars, motionless in a stationary abode, so also is the Earth [in appearance], with respect to the motion of the Moon, motionless at the centre of the *quasi*-motionless sphere of the Sun.[42] In the same way that the region of the moving bodies is ordered about the Sun, so also is the sphere of the Moon traced about the Earth; in the former instance the fixed stars constitute the boundary of the planets, whereas in the latter, the Sun [provides one] for the Moon, to which boundary it returns each month, after having completed all its phases.

'Accordingly, it is conformable to reason, that, as the sphere of the Moon has been made a mean proportional between the apparent sphere of the Sun[43] and the body of the Earth [situated] at its centre, so the region of the moving bodies or the extreme path of Saturn will, in its turn, be a mean proportional between the outermost sphere of the fixed stars and the body of the Sun at the centre of the Universe.

'In fact, we come to this conclusion by considering the great Universe itself and without referring to the small one. For, on the one hand, seeing that the moving bodies aspire to the immobility of the ambient body, which procures them their place,[44] on account of which they resist motion and do not move with all the speed that the driving force tends to confer on them; and, on the other hand, seeing that they receive [in themselves] in some measure the motion of the driving force, and consequently that the motion of the driving force and the repose of the localizing body are combined in the moving bodies, it follows, if it be permitted to enunciate a physical matter in mathematical terms, that the moving bodies may be very justly considered as mean proportionals between the body which is the source of motion and the motionless body which procures them their place. Seeing that this is true physically as well as locally (the source is inside, that which locates is outside, the moving bodies are between), it is very probably also true geometrically, namely, that the semi-diameter of the region of the moving bodies is a mean proportional between the semi-diameter of the Sun and the semi-diameter of the sphere of fixed stars, that is to say, that the ratio of the globe of the Sun to the sphere of all the planets is equal to that of the latter to the spherical body of the whole Universe bounded by the region of the fixed stars.'[45]

Seeing that the angular size of the Sun's disc as seen by us is very nearly one-half degree, it follows that the distance between the Earth and the Sun is equal to 229 solar semi-diameters; and as the diameter of the terrestrial sphere is slightly more than one-tenth of the diameter of the sphere of Saturn, the latter is almost equal to 2 000 solar diameters. From which, by assuming, in accordance with the preceding argument, that the diameter of the sphere of Saturn is to that of the sphere of the fixed stars as the diameter of the Sun is to that of the sphere of Saturn, we come to the conclusion that the diameter of the sphere of the fixed stars contains this diameter 2 000 times; that is to say, it is 4 000 000 solar diameters, and, therefore, if we accept that ratio between the dimensions of the Sun and the Earth established by the Ancients, the value becomes 26 000 000 terrestrial diameters; and if we use instead the value assigned by Kepler, we have 60 000 000 terrestrial diameters.

What a contrast with the minute universe of Tycho Brahe! and even with that of the Ancients! The ball of the Universe enlarges and extends like a soap-bubble before bursting among Kepler's successors. However, to him, the sphere of the fixed stars was still very real, and even very solid, though it had become very thin; for, though he had enlarged the dimensions of the Universe, he had reduced the thickness of its shell and decreased the size of the stars.[46]

The reasoning by which Kepler arrived at an estimate of the thickness of the celestial vault is very interesting, and most significant: it is based on his conviction, (which he had expressed many times), of the *fundamental unity of the matter* constituting the bodies of the Universe, which unity underlies the differences in density by which they are characterized. If the matter throughout the Universe be one, it is reasonable to suppose that neither its partition between the principal parts of the Universe, nor the attribution to the body of each of them of the particular density with which it is endowed, are the results of chance, but are conformable to rational relationships. Therefore, to the question[47]:

> '*In your opinion, what is the proportion of density between the bodies of the Sun, of the aetheric aura which permeates the whole Universe, and of the sphere of the fixed stars which encloses all from outside?*'

Kepler replies:

> 'As these three bodies are analagous to the centre, to the spherical surface and to the interval, symbols of the three persons of the Holy Trinity, we may believe that there is as much matter in the one as in the other two,[48] in other words, that one third of the matter in the Universe is concentrated in the body of the Sun, even though in ratio to the dimensions of the Universe it must be very small; similarly, that another third of the matter is stretched and attenuated throughout the whole vast space of the Universe in such sort that the Sun has as much matter in its body as is found to be brightened by its light and penetrated by its [motive] rays; and similarly, that the last third of the matter forms a sphere which is placed like a wall round about the Universe. In order that we may have some idea of these relationships (though we can never possibly understand them fully), by comparing them with known things, we can imagine that the body of the Sun is entirely of gold, that the sphere of the fixed stars is aqueous, vitreous or crystalline, and that the intermediate space is filled with air.'

Knowing the dimensions of the sphere of the fixed stars and the quantity of matter it contains, its thickness is readily obtained[49]:

> '*What is the thickness of the sphere of the fixed stars?*—Seeing that we have allotted to it as much matter as there is throughout the full extent of the Universe which it encloses, but neglecting the matter contained in the very small globe of the Sun; and seeing that the matter in the sphere of the fixed stars cannot possibly be of the same density as that of the region of the moving [bodies], but must be a mean proportional between the density of the aetheric aura and that of the body of the Sun, it follows that its volume must similarly be a mean proportional between the volume of the body of the Sun and that of the aetheric aura. The ratio between the diameter of the Sun and that of the aetheric aura has been [established] above as being equal to that of 1 to 4 000 000; the ratio of the volumes is therefore treble, namely, 1 to 64 000 000 000 000 000 000. The mean proportional between these numbers is 8 000 000 000. Consequently, this number of volumes of the body of the Sun will equal the space comprised between the concave and convex surfaces of the sphere of fixed stars. Therefore, [as regards its matter], the whole Universe, its three parts taken together, is represented by the number 64 000 000 008 000 000 001, whose cube root, 4 000 000 and one six-thousandth, shows that the sphere, whose thickness is one six-thousandth part of the semi-diameter of the body of the Sun and which is placed round the aetheric aura, contains in itself 8 000 000 000 volumes equal to that of the body of the Sun. Such, then, is this skin or mantle of the Universe; [this] super-celestial crystalline sphere, on account of its vast expansion, [is] of such subtlety that if it were coagulated into a single spherical body it would have a semi-diameter 2000 times greater than that of the body of the Sun; but its thickness is only one six-thousandth of the semi-diameter of the body of the Sun, that is to say, a little more than 2000 German miles.'

It seems superfluous to compare Kepler's deductions with the reality of astronomy as revealed by his successors. His universe followed after the universes of Aristotle, Ptolemy, and Copernicus—it vanished in the infinite space of Newtonian science. The mind of the Divine Mathematician, which we have not finally unravelled, has proved to be infinitely more complex and richer than Kepler ever imagined; and it is quite free from musical inspiration.

However, we must not complain of Kepler's mistakes, nor of the strange detours in his thought. We must remember that the path along which the human mind advances towards truth is even stranger and more wonderful than truth itself.

Conclusion

Kepler's undertaking may be regarded as completed with the deduction of the dimensions of the Cosmos and of the structure of the sphere of the fixed stars. The path along which Kepler had set out a quarter of a century before with all the keenness of youth, a long, difficult path full of snares, a path from which he often wandered and to which he always returned through the infallible care of Divine Providence, was traversed to the end. The mind of the terrestrial mathematician was united with the mind of the Divine Artist; the Universe revealed its *λόγος* to the mind of the creature of meditation.

We can easily understand how Kepler, when revising his *Mysterium Cosmographicum* in 1621, was able to say, after glancing back into the past, that 'never before has it been given to anyone to embark in such a fortunate manner on such a wonderful work, and so worthy of the object to which it had been devoted'. For, as though it had been dictated to him by a celestial oracle, everything that he had stated there (and his later works provide ample evidence thereof) *was true*: the created Universe was an expression of the Creative Trinity, and the whole was subjected to the laws of Harmony and Proportion.[1]

Kepler was right not to count upon rapid success. In his lifetime, he had no follower. His contemporaries rejected unanimously not only his harmonic cosmology—in which rejection they have been followed by posterity—but also his more profound and more fruitful ideas, especially his 'celestial physics' and his most trustworthy and precious discoveries, namely, his three laws. Thus, even three years before the publication of the *Astronomia Nova* (in 1607) we have seen Fabricius, reviving Tycho Brahe's objections against Copernicus, assert that abandonment of circularity destroyed the constancy of celestial phenomena[2]; seven years after its appearance (in 1616), and on the eve of publication of the *Epitome*, Maestlin, to whom Kepler had announced the project, rejected the very idea of 'celestial physics' as contrary to good sense and healthy scientific philosophy.[3]

Needless to say, official and academic astronomy completely neglected the revolutionary theories of the Imperial Mathematician. I have already pointed out that the great innovators, and creators of modern science, Galileo and Descartes, completely ignored him, too.[4] It would seem that the one face of Janus frightened some, and that the other face, repelled the others.

Gassendi and Boulliau are about the only ones who cite him in the first half of the century. Nevertheless, even as Gassendi noted his observations and predictions, including that of the transit of Mercury over the solar disc, and also gave an account of Kepler's 'magnetic' concepts, he seems not to have understood the theory of planetary motions, on which it was based and from which it was derived. As for Boulliau, he borrowed Kepler's notion of elliptical orbits, but completely transformed it by rejecting Keplerian dynamics in favour of returning to a purely kinetic concept of astronomy, and replacing Kepler's law of areas by the Ptolemaic concept of the equant, assuming that the motion of a planet on its elliptical orbit is uniform with respect to its second, unoccupied focus.[5] In England, Kepler was more fortunate. Jeremiah Horrocks, the only real follower Kepler ever had, declared his belief and defended him against the attacks and criticisms of Landsberg[6]; he accepted not only the ellipticity of planetary orbits but also the mechanism of libration conceived by Kepler and its explanation by the action of magnetic forces.[7] He even accepted, at least partially, the harmonic considerations, although they had not enabled Kepler to obtain more exact values for the eccentricities than those of his predecessors, or even those of Landsberg.[8] Unfortunately, Horrocks died very young in 1641, and his works remained unpublished. They did not appear until 1673, when they were edited by Wallis.[6]

In 1653, Seth Ward in his *Inquisitio Brevis* of the foundations of Boulliau's astronomy,[9] and, in 1656, in his *Astronomia Geometrica*, adopted 'the astronomy of the ellipse', and defended Kepler against those who criticized it. Whilst rejecting the scheme of the French astronomer, he borrowed from him his use of the equant (simple elliptical theory) in place of the law of areas.[10] Furthermore, he said nothing about the dynamic substructure of Keplerian astronomy. Obviously, 'celestial physics' had no attraction for Seth Ward. It seems that J. A. Borelli was the first to understand the prime importance of the Keplerian revolution.[11] Not only did he borrow the ellipticity of planetary orbits from Kepler, but he also adopted his general conception of celestial physics, though he modified it in an authoritative manner, particularly by substituting the interaction of centripetal and centrifugal forces for the 'libratory' action of the Sun.

Unfortunately for him, he borrowed also the incorrect dynamical law, velocity is inversely proportional to distance, which he regarded as equivalent to the 'simple elliptical theory', and good Galilean that he was, he did not know how to reveal in the Keplerian concept of gravitation-attraction—(which Kepler never used)—the essential principle of unity of the Cosmos. Hooke had the credit of this, for he was able to resolve the law of change in the force of attraction by making it inversely proportional to the square of the distance.[12] However, like Borelli, Hooke accepted the incorrect Keplerian law of velocities, and, consequently, found himself in a dilemma.

Astronomy could have remained in this state for some time. Kepler's dynamical law is so plausible that it was still accepted by Leibniz.[13] However, Providence decided otherwise, and created a worthy successor to Kepler in the person of Newton. The genius of Newton was able to correct Kepler's mistake; also, starting from Kepler's three descriptive laws, together with the concept of attraction, Newton was able to accomplish in his *Philosophiae Naturalis Principia Mathematica* that synthesis of terrestrial with celestial physics, which had been the dream of the author of the *Astronomia Nova.*

Appendix I

TABLE OF ASTRONOMICAL TERMS*

P_1—aphelion
P_2—perihelion
Q—*punctum aequans*
O—centre of planet's (= spheres) eccentric circle
S—true Sun
ψ—physical equation
φ—optical equation

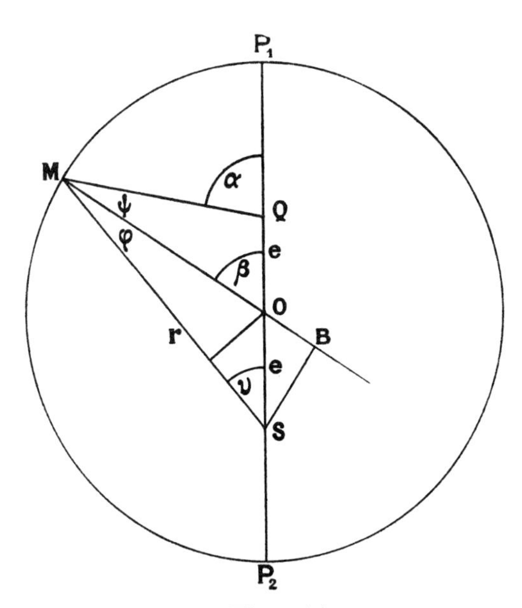

Figure 34.

* Cf. Max Caspar, *Die Neue Astronomie*, München–Berlin, 1929, Vorbericht, p. 44.

α —mean anomaly, proportional to time
β —eccentric anomaly
v —true (or equated) anomaly
e —eccentricity (bisected)

$$\beta = \alpha - \psi$$

$$\sin \psi = e \sin \alpha$$

$$\text{tg} \frac{v - \varphi}{2} = \frac{1 - e}{1 + e} \text{cotg} \frac{\beta}{2}$$

$$\frac{v - \varphi}{2} \quad \text{and} \quad \frac{v + \varphi}{2} = \frac{\beta}{2}$$

$$\tau = \cos \varphi + e \cos v$$

$$\tau = \sqrt{1 + 2 e \cos \beta + e^2}$$

Appendix II

On the duration of motion of a planet along any arc.*

What is the meaning of the term 'anomaly'?

Although, properly speaking, Anomaly (inequality) is an affection of the planet's motion, nevertheless, astronomers use this term for the motion itself in which the inequality is present. Seeing that three measurable things are comprised in motion, namely: (1) the distance to be traversed; (2) the time required [to traverse] the distance; (3) the apparent extent of the distance [traversed]: the term 'anomaly' is applied to all three. Furthermore, with regard to time, the word is used in two [different senses]. In fact, Ptolemy uses it, *primo*, [to designate] the total time taken by a planet for its inequality to return to the starting point [period of revolution]; and he reckons as many anomalies as there are such [periods]; *secundo*, portions of this time (period) are commonly called anomalies from the fact that Ptolemy speaks of the motion of the anomaly, having in mind the parts comprising the total anomaly.

How many anomalies are there in the total anomaly?

Three anomalies are distinguished for each position of the planet: (1) the mean anomaly; (2) the eccentric anomaly; (3) the true anomaly.

What is the mean anomaly?

It is the time taken by the planet [to describe] any portion of its orbit starting from the apside; [this time is] expressed in degrees and minutes, and the total anomaly equals 360° in the logistic or astronomical numeration.

Why is it called 'mean'?

Not because it would be a mean, in respect of quantity, between the related [anomalies], as will be indicated below; but it is called 'mean' in imitation

* *Epitome Astronomiae Copernicanae*, Lib. v. pars II, 3 (*G. W.*, vol. VII, p. 386 f.; *O. O.*, vol. VI, p. 419).

of ancient astronomy, which is accustomed to speak of mean anomaly instead of mean motion (*i.e.*, uniform) of the anomaly; for the time, thus reduced to a logistic denomination, expresses by the number of degrees and seconds the size of the arc [which would be] traversed by the planet if it moved with a uniform motion, and a mean between the slowest and quickest motions, during the whole time that we call 'mean anomaly'.

In what way should the mean anomaly be defined or measured in our diagrams in accordance with ancient astronomy?

The line *BL*, equal to the eccentricity *AB*, having been drawn on the line of apsides *BP* (as was explained in the first part of this Book V): the mean anomaly, according to ancient astronomy, would be the arc of the equant circle described in the direction of the signs [of the Zodiac], starting at *L*

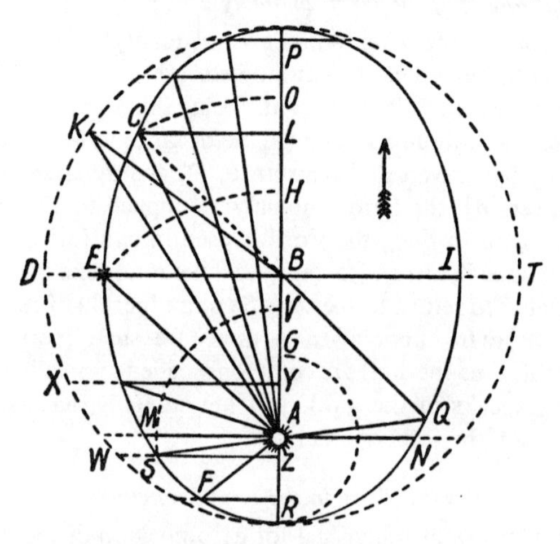

Dic regulam de affectione horum Triangule

and contained between two lines drawn from *L*, one to the apside *P* and the other to the body of the planet *C*. Or, it would be the angle [at the apex *L*] between these two lines, or the complement of this angle to four right angles (360°). Thus, if *C* were the planet, the angle *PLC* could be taken, approximately, as the mean anomaly.

Define the line of mean motion and the mean position of the planet in accordance with the hypothesis of the equant.

This would be the line drawn from the centre of the Sun to the sphere of fixed stars, parallel to the line drawn from the centre of the [circle of the] equant, *i.e.*, from the other focus of the ellipse, through the body of the planet; each of them would show the mean position of the planet on the sphere of fixed stars. In our diagram, if *G* be the planet, and *AM* be parallel to *LC*, then *AM* is the line of this mean motion. In the new astronomy, which takes no account of the equant circle, the mean anomaly is measured by the sector of the [auxiliary] circle fixing the arc of the orbit (*APK*), or that of the ellipse.

What is the eccentric anomaly?

It is the arc of the eccentric, measured *in consequentia* (westwards), intercepted between the lines of apsides and a line perpendicular to this, passing through the body of the planet, or through some point on the orbit. Thus, for the point *C*, or for a planet in this position, a perpendicular to *PAR*, *i.e.*, *KCL*, will cut the [eccentric] circle in *K*, and the arc *PK* will be the eccentric anomaly.

In ancient astronomy, the arc *PK* is the mean anomaly.

What is the true anomaly?

It is the arc of the great circle in the latitude of the Zodiac from the apside to the planet. . . . Or, what is the same thing, it is the angle which the arc of the true orbit subtends at the centre of the Sun, or the complement of this angle to four right [angles].

Thus, for *C*, the true anomaly is the angle *PAC*, and for $Q = 2$ right angles $+ RAQ$.

What is the equation for the prosthaphaeresis, and what is the origin of this name?

It is the difference between the number of degrees and minutes in the mean anomaly and [that of the number] of degrees and minutes in the true anomaly. Or, according to the view of ancient astronomy, it is the angle whose apex is in the Sun, and which is measured by the arc of the great circle of the fixed stars intercepted between the planet's lines of mean motion and eccentric motion. This angle must be subtracted from one semi-circle of the orbit, and added to the mean anomaly in the other in order to obtain the true anomaly: whence the composite name προσθαφαίρεσις by which it is called.

As for the term: equation, it derives from the fact that by its addition to, or subtraction from, the true anomaly, which divides the equal arcs and times into unequal parts, we obtain the mean or uniform anomaly.

What are the parts of this equation, and what is the measure of each?

There are two parts; one, physical; the other, optical. One derives from the inequality which affects the motion of the planets as a result of physical causes; the other derives only from the apparent, or *quasi*-apparent, inequality, that is to say, from the greatest, or smallest, distance of the arc of the true orbit from the Sun. Both can be discerned after a fashion in the same triangle, which, consequently, is called equatorial here.

If the ends of the eccentricity *A* and *B* be joined to the body of the planet *C*, then the physical part of the equation is measured by the area *BAC* (or by equivalence, by the area *BAK*); as for the optical part of the equation, it is equal to the angle *BCA*, but the angle *BKA*, which is always a trifle smaller, is easier to calculate.

What is the use of this equation and its parts?

In the restored form of astronomy, the use of the whole equation comprising both parts is not necessary, nor very useful. For the anomalies are not computed from this equation, but, on the contrary, we determine this equation, when we want to use it, by comparing the true anomaly (which is calculated first of all) with the mean anomaly.

Appendix III*

The most ingenious Kepler wanted the *planets* to be at rest [of their own accord] and on account of their natural inertia, but that they should be seized by the solar light, as though by an agent of motive force, and thereby led round [about the Sun], as he has described in chapter XXXIII of his *Commentaria de Motibus Stellae Martis*, and in other terms in Book IV of the *Epitome Astronomiae Copernicanae*, part iii, question 5: 'The Sun, in order to carry the planets around, makes use, as though they were arms, of the force of its body emitted in straight lines throughout the whole extent of the Universe'; and in defence of this proposition he puts forward the following reasons:

The *first* of these reasons is that the planet is subjected to a retardation during one part of its revolution, where it is, in fact, slower; but in another [part] it is quicker. Now, it is slower, where it is farther from the Sun, and quicker, where it is nearer to the Sun. However, these retardations and accelerations must not be ascribed to [the motive force] of the planet, for in that case, it would tire and weaken with age.

The *second* adds [the consideration drawn] from the motion of planets compared amongst themselves: they complete their circuit more slowly when they are more distant from the Sun, and more quickly when they are closer.

The *third* [rests] on the dignity and aptitude of this body which is the source of life and heat, and from which all vegetative life originates; in such a manner that the heat, as well as the light, may be regarded to some extent as the agents [of its action].

The *fourth* is drawn from the revolution (rotation) of the Sun on its axis, which takes place in the same direction as all the others.

He would, furthermore, endow the Sun with a soul which subdues and enflames this great mass of matter, producing therein divers changes subjected to mutations, whence it appears that there is not a continuous, perpetual and uniform energy in the body of the Sun; he proves [the animation of the Sun] from the very existence of light which he claims is

* Ismaeli Bullialdi..., *Astronomia Philolaica*, Paris, MDCXLV, pp. 21 f.

related to the soul. As for the force by which the Sun lays hold of the planets, it is corporeal and not animal, nor mental; for which reason, namely, because it is an immaterial kind of body, the force turns together with the Sun in the manner of a very rapid whirlpool, covering the whole extent of the path which it fills with the same velocity as the Sun turns about its own centre in its own small space.

He puts forward, also, the analogy of the magnet which attracts iron; he compares [its action] to the magnetic force of the Sun in the following way: the magnet, according to the position of its parts [poles], attracts [the iron] on one side only and repels [the iron] on the other side [only], but the Sun possesses throughout all its parts the active and energetic faculty to attract, repel and retain the planet.

He maintains that one part [pole] of the planets is friendly, and that the other is hostile towards the Sun; consequently, the Sun attracts them by the one part and repels them by the other; and if there were no motion [of rotation], it would have attracted them and united them with itself; but because the Sun turns on its axis, therefore its *species* turns with it, together with the motive force that moves the planets.

In reply to these premises, I shall say, as regards the first, that it does not prove that the planets are moved by the Sun, although it states a truth, namely, that the planets move more quickly the nearer they are to perihelion, and more slowly when they are farther away from it. Indeed, the Sun does not cause this velocity and this state of motion, but they depend entirely on the system of planetary motions and the manner in which it has been ordered, as we shall show shortly.

To the second, I reply that it is true that the superior planets (those which are more distant from the centre) complete their revolutions more slowly, because of the distance to be covered, as well as the form; it would seem [that in order to provide an explanation] that the two causes must be combined. However, it is not the distance from the Sun, nor the attenuation of the motive force that cause this retardation: for the greater distance of the planet [reckoned] from the Sun only makes its revolution [circuit] more extended, and for that reason it needs a longer time to cover the circuit.

As for the third, I say that it merely proves that the rays from the Sun, as well as its motive force, act on the planet through light and heat, not by causing it to move, but by producing changes and, if such be its nature, generations and corruptions of the kind produced on Earth.

The fourth, in my opinion, proves nothing, unless it be that it is fitting for all bodies in the Universe to move, either locally, or by orbiting. I willingly admit this.

However, I should like to ask Kepler why he would endow the Sun with a motive soul and a corporeal life, whereas the other planets are stupid and so inert with respect to motion that they have a resistance thereto within themselves? Is it because mutations such as we see on the Sun are never produced in the planets? For, if sunspots and the other things that show changes, mutations and activity in [various] parts of the Sun persuade us that it is endowed with the semblance of life, is not our Earth, the planet which we inhabit, subject, perhaps, to even more frequent and still graver mutations? Our planet emits flames, vapours, waters; it causes trees to fall from great heights, sends clouds into the air. This proves the activity and changes in generation and corruption in various parts of the Earth: what, then, prevents it from being animated by a motive soul, and consequently moving of its own accord?

Then, how could the following propositions be true [at the same time]: [that which asserts that] *the Sun moves the Earth which, like the other planets, possesses a certain inertia and a certain resistance to motion;* and this other one, on page 175 of the *Commentaria de Motibus Stellae Martis*, which says that *the motion of the chief, monthly Moon* [derives] *entirely from the Earth* as its source; and a little further on, *To be sure, such is the strength of the Earth's immaterial species, and doubtless, too, the rarity of the lunar body is equally great, and the dislike of this body [for motion] is equally feeble.*

If the Sun move the Earth by means of its immaterial *species*, and if the Earth be *inert* with respect to motion like the other planets, how is it able to impress a monthly motion on the Moon? This is absurd and impossible. Perhaps, Kepler means that the Earth, impelled [in the first place] by the Sun, moves the Moon in a subordinate manner—but it is not difficult to close this loop-hole. If the planets be seized and carried on their paths by the light of the Sun in the guise of instrument of the motive force, then the Earth in the grip of this [motive force] would not be able to lay hold of the Moon and carry it along except by a repercussion [an effect] of the solar light, in so far as it (the Earth) reflects light from the Sun to the Moon. In that case, the Earth would cease to do so at the time of Full Moon, and the force of the Earth would increase from Full Moon up to New Moon, at which time the Earth would move the Moon very rapidly. Now, nothing of the kind is revealed by the phenomena.

As for the force by which the Sun seizes or grips the planets, and which, being corporeal, functions in the manner of hands, it is emitted in straight lines throughout the whole extent of the Universe, and like the *species* of the Sun, it turns with the body of the Sun; now then, seeing that it is corporeal, it becomes weaker and attenuated at a greater distance or interval, and the ratio of this decrease [in strength] is the same as in the case of light, namely, the duplicate proportion, but inversely, of the distances. Kepler does not deny this, and yet he claims that the motive force decreases only in the direct proportion of the distance; furthermore, he says that this attenuation of the motive force produces a weakening of the force only in longitude, because local motion impressed by the Sun on the planets (which motion similarly animates the corporeal parts of the Sun itself) occurs only in longitude, and not in latitude; consequently, he offsets the inadequacy of this analogy by increasing the [amount of] matter in the slower planets.

This reply of Kepler's carries little weight. For, if he consider the amount of this motive force with respect to area, then he must necessarily decrease it according to the duplicate proportion of the distance; if, on the other hand, he regard it only as lines, then he contradicts the previous proposition which states that this motive force is corporeal: for, if it were such, it would not be able to dwell in lines only. Furthermore, this force acts by contact of the Sun's *species* which emanates from the Sun together with the motive force; now, this *species* touches the body of the planet in the same way that one surface [touches] another surface; consequently, the force touches it in the same manner, seeing that it emanates from the Sun in the same manner. It follows that the force must decrease in the duplicate proportion of the distance, as does the *species*. However, if we were to accept this argument, Saturn would make one revolution only whilst Jupiter would make three revolutions and part of the next in the same time, Mars 39, Venus 173 and Mercury nearly 557; as for the Earth, it would make 90. This is obviously contrary to the phenomena. In order to show agreement among the ratios between the revolutions, he appeals in vain to the quantity of matter by compensating [balancing] the distance of one by the *moles* of the other. For example, Jupiter at its mean distance from the Sun is three and one-eighth times further away from the Sun than is ♂ [Mars]; for this reason, if the proportion between the periods of revolution were as the lines of distance, Mars would complete only three and one-eighth revolutions [whilst Jupiter completed its revolution]; consequently, Kepler had to increase the corporeal *moles* of Jupiter, or else decrease that of Mars, so that the *moles* of Mars being less, the *species* would show less reluctance [resistance] to move it [Mars], and therefore this planet would be accelerated to the degree where ♂ would complete nearly seven revolutions. However, observations show that Mars is less than one-thousandth of the size of Jupiter; where, then, is the proportion between motions? For the motion of Mars should have exceeded that of Jupiter beyond all measure; therefore, no compensation between the matter and *moles* [of the planet] and the distances is permissible. Also, Kepler cannot avoid [objections] by saying: *all the fibres of the globe taken together are moved* by all the circles of the motive force taken together in simple proportion, in the same manner as the particular (individual) lines, or the fibres of two isolated planetary bodies would be moved by the particular (individual) circles *of the motive force in the simple proportion of the distances* as if the corporeal force could dwell in lines. When diminution founders on these reefs, then the astronomy of fibres is included in the shipwreck.

Comparison with the magnet, as put forward [by Kepler] does not agree very well with the Sun's energy as postulated by him: for a magnet attracts with one of its poles, and repels, as it were, with the other, but that results from the nature of its parts, which cannot be changed. Now, the Sun has no distinct parts as are found in the planets, though Kepler says so, without giving proof. A magnet attracts those parts of iron which are cognate to it, and does not make them turn [round about it]; and if they are equidistant from the poles, they are attracted by neither. The Sun, however, makes the planets to turn [round about it], and attracts them by one part, whilst it repels with the other. When the planet's fibres are in equilibrium, then the Sun either attracts the planet when at aphelion, or repels it at perihelion. On the other hand, a magnet never repels, but only tends to cause cognate parts to unite with it, and to set them in accordance with the direction of its poles. If the fibres, when at the apsides, approve and reject [attract and repel] uniformly, ought not the Sun to propel the planet by the one, and attract it to itself by the other, because the force and the body which is moved are in equilibrium? The planet ought, on this account, to remain stationary; nevertheless it continues to move. Finally, I ask, how can he prove that the Sun would draw all the planets to itself, if it did not turn on its axis, and why, seeing that it does turn, it does not attract them, but makes them turn round about itself. I fear that Kepler's devices are merely phantasies engendered by his extremely lively, and most ingenious, mind which imagines the cause of things where the real cause is hidden.

Notes

INTRODUCTION

1. Cf. my *From the Closed World to the Infinite Universe*, Baltimore, 1957.

2. Kepler's works have been edited by Ch. Frisch, *Johanni Kepleri Astronomi Opera Omnia*, 8 vols., Frankfurt and Erlangen, 1858–91: and more recently by W. von Dyck and Max Caspar, Johannes Kepler, *Gesammelte Werke*, 18 vols., München, 1938–59. The *Harmonice Mundi* has been translated into German by Otto J. Brijk (*Die Zusammenklänge der Welten*, Jena, 1918); and the *Mysterium Cosmographicum* by Max Caspar (*Das Weltengeheimnis*, Augsburg, 1923), who has translated also the *Astronomia Nova* (*Die Neue Astronomie*, München, 1929). Book IV of the *Epitome Astronomiae Copernicanae* and Book V of the *Harmonice Mundi* have been translated into English (Britannica Great Books, vol. 16, 1952). Quotations from Kepler are made as far as possible from Caspar's edition and are designated by *Gesammelte Werke* or *G.W.*, with volume and page; quotations from Frisch's edition are designated by *Opera Omnia* or *O.O.*, followed by volume and page; in some instances both texts are quoted. The best works on Kepler are still some of the old ones: E. F. Apelt, *Johann Keplers Astronomische Weltansicht*, Leipzig, 1849; *idem*, *Reformation der Sternkunde*, Leipzig, 1852; an excellent *Vita Johannis Kepleri* which contains a valuable *Historia Astronomiae Seculo XVI* appears in vol. VIII (2) of Ch. Frisch's edition of the *Opera Omnia*, Frankfurt, 1891. The older works still provide the best exposition of technicalities: J. B. Delambre, *Histoire de l'Astronomie Moderne*, Paris, 1821, vol. I; and especially R. Small, *Account of the Astronomical Discoveries of Kepler*, London, 1804. Amongst the more recent works, mention must be made of the admirable *History of the Planetary System*, by J. L. E. Dreyer (London, 1905; 2nd edition, New York, 1953), and the very valuable introductions and notes in Caspar's edition which often provides 'translations' of Kepler's arguments in modern mathematical language. There are in addition: E. Goldbeck, *Keplers Lehre von der Gravitation* (*Abhandlungen zur Philosophie und ihrer Geschichte*, Halle, 1896, vol. VI); A. Speiser, *Mathematische Denkweise*, Bâle, 1945; W. Pauli, *Der Einfluss Archetypischer*

Vorstellungen auf die Bildung naturwissenschaftlicher Theorien bei Kepler in C. G. Jung and W. Pauli, *Naturerklärung und Psyche*, Zürich, 1952; and the excellent article by G. Holton, 'Johannes Kepler's Universe: Its Physics and Metaphysics', *American Journal of Physics*, 1956, vol. XXIV. Finally, 200 very interesting pages are devoted to Kepler by A. Koestler in *The Sleepwalkers*, London, 1959.

3. However, there is an excellent recent biography of Kepler: Max Caspar, *Johannes Kepler*, Stuttgart, 1948; 2nd edition, 1950. There is a very valuable bibliography by the same writer: Max Caspar, *Bibliographia Kepleriana* . . ., München, 1936; 2nd edition, 1968. The reader is referred to these for all biographical and bibliographical details.

4. In fact, Newton deduced his law of universal gravitation using Kepler's laws as his starting point—particularly the third law, which establishes a relationship between the distance of the planets from the Sun and their period of revolution. Delambre quotes with approval (*Histoire de l'Astronomie Moderne*, vol. I, p. 390) the note made by Samuel Koenig, professor of philosophy at Franeker, in his copy of *Astronomia Nova*:

> '*Nunquam Newtonus principia Philosophiae naturalis scripsisset, nisi Keplerianos maximos conatus circa clarissimos sui libri locos, multum diuque considerasset.*'

5. Not one of Tycho Brahe's collaborators, who with Kepler had the task of elaborating his observations, had ever had the idea of seeking a solution to the difficulties which stood in the way of establishing planetary theories after abandoning the dogma of circular motion. Furthermore, not one of them adopted it, not even amongst Kepler's contemporaries (with the single exception of Horrocks). Fabricius (cf. *supra*, pp. 263 f.) protested strongly; Galileo and Descartes took no account of it; Ismaël Boulliau distorted the arrangement; and the importance of his work was only recognized later by Seth Ward, who however adopted it only in part (*Astronomia Geometrica*, London, 1656), and by J. A. Borelli (*Theoricae mediceorum planetarum ex causis physicis deducta*, Florence, 1666); cf. *infra*, p. 476 and p. 518, n. 14.

6. Even this, will be considered only in its broad outlines, without entering into technical details, however interesting (for these, reference should be made to the works of Delambre and Small quoted in note 2). In so doing, Kepler himself will be taken rather as a guide; he offered his reader the choice: '*vel perlegendi et percipiendi demonstrationes ipsas labore maximo; vel mihi professione Mathematico super adhibita sincere et Geometrica methodo credenti*'; cf. *Astronomia Nova, G.W.*, vol. III, p. 19. The occasional use of technical terms cannot be avoided; the reader will find them explained in Appendix I.

7. Letter of Kepler to Georg Brengger, 4 October 1607; (*G.W.*, vol. XVI, p. 54):

'*Trado enim unà philosophiam seu physicam Coelestem pro Theologia coelestij, seu Metaphysica Aristotelis.... In qua physica simul novam Arithmeticam doceo, computandi non ex circulis sed ex facultatibus naturalis et magneticis.*'

8. Letter to Herwart von Hohenburg, 10 February 1605; (*G.W.*, vol. xv, p. 146):

'*Scopus meus hic est, ut Caelestem machinam dicam non esse instar divinj animalis, sed instar horologii (qui horologlum credit esse animatum, is gloriam artificis tribuit operj) ut in qua penè omnis motuum varietas ab una simplicimi vi magnetica corporalj, utj in horologio motus omnes a simplicissimo pondere, Et doceo hanc rationem physicam sub numeros et geometriam vocare.*'

Comparison of the 'celestial machine' with a clock became usual at a time when there were numerous large astronomical clocks. Although the comparison had already been made by Buridan, as well as by Rheticus, definite identification of the one with the other does not seem to have been made before Kepler.

9. As will be seen later (p. 151), Kepler, too, started by accepting the animistic explanation, and only gradually dropped it.

10. Cf. *supra*, p. 53. Copernicus in setting the moving planets against the fixed stars had only the celestial bodies in mind. Kepler added space to them.

11. Cf. *supra*, p. 153 and p. 284. The Sun will remain stationary only in so far as it always occupies the same central position in the Universe; but whilst remaining in position, it will acquire a rotational motion.

12. Copernicus was very insistent upon the systematic nature of the heliocentric Universe, and Rheticus was perhaps even more so than his teacher. However, neither of them was sufficiently bold to seek the reason for it; cf. *supra*, p. 53.

13. Letter of Kepler to Maestlin, 2 August; (*G.W.*, vol. xiii, p. 27):

'*Quaero an unquam audiveris vel legeris fuisse aliquem, qui rationem dispostionis planetarum iniverit? Creator nil temerè sumpsit. Erit igitur causa cur* ♄ [Saturn] *duplo ferè altior sit Jove,* ♂ [Mars] *paulo plus Terra, et haec Venere, Jupiter vero plus triplo quam Mars. Idem de motuum proportione dictum volo.*'

14. The Universe was created in order to give expression to the Divine Spirit, and for the purpose of being understood by Man, a creature of reason and meditation, made in the image and likeness of God; cf. *infra*, p. 345 and p. 349.

Kepler was deeply imbued in Neo-Platonism; especially in the epistemological concepts of Proclus: cf. Speiser, *Mathematische Denkweise*, Bâle, 1945.

15. For Kepler, as well as for Rheticus or Copernicus, God 'geometrizes eternally', ὁ Θεός ἀεί γεωμετρεῖν (cf. *supra*, p. 145). For example, he wrote to Maestlin, 9 April 1597; (*G.W.*, vol. XIII, p. 113):

> '*Deo Conditorj devotè gratias ago. Faxit is . . . ut fides de rerum creatione hoc externo adminiculo* [the Mysterium Cosmographicum] *confirmetur. Conditricis mentis natura cognoscatur, majorque nobis quotidiè fiat inexhausta illius sapientia; atque adeò ut vel tandem homo suae mentis vires justo modulo metiatur, et intelligat, Cum Deus omnia ad quantitatis normas condiderit in toto mundo: mentem etiam hominj datam, quae* TALIA *comprehendat. Nam ut oculus ad colores, auris ad sonos, ita mens hominis non ad quaevis, sed ad* QUANTA *intelligenda condita est, remque quamlibet tanto rectius percipit, quanto illa propior est nudis quantitatibus, ceu suae originj: ab his quo longius quidlibet recedit, tantò plus tenebrarum et errorum existit. Affert enim mens nostra suapte naturâ secum ad divinarum rerum studia notiones suas in praedicamento quantitatis extructas: quibus si spolietur, nihil nisi meris negationibus definire potest.*'

Letter to Herwart von Hohenburg, 9 April 1599; (*op. cit.*, p. 309):

> '*Desinamus igitur super caelestia et incorporea plus, quam Deus nobis revelavit scrutarj. Haec sunt intra captum judicij humani, haec nos scire deus voluit, dum ad suam nos imaginem condidit, ut in consortium earumdem secum ratiocinationum veniremus. Quid enim est in mente hominis praeter numeros et quantitates? Haec sola rectè percipimus, et, si pie dici potest, eodem cognitionis genere cum Deo, quantum quidem in hac mortalitate de ijs percipimus.*'

Galileo said the same, but drew quite different conclusions. Cf. my *Études Galiléennes*, III. *Galilée et la loi d'inertie*, Paris, 1939, p. 125.

16. The *Harmonices Mundi Libri* V, which was published at Linz in 1619, is in some measure encompassed by the *Epitome Astronomiae Copernicanae*, the first part of which appeared in 1618, the second in 1620, and the third in 1621; the first two parts were published at Linz, the third at Frankfurt.

PART I. CHAPTER I. *MYSTERIUM COSMOGRAPHICUM*

1. The official documents are full of eulogies bestowed on Kepler. Nevertheless, it is probable that he was already regarded with suspicion by orthodox Lutherianism on account of his avowed and enthusiastic partisanship of Copernican cosmology, and that the university authorities sent him to Graz to a post socially inferior to that of a pastor in order to get rid of him; cf. *infra*, p. 389, n. 4.

2. Once he had embarked on a scientific career, Kepler was never able, in fact did not wish, to forsake it. He, for his part, considered his settlement

at Graz, which diverted him from theology to the profit of astronomy, was decided by Divine Providence. He always attributed his discoveries to the particular intervention of the same Providence; cf. *supra*, p. 163.

3. Kepler's first 'prognostication' in which he foretold a very cold winter, peasant revolts and war against the Turks was realized in every detail; and so Kepler's reputation as an astrologer was established once and for all. Consequently, important personnages asked him to cast horoscopes, which he was not always able to refuse to do. Kepler, in spite of his rejection of judicial astrology, believed in the influence of the celestial bodies, and some of his horoscopes, such as that of Wallenstein (*O.O.*, vol. VII, pp. 343 f.) as well as his own (*ibid.*, vol. V, pp. 476 f.) are masterpieces of psychological analysis. Cf. the analysis of the latter by Koestler in *The Sleepwalkers*, pp. 231 f.

4. The full title of this work by Kepler is: *Prodromus Dissertationvm Cosmographicarvm continens Mysterivm Cosmographicvm, De Admirabili Proportione Orbium Coelestium: deque causis coelorum numeri, magnitudinis, motuumque periodicorum genuinis et proprijs, demonstratvm, per quinque regularia corpora Geometrica*, Tübingen, 1596. In Kepler's opinion, the *Prodromus* would serve as a suitable introduction to a series of cosmographical treatises (see his letter to Herwart von Hohenburg, 26 March 1598; (*G.W.*, vol. XIII, p. 190)) dealing particularly with: (a) the Universe, mainly the motionless parts, the Sun and fixed stars; (b) the planets, a study of their motion according to Pythagoras, the music of the spheres . . ., etc.; (c) the celestial bodies, the Earth, origin of mountains and rivers, etc.; (d) the relationships between the Sky and Earth, light, aspects, physical principles of meteorology and astrology. The first three books would correspond, as Kepler mentioned, to Aristotle's *De Coelo*, and the fourth to the same author's *De Generatione*.

5. Maestlin not only supervised the printing of Kepler's work, but also helped him to compile it by carrying out a series of difficult calculations (determination of the distances of the planets from the Sun, cf. *supra*, p. 149) which were published together with some very fine drawings and diagrams as an appendix to the *Mysterium Cosmographicum* (cf. pp. 62 f.). He was responsible also for securing for Kepler the approval of the University of Tübingen by stressing in his report to the Pro-rector, Matthias Hafenreffer (end of May 1596; (*G.W.*, vol. XIII, pp. 84 f.)) the interest and exceptional importance of the work, the author of which had been the first to consider the problem of the arrangement of the celestial spheres, and to have had the good fortune to resolve it. He was careful to point out that the work, which had been hastily composed and presupposed some mathematical and astronomical knowledge on the part of the reader, especially of the Copernican system, should be completed by an exposition of this system and made clearer by re-editing certain very obscure passages. These, and other, recommendations were imparted to Kepler also by

Maestlin. The advice was accepted, at least in part, for Kepler suppressed (or allowed Maestlin to do so) the chapter dealing with the non-contradiction between Copernican astronomy and the Holy Scriptures: he reverted to this most dangerous subject in the preface to *Astronomia Nova*. If he added to his original text the admirable exposition of Copernicanism which is contained in the first two chapters; and if he agreed to rewrite the obscure passages, or suppress others as suggested by Maestlin, then he refused to do anything more. Otherwise, he said (cf. letter to Maestlin, 11 June 1596; (*G.W.*, vol. XIII, pp. 89 f)), it would be necessary to rewrite the whole work completely.

It was for this very reason that Maestlin, who considered Kepler's corrections and alterations to be inadequate, made others of his own; and as he regarded his own contribution (the appendix mentioned above) to be too difficult for the generality of readers, he added to Kepler's text a reprint of the *Narratio Prima* by Rheticus, to which Kepler had made reference in his work. Kepler was not very pleased on this account, because the cost of printing was considerably increased. All told, the correction of proofs, preparation of tables and drawings, cost Maestlin much time and trouble, a fact which he did not fail to tell Kepler more than once (cf. letters to Kepler, 15/16 November 1596 and 10 January 1597; (*ibid.*, pp. 95 f., 101 f.)). By way of compensation Kepler finally made him a present—a silver gilt cup and six silver thalers.

6. Cf. *Mysterium Cosmographicum*, preface *ad lectorem*; (*G.W.*, vol. I, p. 9). Just as significant is the letter of dedication to Sigismund, Freiherr von Heberstein, and to the magistrates of Styria, in which Kepler without false modesty proclaims the value of his work (*ibid.*, p. 5):

'*Opus . . . exigua mole, labore modico, materia vndiquaque mirabili. Nam siue quis antiquitatem spectet; tentata fuit ante bis mille annos à Pythagora; siue nouitatem, primùm nunc à me inter homines vulgatur. Placet moles? Nihil est hoc vniuerso mundo maius nec amplius. Desideratur dignitas? Nihil preciosius, nihil pulchrius hoc lucidissimo Dei templo. Lubet secreti quid cognoscere? Nihil est aut fuit in rerum natura occultius.*'

7. It is interesting to note that, right from the start, Kepler was preoccupied with *physical* reasons which militate in favour of Copernicanism and was not satisfied with purely 'mathematical' reasons.

8. There is probably a relic of it in the early chapters of the *Epitome Astronomiae Copernicanae*.

9. *Mysterium Cosmographicum*, cap. I, p. 14.

10. Allusion to one of Aristotle's propositions.

11. Except for the first part of the first book, *De Revolutionibus Orbium Coelestium* is a very difficult technical work.

12. *Mysterium Cosmographicum*, cap. I, p. 15. Kepler's aversion from

'hypotheses', that is to say, groundless, arbitrary and even false suppositions put forward as the basis of astronomical calculations (this is strangely reminiscent of Newton's aversion from Cartesian 'hypotheses'), and Kepler's conviction of the truth of his own system were such that he wrote to his teacher (cf. letter to Maestlin, October 1597; (*G.W.*, vol. XIII, pp. 140 f.)) saying, that through his *Mysterium Cosmographicum* he had acquired the chair—or at least the right to the chair—of Ramus at the Collège Royal, who had promised it to anyone who could produce an astronomy *without hypotheses.*

> '*Non possum te Clarissime domine praeceptor, non certiorem reddere, mihi pro edito libello hunc honorem a Gallis haberj, ut Regius professor Lutetiae Parisiorum ex pacto constitutus ego sim, aut certe illum, qui haec ad me scripsit stultum fuisse necesse sit. Quia verò is qui mihi cedere volebat intera mortuus est; et procul dubio diversae sententiae successorem jam reliquit: ideò decrevi, quoad deo placuerit spreta regia professione in Styria manere. Exemplar epistolae ad me missae tibi transmitterem, nisi mihi constaret te id pridem habere. Scripta est anno 1569 biennio ante me natum. Author Ramus est in scholis suis (non Geometria) Geometricis fol. 49 et 50. Quo loco Ramus praemium conformatae absque hypothesibus astrologiae spondet, suae professionis cessionem. Si Ramus illas exterminatus cupit hypotheses, quae ut credantur, postulantur non probantur, et si hanc absque hypothesibus astronomiam laudat, quae solius naturae apparatu orbium coelestium contenta est, quod quidem ante et post omninò videtur: vicimus vel Ego, vel Copernicus, vel uterque simul, nobisque professio debetur Ramea. Si autem Ramus omnes omninò hypotheses rejicit seu veras et naturales, seu falsas, tum id est, quod supra dixi: Stultus nempe, idque ut opinor vel te judice. Verum ut utriusque honorj consulatur malo me Professorem Regium, quam Ramum stutltum appellare.*'

This letter of Kepler's is certainly not to be taken seriously; but it undoubtedly expresses his firm belief. Moreover, the claim was repeated, seriously this time, in *Astronomia Nova*; cf. *infra*, p. 395, n. 1, and *supra*, p. 95, n. 7.

13. *Mysterium Cosmographicum*, pp. 15 and 16.

14. *Ibid.*, p. 16.

15. This is an exaggeration of the simplicity of the Copernican system, as Kepler, better than anyone, was well aware; but since the time of Copernicus, this had been an accepted manner of speaking.

16. Cf. *supra*, p. 33 and p. 43 for similar claims by Copernicus and Rheticus, which do not agree with reality.

17. *Mysterium Cosmographicum*, pp. 17 f.

18. The three motions mentioned by Copernicus are: (a) motion of the Earth round the Sun: (b) rotation of the Earth on its axis; (c) motion of this axis as a result of which it remains parallel to itself. Kepler added to these

the motion which produces libration in latitude of the planets in relationship to the Earth. Copernicus attributed this motion to the planetary circles. On the other hand, Copernicus admitted as a fourth motion the 'trepidation' of the Earth's axis (cf. *supra*, p. 88, n. 58). In a note to the second edition of the *Mysterium Cosmographicum* (Frankfurt, 1621) Kepler makes a correction: there are in fact only two motions; the third is not motion, but rest.

19. It was from these inequalities that Copernican astronomy 'liberated' nature.

20. Cf. *supra*, p. 59. The variation in the Earth's (Sun's) eccentricity was accepted by Tycho Brahe. According to Copernicus it resulted from the (double) motion of the centre of the terrestrial orbit round the Sun. Cf. *supra*, p. 113.

21. Kepler has forgotten—or neglected—to say, that Copernicus, by rejecting the equant, was obliged to reintroduce one epicycle per planet in his model; consequently, the gain was not ten, but only five, circles; cf. *supra*, p. 43.

22. The term 'prosthaphaeresis' has several meanings which it is unnecessary to describe in detail. Here, it means the relation of the deferent to the epicycle, or, in other words, the angle under which the planet's epicycle is seen from the Earth, this angle being equal to that under which the Earth's circle is seen from the planet in question.

23. *Mysterium Cosmographicum*, p. 9. Cf. *supra*, p. 153.

24. *Ibid.*, p. 10.

25. *Ibid.*, p. 10; cf. chap. x *De origin enumerorum nobilium* (pp. 36 f.). In the second edition of *Mysterium Cosmographicum*, Kepler added (*O.O.*, vol. I, p. 134):

> '*Omnis numerorum nobilitas (quam praecipue admiratur theologia Pythagorica, rebusque divinis comparat), est primitus a geometria. . . . Non enim ideo numerabiles fiunt anguli figurae, quia praecessit conceptus illius numeri, sed ideo sequitur conceptus numeri, quia res geometricae habent illam multiplicitatem in se, existentes ipsae numerus numeratus.*'

Words to the same effect are found in the *Harmonice Mundi*; (*G.W.*, vol. VI, p. 222; *O.O.*, vol. V, p. 221):

> '*De numeris quidem haud contenderim, quin Aristoteles rectè refutaverit Pythagoricos; sunt enim illi secundae quodammodò intentionis, imo et tertiae et quartae.*'

26. Cf. Georgius Johachimus Rheticus, *De libris Revolutionum . . . Narratio prima*; (*G.W.*, vol. I, p. 105).

Cf. *supra*, p. 52. One could put forward in favour of the number six the fact that the Universe was created in *six days*. It is rather surprising that neither Rheticus nor Kepler made use of it.

27. *Mysterium Cosmographicum*, p. 11. It is characteristic of Kepler to speak of *vis motus* and *virtus ad motus*: but a virtue or infinite motive force is impossible, even though at first sight it seems to 'conform' to the Sun, the source of motion, for it would make all the motions infinitely rapid. Kepler, of course, considered the velocities to be proportional to the motive forces: it was for this reason that he was able to replace an estimate of the latter for the former.

28. *Ibid.*, pp. 11 f.

29. In his letter of 2 August 1595 to Maestlin (*G.W.*, vol. XIII, p. 28) Kepler gives the date as 20 July:

> '*Tandem die 20 Julii cum lacrumarum copiosa profusione* (*instar illius qui clamavit Εὕρηκα*) *reperi modum et causam numerj orbium senarij simul et distantiae talem, ut censeam vel solius illius retinendae causa* (*tam est probabilis et concinna, deoque creatore imprimis digna*) *mutandas esse προσθαφαιρέσεις, ut compensetur, quod in distantia Copernicana adhuc ab ea discrepat*.'

In a letter of 1 October 1602 to Fabricius, Kepler gives 17 July. The correct date is probably 20 July.

30. *Ibid.*, p. 12. At a later date (cf. *supra*, p. 345) Kepler identified light with the Sun, whose creation he then placed as having been made on the first day.

31. Kepler considered it appropriate for the most regular solid (the cube) to follow immediately after the sphere of the fixed stars, and that it should be followed in its turn by the tetrahedron; cf. chaps. IV to VIII, pp. 30–36; cf. n. 57.

32. *Mysterium Cosmographicum*, cap. II, p. 23. The restriction of quantity to the body recalls the views of Descartes on bodies or extended things.

33. *Ibid.*; Cicero, *Timaeus*, 3.

34. *Mysterium Cosmographicum*, pp. 23 f.

35. Even Descartes, who had started by trying to solve the problem of the fixed stars, retreated in face of the difficulty.

36. *Mysterium Cosmographicum*, p. 26.

37. The truth was not acquired in Antiquity but in our own 'age'.

38. *Ibid.*, p. 13.

39. The spheres referred to here are not the solid spheres of Aristotelian cosmology, but merely the spherical envelopes within which the planetary orbs are located. In fact, in chap. XVI of the *Mysterium Cosmographicum* (p. 56) Kepler writes:

> '*Nam absurdum et monstrosum est, corpora haec* [the celestial bodies] *materia quadam vestita, quae alieno corpori transitum non praebeant, in coelum collocare. Certè multi non verentur dubitare, an omnino sint in coelo eiusmodi Adamantini orbes; an divina quadam virtute moderante cursus intellectu proportionum Geometricarum, stellae per campos et auram aetheream liberae istis orbium compedibus transportentur*.'

The Copernican spheres or circles (they were 'destroyed' by Tycho Brahe) consequently disappeared from the sky, which was left filled only with aether. Cf. Letter to Maestlin, 3 October 1595; (*G.W.*, vol. XIII, p. 43).

40. Kepler's famous diagram shows the model of a silver bowl which he suggested to the Duke of Württemberg should be made by the goldsmiths of Stuttgart: the making of this bowl giving a true representation of the Universe would contribute greatly to the glory of the Duke, and would be a tribute to the glory of God (cf. letter to Friedrich, Duke of Württemberg, 17 February 1596; (*G.W.*, vol. XIII, pp. 50 f.)). Strangely enough, the Duke agreed to the scheme, but stipulated that a model in copper should first be made (*ibid.*, p. 52), and that Maestlin's opinion should be asked. After he had received the latter (12 March 1596; (*ibid.*, p. 67)), the Duke decided to have the bowl made. Unfortunately, the goldsmiths were not sufficiently skilled, and the project was finally abandoned.

41. If, instead of taking the radius of the sphere inscribed in the octagon of Mercury as the term for comparison, one takes the radius of the circle inscribed in the square base of the two pyramids of which it is comprised, then the value 707 is obtained.

42. Cf. Letter of Maestlin to Kepler quoted in the *Mysterium Cosmographicum*, cap. XIX, p. 67, and Maestlin, *Appendix*, p. 143.

43. *Mysterium Cosmographicum*, cap. XV, p. 50. In fact, the centre of the terrestrial orbit corresponds to the mean Sun of Ptolemy: according to Copernicus, the Earth's orbit is consequently exactly that of the Sun according to Ptolemy.

44. In this connection, Kepler quotes the well-known preface of Rheticus to his *Ephemerides* for 1550, where the latter relates that Copernicus told him that he would be just as happy to have an agreement of about 10′ with observed phenomena as Pythagoras was on the discovery of his famous theorem. Cf. *supra*, p. 81, n. 37.

45. Apelt has pointed out (*Johann Keplers Astronomische Weltansicht*, p. 93) that Bode's Law could have been discovered by Kepler, for it differs by a single constant only from one of the relationships examined and discarded by the latter.

According to Bode's Law the distances of the planets from the Sun stand in the following proportions:

☿ $= \ldots 0 + 4 = 4$
♀ $= 2^0 \times 3 + 4 = 7$
♁ $= 2^1 \times 3 + 4 = 10$
♂ $= 2^2 \times 3 + 4 = 16$
$= 2^3 \times 3 + 4 = 28$ (Gap between Mars and Jupiter occupied by Asteroids)
♃ $= 2^4 \times 3 + 4 = 52$
♄ $= 2^5 \times 3 + 4 = 100$

In other words, the distances *minus the constant 4* are in the ratio of 3:6:12:24:48:96.

An excellent account of modern theories of the structure of the solar system is given by E. Schatzmann, *L'origine et l'évolution des mondes*, Paris, 1957.

46. Cf. *supra*, p. 34 and p. 135.

47. *Mysterium Cosmographicum*, p. 69. It will be noted that Kepler's dynamics, to which I shall refer again, is purely Aristotelian: velocities are strictly proportional to the motive forces.

48. In the second edition of *Mysterium Cosmographicum*, which appeared in 1621 with numerous notes and corrections, Kepler wrote (cf. note *c* of cap. XX, *Opera Omnia*, vol. I, p. 176):

> '*Si pro voce anima vocem vim substituas, habes ipsissimum principium ex quo physica coelestis vis Comment. Martis est constituta et lib. IV Epitomes Astr. exculta. Olim enim causam moventem planetas absolute animam esse credebam, quippe imbutus dogmatibus J. C. Scaligeri, de motricibus intelligentiis. At cum perpenderem, hanc causam motricem debilitari cum distantia, lumen Solis etiam attenuari cum distantia a Sole: hinc conclusi vim hanc esse corporeum aliquid, si non proprie saltem aequivoce; sicut lumen dicimus esse aliquid corporeum, id est, speciem a corpore delapsam, sed immateriatam.*'

Cf. J. C. Scaliger, *Exercitationes Esothericae*, VIII (2), 673; cf. *Astronomia Nova*, cap. XXXIII; *supra*, p. 203.

49. *Mysterium Cosmographicum*, cap. XX, *G.W.*, vol. I, p. 70; cf. *ibid.*, cap. XXII, *G.W.*, vol. I, p. 77:

> '*Nempe mundus totus animâ plenus esto quae rapiat, quicquid adispicitur stellarum siue cometarum, idque ea pernicitate, quam requirit loci à Sole distantia et ibi fortitudo virtutis. Deinde esto in quolibet Planeta peculiaris anima, cuius remigio ascendat in suo ambitu.*'

50. All the forces or virtues emanating from the Sun should have the same law of propagation; but, as will be seen later (pp. 210 f.), the diminution in motive force and of light cannot take place in the same way, for the Keplerian concept of motion, which in the last analysis is Aristotelian, requires a motive force proportional to the velocity (and not to the acceleration) of the moving body. A motive force which would be attenuated in the same degree as light, that is to say, in the inverse ratio of the *square* of the distance from the source, as Kepler himself proved at a later date, would be too weak to endow the planets with their observed velocities; cf. my *Études Galiléennes*, III, pp. 31 f.

51. Cf. Appendix III.

52. *Mysterium Cosmographicum*, cap. XX, *G.W.*, vol. I, p. 71.

53. Cf. *ibid.*, cap. XX, *G.W.*, vol. I, p. 71: *Dimidium incrementi*

additum periodo minori exhibere debet proportionem veram distantiarum.' In modern notation:

$$\frac{t_2 + t_1}{2t_1} = \frac{r_2}{r_1}.$$

Cf. Letter to Maestlin, 3 October 1595; (*G.W.*, vol. XIII, p. 38):

> '*Motrix anima ut dixi, in Sole. Si motus aequalitas et idem vigor veniret a Sole in omnes orbes, tamen unus alio tardius circumiret propter inaequalitatem ambitus. Tempora periodica essent ut circulj. Nam quantitas motum metitur. Circulj autem ut semidiametrj, scilicet ut distantiae. Sic facilime constitueremus ex motibus medijs certo cognitjs, medias etiam distantias. Sed accedit alia causa quae tardiores efficit remotiores. Capiamus a luce experimentum. Nam lux et motus utique ut origine sic etiam actibus conjunctj, et forsan ipsa lux vehiculum motus est. Igitur in parvo orbe, et sic etiam in parvo circulo prope Solem tantum est lucis, quantum in magno et remotiore. Tenuior igitur lux in magno, in angusto confertior et fortior. At haec fortitudo rursum est in circulorum proportione, sive in distantiarum, Si jam eadem ratio motus est (qua quidem re nihil concinnius fingi potest) sequetur distantias bis facere ad motus tarditatem inducendam. Quare dimidia differentia motuum duorum adjuncta minorj motuj erit distantia remotioris: motus ipse minor erit distantia propinquioris.*'

54. The *Mysterium Cosmographicum* was republished in 1621; his essential propositions were corrected and amplified in the *Harmonice Mundi* as well as in the *Epitome Astronomiae Copernicanae*.

55. The problems of 'harmony' and the establishment of relationships between the musical scale and celestial motions had already intrigued Kepler at the time of the *Mysterium Cosmographicum* (cf. cap. XII, pp. 39 f.); he mentioned the matter in his letter to Herwart von Hohenburg quoted on p. 380, n. 4; and in subsequent years it was the subject of letters to Edmund Bruce (18 July 1599; (*G.W.*, vol. XIV, pp. 1 f.)), to Herwart von Hohenburg (6 August 1599; (*ibid.*, pp. 27 f.)), to Maestlin (19/24 August 1599 (*ibid.*, pp. 43 f.)), to Herwart von Hohenburg again (14 September 1599; (*ibid.*, pp. 63 f.)), etc. Even the plan of the *Harmonice Mundi* was already clearly formed in his mind by 1599, when he informed (14 December 1599; (*ibid.*, p. 100)) Herwart von Hohenburg that he had formulated the method and plan of a work, *Harmonice Mundi*, which would consist of five parts: (1) geometrical, dealing with the figures that can be constructed; (2) arithmetical, dealing with the relationships between polyhedral solids; (3) musical, dealing with the causes of harmony; (4) astrological, dealing with the causes of aspects; (5) astronomical, dealing with the causes of periodic motions. This was exactly the plan of the work published in 1619.

56. Cf. *supra*, pp. 201 f. In the *Tertius Interveniens* (no. 51), Kepler says that the planets are magnets, and are pushed round the Sun by magnetic forces, but that the Sun alone is alive (*Opera Omnia* (Frisch edition), vol. I, p. 554). There is not a complete identification of the magnetic forces with the motive forces; Kepler says that the planets are moved *per speciem immaterialem Solis* (*ibid.*, p. 590).

57. Letter to Maestlin, 3 October 1595; (*G.W.*, vol. XIII, pp. 35 f.). Kepler continued:

'*Conclusi, ut quae sententia eundem ordinem in motu et distantijs servaret, vera esset; quae non, falsa. Exinde probavi corpora haec* [regular polyhedra] *duorum esse ordinum, tria in uno, duo in reliquo ordine: Cubum esse primum omnium et sui ordinis etiam, post Pyramida, ultimum primorum dodecaedron. Secundariorum primum esse Octaedron, ultimum Icosaedron: Hinc constitit ratio quare res totius mundj praecipua, terra nempe cum imagine Dej in homine, distingueret inter ordines. Probavi igitur quod Primaria debeant extra terrenj orbis complexum esse, Secundaria intra: quod Caput exterioris ordinis debeat ad fixas vergere, caput interioris ad Solem. Atque hic explicandae fuerunt omnes proprietates et cognationes horum corporum. His ita constitutis accensui corporibus planetas, contentis a terrae orbe, contentos a se; exclusis exclusos. Sic* ♄ *cubum sortitis est,* ♃ *Tetraedrum,* ♂ *Dodecaedrum,* ♀ *Icosaedrum,* ☿ *Octaedrum. Digressus igitur in Astrologiam demonstravj cognationes et proprietates planetarum plerasque ex horum 5. corporum affectionibus mathematicis.*'

58. *Divinus mihi Cusanus*, said Kepler; (*Mysterium Cosmographicum*, cap. II, p. 23). On the connection between Kepler and Nicholas of Cusa see D. Mahnke, *Unendliche Sphaere und Allmittelpunkt*, Halle, 1937.

59. This modification—shifting the common point of the planetary circles from the centre of the Earth's circle to the body of the Sun—deprived the Copernican system of its greatest kinematic advantage, namely, suppression of the equants, *i.e.*, those points, which are eccentric with respect to the centres of the planetary circles, but with respect to which the planets have constant angular velocity whilst moving with variable velocity over their circles. In fact, as a result of referring the planetary motions, including that of the Earth, to the Sun, Kepler was forced to reintroduce the very notion from which astronomy had been (in the words of Rheticus) 'freed' by Copernicus. Now, it was precisely this reconversion to uniform circular motions that was responsible for the failure of Copernicus; and it was their abandonment, which requires a dynamical explanation of the non-uniform motions of the celestial bodies, that led Kepler to his greatest triumph—rejection of circularity. However, Kepler's first step, by which he reintroduced the equant and got rid of the

Copernican epicycles, was the *triumph of circularity in astronomy*. Cf. *supra*, p. 170 and n. 34.

60. Cf. *supra*, p. 149 and n. 43.

61. Thus, Kepler distinguishes between 'Copernicus who speculates' and 'Copernicus who computes'.

PART II. CHAPTER I. KEPLER AND TYCHO BRAHE

1. Meanwhile, Kepler had published, besides *Calendars* and *Prognostications*, *De fundamentis astrologiae certioribus* (Prague, 1601), *Ad Vitellionem Paralipomena quibus Astronomiae Pars Optica traditur* (Frankfurt, 1604), *De stella nova* (I, *De stella nova in pede Serpentarii* (Prague, 1606); II, *De stella tertii honoris in Cygno* (Prague, 1606); III, *De stella nova in pede Serpentarii, pars altera* (Frankfurt, 1606); IV, *De Jesu Christo Salvatoris nostri vero Anno natalitio* (Frankfurt, 1606)).

2. Maestlin enthusiastically accepted the change of Copernicanism into *a priori* doctrine. Cf. *supra*, p. 127, and Maestlin's preface to his edition of the *Narratio prima*; (*G.W.*, vol. I, p. 82): *Michael Maestlin Goeppingensis*, candido Lectori S.

> '*Magna sanè sunt, quae Artifices Astronomi huc vsqe inuenerunt: Astronomiam tamen hactenus omnes non nisi à tergo adorti sunt, et tam motus, quàm magnitudines et distantias ex solis obseruationes indagare docuerunt. An autem à priori, siue à fronte vllus ista dimetiendi pateat aditus, vel annè vlla alia, praeter obseruationes, geometrica Norma, inuentos motuum et quantitatum numeros examinandi, haberi possit, nulli ne peritissimo quidem Artifici hactenus vel per insomnium in mentem venit. Iam vero Keplervs noster solertissimo ex Geometria inuento orbium seu spherarum coelestium certum finitumque numerum et ordinem, atque quod maximum est, certam magnitudinem, sicut et motuum, ad se mutuo proportionem tradit; et paulò altius sumpto initio ostendit, quod Creator Deus Opt. Max, in Mundi creatione, iuxta quinque regularium Corporum geometricorum, aliàs omnibus Geometris notissimorum, proportionem, sphaeras coelestes mobiles fabricauerit, extenderit, disposerit, adornauerit, et ordinauerit.... Ab hoc igitur tempore, qui coelorum motus pleniùs inquirere, et quae in Astronomia adhuc manca sunt, reficere et redintegrare volet, habet iam à priori patentem ianuam, qua ingrediatur, habet rectissimam normam, ad quam, ceu ad Lydium lapidem, omnes suas obseruationes, totumque calculum examinet.*'

3. Cf. *ibid.*, pp. 84 f.

4. Matthias Hafenreffer, Rector of the University of Tübingen, who had received a copy of the *Mysterium Cosmographicum* from Kepler, wrote to the latter on 11 April 1597 (*G.W.*, vol. III, p. 203) enjoining him not to

stress the truth of Copernicanism, but to treat it only as an 'hypothesis'. Galileo (*ibid.*, p. 130) sent an enthusiastic letter (4 August 1597). It was written, he said, no more than several hours after the receipt of the volume, of which he had so far read only the preface. In this letter he congratulated himself on having found such a *socius* in the pursuit of truth (Copernican), a truth which would undoubtedly ensure its author (Copernicus) immortal glory in certain quarters, though it was nevertheless ridiculed and mocked by the majority (for the number of idiots is infinite). He thanked Kepler for having thought him worthy of this token of friendship. However, when Kepler replied asking for Galileo's approval, and inviting him publicly to declare himself a follower of Copernicus, Galileo, who had probably read the book in the meanwhile, made no answer. It is obvious that Kepler's ideas could not secure a sympathetic reception from Galileo; on the contrary, they must have provoked considerable distaste.

5. Cf. *G.W.*, vol. XIII, p. 154:

'*Cum te, Vir amplissime, Mathematicorum omnium, non huius tantum aetatis, sed totius aeui Monarcham constituerit incomparabilis doctrina, iudicijque praestantia: iniquum fecero, si opusculo de proportione caelorum nuper in lucem dato (sub titulo: Prodromus dissertationum Cosmographicarum) vllam gloriam, tuo iudicio et commendatione neglectâ, aucuper.*'

6. Tycho Brahe to Kepler, Wandsbeck, 11 April 1598; *ibid.*, p. 197.

7. Tycho Brahe wrote: *Apollo*—as did Kepler also (cf. *infra*, p. 399)—but he certainly had *Apollonius* in mind.

8. Tycho Brahe to Maestlin, May 1598; *ibid.*, p. 204.

9. Tycho Brahe alludes to Maestlin's preface to the *Narratio Prima*.

10. They seemed to him to have just as little value as Tycho Brahe's objections against the Copernican system. In Kepler's view, it was not the Copernican system, but the Tychonian, which made the Sun together with the five planets revolve round the Earth; which is quite absurd. In his opinion, Tycho Braye should do one thing only, namely publish his observations, and *in toto*, in order to allow those who are capable of interpreting them (Kepler, for example) the opportunity to do so. As for Tycho Brahe himself (cf. Letter of Kepler to Maestlin, 16/26 February 1599; (*G.W.*, vol. XIII, p. 289)): '*Ego sic censeo de Tychone: Divitijs abundare, quibus non rectè utatur, more plerorumque divitum. . . .*' One only of Tycho Brahe's objections against Copernicus did impress him; it was the immensity of the Copernican Universe coupled with the lack of connection between the eighth sphere (that of the fixed stars) and the rest of the Universe. Nevertheless, Kepler thought that there must be some essential difference between the mobile Universe and the immobile Universe. As for the size of the Universe, it is not the Universe which is large—at least for God—it is we who are extremely small: '*nos mundo parvi sumus, mundus Deo non est magnus.*'

11. Tycho Brahe to Kepler, 9 December 1599; (*G.W.*, vol. XIV, pp. 94 f.)). This is a very interesting letter because it reveals (a) Tycho Brahe's belief that the 'destruction' of the solid spheres *condemned* Copernicanism by excluding the possibility of the Earth's orbital motion; and (b) the power of the obsession with circularity on a mind of his quality; and, on the other hand, the magnitude of Kepler's liberating effort.

12. Basically, they did not get on together at all. Tycho Brahe was looking for a meek collaborator; a mathematical computator who would work under him and in accordance with his instructions—a 'technical assistant'. Now, this was precisely what Kepler, who was well aware of his own worth, could not become at any price. What Kepler wanted, was access to Tycho Brahe's material in order to use it in his own way in order to develop his 'harmonious' theory of the Universe (cf. *supra*, p. 67 and Kepler's letter to Magini, 1 June 1601 (*G.W.*, vol. XIV, p. 173)): '*Harmonicen Mundi, perficere, nisi restaurata per Tychonem Astronomia, non possum.*' However, Tycho Brahe had no intention of delivering up to him his treasury of observations. Hence, we have Kepler's complaint (*ibid.*): '*Observationes quidem lectissimas porrigit, non tamen aliter quam intra suos parietes. Labora, inquit, tu quoque; credo quod Copernicanae hypotheseos defensorem, alius ipse sententiae, spectare constituit.*' Kepler was all the more discontented on this account because he thought he could have solved very quickly the theoretical problems which were holding up Tycho Brahe and his assistants. Experience revealed to him his error, and caused him to realize the great value of the empiricism of the great Danish astronomer to whom he was to show his faithfulness in the *Astronomia Nova*.

13. Kepler relates the history of his relations with Tycho Brahe in the preface to the second edition of the *Mysterium Cosmographicum*. For the true account of these relationships, see the biography of Kepler by Max Caspar, and the relevant chapter in Koestler's *The Sleepwalkers*, p. 377.

14. *Tabulae Prutenicae*, published by Erasmus Reinhold in 1551.

15. Cf. *supra*, p. 178. Kepler finally compiled these tables, and published them under the title of *Tabulae Rudolphinae* (Ulm, 1627).

16. Initially in collaboration with his pupil Longomontanus; Kepler continued the work alone, when Longomontanus left Prague to return to Denmark as professor at the University of Copenhagen at the beginning of 1601.

17. One might go even further and regard the meeting between Tycho Brahe and Kepler as only the last but one of the steps taken by Providence, the last step being the providential death of the former who put the latter in possesion of the documents he—or astronomy—needed.

18. I refer to Tycho Brahe's loss of his observatory at Hven, and the expulsion of the Protestants from Styria by the Archduke Ferdinand in 1601.

19. Cf. *Astronomia Nova*, cap. VII; (*G.W.*, vol. III, p. 108;*O. O.*, vol. III, p. 209).

20. Cf. *supra*, p. 149 and n. 43.

21. Cf. *infra*, p. 434, n. 36.

22. The eccentricity of the orbit of Mars is 0·093; of the Earth, 0·016; of Mercury, 0·2. Mercury is difficult to observe, and even Tycho Brahe did not succeed in doing so.

23. *Astronomia Nova, Argumentum singulorum capitulum* (*G.W.*, vol. III, p. 36; *O.O.*, vol. III, p. 160):

> '*Cum alia sit Methodus, quam Natura rei docet; alia, quam cognitio nostra requirit; utraque artificialis: neutram a me lector sinceram expectare debet. Mihi enim scopus non hic praecipuus est: explicare motus coelorum, quod fit in libellis Sphaericis et Planetarum Theoriis: neque tantum, docere lectorem, et perducere a primis et per se notis ad ultima; quam viam Ptolemaeus ut plurimum observavit; sed accedit tertium aliquid, commune mihi cum Oratoribus; ut quia nova multa trado, id coactus fecisse manifestus sim; itaque demeream et retineam assesum lectoris, et amolior suspicionem de studio novandi.*
>
> '*Nil igitur mirum, si methodis superioribus admisceam tertiam Oratoribus familiarem, hoc est, historiam mearum inventionum: ubi non de hoc solo agitur, quo pacto lector in cognitionem tradendorum perducatur via compendiosissima: sed de hoc potissimum, quibus Ego author seu argumentis seu ambagibus seu fortuitis etiam occasionibus primitus eodem devenerim. Quod si Christophoro Colvmbo, si Magellano, si Lusitanis non tantum ignoscimus, errores suos narrantibus, quibus ille Americam, iste Oceanum Sinensem, hi Africae Periplum aperuerunt; sed ne vellemus quidem omissos, quippe ingenti lectionis jucunditate carituri: nec igitur mihi vitio vertetur, quod idem eodem lectoris studio per hoc Opus sum secutus. . . .*'

24. In fact, the *Astronomia Nova* is not even a history: it is, for the most part, a diary written as Kepler's ideas progressed, and was not revised till afterwards.

25. The best commentary on the *Astronomia Nova* is provided by Kepler's correspondence during the years the work was being compiled. This was fully realized by Frisch, whose edition of this work (*O.O.*, vol. III) is consequently extremely useful.

26. References to the *Astronomia Nova* will be made to both Caspar's edition and that of Frisch.

27. *Astronomia Nova*; (*G.W.*, vol. III, p. 19; *O.O.*, vol. III, p. 147).

28. *Ibid.*; (*G.W.*, vol. III, p. 19; *O.O.*, vol. III, p. 147).

29. I.e., Ptolemaic, Copernican and Tychonian.

30. My italics. Cf. *supra*, p. 120.

31. Cf. *supra*, Introduction, n. 7 and 8.

32. Cf. *supra*, p. 154.

33. Cf. *Mysterium Cosmographicum*, cap. XXII, *G.W.*, vol. I, pp. 75 f.: *Planeta cvr super aequantis centro aequaliter moveatur.*

... Maestlin ... '*me de superiorum epicyclijs monuit, quos Copernicvs loco aequantium introduxit, quique duplo maiorem efficiant orbi spissitudinem, quàm Planetae ascensus descensusque requirit.... Existimauit igitur, eam orbibus relinquendam esse spissitudinem, quae motibus demonstrandis sufficiat. Cui respondi, primùm, deserendum esse totum negoteium, si duplo crassiores fiant orbes: nam nimium προσθαφαιρέσεων ademptum iri: Deinde nihil decedere nobilitati miraculosae huius machinationis, si modò viae ipsae, planetarum descriptae globulis, retineant hanc proportionem: quibuscunque illi agitentur orbibus, magnis an paruis. Et addidi, quae cap, XVI habes, de materia figurarum, quae nulla sit; atque inde non absurdum esse, corpora cum orbibus eodem loco includere. Imò verò vel sine orbibus hanc viae inaequalitatem defendi posse. In qua sententia video Nobilem et excellentissimum Mathematicum Tychonem Brahe, Danum, versari. Causam tamen et modum haec nostra disertiùs indicant. Nempe si eadem sit causa tarditatis et velocitatis in singulorum orbibus, quae suprà cap. XX fuit in vniuerso mundo, hoc modo: Via Planetae eccentrica, tarda superiùs est, inferiùs velox. Ad hoc enim demonstrandum assumpta Copernico epicyclia, Ptolemaeo aequantes. Describatur igitur concentricus aequalis viae Planetariae eccentricae; cuius motus undiquaque aequalis erit, quia aequaliter ab origine motus distat. Ergo in medietate viae eccentricae supra concentricum eminenti tardior erit Planeta, quia longiùs à Sole recedit, et à virtute debiliori mouetur: in reliqua celerior, quia Soli vicinior, et in fortiori virtute.*'

Addition in the 1621 edition (cf. Frisch, *O.O.*, vol. I, p. 183):

'*Si quae causa efficit, ut Saturnus altus sit tardior Jove humiliori et Soli viciniori, eadem efficiat, ut Saturnus altus et apogaeus sit tardior se ipso perigaeo et humili.*'

34. Kepler had already written in 1600 to Herwart von Hohenburg saying that all the planets describe identical orbits, namely, *perfect circles*; and in 1601 he informed Magini that for the theory of the Sun (or of the Earth) one single eccentricity in the orbit was not enough, and that it was necessary to assume in fact some inequality in orbital velocities—or an equant. Cf. Letter to Magini, 1 June 1601 (*G.W.*, vol. XIV, p. 176):

'... *omnium septem Theoriarum, quod motus siderum reales attinet, formam esse plane eandem, eamque simplicissimam; quilibet enim in una revolutione constantissimum exactissimumque circulum decurrit, tardius supra, velocius infra, hoc est, prope Solem, idque non per φαντασιαν, sed re vera. Nam Tycho etiam in Luna aequantem adhibuit. Ex quo concinnitate, et simplicitate, hoc est, perfectione motuum caelestium, quantum Copernico roboris accedat, facile perspecis.*'

35. Cf. *Astronomia Nova*, cap. XXII; (*G.W.*, vol. III, p. 191; *O.O.*, vol. III, p. 267):

> '*In Mysterio Cosmographico cap. XXII cum Physicam causam aequantis Ptolemaici vel secundi epicycli Copernico Tychonici redderem, mihi ipsi objeci in fine capitis: quod si causa a me allata genuina esset, omnino per omnes Planetas valere debuerit. Cum autem Tellus, una ex sideribus* (*Copernico*), *vel Sol* (*reliquis*), *aequante hoc hactenus non indiguerit, speculationem illam incertam esse volui, quoad Astronomis amplius liqueret. Suspicionem tamen concepi, fore et huic theoriae suum aequantem.*'

36. *Mysterium Cosmographicum*, cap. XXII, p. 77. In the passage which has just been quoted from the *Astronomia Nova*, it seems to me that Kepler gives more precision to his thought than is justified. He says, in effect, that from that time onwards he suspected the uniformity of the nature of planetary motions, particularly in view of the fact that the mechanism (dynamism) of his concept made it a necessary condition for all the planets. All the same, he left the matter open, seeing that the Earth (according to Copernicus), and the Sun (according to other astronomers), did not require an equant. In the second edition of the *Mysterium Cosmographicum* (1621) Kepler added to the passage in question from chap. XXII a note in which he says: 'undoubtedly [the Earth] did not need one according to Ptolemy, but did according to Copernicus. *At ego in Comment. Martis praecipuorum libri membrorum hoc unum feci et velut angula rem lapidem in fundamento posui, imo clavem astronomiae merito appellavi.*'

37. J. Dreyer, *A History of Planetary Systems*, 2nd edition, New York, 1950, p. 381. Dreyer, following Kepler, is of the opinion that Tycho Brahe must have suspected the true explanation of the phenomenon he had observed, and that Kepler was alluding to this in his letter on 12 July 1600 to Herwart von Hohenburg when he wrote as follows on the subject of the Sun's equant: '*Et hoc est quod Tycho quasi sub enigmatis involucro* (*ut interdum solet*) *ad me prescripserat de variabili quantitate orbis annui*' (*G.W.*, vol. XIII, p. 131; *O.O.*, vol. III, p. 24). In fact, in the letter quoted *supra*, p. 160 (April 1598; (*G.W.*, vol. XIII, p. 198)), Tycho Brahe started by setting forth the inaccuracies on Copernican astronomy, to wit:

> '*Eccentricitates ipsas utriusque illius, ut sic nunc loquar, Eccentricj, quas Copernicus paulò aliter excusavit, non habere eas ad invicem proportiones; ut una sit alteriùs tertia pars; veluti ex Ptolemaeo deduxit: Neuter verò eorum demonstratum reliquit. Sed haec aliam inter se habent rationem, et in singulis variant. Neque enim tres acronychiae observationes ut ut inter se in Eccentrico* (*quod praestat*) *remotae sufficiunt, ad trium superiorum Planetarum Apogaea . . . et Eccentricitates ambas ad amussim extricandas; sed per totum orbis circuitum in plurimis locis haec experienda veniunt: veluti à nobis quoque factum est.*'

Then he added:

> '*Vt vero de Marte aliquid exemplj loco addam, cùm is maximam ingerat varietatem, atque inter Coelum et Tabulas discrepantiam caeteris evidentiorem, scias, tantum abesse, ut eius Eccentricitas sit minor reddita,* (*prout voluit Copernicus, quo suae imaginationj circa quietam Solis fidem faceret*); *ut potiùs nunc aliquanto maior sit, quàm ea, quae à Ptolemaeo prodita est, licet discrimine non magni momentj, Apogaeo ejus 5° Copernicj terminos anticipante. Quin et Orbis annuus juxta Copernicum, aut Epicyclus secundum Ptolemaeum non videtur eiusdem semper magnitudinis, quoad ipsum Eccentricum collatione facta; sed alterationem introducit in omnibus tribus superioribus sensibilem; adeò ut angulus differentiae in Marte ad 1° 45 excrescat. Quomodo haec cum tuâ speculatione conciliandasint, tute videris.*'

CHAPTER II. FIRST ATTACK UPON THE THEORY OF MARS

1. *Astronomia Nova*, Introduction; (*G.W.*, vol. III, p. 34; *O.O.*, vol. III, p. 156):

> '*Coepi dicere, me totam Astronomiam non Hypothesibus fictitiis, sed Physicis causis hoc opere tradere: ad hoc vero fastigium me contendisse duobus gradibus; altero, quod deprehenderam, in corpore Solis concurrere Planetarum Eccentricos; reliquo, quod in Theoria Telluris intellexerim inesse circulum Aequantem, ejusque Eccentricitatem bisecandam. Igitur hic sit tertius gradus, quod . . . certissime demonstratum fuit, etiam Martialis Aequantis Eccentricitatem bisecandam praecise, quod Brahevs diu et Copernicvs dubium effecerunt.*'

On the verso of the title-page of the *Astronomia Nova*, after having reproduced the passage (Lib. II, p. 50) from the *Scholarum Mathematicarum Libri XXXVI* by Petrus Ramus, which contains his condemnation of Hypotheses and the promise to vacate his professional chair to anyone who should produce an Astronomy (Astrology) *absque hypothesibus*, Kepler wrote: '*Commodum Rame vadimonium hoc destruisti, vita digressus et professione: quam si nunc retineres, mihi quidem illam ego jure meo vindicarem*'; cf. *supra*, p. 96. Kepler published his denunciation of Osiander in the same place, cf. *supra*, p. 96, n. 7.

2. A letter to Herwart von Hohenburg, dated 12 July 1600 (*G.W.*, vol. XIV, pp. 128 f.), gives an admirable description of the working of Kepler's mind, as well as the transformation which took place within him when confronted with the real work of an astronomer. He came to Tycho Brahe in the hope of obtaining from him the *results* (distances and eccentricities) by means of which he would be able to confirm the ideas put forward in the *Mysterium Cosmographicum*, or, in a wider sense, his theories on the

Harmony of the Universe; but he had the unpleasant surprise of discovering that Tycho Brahe had the observational data only, and that the task of working them out and building up theories remained to be done. Certain in his own mind of his own superiority, he thought at first that it would be possible to complete this task in a very short time: eight days were quite enough to put his theory of Mars in order—so he told Longomontanus. He even made a bet with him. He then had his second surprise, which taught him a lesson. He learnt how difficult it was with really accurate data, such as those of Tycho Brahe, when one wanted to secure agreement with theory. He wrote (*ibid.*, p. 130):

> 'My dissertation on the Harmony of the Universe would have been brought to a satisfactory conclusion ere now had not the Tychonian Astronomy taken up my time to the point of sending me almost crazy.... Indeed, one of the main reasons for my visit to Tycho Brahe was [the wish] to obtain from him more accurate values (ratios) for the eccentricities by means of which I might consider my *Mysterium* and the Harmony which I have just mentioned, for speculations made *a priori* should not disagree with plain evidence, but must agree with it. Now, Tycho Brahe did not deliver up to me his entire collection of observations. At the very most, during meals, or in between times, he spoke sometimes of the apogee of one [of the planets] and sometimes of the nodes of another. However, when he saw that I possessed a challenging mind, he decided that the best way of treating me was to give me the observations relating to one single planet, namely Mars, to deal with them as I intended. All my time was taken up with them, and I was not concerned with the observations relating to other planets. Every day I hoped to finish with the theory of Mars; and then [I said to myself] I shall certainly receive the others.'

However, time was passing, and Kepler was always far from his goal. All the same, he made some definite progress:

> 'For Mars, as far as I was able to discover from Tycho Brahe's observations, started to sound the major third which I had assigned to it. It confirmed also my *Mysterium* in a remarkable way in two points: in fact, I had referred the eccentricities of the planets to the true Sun, and I was very much afraid that Tycho Brahe would refer them to the mean Sun, as Copernicus had done. Now, Mars obstinately refused any point other than the centre of the body of the Sun. Tycho Brahe was satisfied with my attempts; for, he told me, that he himself had long been of the same opinion, but had wanted to avoid making the complicated calculations, and therefore wished to know the opinions of others besides. Furthermore, at the end of my *Mysterium* I have mentioned the equant of the Sun, and expressed my regret that it was denied to the Sun (or the Earth)

alone: now, Mars has most clearly given evidence of this, that the equant is related to the body of the Sun.'

3. The 'acronical ' positions, *i.e.*, the positions of planets when at opposition with the Sun, are of special interest, because their longitudes then coincide with those of the centres of their epicycles (in Ptolemaic theory), and with their heliocentric longitudes (in Copernican theory).

4. On 20 December 1601, Kepler made a report on the progress of his labours to Maestlin (*G.W.*, vol. XIV, pp. 203 f.) saying, with a certain amount of exaggerated optimism, that he had established a theory for the five planets, even during Tycho Brahe's lifetime.

> 'I have built up a theory of Mars such that there is no difficulty about agreement between calculation and the accuracy of observational data. The reason why its motion had been regarded as less certain is not peculiar to that planet, but is common to them all; it is merely more apparent in that case. In the first place, hitherto, the line of apsides did not divide its circle into two halves, because the line was made to pass through the centre of the equant [circle] and the centre of the *orbis magnus* [the path of the Earth's annual revolution round the Sun]. However, in fact, it passes through the centre of the equant [circle] and the Sun itself. Hence, the centre of the eccentric (taken after the manner of Ptolemy) is between the equant point and the Sun; whereas, [hitherto], it was placed on another line between the equant point and the centre of the *orbis magnus*. . . . With regard to the libration (oscillation) of the plane [of the orbit] and variation of its inclination, I have found nothing of the kind. Consequently, the theory of Mars becomes quite simple; it consists of one circle only. . . . The theory of the Sun, or of the Earth, becomes exactly similar to it, and includes an equant also. In both cases, the proof and the exigency of numbers obliged us to divide the compound eccentricity into two, as did Ptolemy. . . . This having been proved [by studying observational data] for two planets, consideration of the matter showed the cause of the equant to be of a purely physical nature, but that it finds expression in the geometrical relationships. As [the ratio] of any one distance is to another, so is [the ratio] of the time that the planet remains in the position corresponding to the first distance to the time it remains in the position corresponding to the second distance. Now, as everything has succeeded so well for two planets . . . I have every hope that the others will conform to the same law.'

In this same letter, Kepler gives Maestlin his opinion of Tycho Brahe . . . and of himself:

> 'What Tycho Brahe has accomplished, he had accomplished before 97. Since that time his affairs took a bad turn; he was beset by terrible worries, and started to become childish. He was overwhelmed by the

inconsiderate way his native land had abandoned him. The court, here, was definitely his undoing. He was not a man who could live with anyone without serious conflicts; he could not do so especially with persons of such eminence, counsellors to kings and princes, conscious of their importance. Tycho Brahe's most important work consists of his observations, which fill as many volumes as years devoted to this labour; then, the *Progymnasmata*, in which he treats of the fixed stars and the motion of the Sun and the Moon in our time. The work exhales the pure fragrance of ambrosia. I hope to publish it in time for the next fair....

'Tycho Brahe also wanted to write a new book on comets; and he had made very learned and diligent investigations (commentaries) on all the planets; but rather in the manner of Ptolemy, *mutatis mutandis*, as Copernicus had done, too. You can see [thereby] in what manner God disposes of his gifts; one man cannot do everything. Tycho Brahe has done what Hipparchos did; he has laid the foundations of the edifice, and has accomplished an enormous amount of work. Hipparchos had need of a Ptolemy who built thereon [the theories] of the five planets. I have done as much whilst he was still alive.'

It is interesting to compare Kepler's verdict on Tycho Brahe with the one he made on himself some years later (Letter to Maestlin, 5 March 1605; (*G.W.*, vol. xv, p. 170)):

'High positions and great honours do not exist for me. I live here on the world's stage as a private gentleman. If I am able to extract part of my salary from the court, I am content not to be obliged to live entirely at my own expense. Howbeit, I conduct myself as though I did not serve the Emperor, but the whole of mankind, as well as posterity. With this assurance, I have kept a secret pride in despising all honours and positions, and, if necessary, even those who confer them. The only honour for me is to have been entrusted by Divine Providence with the observations of Tycho Brahe.'

5. Eliminating the *punctum aequans* whilst assuming that the Sun (or the Earth) moves uniformly about the centre of its circle, comes to exactly the same thing as identifying this centre with the *punctum aequans*.

6. Tycho Brahe, however, in his letter to Kepler quoted *supra* (p. 394, n. 37) points out that three observations are not enough, because of the change in dimensions of the Sun's orbit.

7. *Astronomia Nova*, caps. XVI and XXII. In my account I follow that of J. L. E. Dreyer, *History of the Planetary Systems*, pp. 383 f.

8. The distances of Mars from the *punctum aequans*, namely, the sides AF, AE, AD and AG, may be calculated similarly.

9. A similar argument and calculation are applicable to triangle CGS;

and the whole deduction can be repeated using each side of the quadrilateral FEDG. Kepler actually made them.

10. Cf. *Astronomia Nova*, cap. XVI; (*G.W.*, vol. III, p. 156; *O.O.*, vol. III, p. 245). The semi-empirical method followed by Kepler, and copied from that used by Copernicus to calculate the circle from three positions of the planet, is obviously very laborious and defective. Consequently, when he started his calculations, Kepler wrote to Herwart von Hohenburg (12 July 1600), postscript to the letter quoted *supra*, p. 395, n. 2 (*G.W.*, vol. XIV, pp. 132 f.) begging him to ask Viète (the *Apollonius Gallus*, whom Kepler had seen at Tycho Brahe's) if he could not come to his aid and solve this problem, as he had done for Copernicus. This is what was to be asked of Viète:

> 'From four acronical positions (oppositions) of planets in the Zodiac and the time intervals, to determine the true mean longitude, the apogee, the eccentricity and the proportion of the eccentricity in the equant to the eccentricity of the path of the planet . . . I have tried hard to solve this problem, but without success. In four months, I have not completed my one attempt [case under consideration]. It is necessary in effect to make use of a double false position, or even quadruple; that is to say, one must in fact make use of the doctrine of chances, and ἀτεχνία, to use Viète's expression in the proof of the Copernican problem dealing with three observations ἀχρονυχίοις. Viète's proof has given me hope that my problem could be similarly solved by him. If I should find a proof before him, then I shall communicate it to him. Up to now, I have sought it in vain; I think it is because I am not sufficiently skilled in this kind of problem.'

I do not know if Herwart von Hohenburg passed this naive, yet touching, request on the part of Kepler to Viète: the correspondence is silent on the matter. However that may be, Viète never dealt with the problem; and Kepler was never able to solve it. Thus, he wrote in the *Astronomia Nova* (*G.W.*, vol. III, p. 156; *O.O.*, vol. III, pp. 245 f.):

> 'Shrewd minded geometers, such as Viète, will consider it a fine thing to demonstrate the non-scientific nature of my procedure (ἀτεχνίαν). The same charge was made . . . against Ptolemy, Copernicus and Regiomontanus by Viète. Let them set about the problem, and let them find a geometric solution; in my opinion they will then be great Apollos.'

The passage from Viète to which Kepler alludes occurs in: *Francisci Vietae Apollonius Gallus.* Seu *Exsuscitata Apollonii Pergaei περι 'Επαφῶν Geometria*, Paris, 1600, fol. 11. *Appendicula. De problematis quorvm factionem geometricam non tradunt Astronomi:*

> '*Ptolemaeus ipse, et Ptolemaei paraphrastes Copernicus, cum ex tribus Epochiis mediis, et totidem adparentibus exquirunt summarum absidum*

loca, et Eccentro tetas vel Epicyclorum semidiametros, Geometras non se produnt, adsumentes opus tanquam confectum, quod ideo resolvunt infeliciter. Immo vero Copernicus ἀτεχνίαν non solum profitetur, sed docet. . . . Jubet enim non jam artis, sed aleae magister, circulum tandiu revolvi, donec error, quem ex sua ἀγεωμετρησία nasci agnoscit, tandem si sors dederit compensetur.'

Cf. *G.W.*, vol. III, p. 464; *O.O.*, vol. III, p. 478. It is interesting to note the contemptuous term used by Viète in reference to Copernicus: *paraphrastes Ptolemaei. . . .*

11. Cf. *supra*, p. 150 and p. 396. Kepler had already stated the necessity for seeking perfect agreement between theory and data in the *Mysterium Cosmographicum.* It seems quite possible that his passion for accuracy and strict agreement was intensified by his dealings with Tycho Brahe and the very nature of the work he had to perform.

12. Cf. *De fundamentis Astrologiae certioribus*, Thesis XX; (*O.O.*, vol. I, p. 423):

'*Primam contrarietatem Aristoteles in metaphysicis recipit illam, quae est inter idem et aliud: volens supra geometriam altius et generalius philosophari. Mihi alteritas in creatis nulla aliunde esse videtur, quam ex materia aut occasione materiae, aut ubi materia, ibi geometria. Itaque quam Aristoteles dixit primam contrarietatem sine medio inter idem et aliud, eam ego in geometricis, philosophice consideratis, invenio esse primam quidem contrarietatem, sed cum medio, sic quidem, ut quod Aristoteli fuit aliud, unus terminus, eum nos plus et minus, duos terminos dirimamus.*'

13. At a later date (cf. *infra*, p. 183) Kepler was to realize that Ptolemy had good reasons for adopting it.

14. *Astronomia Nova*, cap. XVI, p. 242.

15. Cf. *supra*, p. 52. The same result can be obtained—as Copernicus himself pointed out—by using one concentric circle + two epicycles instead of one eccentric + one epicycle. Consequently, Kepler frequently describes the system of Copernicus as concentro-biepicyclic. Cf. *Astronomia Nova*, cap. IV.

16. Cf. *supra*, p. 172.

17. In the *Astronomia Nova*, cap. IV; (*G.W.*, vol. III, p. 182; *O.O.*, vol. III, p. 182) Kepler goes further and states that it is absolutely impossible. However, the passage was certainly written during the final editing of the *Astronomia Nova* and relates to a period of Kepler's thought subsequent to the great discoveries in 1605 and 1606, namely, the ellipticity of planetary paths and the physical mechanism producing them. In fact, as we shall see later, it was only after much trouble and effort that Kepler decided to abandon the epicycle arrangement.

18. Cf. *Astronomia Nova* (*ibid.*): 'If the solid spheres are cleared away,

as Tycho Brahe rightly does, then this hypothesis implies something which is well nigh impossible. For, apart from the fact that it ascribes three motive forces endowed with reason to one single planet, there is in addition the fact that each of these [motive forces] will be affected by the motions produced by the other two, as well as the proximity of the central body. It is impossible to imagine that each of them fixes its central point, which is not distinguished by any body, and furthermore is mobile.' Cf. *supra*, where Kepler uses the same argument. It is amusing to recall that Dante in his *Convivio* saw no objection in ascribing several angelic motive forces to the planets.

19. Cf. *Astronomia Nova*, caps. XVIII, XIX.

20. *Ibid.* When Kepler came to write the *Astronomia Nova*, he made use also of his own observations of 1602 and 1604.

21. In fact, as was mentioned *supra* (p. 84, n. 54) and as Kepler subsequently recognized (cf. *supra*, pp. 225 and 267), Ptolemy was *absolutely* right: the *punctum aequans* of a planet's motion being at the focus of its elliptical orbit, it can be bisected only by the centre of the latter.

22. Observational astronomy was taken on to another stage by Tycho Brahe; it was the one concerned with 'phenomena'. We could say that even the phenomena of astronomy were modified by Tycho Brahe. Cf. *Astronomia Nova*, cap. XIX; (*G.W.*, vol. III, pp. 177 f.; *O.O.*, vol. III, p. 258):

> '*Atque ex hac tam parva differentia octo minutorum patet causa, cur Ptolemaevs, cum bisectione opus habuerit, acquieverit puncto aequatorio stabili. Nam si aequantis eccentricitas, quantam indubie poscunt aequationes maximae circa longitudines medias, bisectur, vides omnium maximum errorem ab observatione contingere VIII minutorum, idque in Marte, cujus est eccentricitas maxima; minorem igitur in caeteris. Ptolemaevs vero profitetur, se infra X minuta seu sextam partem gradus observando non descendere. Superat igitur observationum incertitudo seu (ut ajunt) latitudo hujus calculi Ptolemaici errorem.*
>
> '*Nobis cum divina benignitas Tychonem Brahe observatorem diligentissimum concesserit, cujus ex observatis error huius calculi Ptolemaici VIII minutorum in Marte argitur; aequum est, ut grata mente hoc Dei beneficium et agnoscamus et excolamus. In id nempe elaboremus, ut genuianm formam motuum coelestium (his argumentis fallatcum suppositionum deprehensarum suffulti) tandem indagemus. Quam viam in sequentibus ipse pro meo modulo allis praeibo. Nam si contemnenda censuissem* 8 *minuta longitudinis, jam satis correxissem (bisecta scilicet eccentricitate) hypothesim cap. XVI inventam. Nunc quia contemni non potuerunt, sola igitur haec octo minuta viam praeiverunt ad totam Astronomiam reformandam, suntque materia magnae parti hujus operis facta.*'

Cf. Letter to Longomontanus, beginning of 1605 (*G.W.*, vol. XV, pp. 134

f.): 'Had I wanted to ignore 8 minutes, I should not have given myself the enormous effort of 1604.'

The method used by Kepler (cf. Dreyer, *History of the Planetary Systems*, p. 390) to determine the distance of Mars from the Sun is roughly as follows: the *hypothesis vicaria* enables the heliocentric longitude to be found with sufficient accuracy, *i.e.*, the direction of the line (radius vector) connecting the planet and the Sun; but it does not give the distances directly. According to the hypothesis, the eccentricity of Mars should not be bisected, but divided in the ratio of 0·072 32 to 0·113 32. On the line of apsides, *IH*, mark the point *C* (centre of the eccentric orbit), and place *S* (Sun) and *A* (equant point) at distances such that $AC = 0{\cdot}07\,232$ and $SC = 0{\cdot}113\,32$. The line *AM* is then drawn so that angle *HAM* is equal to the mean anomaly; *CM′* is then drawn from *C* with a length equal to the mean distance of Mars; *SM′* will then be the true heliocentric direction of the planet (in accordance with the *hypothesis vicaria*). Now bisect *SA* with *B* as the mid-point. Draw *BP* parallel to *AP*: the angle *HBP* will then be equal to the mean anomaly. On *BP* mark the point *M″* so that $BM'' = CM'$, *i.e.*, so that *BM″* is equal to the mean distance of Mars (or radius of its orbit). *SM″* will then be the true length of the radius vector of Mars (its distance from the Sun). With *S* as centre and radius *SM″* draw a circle to cut *SM′* (or find the point *M‴* on *SM′* so that *SM‴* is equal to *SM″*); this point will be the true position of the planet.

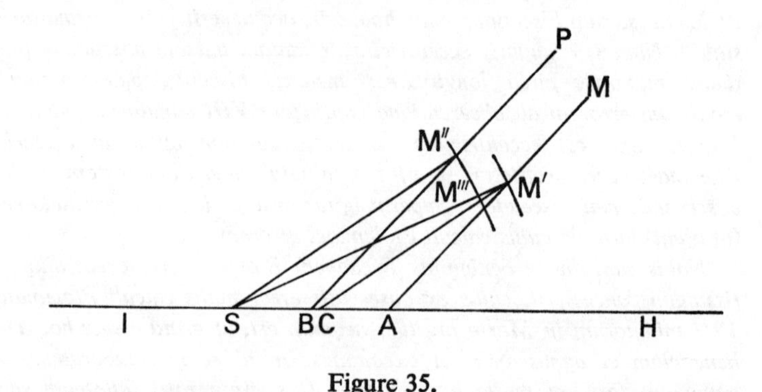

Figure 35.

23. It may be mentioned in passing that, from Ptolemy's point of view, it was quite acceptable for the *punctum aequans* to be mobile, and that he had assumed it to be so in his very complicated theory of the motion of Mercury. If this view were unacceptable to Kepler, it was simply because he sought a natural, *i.e.*, physical, explanation of celestial phenomena, whereas Ptolemy did nothing of the kind.

24. *Astronomia Nova*, cap. xx; (*G.W.*, vol. III, p. 182; *O.O.*, vol. III, p. 262).

CHAPTER III. STUDY OF THE EARTH'S MOTION

1. The motion of the Sun in Ptolemy's view—or Tycho Brahe's—was equivalent to that of the Earth for Copernicus. In each case the motion is uniform. Futhermore, pre-Copernican astronomy studied only the angular motion of planets (cf. *supra*, p. 106, n. 14) and had no means for determining distances; so, a calculation of the orbit, *i.e.*, the true path on which a celestial body moves, was not feasible; and, what is more, was of no interest.

2. Cf. *Astronomia Nova*, cap. XXIV; (*G.W.*, vol. III, p. 198; *O.O.*, vol. III, p. 273). The four observations of Mars selected by Kepler were those of 5 March 1590, 21 January 1592, 8 December 1593 and 15 October 1595.

3. Cf. Max Caspar, Introduction to *Die Neue Astronomie*, München, 1929, pp. 43 f.

4. Cf. *Astronomia Nova*, cap. XXV; (*G.W.*, vol. III, p. 203; *O.O.*, vol. III, p. 277).

5. Cf. *supra*, p. 169.

6. Cf. *Astronomia Nova*, caps. XXVI–XXVIII.

7. Table of the distance of Mars from the Sun, the radius of the Earth's orbit being taken as unity. Cf. Dreyer, *History of the Planetary Systems*, p. 389.

Date	Distance from aphelion	According to circular hypothesis (a)	By observation (b)	Difference (b–a)
31 Oct 1590	9° 37′	1·666 05	1·662 55	−0·003 50
31 Dec 1590	36° 43′	1·638 83	1·631 00	−0·007 83
25 Oct 1595	104° 25′	1·485 9	1·477 50	−0·007 89

It should be remembered that all Kepler's distances are wrong, seeing that the essential piece of information, the distance from the Earth to the Sun, is twenty times greater than the value accepted by him (cf. *supra*, p. 107, n. 24). However, that does not matter greatly, so long as one keeps to relative values; and, in spite of wrong data, Kepler was able to establish the quantitative laws of planetary motion.

8. *Paralipomena ad Vitellionem, Astronomiae Pars Optica.* The writing of this work on optics was envisaged in Kepler's programme of work. However, like many others, both before and after him, he underestimated the time required for its completion. He wrote to Herwart von Hohenburg as follows on 12 November 1602 (*G.W.*, vol. XIV, p. 299):

'I am engaged on two works. One which ought to be ready about Easter is

> *Commentaria de Motibus Stellae Martis* (or whatever the title may be), being the key to the whole of astronomy, and containing many very interesting investigations on the planets, [a work] based on Tycho Brahe's observations; the other—*Astronomiae Pars Optica*—should be ready in eight weeks' time.'

The *Astronomiae Pars Optica* did not appear until 1604; as for the *Commentaria de Motibus Stellae Martis* they occupied Kepler until 1606. He explained to Longomontanus in 1605, when reproached with tardiness, that he was not entirely free: he had 'to dance attendance' in order to get his salary; cast horoscopes; attend court, and so on. Nevertheless, he had not been idle, and had not wasted his time (cf. Letter to Longomontanus, beginning of 1605; (*G.W.*, vol. xv, pp. 134 f.)):

> 'In September 1602 I announced the *Astronomiae Pars Optica* for Christmas, and the investigations on the motion of Mars . . . for Easter. I put these on one side for the time being, and devoted myself to the *Optics*. However, whilst I was engaged in arranging it for publication, I decided upon a revision, which occupied me up to Christmas 1603, instead of from September 1602 till Christmas 1602. Having completed this work, I was then concerned with the printing of it during nearly the whole of the following year, 1604. Consequently, I gave hardly any attention to Mars during 1603. However, I prepared the Ephemerides for Mars. . . . March and April were taken up by the lunar tables. . . . Since the beginning of 1604, I have resumed my work on Mars and the commentaries. . . . At the same time I computed general tables for the prosthaphaereses during the annual circuit for all paths of Mars, as well as the equations according to the physical hypothesis. Believe me, I had to make 181 calculations with the same equation in more than 40 instances, for the accuracy of the problem did not permit the progression to be made in steps of 10°, or even 6°. All this is explained in 51 chapters.'

9. If r be the radius of the eccentric circle (orbit) of the planet, and $2e$ be the total eccentricity, then the velocity at points on the orbit located on a perpendicular to the line of apsides drawn through the central body is equal to $\sqrt{r^2 + 3e^2}$, according to Ptolemy, and equal to $\sqrt{\frac{r^2 + e^2}{r^2 - e^2}}$, according to Kepler.

CHAPTER IV. *A QUO MOVENTUR PLANETAE?*

1. Let us be quite clear on the matter: Kepler's view, according to which the velocity of a planet on its eccentric (circular) orbit is inversely proportional to the radius-vector, and Ptolemy's view, according to which the planet moves with uniform angular velocity about the *punctum aequans*, are not equivalent, but incompatible. This has been frequently pointed out

(consult, for example, Max Caspar's fine introduction to his German translation of *Astronomica Nova—Die Neue Astronomie*, München, 1929, p. 33). Ptolemy's view implied for Kepler—wrongly—a proportionality of orbital velocity to the distance from the *punctum aequans*; consequently, it seemed to him, as it did later to Boulliau and Borelli, to be the counterpart of his own. Only at a much later date did he realize that the two views were not equivalent; cf. *supra*, p. 318.

2. For this reason, instead of embarking on a search for the law of planetary motion starting directly from the velocities, Kepler adopted the roundabout way involving motive forces, although, in fact, he knew nothing more about these motive forces than their proportionality to the velocity of the moving objects (planets).

3. Kepler's celestial dynamics is (almost) purely Aristotelian: velocity is proportional to the motive force, and in the absence of any such force the motion changes to rest. Cf. my *Études Galiléennes*, III, pp. 26 f.

4. *Astronomia Nova*, cap. XXXII; (*G.W.*, vol. III, p. 233; *O.O.*, vol. III, pp. 297 f.): 'The force which moves a planet on its orbit decreases as the distance from the source increases.'

5. It is rather strange to learn from Kepler that Copernicus had copied Ptolemy by adopting the bisection of the eccentricity of the three superior planets. . . . We know, and Kepler knew better than anyone, that Copernicus did nothing of the kind; on the contrary, he explicitly rejected it. Nevertheless, Kepler was of the opinion, and rather justifiably so, that by accepting the value for the eccentricity given by Ptolemy—even though he assigned part of it to the epicycle—Copernicus thereby accepted the views of the latter.

6. The date of the first writing of this passage is therefore 1603.

7. The cause of the equant = the cause of truly non-uniform planetary motion.

8. In fact, it is not the planet, but the centre of its epicycle which moves in this manner on the deferent.

9. The arc of the path = the arc traversed by the planet, or more precisely, the centre of its epicycle on its orbit.

10. Two equal arcs are therefore traversed in different times.

11. *Astronomia Nova*, cap. XXXIII; (*G.W.*, vol. III, p. 234; *O.O.*, vol. III, pp. 300 f.): 'The force which moves the planet resides in the body of the Sun.'

12. This statement is significant for Kepler's dynamics—and its limitations. Indeed, it was precisely 'the strength or weakness of motion in longitude' that was later put forward by J. A. Borelli as the cause of distance from the centre; cf. *infra*, p. 507.

13. Motion presupposes distance; but distance does not imply motion; therefore, logically and ontologically it is antecedent to motion, and consequently motion cannot be the cause of distance.

14. Distance is not something as such, but merely the distance *between* one thing and another; therefore, it arises *subsequently* to things, and distance *qua* distance cannot be the cause of something that happens to them.

15. Curiously enough, this was the solution put forward by Borelli. Naturally, the body of the planet does not become heavier *in itself:* nevertheless, it becomes so in relation to the motive forces, and consequently will offer greater resistance to them; cf. *infra*, p. 500.

16. It would be ridiculous to say that a *variable* 'animal' force does not suffer from the effects of age. On the other hand, a *constant* 'animal' force has no reason to be so affected.

17. Kepler's objection to the theory of planetary motive souls was all the stronger and more resolute, seeing that he had adopted it himself in his younger days. He criticized it in detail (cf. *supra*, p. 223) showing: (a) that a purely motive soul would not suffice, and that an *intelligence* would be absolutely indispensable for the purpose of calculating the changes in velocity along the path; (b) that this intelligence would have a very difficult, if not impossible, task to perform, seeing that it would be able to base the calculations only on the changes in apparent diameter of the solar disc; (c) that such an intelligence, even with the assistance of a soul, would not be able to confer motion of translation to the body of the planet, though it would be able to make it perform a *rotatory* motion. In the latter instance, an intelligence is superfluous: a soul is all that is needed; consequently it could be assumed that souls gave motion to the solid spheres of Copernicus, provided always, of course, that the motion were uniform—hence the importance of their destruction by Tycho Brahe; similarly, we can, or even must, assume that the Sun's rotation is to be explained by the action of a soul.

18. Cf. *supra*, p. 177, and n. 18 of chap. II.

19. *Astronomia Nova*, Introduction; (*G.W.*, vol. III, p. 23; *O.O.*, vol. III, p. 149). It is rather interesting to note that Aristotle in *De Coelo* (284a) had already raised a similar objection against the notion of a soul in the Universe whose function was to make the celestial sphere turn round; this soul never being able to stop or rest would seem to have shared the fate of Ixion.

20. The law of the lever, the first and for centuries the only example of a mathematical law capable of being realized in practice, played an enormous, and often baneful, rôle in pre-Galilean physics. It was so, even later, as will be seen in my study of Borelli (cf. *infra*, p. 500).

21. In fact, the resistance of the planet does not increase; it is the motive force that decreases. In Borelli's view, on the contrary, the resistance increases.

22. In the first (speculative) part of *De Revolutionibus Orbium Coelestium* the Sun is placed at the centre of the Universe (cf. *supra*, p. 59). On the

other hand, in the technical parts of the book, the centre of the terrestial sphere functions as the centre of the planetary system.

23. We shall not adopt Kepler's advice—given, moreover, grudgingly—but shall continue methodically and without breaking the continuity; especially as we can treat as established those matters which Kepler means to prove for the benefit of his readers.

24. Kepler gave this proof *a priori* in both the *Harmonice Mundi* and the *Epitome Astronomiae Copernicanae.*

25. The second inequality, caused in Ptolemy's view by the motion of a planet on its epicycle, and in Tycho Brahe's view by the motion of the Sun round the Earth, was explained by Copernicus as an illusion resulting from the projection of the Earth's orbital motion round the Sun on to the planets.

26. The central position of the Sun with respect to all the planets was affirmed by Tycho Brahe, but Kepler insinuates that Tycho Brahe and his supporters forsook the logic of their own system in order to affirm the immobility of the Earth and to conform to the appearances.

27. *Astronomia Nova,* Introduction; (*G.W.*, vol. III, p. 23; *O.O.*, vol. III, pp. 149 f.)

28. Immobility, here, does not exclude *rotation in situ.* Indeed, in the celestial physics of the *Astronomia Nova,* if the Sun did not rotate on its axis, the planets would not move round the Sun.

29. The planets are endowed with different 'natures' and are distinguished from each other qualitatively; their difference from the Sun is even greater, though they are all material—the Sun also—and as such possess different quantitative properties of volume and density.

30. The disappearance of 'natural' motions implies, paradoxically, an extension of Aristotle's axiom—*quodquod movetur, ab alio movetur*—to all motion, both terrestrial and celestial.

31. Cf. *Astronomia Nova,* Introduction; (*G.W.*, vol. III, p. 23; *O.O.*, vol. III, pp. 149 f.).

32. My italics.

33. The Earth in itself is not 'heavy' in the Aristotelian sense; it would be so only if it were attracted by a neighbouring body, for example, another larger Earth. Cf. Letter to Herwart von Hohenburg, 28 March 1605 (*G.W.*, vol. XV, p. 184):

> '*Si autem ad Tellurem quocunque in loco quiescentem applicaretur Tellus alia, et major, tunc illa sane fieret gravis respectu majoris, attraheretur enim ab illâ, planè uti haec Tellus lapides attrahit, etc. Itaque gravitas non est actio sed passio lapidis, qui trahitur, principium inquam ejus.*'

Seeing that the Earth, in fact, is attracted only by the Moon, which is smaller, the former consequently is not 'heavy'. It should be pointed out that gravific attraction is always mutual: cf. Letter to Fabricius, 11

October 1605 (*G.W.*, vol. xv, p. 241): '*Non tantum lapis ad Terram eat sed etiam Terra ad Lapidem.*'

34. The 'other force, equivalent to animal virtue' which prevents the Moon from falling on to the Earth *is not the centrifugal force.* The latter is not produced by the motion of celestial bodies, and consequently, as will be seen later (p. 215), the planets have no tendency to move away from the Sun. Only Descartes and Borelli subscribed to this view.

35. In fact, the substance of the Earth is not the same as that of the Moon: that of the Moon is more 'rarefied'; cf. *supra,* p. 207.

36. *Moles* is not yet the same as 'mass'—though it is very close to the latter, seeing that it involves the volume and the density of the body in question. Several years later Kepler reached the clear concept of mass = quantity of matter in a body (cf. *supra,* p. 353). The term *moles* then came to mean the *volume* of a body.

37. In Kepler's view all terrestial bodies are 'heavy', and 'light' ones are merely less heavy = less dense than the others. This was nothing new, for it had already been expounded most clearly by J. B. Benedetti in his *Diversarum Speculationum . . . liber,* Turin, 1585. Cf. my *Études Galiléennes,* I, pp. 41 f.

38. Kepler's view of *inertia* = resistance to motion or tendency to rest, although derived directly from Aristotle's physics, is, however, not a strictly Aristotelian concept. Indeed, in dynamics, Aristotle envisaged only resistance *external* to bodies in motion; consequently, the velocity of such a body, being, in his opinion, proportional to the motive force and inversely proportional to the resistance, would become infinite in a vacuum. On the other hand, *Kepler's inertia* is an internal resistance and would make itself apparent in a vacuum just as much as in a plenum; consequently, the velocity of the planets is proportional to the motive force acting on them, and inversely proportional to their 'natural inertia', which is a function of their *moles.* It is unnecessary to point out that Kepler's inertia (he was responsible for the term) is quite different from Newton's inertia, which is not *resistance to motion,* understood as a process and the reverse of rest, but is eternal persistence of motion or of rest, each being understood to be a state. Cf. my *Études Galiléennes,* III, 'Galilée et la loi d'inertie' and 'J. B. Benedetti, critique d'Aristote', *Mélanges E. Gilson,* Paris, 1959.

39. Cf. my *Études Galiléennes,* III, pp. 23 f.

40. In the view of both Kepler and Aristotle rest and motion are related to each other as absence to presence, darkness to light.

41. Once more, the notion of inertia is not Aristotelian. Nevertheless, it must be pointed out that the Aristotelian conception of the process of motion implies almost of necessity the admission of a natural tendency to rest, at least in the case of 'heavy' bodies. Certain medieval commentators on Aristotle have laid down the existence of an *inclinatio ad quietem, i.e.,*

an internal resistance to motion. This view, already used by Benedetti, is the basis of Kepler's inertia.

42. Inertia of celestial bodies of necessity implies a motive force which acts on them and gives them motion; consequently, even the physical concepts of Aristotelian physics are applicable to the planets, and 'terrestrial physics' extends to the heavens.

43. Cf. *Astronomia Nova*, Introduction; (*G.W.*, vol. III, p. 24; *O.O.*, vol. III, p. 150).

44. *Ibid.*; (*G.W.*, vol. III, p. 34; *O.O.*, vol. III, p. 156).

CHAPTER V. THE MOTIVE FORCE

1. *Astronomia Nova*, pars III, cap. XXXIII; (*G.W.*, vol. III, p. 239; *O.O.*, vol. III, pp. 302 f.).

2. Kepler's text certainly shows the hesitation and changes in his thought. At times he seems to identify the motive force purely and simply with light, and at others with magnetic force; in fact, he distinguishes it from these two phenomena, as will be seen later, p. 268.

3. Cf. *G.W.*, vol. III, pp. 233 f.; *O.O.*, vol. III, pp. 296 f.

4. Cf. *Astronomiae Pars Optica*; (*G.W.*, vol. II, pp. 18 f.; *O.O.*, vol. II, pp. 130 f.).

5. The relationship of motive force to light could be similar to that between light and heat. Heat is not identical with light, but is transported as it were by the latter. In the *Epitome Astronomiae Copernicanae* (cf. *infra*, p. 440), light from the Sun has the function of exciting the motive force of the planets.

6. It is strange that Kepler, who was more Aristotelian in his celestial physics than in his terrestrial physics, was embarrassed by the objection of occultation instead of disposing of it by ascribing to the planets an *impetus* acquired during their earlier motion.

7. That is to say, throughout the whole sphere and in all directions, whereas the motive force is propagated only in a plane.

8. The motive force or its *species* is not propagated spherically like light, but in a plane; consequently, the force received by the moving body is not in ratio to that portion of the spherical surface which it occupies, but is in ratio to that portion of the path (plane).

9. Kepler does not mean that the motive *species* is propagated like light in that it spreads outwards over a spherical surface; he means that it has no intrinsic thickness.

10. Unlike other *species* or forces—for example, attraction or magnetic force which is propagated only to a certain distance (their *orbis virtutis* is essentially finite), the *effluxus* of light has no intrinsic limits.

11. The *species* or motive force, like light, is propagated instantaneously.

For this reason, when viewed from the moving body, or the illuminated body, it (the force) appears to have passed through all the intermediate points.

12. *Astronomia Nova*; (*G.W.*, vol. III, p. 240; *O.O.*, vol. III, p. 303).

13. The motive force is more concentrated—denser—and therefore stronger in the paths nearer to the Sun than those which are more remote. Consequently, its action is stronger in the former, and weaker in the latter; thereby causing the greater, or lower, velocity of the planets. It is interesting to note that Kepler uses here the terminology employed by the school of logicians at Paris and Oxford: *intensio* and *remissio*.

14. Cf. n. 10.

15. Resistance from the medium—a reminiscence of Aristotelianism—was completely neglected by Kepler in favour of internal resistance.

16. Cf. *supra*, p. 194.

17. Kepler's celestial physics, as I have already said, remained faithful to the Aristotelian principle: *no motion without a motive force*, not even external motive force in the case of translatory motion. Animated bodies themselves do not escape from this necessity of being moved: a soul cannot transport them from one place to another; the most that it can do is to make them rotate in a given position.

18. The period of revolution of the planets is therefore no longer a function of two factors, as in the *Mysterium Cosmographicum*, but of three: (a) length of path; (b) intensity of the motive force; (c) intensity of the resistance to motion (inertia). If the last mentioned factor were not present, all the periodic times would be inversely proportional to the square of the distance of the planets from the Sun. The fact that this is not the case is explained therefore by the difference in inertia (function of the *moles* and density) of the planets. Conversely, the different periodic times enable conclusions to be made concerning the differences in their *moles*. This is a most important deduction, which marks the beginning of astrophysics, and from which Kepler finally derived the physical basis of his 'third' law; cf. *infra*, pp. 353.

19. *Astronomia Nova*, pars III, cap. XXXIV; (*G.W.*, vol. III, p. 242; *O.O.*, vol. III, p. 304): 'The body of the Sun is magnetic and rotates on itself in space.'

20. Kepler endowed the Sun with a soul. Though, in the last analysis, the motive soul of the Sun provides the explanation of its rotatory motion, the *species motrix* does not derive from this soul, but from the body of the Sun.

21. Kepler is referring here to the theory of motion put forward by Nicholas of Cusa (cf. P. Duhem, *Études sur Léonard de Vinci*, vol. II, pp. 186 f.) as well as to the hypothesis of planetary animation he put forward in the *Mysterium Cosmographicum* as one of the two possible explanations of their motion; cf. *supra*, pp. 151 f.

22. Planetary motion is therefore mechanical, and not animal. It is

strange that Kepler did not pursue the analogy with the ballista which confers an *impetus* on the projectile allowing it to continue its motion after separation from its launching force. Keplerian planets receive no *impetus* from the *species*.

23. The *species* partaking, so to speak, of the properties of its source, a motionless body could not confer motion on the planets.

24. A good example of Kepler's genius. Starting from false premises, and using an absurd argument, he arrives at a correct conclusion, which is of first-rate importance. All the same, it should be pointed out that Kepler was not the only one to have assumed that the Sun rotates on its axis; Edmund Bruce had had the same idea: cf. Letter of Edmund Bruce to Kepler, Venice, 5 November 1603 (*G.W.*, vol. XIV, p. 450):

> '*Multas habeo in Astronomia dubitationes in quibus te vnicus me certior facias; nam ego opinor mundos esse infinitos; vnusquisque tamen mundus est finitus sicut Planetarum in cuius medio est centrum Solis. Et quemadmodum tellus non quiescit sic neque Sol; Voluitur namque velocissimè in suo luoco circa axem suum; quem motum sequuntur reliqui Planetae: in quorum numero Tellurem existimo; sed est tardior vnusquisque quo ab eo distat longior. Stellae etiam sic mouentur vt Sol; sed non illius vi sicut Planetae circumaguntur; quoniam vnaquaeque earum Sol est, in non minori mondo hoc nostro Planitarum. Elimentalem mundum nobis proprium et particularem non puto; nam aer est et inter ipsa corpora: quae stellas vocamus; per consequens et ignis et aqua et terra: Terram autem quam calcamus nostris pedibus, nec rotundam nec globosam esse credo; sed ad ovalem figuram propius accedere. Nec Solis nec stellarum lumen ex materia sed potius ex eorum motu procedere et demanare iudico: Planetae vero a Sole eorum lumen assumunt: quia tardius mouentur et proprijs motibus impediuntur.*'

It is curious to find, as Kepler himself admitted, that on receiving this letter he paid no attention to Bruce's theory, which is almost identical with his own, no doubt because Bruce maintained the boundless extent of the Universe, the identity of the fixed stars with the Sun, and the figure of the Earth to be oval . . . in addition to rotation of the Sun. In 1610 (5 April) Kepler came across Bruce's letter again and noted down:

> '*Quid potius mirer? Stuporemne meum, qui patefacta mihi Naturae penetralia his literis, cum illas accepissem, introspicere contempsj; adeoque oblivione sepelivj, ut ne postea quidem cum clavem eandem ad haec penetralia quaererem et invenissem, literarum harum fuerim recordatus. An potius mirer Vim veritatis, quae duobus sese non una via aperuit?*'

25. Kepler admitted that this idea was very confused. As a result, the argument in the *Astronomia Nova*, as well as in the *Epitome Astronomiae*

Copernicanae (cf. *infra*, p. 300), is quite false. No doubt, if we assume that the *species* or motive force be propagated 'circularly', we are justified in concluding that it becomes rarer and weakens proportionally with the distance. On the other hand, Kepler seems to have forgotten to take into consideration the fact that it is all the more rapid. As a result, a planet will receive the same amount of *species*, in a given *time*, whether it be near to, or remote from, the Sun; and consequently it will move with the same velocity.

26. *G.W.*, vol. III, p. 243; *O.O.*, vol. III, p. 304.

27. In the history of astronomical thought, this notion of Kepler's had a certain influence on J. A. Borelli; cf. *infra*, p. 485.

28. *Ibid.*

29. If the motive *species* emanated from an immaterial entity, it would have an infinite velocity of translation, for its source, having no inertia, would be able to rotate only with infinite velocity.

30. With the same angular velocity.

31. Cf. *supra*, p. 195.

32. Cf. *supra*, p. 190. The weakening of the motive force as the distance increases does not in itself explain the slowing down of the planets in their orbital motion: a progressive increase in their *moles* must be taken into account; cf. n. 18.

33. Strictly speaking, it is not a question of a 'force', but of inability = resistance = inertia.

34. Cf. *G.W.*, vol. III, p. 244; *O.O.*, vol. III, p. 305.

'*At cum in meo* Mysterio Cosmographico *monuerim, eandem fere proportionem esse inter semidiametros corporis Solis et orbe Mercurii, quae est inter semidiametros corporis terrae et orbis Lunae; hinc non absurde concluseris, sic esse periodum orbis Mercurii ad periodum corporis Solis, ut est periodus orbis Lunae ad periodum corporis terrae . . . adeo ut verisimile sit, Solem triduo circiter gyrari.*'

Kepler was disagreeably surprised when Scheiner fixed the period of rotation at 25 days as a result of his observations of sunspots. However, he accepted the fact without protest. Cf. p. 440, n. 18.

35. *G.W.*, vol. III, p. 245; *O.O.*, vol. III, p. 306.

36. The 'example of the Earth and the Moon' is a very good one. It enables us to define more accurately the nature of the motive force, and, seeing that the Earth is not a luminous body, to distinguish this force from light.

37. The distance from the Moon to the Earth being estimated (approximately) at 60 terrestrial semi-diameters, its path is 60 times longer than the circle on the equator.

38. In a letter to Herwart von Hohenburg (28 March 1605; (*G.W.*, vol. XV, pp. 180 f.)), Kepler say that the *moles* of the body of the Moon is

one-fortieth that of the Earth. It is difficult to say how Kepler arrived at this figure; in fact, according to his Aristotelian principles, a force one-sixtieth of that which causes the Earth to rotate, moving a body (the Moon) at twice the velocity, could do so only if the resistance to motion of the latter body were 1/120 of that of the former.

39. Kepler was much taken up with magnetism. In a letter to Herwart von Hohenburg (12 January 1603; (*G.W.*, vol. XIV, pp. 347 and 351)) he says:

> 'As is known to Your Grace, I have been much occupied with magnetism. I have studied most of Porta's experiments, both as regards their purpose and performance, as well as I was able with an indifferent lodestone. I have found some of the errors which Gilbert also has particularly noted somewhere. Indeed, Gilbert has treated this subject so amply, clearly and fully, and he has taken such care to be on his guard [against error] by most noble experiments [that] he has silenced all gainsayers, [and] my speculations on a slow displacement of the pole [axis] of the Earth's diurnal motion from the places assigned to it at the creation have completely vanished. . . .
>
> 'As regards the Earth's motion, you can easily imagine that I heartily approve of what [cf. Gilbert, *De Magnete*, Lib. I, cap. VI] he urges in support of the Copernican theory, for I myself am a Copernican. However, I should like you to realize that he [Gilbert] asserts only a probability. His argument is somewhat as follows: iron is moved by a magnet; the magnet is moved by the Earth by means of a certain, almost corporeal, force, seeing that it [the Earth] assumes corporeal dimensions. Why then should not the Earth itself be moved by an animal force? Furthermore, it is fitting that motion be given to that which possesses natural, and not imaginary, elements of circular motion. The Earth has natural poles, a natural axis, and a natural equator set out by an evident natural force: that is what is required for motion. Hence, it is fitting that the Earth should move. As for the singularities of the Earth's motion, they are explained in a very judicious manner. If the Earth's axis were always pointed towards the same fixed stars, one could believe that it tended towards something akin to itself, in the same way that our magnet turns towards the North, without [the action of] any soul [being implied thereby], and simply by the natural force of union. However, seeing that the Earth's axis does not remain entirely motionless with respect to the fixed stars, but describes a small circle round the poles of the ecliptic, it appears that the Earth's axis is not controlled by these parts of the sky, but by a certain soul in the Earth itself; otherwise the axis would always point in the same direction.'

In this passage, Kepler, as well as Gilbert, was concerned with the Earth's rotatory motion, and not its orbital motion.

40. Kepler and everyone else at that time—and even later—believed that magnetic action extended only to a relatively short distance from the magnet.

41. If a small strip of iron be placed on the equator of Gilbert's *terrella*, it will adopt a fixed direction with respect to the *terrella*, one end pointing to one pole, the other end to the other pole; it will return to this position if displaced, but it will not move to either of the poles. Kepler concluded, not that the iron is attracted equally to both poles, but that it is not attracted at all; that it is subjected only to a *directive* force, and not to any *attractive* force by the *terrella*.

42. Therefore, the Sun does not attract the planets; neither in virtue of a relationship comparable to that between the Earth and the Moon (there is a *mutual* attraction between the Moon and the Earth) which presupposes a common, or similar, nature that does not exist between the planets and the Sun: nor in virtue of a magnetic force; there is no attraction at all. The Sun's action is limited to urging the planets along their orbits: this was pointed out by E. F. Apelt in *Johann Keplers Astronomische Weltansicht*, Leipzig, 1849, p. 72:

> '*Diese Idee einer Centralbeschleunigung liegt den physicalischen Ansichten Kepplers durchaus fern. . . . Diese von der Sonne ausströmende Kraft ist also etwas ganz anderes, als die Schwere, mit der man sie oft verwechselt hat. Die letztere ist eine Anziehungskraft, die erstere dagegen eine Umdrehungskraft, eine wahre Tangentialkraft, wenn man die Vorstellungen Kepplers auf unsere mechanischen Begriffe bringen will.*'

Cf. also E. F. Apelt, *Die Reformation der Sternkunde*, Jena, 1852, vol. II, p. 275:

> '*Man muss sich daher wohl hüten diese Keppler'sche Centralkraft der Sonne mit Newton's allgemeiner Gravitation zu verwechseln.*'

Apelt is in error in speaking of tangential force: Kepler's force is 'circular'.

Still, that is what it is in the *Astronomia Nova*; in the *Epitome Astronomiae Copernicanae*, the Sun attracts and repels the planets by an action peculiar to the *species* or magnetic force. Yet, a note in the *Somnium seu de Astronomia Lunari* (*Opera Omnia*, vol. VIII, p. 61, n. 202) seems to assume attraction by the Sun: Kepler explains the high tides at the time of the Full Moon and the New Moon as resulting from the conjoint action of the Sun and the Moon.

43. Consequently, the Sun, like magnets, has two kinds of lines of force: straight and circular.

44. Gilbert was a strenuous supporter—against Tycho Brahe—of the Earth's diurnal rotation. He never made any claim for orbital motion. Cf. my *From the Closed World to the Infinite Universe*, pp. 73 and 284 f.

45. There is no question here of orbital velocity.

46. Cf. Letter to Maestlin dated 5 March 1605 quoted later, p. 252.

47. *Astronomia Nova*, cap. XXXV; (*G.W.*, vol. III, p. 247; *O.O.*, vol. III, p. 309).

48. *Ibid.*

49. The forces or immaterial *species* are one genus of which each is one species. Hence, they are similar in that they belong to one genus, but nevertheless they have specific properties which distinguish them.

50. *Astronomia Nova*, cap. XXXV; (*G.W.*, vol. III, p. 246; *O.O.*, vol. III, pp. 307 f.).

51. There are transparent substances.

52. This naturally implies a qualitative, as well as a quantitative, difference. The body of the Sun is not only the densest one in the whole Universe, but is essentially different. It is precisely for this reason, that the Sun does not attract the other planets, whereas the Moon, although not so dense as the Earth, is attracted by the latter. All the 'similarities' that we can establish between the nature of the Sun and other bodies should not cause us to lose sight of the fact that celestial physics and terrestrial physics comprise only one physics; the Sun nevertheless retains its uniqueness.

53. *Astronomia Nova*, cap. XXXVI; (*G.W.*, vol. III, p. 248; *O.O.*, vol. III, p. 309).

54. Cf. *supra*, p. 199.

55. Cf. *Appendix* III.

56. Kepler seemed to believe that the effects of the lines [of force] stretching towards the poles and originating from different parts of the Sun (those which face the planet and those which face the opposite side of the sky) are in equilibrium, and that their total action will be nil. Kepler's line of argument could also be interpreted as supposing a summation of the impulses from the radii vectores reaching the body of the planet from different parts of the Sun's body: each radius vector will bring only a force attenuated according to the law of proportionality to the inverse square of the distance; but the combined effect would be proportional only to the distance. However, the same argument should apply in the case of light.

57. *Astronomia Nova*, cap. XXXVI; (*G.W.*, vol. III, pp. 250 f.; *O.O.*, vol. III, pp. 310–311).

58. It could be said that he confirmed, instead of invalidating, the objection that the strength of the *species* should be attenuated in the same manner as light in the duplicate proportion (proportionally to the square of the distance), even in the plane of the ecliptic (rotation of the Sun) seeing that the *species* sends out lines [of force] from the Sun in all directions (orbicularly); and that the attenuation should be still greater in regions outside this plane.

59. *Astronomia Nova*, cap. XXXVI; (*G.W.*, vol. III, p. 250; *O.O.*, vol. III, pp. 310 f.).

60. *Ibid.*, Introduction; (*G.W.*, vol. III, p. 34; *O.O.*, vol. III, p. 156).

CHAPTER VI. THE INDIVIDUAL MOTIVE FORCES

1. Cf. *supra*, p. 194 and p. 264. It was an equally serious matter for Galileo. Kepler, if he had thought of it, could have explained the absence of centrifugal force by the fact that the *species motrix* emanating from the Sun being a rotatory motion, could only impress a motion of revolution on the planets; or at most, a rotatory motion; but in no case a motion in a straight line.

2. This is precisely why there is no need of a soul to maintain the motion: any *physical* motive power would be gradually exhausted in producing it; cf. *supra*, pp. 295 f.

3. Cf. *Astronomia Nova*, cap. XXXVIII; (*G.W.*, vol. III, pp. 254 f.; *O.O.*, vol. III, pp. 313 f.).

4. The solid spheres of Copernicus (or of pre-Copernican astronomy) turn on themselves and consequently can be located both concentrically and eccentrically with respect to the Sun (or to the Earth). Kepler thought otherwise.

5. This is a curious statement which seems to contradict Kepler's previous argument. In fact, if the force of the *species* set the aetheric aura in motion, ought not it to become weaker and exhaust itself *before* reaching the planets ? Yet, Kepler thought it probable and repeated it in the *Epitome*; cf. *infra*, p. 298

6. In *Astronomia Nova*, cap. LVII; cf. *infra*, p. 257.

7. *Astronomia Nova*, cap. XXXIX; (*G.W.*, vol. III, p. 256; *O.O.*, vol. III, pp. 314 f.). Kepler wrote: *Qua via et quibus mediis movere debeant virtutes planetis insitae, ut circularis planetae orbita, qualem vulgo credunt, per auram aethereum efficiatur*. This is a curious expression on the part of Kepler: in fact, belief in the circularity of orbits, *i.e.*, the effective paths of the planets, was quite unusual and nobody except Kepler himself accepted it (cf. *supra*, pp. 170 f.), at least not since Hipparchos.

8. *Ibid.*

9. Kepler makes a correction here of a mistake in the *Mysterium Cosmographicum*; cf. *supra*, pp. 152 f.

10. *Astronomia Nova*, Introduction; (*G.W.*, vol. III, p. 34; *O.O.*, vol. III, p. 156).

11. Cf. *supra*, p. 161. Letter of Tycho Brahe.

12. *Astronomia Nova*, cap. XXXIX; (*G.W.*, vol. III, p. 257; *O.O.*, vol. III, p. 315).

13. Cf. *supra*, pp. 171 f.

14. *Astronomia Nova*, cap. XXXIX; (*G.W.*, vol. III, p. 260; *O.O.*, vol. III, p. 318).

15. He admits it himself; cf. *Astronomia Nova*, cap. XLIV and cap. XLV; (*G.W.*, vol. III, pp. 285 and 288; *O.O.*, vol. III, pp. 335 and 337).

16. Cf. *Astronomia Nova*, caps. II–IV.

17. This is almost the same as the view held by Copernicus.

18. The planet's motion on the diameter (radius) $\alpha\gamma$, consequent upon—or corresponding to—its motion on the epicycle, is rather complicated, and Kepler remarks that in this instance it is easier to say what does not occur rather than what does occur. In order for the path to be a perfect circle, it would be necessary (*G.W.*, vol. III, p. 239; *O.O.*, vol. III, p. 317) 'for the planet, when placed by the Sun [through the motive force emanating from it] on the [straight] lines drawn from *A* to *C*, *D*, *E*, *F*, *G*, *H*, to proceed of its own accord from the point γ to the points ι, λ, ζ, λ, ι. Consequently, equal arcs *CD*, *DE*, *EF* of the eccentric would correspond to unequal 'falls' on the diameter (radius), namely $\gamma\iota$, $\iota\lambda$, $\lambda\zeta$, and in an irregular manner: the highest [the farthest from the Sun] would not be the least, and the lowest [the nearest to the Sun] would not be the greatest, but those in the middle, $\iota\lambda$, would be the greatest, and the extremes, $\gamma\iota$ and $\lambda\zeta$, would be the least; furthermore, the highest, $\gamma\iota$, would be slightly less than the lowest, $\lambda\zeta$. In fact, $\gamma\kappa$ and $\mu\zeta$ are equal and $\gamma\iota$ smaller than $\gamma\kappa$, whilst $\lambda\zeta$ is greater than $\mu\zeta$.' Consequently, the 'fall' of the planet on the diameter $\gamma\zeta$ would correspond neither to the time of motion on the eccentric, nor to the distance covered on it, nor to the angles [between the radii vectores] having their apices at the Sun. In fact, they would not correspond to any physical magnitude, but to some abstract magnitude, namely, the *arc sine* of $\gamma\delta$, $\delta\epsilon$, $\epsilon\zeta$, etc., of the epicycle, as shown by Kepler later on.

19. *Ibid.*

20. Cf. *supra*, p. 195 and p. 252.

21. *Astronomia Nova*; (*G.W.*, vol. iii, p. 254; *O.O.*, vol. iii, p. 316).

22. This radius vector, though quite imaginary, condenses, so to speak, the motive force of the *species*.

23. If the eccentric and the epicycle move in opposite directions with the same angular velocity, the radius *ND* will always remain parallel to itself. Conversely, if it be kept in this position, the point *D* describes an epicycle.

24. *Astronomia Nova*, cap. XXXIX; (*G.W.*, vol. III, p. 258; *O.O.*, vol. III, p. 316); cap. II; (*G.W.*, vol. III, p. 69; *O.O.*, vol. III, p. 178); cap. IV; (*G.W.*, vol. III, p. 73; *O.O.*, vol. III, p. 181).

25. The multiplicity of arguments, the length and persistence of the discussion show that we are confronted by a serious problem; cf. *supra*, pp. 189 f.

26. Cf. *Astronomia Nova*, cap. XXXIX; (*G.W.*, vol. III, p. 260; *O.O.*, vol. III, p. 318).

27. *Ibid.*

28. *Ibid.* Kepler's figures are wrong: the Sun is twenty times further off than he thought. However, it is irrelevent to the build-up of his argument.

29. Kepler, as we have seen, believed in the animation of celestial bodies, including the Earth; though he was opposed to the judicial astrology of his times, he firmly believed in the influence of 'aspects' on events occurring on Earth (climate, earthquakes, volcanic eruptions, etc.), and even their influence on the physical and mental make-up, besides the destiny, of human beings. Therefore, having asserted the possibility that changes in dimension of the solar disc could be apprehended by the planets, he wrote (*Astronomia Nova*, cap. xxxix; (*G.W.*, vol. III, p. 261; *O.O.*, vol. III, p. 319)):

> 'Well then, Kepler, do you claim that the planets possess a pair of eyes? Not at all. It is not necessary, any more than it is necessary to give them feet and wings so that they may move. Tycho Brahe has already got rid of the solid spheres; and our speculation has not exhausted all the treasures of nature, so we are able by our knowledge to ascertain how much reason there is in it. Some good examples immediately come to mind.
>
> 'Say then, O philosopher, by what eyes do the animal faculties of sublunary bodies perceive the positions of celestial bodies in the Zodiac, so that by an harmonious arrangement (which we call aspect) they come to agreement amongst themselves and blaze forth to accomplish their work? Was it by means of eyes such as these that my mother noted the positions of the stars, in order that she might know that she was born in the configuration of Saturn, Jupiter, Mars, Venus, Mercury in sextile and trine; and that she gave birth to her children, particularly me, the first-born, under very nearly the same conditions where most of these aspects, especially those of Saturn and Jupiter, had returned, or where most of the earlier positions of these bodies were in square or in opposition.'

30. *Ibid.*

31. *Ibid.*

> 'Finally, I must say some few words on the animal faculty of moving planetary bodies: anyone who says that the planets are carried around by their own power is not making a likely statement. We have denied this from the start. Nor, on the other hand, can this power be simply transferred to the Sun. In fact, the same force which attracts the planet, will repel it also; which implies a contradiction of the simplicity of the solar body. However, he who, for any particular reason, sets this translatory motion in the mutual *consensus* of the body of the Sun and of the planet is otherwise informed in this chapter, and chapter LVII is specially devoted to this question.'

Later, in the *Epitome Astronomiae Copernicanae*, Kepler dropped this objection, and ascribed to the Sun itself the ability to attract and repel the planets, as well as to urge them on their path.

32. Cf. *infra*, p. 475.

CHAPTER VII. FROM THE CIRCLE TO THE OVAL

1. Cf. *supra*, p. 218.
2. The *oval* came before the law of areas: but the *ellipse* came after.
3. *Astronomia Nova*; (*G.W.*, vol. III, p. 34; *O.O.*, vol. III, p. 156).
4. As such, he used it until he discovered the law of areas.
5. This would explain the set-back to the *hypothesis vicaria.* At this time, Kepler still believed that the Earth was an exception to the rule: in fact, a study of the Earth's orbit (cf. *supra*, pp. 182 f.), made after that of Mars, had convinced him that the latter was a perfect circle. Still, he did not persist in this belief, and after 1604 he reverted to his first intuition—the dynamic structure of planetary motion being the same for them all, their geometric structure must be the same also. This means, that if Mars move on an oval, then the other planets, including the Earth, must do the same. This is why his analysis of the physical mechanism of celestial motions (cf. *supra*, pp. 217 f.) is set forth in quite general terms. So, having established the ellipticity of the orbit of Mars, both by observation and calculation, he applied his discovery to all the other planets, including the Earth, for purely theoretical reasons.
6. Cf. *supra*, p. 45. The orbit of Mars bristles with nodes; the moving deferent of Mercury describes an oval; that of Venus, describes an even more complicated curve. This was not of great importance for ancient astronomy seeing that it was not the *orbits* of the planets that mattered, but only the angular distances between the 'phenomena'. There was a good reason for this: computation of the orbit presupposes a knowledge of the linear distances of the planet (from the Sun), which ancient astronomy did not, and could not, have. We could go even further and say that the orbit of a planet had no real existence for pre-Copernican astronomy. It interested no-one. A study of the orbit, and a determination of the distances of a planet from the Sun were radical innovations.
7. The two great discoveries that Kepler prided himself on having made were: (1) circular motion; and (2) the need to reintroduce the equant; cf. *supra*, pp. 169 f.
8. The unlikely nature of such a motion was admitted by Kepler when the resultant had to be a circle. This did not prevent him from re-adopting it for his description of the oval.
9. Letter to Fabricius, 1 October 1602; (*G.W.*, vol. XIV, pp. 277 f.; *O.O.*, vol. III, pp. 63 f.):

'*Accipe hic Schema Theoriae Martis et Solis. . . . A Sol, AB radius orbis,*

Martis, BC verissima eccentricitas ♂. Quodsi Sol non raperet Martem Mars a C per I, H, G, in F circellum describeret, cuius eccentricitas 9165 qualium AB = 100.000. . . . Jam leges motus puncti B explicabo. Imaginare itaque, sicut est AF ad AE, AB, AD, AC, sic esse moram ♂ [Martis] in 1° vel 1′ circuitus circa ☉ [Solem], quando est in F, ad moram in G, H, I, C. Nam virtus Solis circularis quidem est, et si planeta non descenderet ex C, raperetur perfecto circulo per C descripto. Ita si in I maneret, raperetur circulo per DI descripto, sed breviori tempore rediret eodem, quia virtus in angustioribus circulis stipatior est et densior ideoque et in effectu fortior, quam in laxioribus et superioribus. Itaque Martis in C tardus est motus circa ☉ [Solem], in H mediocris, in F velox. Interim tamen aequalis manet motus circa B centrum. Hoc modo fit illa, quam superioribus literis innui, figura ovalis. *Nam citius in H venit facitque distantiam AH, quam circa ☉ in locum a B quadratum: breviores itaque ad latera seu in longitudinibus mediis fiunt distantiae, quam ratio perfecti circuli eccentrici exigit, et differentia versus perigaeum major est quam versus apogaeum; quae justa definito figurae ovalis est. . . .*

'*Si vis scire, in quo gradu celeritatis sit ♂, quaere per anomaliam simplicem ejus a ☉ [Sole] distantiam. At si vis locum scire, oportet scire, quantum accumulatae omnes ab apogaeo distantiae in iis temporibus, in quae singulae inciderunt, in motu efficerint. At cum infinitae sint distantiae, quia infinita F, G, H, I, C puncta, et illae non aequaliter sparsae per circuitum circa ☉, confertiores enim quae longiores, quia ibi ♂ [Mars] tardus, sparsiores quae breviores, hic vides difficultatem computandi loci eccentrici ex hypothesi vera et physica.*

'*Modus tamen iste est: divido circuitum in 360° et fingo planetam moveri aequaliter quamdiu in uno est. Ita ordine ab apogaeo omnium graduum morae computatae dant cujusque anomaliam. Idem intellige per omnia (in depicto schemate Terrae) Haec ergo omnis hypothesis ♂ [Martis], Proportio orbis ♁ [Terrae] ad orbem ♂ = 100.000:152518, circelli ♁ [Terrae] semidiameter 180, ⊕ in 6° ♋; ♂, in 29° ♌.*'

The figure in the text is taken from Caspar's edition; the one on page 421 for this note comes from the *Opera Omnia.* The reference to the Earth's *circellum* shows that Kepler regarded the Earth's motion as being like that of Mars.

10. These letters have not survived.

11. It is interesting to note the part played by distances (*radii vectores*).

12. Letter to Fabricius, 7 February 1604 (*G.W.*, vol. xv, pp. 17 f.):

'I continue computation of distances [of Mars from the Sun]. When I shall have calculated a large number of such distances over the whole length of the orbit, the position of apogee will be easily found; and if the orbit have an oval figure, then it will be necessary to find an hypothesis which accounts for all these distances.'

13. Letter to Fabricius, 4 July 1603 (*G.W.*, vol. XIV, pp. 409 f.):

'*Mars circulari lege Soli appropinquat aequales Epicycli arcus conficiens temporibus aequalibus. At singula verò momenta sui accessus ad Solem mutat modulum suae celeritatis. Nam in apogaeo tardè (circa proprium Eccentrici centrum) volvitur. Et tamen ad singula momenta, distantiâ suâ*

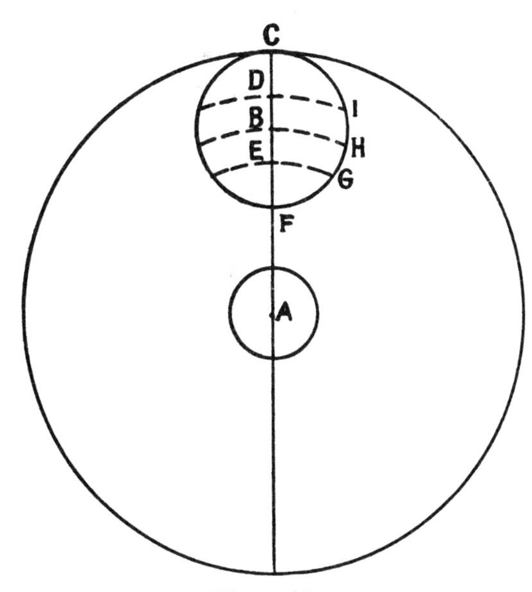

Figure 36.

à Sole, monstrat certum suae celeritatis modulum. Collectione igitur omnium distantiarum quae sunt infinitae, habetur virtutis effusae certo tempore, summa; igitur et emensi circa centrum Eccentrici (adeoque et circa Solem) itineris. Ito tu jam, et vel infinita ejus via puncta, vel arcus in minutissima sectos, puta in dena scrupula, computa: invenies idem, quod ego invenio ex hypothesi vicaria, vix dimidio scrupulo deficiente. Ego vero citius ero expeditus: tibi ab apogaeo incipiendum et per minima erit eundum, quare immanem laborem hauseris. Ad compendiosam vero solummodo calculationem genuinae hypotheseos et cujuslibet loci eccentrici seorsim sine diductione per minima usque in apogaeum aliquid mihi deest: scientia Geometrica Generationis viae Ovalis seu facialis μετωποειοῦς *ejusque plani sectionis in data ratione. Si figura esset perfecta Ellipsis jam, Archimedes et Apollonius satisfecissent.*'

Ibid., p. 411: '*Martis motum esse figuram Ovalem . . .*'; cf. *supra*, p. 246.

14. *Ibid.*, p. 429:

'*Nihil nobis, mi Fabrici, praeter geometriam deest. Doce me Geometricè constituere, quadrare, secundùm datam rationem secare Ellipoides, statim docebo te ex genuina hypothesi calculare. Adeoque excita illa tua Belgica ingenia, ut hic me juvent. Eo usque perveni, ut sciam circelli seu Epicycli plano summam contineri defectus. Quomodò verò sit hoc spaciolum distribuendum, ignoro; ut cuilibet portioni Eccentrici plani suus defectus adjudicetur.... Possumus tamen inire rationem calculi distributa Ellipoide orbita in partes 360 aequales, seu potius ex termino cujusque arcus educta linea ad centrum A* [Sun]. *Tunc enim fient 360 aequebasia triangula, verticibus in A Sole coeuntia. Sciturque primi altitudo AC* [distance at aphelion], *fingitur alterum aequealtum, quia parum deest. Datur igitur area trianguli primi in proportione areae CKH, seu CIH subtracto defectu. Ex quantitate hujus trianguli datur tempus seu anomalia, ex anomalia altitudo secundi trianguli, ex altitudine area, ex area tempus et anomalia, quae juncta priori anomaliae ostendit altitudinem tertii trianguli, et sic consequenter usque ad 360. Praecipere equidem possum, tu exsequere.*'

15. Chapter XL was written in 1604.

16. *Astronomia Nova*, cap. XL; (*G.W.*, vol. III, p. 263; *O.O.*, vol. III, p. 320).

17. Literally: 'has been a thief of my time'.

18. *Ibid.* The Sun's orbit: the Earth's orbit.

19. Cf. *Astronomia Nova*, cap. XXXIX; (*G.W.*, vol. III, p. 358; *O.O.*, vol. III, p. 317). It is true that it was then a question of the knowledge that the planet should have had—though it could not have—to be able to determine its path; but the impossibility of knowing the infinite was a fundamental conviction with Kepler. The infinite was 'irrational' and 'ineffable'; even God himself was unable to *know* it, in the strict meaning of the term: knowledge extends only to that which is rational, and in geometry to that which is capable of 'construction'; cf. *infra*, n. 33.

20. *Ibid.*, cap. XL; (*G.W.*, vol. III, p. 263; *O.O.*, vol. III, p. 320). See diagrams on p. 228 and p. 421.

21. In other words, Kepler replaces the planet's continuous path by a discontinuous path. Cf. *supra*, p. 230.

22. At this stage in the development of his thought, Kepler added the *distances* (lengths of the *radii vectores*) and not the areas of the small triangles formed by them.

23. The mean anomaly (cf. diagram in Appendix I, p. 365) is the angle (or arc) at the equant point corresponding to the mean motion of the Sun (or of the Earth) or of the centre of the planetary epicycle. The mean anomaly is therefore a measure of time. The eccentric anomaly is the angle (or arc) at the centre of the deferent circle.

24. Here, it is a question of the time that the planet takes to cover the

smallest portion of the path (one degree), along which the distance from the Sun, and hence the velocity, are unchanged.

25. Cf. diagram in Appendix I.

26. This was the essential step.

27. Cf. *supra*, n. 13; Letter to Fabricius, 4 July 1603.

28. The distances (cf. *supra*, p. 186) determine the times of travel; the latter are then the 'effects' of the distances to which they are directly proportional.

29. It then follows that equal areas correspond to the same time, and conversely. This is a statement of the law of areas.

30. *Astronomia Nova*, cap. XL; (*G.W.*, vol. III, p. 267; *O.O.*, vol. III, p. 323).

31. In the theory of the Sun (or the Earth). The theory of Mars is quite different.

32. Kepler took the radius as being equal to 100 000.

33. Kepler's reasoning is wrong in both places: in effect, he establishes proportionality between non-corresponding lines. In the first part of his argument, he makes a diameter rotate about the point *B*, and draws two lines from the point *A* to the ends of this diameter for each position it occupies; in the second part of his argument, he makes a straight line, which cuts the circle in *E* and *V*, rotate about the point *A*, and draws two radii (= one diameter) of the circle to the chord *EV*. The correct argument should consider the respective distances of a point *N* on the circumference from the points *B* and *A*; it would then be found that the distance of *A* from *N* will be always greater than that of *B* from *N* for the greater part of the circumference; for the remaining, smaller portion of the circumference, the distance *AN* will be always shorter than *BN*; two points will be equidistant.

34. *G.W.*, vol. III, p. 268; *O.O.*, vol. III, p. 324.

35. Kepler ought to have said 'oval', for he believed at first that his method was correct for the oval; but he subsequently realized that the orbit was not an oval, but an ellipse.

36. Letter to Maestlin, 15 March 1605; (*G.W.*, vol. XV, p. 173; *O.O.*, vol. III, pp. 57 f.).

37. Kepler gave Maestlin concrete examples. Following the precedent of Max Caspar in his *Johann Kepler in seinen Briefen*, München and Berlin, 1930, vol. I, pp. 222 f., I have omitted them.

38. Kepler's reasoning foreshadowed that of Cavalieri, who probably used Kepler as his source; cf. my 'Cavalieri et la géométrie des continus' in *Mélanges Lucien Febvre, L'éventail de l'Histoire vivante*, Paris, 1955.

39. In the motion with uniform angular velocity invented by Ptolemy, the planet—or more correctly the centre of its epicycle—is not regarded as being linked to the radius vector rotating about the *punctum aequans*; it is considered to move—in imagination—on the *equant circle* whose centre

is the *punctum aequans*. The diagrams given in modern works often leave out the (quite useless) circle and show only the radius vector; though this provides an exactly equivalent representation, it completely falsifies the intention of the Greek astronomer.

40. And Tycho Brahe, also.

41. No doubt, we could reason as follows—and Kepler certainly did so. The surface of a segment of a circle is equal to the 'sum' of the distances from the point S (when the radii vectores coincide with the radii of the circle). By transferring the point from which the distances are measured from S to S', they are increased in the same proportion that the velocity is decreased, and hence in proportion to the length of the planet's path on its orbit. In other words, the base of the 'triangle' becomes shorter in proportion to the increase in length of its sides, and consequently, of the height: the surface, that is to say, the 'sum' of the distances, therefore remains constant. The reasoning is plausible, but unfortunately, erroneous.

42. On the other hand, the law of areas is valid for Ptolemaic motion (uniform about a *punctum aequans*) on a circle.

43. It is unlikely that Maestlin would have said what Kepler ascribed to him.

CHAPTER VIII. FROM THE OVAL TO THE ELLIPSE

1. It was necessary to determine if this orbit were elliptical or ovoid; cf. *supra*, chap. VI, n. 13.

2. Cf. *Astronomia Nova*; (*G.W.*, vol. III, p. 271; *O.O.*, vol. III, pp. 326 f.).

3. *Ibid.*; (*G.W.*, vol. III, p. 275; *O.O.*, vol. III, pp. 328 f.).

4. *Ibid.*; (*G.W.*, vol. III, p. 276; *O.O.*, vol. III, p. 328).

5. *Ibid.*; (*G.W.*, vol. III, pp. 381 f.; *O.O.*, vol. III, p. 333).

6. Cf. *ibid.*, cap. XLIII; (*G.W.*, vol. III, p. 282; *O.O.*, vol. III, p. 333). *On the error in the equations calculated by means of triangles assuming bisection of the eccentricity and perfect circularity of the planetary orbit.*

7. *Ibid.*; (*G.W.*, vol. III, p. 283; *O.O.*, vol. III, p. 334); more exactly, more than 8′ 21″ in the first octant, and less in the third; cf. *supra*, p. 178, for the part played by these eight minutes.

8. *Ibid.* It is to be noted again that Kepler describes the Copernican system as bi-epicyclic; cf. *supra*, p. 178, for the part played by these eight minutes.

9. *Ibid.*, cap. XLIV; (*G.W.*, vol. III, pp. 285 f.; *O.O.*, vol. III, p. 335). The orbit of the planet through the aetheric aura is not a circle, even if we take into consideration only the first inequality, and disregard the complicated curves which, for Tycho Brahe and Ptolemy, derive from the second inequality.' The passage quoted is on p. 286 in *G.W.*, and on p. 336 in *O.O.*

10. *Ibid.*; (*G.W.*, vol. III, p. 286; *O.O.*, vol. III, p. 326).

11. *Ibid.*; (*G.W.*, vol. III, p. 287; *O.O.*, vol. III, p. 337). The acceleration of the planet's motion in the mean and lower longitudes would enable it to make up for the delay caused by its retardation in the first octant and would provide the necessary gain.

12. *Ibid.*

13. *Ibid.*, cap. XLIV, at the end: '*Orbitum planetae non esse circulum sed figurae ovalis.*'

14. *Ibid.*, cap. XLV; (*G.W.*, vol. III, p. 288; *O.O.*, vol. III, p. 337): *On the natural causes of the deviation from a circle* [*of the orbit*] *of the planet: consideration of the first opinion.*

15. Cf. *supra*, pp. 219 f.

16. Cf. *supra*, pp. 221 f. If the angular velocities of these two motions be equal, the radius *ND* of the epicycle will be parallel to *AC* (line of apsides), and the point *D* (the planet) will describe a circle.

17. *G.W.*, vol. III, p. 289; *O.O.*, vol. III, p. 338.

18. Cf. *supra*, p. 245.

19. Cf. *supra*, p. 231.

20. Cf. *Astronomia Nova*; (*G.W.*, vol. III, p. 289; *O.O.*, vol. III, p. 239). Psychologically, the passage is very interesting.

> '*Si diameter epicycli ND mansisset ipsi AB aequidistans* [that is to say, if the motion on the epicycle were non-uniform], *poteram exuisse hanc meam opinionem erroneam, poteramque, quod est verissimum, omnem promotionem in longitudinem zodiaci, transscribere Soli, solam Planetae librationem in diametro* $\gamma\zeta$ (*see figure p. 219*) *relinquere, ut in parte cap. XXXIX.*' [However, the uniformity of epicylic motion] '*admirabiliter me confirmavit in errore hoc de motu Planetae in ipsa epicycli circumferentia . . . quia non putavi fieri ullo alio medio posse, ut Planetae orbita redderet ur ovalis*'.

21. In the first enthusiasm for his discovery for the mechanism by which the oval was produced, Kepler was not greatly concerned with its agreement with fact. *Ibid.*; (*G.W.*, vol. III, p. 290):

> '*Haec itaque cum ita mihi incidissent, plane securus de quantitate hujus ingressus ad latera, nimirum de consensu numerorum, jam alterum de Marte triumphum egi. Neque mihi difficile videbatur, si quid adhuc numeros esset discordia, id τῷ προσθαφαιρεῖν per minima circumcirca dissipare, ut redderetur insensibile*'.

22. Cf. *supra*, p. 230.

23. Cf. *Astronomia Nova*, cap. XLVII; (*G.W.*, vol. III, p. 297; *O.O.*, vol. III, p. 345).

24. *Ibid.*; (*G.W.*, vol. III, pp. 302 f.: *O.O.*, vol. III, pp. 349 f.).

25. *Ibid.*; (*G.W.*, vol. III, pp. 309 f.; *O.O.*, vol. III, p. 355).

26. *Ibid.*; (*G.W.*, vol. III, pp. 313 f.; *O.O.*, vol. III, p. 359).

27. *Ibid.*, cap. L; (*G.W.*, vol. III, p. 314; *O.O.*, vol. III, p. 359): *De aliis sex modis, equationes eccentrici exstrvendi, tentatis.*

28. Cf. Letter to Maestlin, 14 December 1604; (*G.W.*, vol. XV, pp. 72 f.).

29. *Astronomia Nova*, cap. LIII; (*G.W.*, vol. III, p. 340; *O.O.*, vol. III, pp. 380 f.).

30. Letter of Fabricius to Kepler, 27 October 1604; (*G.W.*, vol. XV, pp. 59 f.; *O.O.*, vol. III, p. 95).

31. *Astronomia Nova*, cap. LV; (*G.W.*, vol. III, pp. 344 f.; *O.O.*, vol. III, p. 384).

32. *Ibid.*, cap. LV; (*G.W.*, vol. III, p. 345; *O.O.*, vol. III, p. 384: '*Itaque causae physicae cap. xlv in fumos abeunt.*').

33. Letter to Fabricius, 18 December 1604; (*G.W.*, vol. XV, p. 79; *O.O.*, vol. III, p. 96:

> '*Sic igitur est, mi Fabrici. Negativa circuli validissimis quidem nititur argumentis, et ovalitas (frusta te concludente contra hanc), sed affirmativa harum distantiarum ex ratiocinatione meâ nudè dependet. Tu vitiosè: Kepleriana ovalitas nimium curtat, ergò nulla planè ovalitas ponatur. Ego aequè vitiosè: Ovalitas est aliqua, ergò haec erit, quam aequabilitas motus epicycli monstrat. In dimensione orbis annui 100 000, circuli perfectio prolongat circiter 800 aut 900 nimis. Ovalitas mea curtat 400 circiter nimis. Veritas est in medio, propier tamen ovalitati meae. Neque tamen infra longitudines medias prolongandae, sed etiam supra etiamnum magis decurtandae sunt differentiae, quàm mea fert ovalis:* omninò quasi via Martis esset perfecta Ellipsis. *Sed nihil dum circa hanc exploravi. Hoc verisimilius, Epicyclum et in Aphelio et in Perihelio accelerari.*'

34. Cf. *supra*, pp. 219 f. The title of cap. LVI (*G.W.*, vol. III, p. 345; *O.O.*, vol. III, p. 384): '*Demonstratio ex observationibvs ante positis, distantias Martis a Sole desvmendas esse qvasi ex diametro epicycli.*'

35. With all due respect to this great thinker, we might even say that it was difficult for Kepler to miss the chance of making a mistake. In a letter, dated 14 December 1604, written to his old teacher Maestlin, Kepler said (*G.W.*, vol. XV, p. 72: *O.O.*, vol. III, p. 55):

> 'If you could but see my work on the motion of Mars, I believe you would say that I often look for difficulties where they do not exist—it is true and you have said as much of the *Optica.*' '*In* Commentarijs de Motibus Martis *si meos labores cerneres; opinor id diceres, quod res, est quodque etiam de Opticis dicere te non dubito; Me scilicet non rarò nodum in scirpo quaerere. Cur ergò non mecum communicas per literas? Saepe mihi non cogitandj inepta multa obveniunt, quae per literas ventilata facile agnoscerem. Omnis meus labor in hoc est, ut jam porrò ex genuinis causis tam aequationes Excentrj justas, quàm distantias extruam. Profecj autem per Dei gratiam eo usque, ut non plus aberrem in uno quam il altero, certusque*

sim, utrumque ab eâdem hypothesi proficisci, ac proinde non posse esse vana, quae de virtutibus motricibus disputo. Cumque toties jam triumphaverim de Marte, hoc tamen etiamnum in causa manet: Si Eccentrici ratio distribuitur in Concentricum et Epicyclum; scis centrum Epicyclj inaequalis motus fierj in concentrico; id est, concentricum super alieno centro aequaliter ire: quia etiam Eccentrus movetur super alieno centro. Quod si ergò motus et Concentricj et Epicyclj simul intenduntur simul remittuntur (id est, si linea ex centro aequalitatis concentricj per centrum Epicyclj educta monstrat apogaeum verum Epicyclj), tunc in Effectu manet orbita planetae, quam corpore transit, perfectus circulus Eccentricus. At observationes testantur, in longitudinibus medijs utrinque planetam ad latera ingredj circiter 900 partes de 152 500. Et ipsae rationes physicae suadent, Epicyclj motum super proprio centro plane aequabilem dicere (Id est lineam veri apogaej epicyclj agere per centra concentricj et Epicyclj). At si hoc facias, planeta deflectet ab orbita circulari per 1300, debuit secundum observata tantum per 900. Quin etiam uti in longitudinibus medijs et versus perigaeum nimius est hic ingressus ad latera, ita versus apogaeum non satis magnus esse videtur. Unde videtur sequi, ipsum Epicyclum non omninò aequabilem esse: neque tamen in motus inaequalitate cum Concentrico convenire, sed exiguo velociorem fieri planetâ tam circa Apogaeum Epicyclicum versante, quàm circa perigaeum Epicyclicum . . . idem Tycho Lunae tribuit, ut sit velox caeteris paribus tam in ☌ quam in ☍.'

36. Letter to Maestlin, 5 March 1605; (*G.W.*, vol. xv, pp. 171 f.; *O.O.*, vol. III, pp. 56 f.).

37. The one dealing with the circular orbit, the other with epicyclic motion.

38. If Kepler had spoken about it 'in his earlier letters' to Maestlin, then they have not survived.

39. We note that Kepler, whilst informing Maestlin that the planet does not really move on an epicycle, and that the equant point has no real existence, used these concepts as approximations; as well as for the purpose of making himself understood.

40. This implies—if the Sun be regarded as a magnet—that its surface constitutes one of its poles, the other being at the centre. This is an impossible arrangement, as was subsequently proved.

41. This 'animal' force which maintains the axis of the planet in a constant direction was ultimately reduced to the magnetic force itself. Kepler had need of it in order to prevent 'the body of the planet' from turning its 'pursuing' pole towards the Sun and then approaching closer and closer until it finally touched the Sun.

42. As before, it is a matter of libration of the planet (Mars) on the diameter of the imaginary epicycle.

43. Cf. *supra*, p. 300. The angle between the (magnetic) axis of the planet

and the radius vector determines the force of attraction—or of repulsion—exerted between the planet and the Sun; the greater the angle, the weaker the force; cf. n. 45.

44. The geometric hypothesis, which explained the motion of Mars by a combination of epicyclic—or oscillatory—motions without recourse to a dynamic explanation.

45. In the apsides (cf. p. 30) the planet's axis is perpendicular to the radius vector—the line of the attracting or repelling force: furthermore, the two poles, or two forces, act with equal power; in the median parts of the path, one or other of the poles is turned more towards the Sun; consequently, its action preponderates, and to an extent which increases as the angle between the axis and the radius vector decreases.

46. It is this unequal action of the couple comprising attraction-repulsion in aphelion and perihelion that determines the ovoid shape of the orbit.

47. Cf. *Astronomia Nova*, cap. LVI; (*G.W.*, vol. III, pp. 345 f.; *O.O.*, vol. III, pp. 384 f.).

48. The optical equation (see diagram, Appendix I) is the angle subtended by the eccentricity at the planet.

49. Cf. Max Caspar, *Johann Kepler, Die Neue Astronomie*, München, 1929, Introduction, p. 34.

50. *Astronomia Nova*, cap. LVI; (*G.W.*, vol. III, p. 346; *O.O.*, vol. III, p. 385). Cf. Letter to Fabricius, 11 October 1605; (*G.W.*, vol. XV, p. 247); quoted *infra*, n. 66.

51. The equivalence assumed here by Kepler does not exist, but he did not become aware of this till several months later.

52. It is interesting to compare the text of the *Astronomia Nova* with that of Kepler's letter to Longomontanus at the beginning of 1605; (*G.W.*, vol. XV, p. 138; *O.O.*, vol. III, p. 34):

'. . . The oval shape of the orbit that Mars describes through the aether is definite. The cause of this orbit has not been discovered with any certainty. Up to now, I have assumed a cause which resulted in Mars deviating [from the circle] towards the [in-]side by 1300/152500 parts. However, on examining this phenomenon more closely, I found that this deviation did not amount to more than 800 or 900 parts. Mars itself performs a libration on the diameter of the epicycle. Furthermore, the planet is carried round by a force which is propagated from the Sun throughout the space of the Universe. The two motions are not uniform. The libration is greater, or less, not only in conformity with the law dictated by one of the two circles as required by Copernicus, but also the two circles move faster, or slower, according to the increase, or decrease, in the apparent size of the Sun's diameter. The greater the apparent diameter of the solar disc is in perihelion than in aphelion, the greater is the diurnal motion in the former case; or, conversely, that much more [time] is required by the planet to traverse one degree of the true anomaly in aphelion than in perihelion. Now, the measure of the period of revolution is the quantity of light which falls on the planet from the Sun at any particular distance, that is to say, the size of the disc of the solar body . . . [as seen from the planet].

'One thing is certain; the Sun sends forth a force which seizes the planet. Everything else is still uncertain.'

53. *Astronomia Nova*, cap. LVII; (*G.W.*, vol. III, p. 348; *O.O.*, vol. III, p. 386). The mechanism referred to by Kepler is that described in cap. XXXIX; cf. *supra*, p. 228.

54. The intervention of an intelligence which would be able to calculate the required motions, and of a soul which would be able to perform them; but, as we have seen, the two operations are very nearly impossible.

55. *Astronomia Nova*, cap. LVII; (*G.W.*, vol. III, p. 348; *O.O.*, vol. III, p. 386):

> '*Modo naturali, qui nititur non aequalitate angulorum DBC, EBD, FBE, sed fortitudine anguli DBC, EBC, FBC, perpetuo crescentis, quae fortitudo fere sequitur sinum Geometris dictum; ubi ascensus continua imminutione sensim in descensus mutatur, probabilius, quam si subito Planeta proram convertere diceretur.*'

56. In 1605 Kepler wrote to Longomontanus as follows; (*G.W.*, vol. XV, p. 139; *O.O.*, vol. III, p. 34):

> '*Hypothesin habeo jam ante quatuor annos constructam, quae mihi planetam Eccentrico debitis locis sistit scrupolosissimè. Sed non placet mihi, quia non est physica, sed vere id, quod dicitur,* Hypothesis.'

Cf. *supra*, n. 52.

57. Letter to Herwart von Hohenburg, dated 10 February 1605, quoted *supra*, Introduction, n. 8; (*G.W.*, vol. XV, p. 145).

58. I have given the physical theories of cap. XXXIX (cf. *supra*, cap. VI) in considerable detail in order not to spend so much time on those in cap. LVII.

59. *Astronomia Nova*, cap. LVII; (*G.W.*, vol. III, p. 348; *O.O.*, vol. III, pp. 386 f.). In cap. XXXIX Kepler had envisaged the possibility of a whirlpool motion throughout the whole 'aetheric sphere' (cf. *supra*, p. 216). Probably, he did not regard it as being sufficiently material to be able to produce a mechanical effect.

60. *Ibid.*; (*G.W.*, vol. III, pp. 350 f.; *O.O.*, vol. III, pp. 387 f.). Cf. *supra*, Letter to Herwart von Hohenburg quoted on p. 412, n. 39.

61. *Ibid.*; (*G.W.*, vol. III, pp. 351 f.; *O.O.*, vol. III, pp. 388 f.). It should be noted, however, that the planet, or its intelligence, would not be able to

discern the true anomaly; cf. *supra*, p. 223, and *G.W.*, vol. III, pp. 359 f.; *O.O.*, vol. III, p. 395.

62. Cf. *ibid.*; (*G.W.*, vol. III, pp. 359 f.; *O.O.*, vol. III, pp. 392 f.).

63. Cf. *G.W.*, vol. III, pp. 354 and 356; *O.O.*, vol. III, pp. 390 and 392:

Mensurat anomaliae { *eccentri coaequatae* } { *sinus versus* } { *Librationem Planetae. Augmentum diametri Solis, ut ea apparitura fuit spectatori in corpore Planetae supposito: et vicissim.* }

64. Cf. *supra*, p. 192.

65. *Astronomia Nova*, cap. LVIII; (*G.W.*, vol. III, p. 364; *O.O.*, vol. III, p. 399).

66. Letter to Fabricius, dated 11 October 1605; (*G.W.*, vol. XV, pp. 247 f.; *O.O.*, vol. III, p. 99):

'*Quae hactenus in meo Marte profecerim accipies. Cum viderem distantias ex perfecto circulo eccentrico exstructas penè tantum peccare in excessu, tam quod seipsas, et earum effectuum in prostaphaeresibus orbis annui, quam quoad aequationes Eccentri: quantum Ellipsis mea (quae perparum ab ouali differt), quam tibi in numeris praescripsi peccabat in defectu; rectissimè fuissem argumentatus in hunc modum. Circulus et Ellipsis sunt ex eodem figurarum genere, et peccant aequaliter in diuersa, ergò veritas consistit in medio, et figuras Ellipticas mediat non nisi Ellipsis. Itaque omninò Martis via est Ellipsis, resectà lunula dimidiae latitudinis pristinae Ellipseos. Erat autem lata lunula 858 de 100 000. Ergò debuit esse lata 429, quae est justa curtatio distantiarum in longitudinibus mediis, ex perfecto circulo extructarum. Hic inquam veritas ipsa est. At vide quomodo ego interea rursum hallucinatus et in nouum laborem conjectus fuerim. Imo vide quam misere trepidem super inventa veritate, secundum illud, qui nunquam dubitat, nunquam certus est de re aliqua. Ellipsis illa pristina cum curtatione 858 habuit causam naturalem hanc, nempe, ut dicatur centrum Epicycli tardè incedere, quando Planeta versatur in Apogaeo Epicycli, velociter infra. Epicyclum verò ipsum aequalibus temporibus incedere aequaliter. Hoc erat mediocriter consentaneum naturae. Jam uerò si Ellipsis esset cum curtatione 429, carebam causa naturali. Nam absurdeum erat, centrum Epicycli incedere inaequaliter, circumferentiam Epicycli nec aequaliter, nec inaequalitatis ipsius centri, sed inaequalitate peculiari, quae esset dimidia saltem inaequalitatis centri. Loquor enim jam tecum non ex meis* Commentariis, *hoc est rationibus naturalibus, sed ex Ptolemaeo et antiqua Astronomia ut me capias*'.

67. See the accompanying diagram by Kepler (Fig. 37, p. 431). *Astronomia Nova*, cap. LVIII; (*G.W.*, vol. III, p. 365; *O.O.*, vol. III, p. 399).

68. Refer to the diagram in chapter LVII of the *Astronomia Nova* (Fig. 10a, *supra*, p. 219). The epicycle *N* being supposed to turn with the same speed as the eccentric circle *CDEF*, the angle *BDN* is always equal to angle *NAC* (eccentric anomaly); and as its radius is equal to the eccentricity the value of the oscillation is equal to *e. versed sine DBC.*

69. The true anomaly would be equal therefore to angle M_2SA. In fact, as Kepler shows in cap. LVII of the *Astronomia Nova*, the position of the planet is to be found at the intersection of the circle traced by the radius SM_2 about *S*, and the perpendicular *MB*; or, if preferred, the planet is to be found on this perpendicular at a distance of M_2S from the Sun; that is to say, at the point M_3, and the true anomaly is the angle M_3SA.

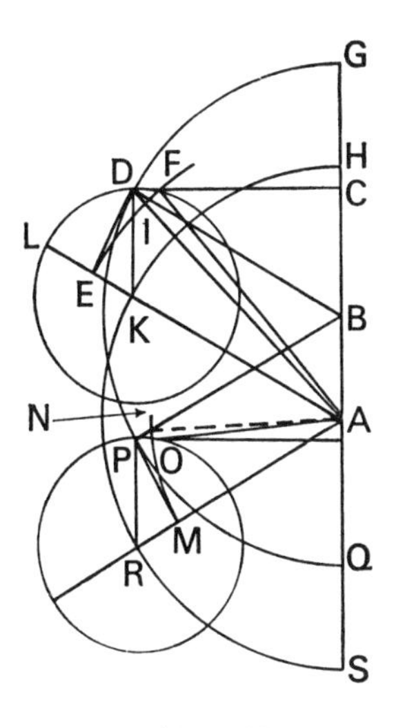

Figure 37.

70. Letter of Fabricius to Kepler, dated 20 January 1607; (*G.W.*, vol. xv, p. 377):

'*Contra verò tuas hypotheses aliquid inferam generaliter. 1. per oualitatem vel ellipsin tuam, tollis circularitatem et aequalitatem motuum, quod mihi imprimis penitius consideranti absurdum videtur. Caelum vt rotundum, est, ita circulares et maximè circa suum centrum regulares et aequales motus habet. Corpora caelestia sunt perfectè rotunda, vt ex* (*Sole*) *et* (*Luna*) *liquet. Ergò non dubium est omnes omnium motus per circulum perfectum, non ellipsin aut excessum etiam fieri, item aequaliter moueri super suis centris, at cum in ellipsi tua centrum non vbique aequaliter distet à circumferentia, certè motus aequalis maximè erit super suo proprio centro inae-*

qualis. Quod si igitur retento circulo perfecto, ellipsin per alium circellum excusare posses, commodius esset, non sufficit saluare posse motus, sed etiam tales hyptheses constituere, quae principijs naturalibus minimè dissentiant.'

71. Letter to Fabricius, dated 1 August 1967; (*G.W.*, vol. XVI, pp. 14 f.).
72. *Astronomia Nova*; (*G.W.*, vol. III, p. 365; *O.O.*, vol. III, p. 400).
73. In fact, they are equivalent. Furthermore, he says:

'*Ipsa veritas et rerum Natura repudiata, et exulare jussa, per posticum se furtim rursum recipit intro, et sub habitu alieno a me recepta fuit.*'

74. *Astronomia Nova*; (*G.W.*, vol. III, p. 366; *O.O.*, vol. III, p. 400):

'*Multo vero maximus erat scrupulus, quod pene usque ad insaniam considerans et circumspiciens, invenire non poteram, cur Planeta, cui tanta cum probabilitate, tanto consensu observatarum distantiarum, libratio . . . in diametro . . . tribuebatur, potius ire vellet ellipticam viam, aequationibus indicibus.*'

75. *Ibid.*:

'*O me ridiculum! perinde quasi libratio in diametro, non possit esse via ad ellipsin. Itaque non parvo mihi constitit ista notitia, juxta librationem consistere ellipsin; ut sequenti capite patescet: ubi simul etiam demonstrabitur, nullam Planetae relinqui figuram Orbitae, praeterquam perfecte ellipticam; conspirantibus rationibus, a principiis Physicis, derivatis, cum experientia observationum et hypotheseos vicariae hoc capite allegata*'.

76. *Ibid.*, cap. LIX; (*G.W.*, vol. III, p. 367; *O.O.*, vol. III, p. 401):

'*Demonstratio, qvod orbita Martis, librati in diametro epicycli, fiat perfecta ellipsis: et qvod area circvli metiatur svmmam distantiarvm ellipticae circvmferentiae pvnctorvm.*'—'Proof that the orbit of Mars, performing libration on the diameter of the epicycle, becomes a perfect ellipse; and that the area of the circle [is a] measure of the sum of the distances of the points from the elliptical periphery.'

CHAPTER IX. ASTRONOMY WITH THE ELLIPSE

1. *Astronomia Nova*, cap. LIX; (*G.W.*, vol. III, pp. 375 f.; *O.O.*, vol. III, p. 407; cf. *supra*, p. 274).

2. The 'geometrical' proof of the law of areas by Newton (*Principia*, Lib. I, sect. II, prop. I) is, literally, astounding.

3. The law according to which the velocities are inversely proportional to the distances. The correct law, established by Newton on the basis of the principle of inertia is: the velocities are proportional to the perpendiculars dropped from the central point (the Sun) on to the tangents to the

point where the moving body (the planet) is located. Cf. *supra*, p. 318, for Kepler's correction to his law.

4. *Astronomia Nova*; (*G.W.*, vol. III, p. 367; *O.O.*, vol. III, p. 401). Kepler explicity quotes Apollonius and Commandinus: '*Ex libro Apollonii Conicorum pag. XXI. demonstrat Commandinus in Commentario super V Sphaeroïdeon Archimedis.*' Kepler is referring to F. Commandinus, *Commentarii in Opera non nulla Archimedis*, Venice, 1552.

5. *Ibid.*; '*Est quinta Spheroïdeon Archimedis.*'

6. *Ibid.*

7. This means that the difference between the arcs of the circle and of the ellipse will be greater when these arcs are situated near to the major axis, and smaller when they are near to the minor axis.

8. *Astronomia Nova*; (*G.W.*, vol. III, p. 368; *O.O.*, vol. III, p. 402); cf. *supra*, p. 312. The term 'focus' owes its origin to Kepler.

9. Cf. *supra*, p. 240.

10. *Astronomia Nova*; (*G.W.*, vol. III, p. 369; *O.O.*, vol. III, p. 402): '*Si circulus dividatur in quotcunque seu infinitas partes; et puncta divisionum connectantur cum puncto aliquo, praeter centrum, intra complexum circuli; connectantur item cum centro: summa earum quae ex centro, minor erit summa earum quae ex alio puncto.*'

11. Note this 'infinite number' of parts.

12. *Astronomia Nova*; (*G.W.*, vol. III, p. 369; *O.O.*, vol. III, p. 402): '*Si autem . . . sumantur distantiae diametrales pro circumferentialibus, ut cap. xxxix et lvii denominatae sunt; tunc summa aequat summam earum, quae ex centro ducuntur.*' Cf. *supra*, pp. 254 f.

13. *Ibid.*; (*G.W.*, vol. III, p. 370; *O.O.*, vol. III, p. 403). On the other hand, as will been seen later, if the ellipse be divided into unequal arcs, and their extremities be joined to an eccentric point, their 'sum' will be equivalent to the 'sum' of the radii of the circle; cf. *supra*, pp. 274 f.

14. Cf. *supra*, pp. 254 f.

15. *Astronomia Nova*; (*G.W.*, vol. III, p. 370; *O.O.*, vol. III, p. 403): '*Patet, quod Area circuli et totaliter et per partes singulas, sit mensura genuina summae linearum, quibus distant arcus elliptici itineris Planetarii, a centro Solis.*'

16. With the focus of the ellipse.

17. *Astronomia Nova*; (*G.W.*, vol. III, pp. 371 f.; *O.O.*, vol. III, p. 404).

18. The circle which has the major axis of the ellipse for its diameter; cf. *supra*, n. 76 in chap. VIII.

19. Max Caspar, *Johannes Kepler, Die Neue Astronomie . . .*, München-Berlin, 1929, p. 412:

> '*Kepler hat den Flächensatz im bisherigen in der Form angeführt und angewandt dass er sagt, die Fläche des Kreises über der grossen Achse und ihre einer gegebenen exzentrischen Anomalie β entsprechenden Teile sind*

der Summe der Abstände aequivalent und geben ein Mass ab für die Zeit α *die der Planet braucht bis seine excentrische Anomalie den Wert* β *erreicht hat, d.h., in der Form:* $\alpha = \beta + e \sin \beta$. *Nun sieht er dass, wenn man von den Endpuncten eines kleinen Bogens jenes Kreises auf die grosse Achse Lote fällt, diese auf der Ellipse ungleiche Bögen ausschneiden. Da erhebt sich für ihn die Frage: Sind diese Ellipsenbögen gerade so gross, dass die Flächenstücke der Ellipse die von jenen Bögen und von den Radien-Vectoren nach ihren Endpuncten umgrenzt werden, die Zeiten messen, die der Planet beim durchlaufen jener Ellipsenbögen braucht? Natürlich folgt dieser Satz ohne weiteres aus der vorigen Formulierung des Flächensatzes und dem Satz den Kepler im vorliegenden Kapitel unter der Ziff. III anführt. Kepler, von früheren Vorstellungen her befangen, durchschaut diesen Sachverhalt nicht. . . .*'

20. *Ibid.*: '*Kepler hätte an der vorliegenden Stelle ausdrücklich sagen müssen, dass sein Radiensatz unrichtig ist.*' By the term: *Radiensatz*, Caspar designates the fundamental dynamical law of Kepler: the velocities are inversely proportional to the distance. Cf. *supra*, p. 318.

21. Cf. *supra*, pp. 261 f., and p. 275.

22. *Astronomia Nova*; (*G.W.*, vol. III, p. 372; *O.O.*, vol. III, p. 404).

23. Cf. *supra*, pp. 204 f.; pp. 212 f.

24. An indivisible as understood in the seventeenth century, and not as understood by Cavalieri; see my article cited above, p. 423, n. 38.

25. *Nova Stereometria Doliorum Vinariorum*, Linz, 1615. A study of Kepler's mathematical work, as well as his work in optics, does not come within the scope of the present undertaking.

26. *Astronomia Nova*; (*G.W.*, vol. III, p. 372; *O.O.*, vol. III, p. 405).

27. As in the case of the circle, the 'sum' of the distances is greater than the surface; cf. Pro-theorem X.

28. Cf. *supra*, p. 239 and p. 245 (analogy of the sausage); it is clear that Kepler regarded the ellipse as a 'compressed' circle; the (lateral) compression produces a shortening of the periphery as well as of the radii; furthermore, as the shortening of the periphery is more pronounced at the apsides, whilst that of the radii is more pronounced in the median parts, the radii 'are condensed' (are closer together) at the apsides, and 'are rarer' (are wider apart) in the mean longitudes.

29. *Astronomia Nova*; (*G.W.*, vol. III, p. 374; *O.O.*, vol. III, p. 406).

30. *Ibid.*; (*G.W.*, vol. III, p. 375; *O.O.*, vol. III, p. 406).

31. *Ibid.*; (*G.W.*, vol. III, p. 376; *O.O.*, vol. III, p. 407): cap. LX:

'*Methodvs, ex hac physica, hoc est genvina et verissima hypothesi, extrvendi vtramqve partem aeqvationis, et distantias genvinas: qvorvm vtrvmqve simvl per vicariam fieri hactenvs non potvit. Argvmentvm falsae hypotheseos.*'

32. *Ibid.*; (*G.W.*, vol. III, pp. 376 f.; *O.O.*, vol. III, pp. 407 f.).

33. *Ibid.*; (*G.W.*, vol. III, p. 381; *O.O.*, vol. III, p. 411).

34. *Ibid.*, cap. LX; (*G.W.*, vol. III, p. 381; *O.O.*, vol. III, p. 411):

> '*Haec est mea sententia. Quae quo minus habere videbitur Geometricae pulchritudinis, hoc magis adhortor Geometras, uti mihi solvant hoc problema: Data area partis semicirculi, datoque puncto diametri, invenire arcum, et angulud ad illud punctum; cujus anguli cruribus, et quo arcu, data area comprehenditur. Vel: Aream semicirculi ex quocunque puncto diametri in data ratione secare.*
>
> '*Mihi sufficit credere, solvi a priori non posse, propter arcus et sinus ἑτερογένειαν. Erranti mihi quicunque viam monstraverit, is erit mihi magnus Apollonius.*'

The problem propounded by Kepler (cf. *supra*, p. 399) has since become known as 'Kepler's problem', and has given rise to numerous studies.

35. *Ibid.; Epistola Dedicatoria*, dated 4 June 1609 in the Dionysian era (that is to say, in the common era—Dionysius Exiguus having established the agreement between the Roman and Christian calendars), but it was written 29 March 1607. Kepler was interested in chronology, and believed that the starting point for the common era, namely, the date on which Christ was born, had been determined in a faulty manner, and that it ought, in fact, to have been moved back two years; cf. *G.W.*, vol. III, pp. 6 f.; *O.O.*, vol. III, pp. 138 f. Cf. G. Bigourdan, *L'Astronomie*, Paris, 1924, pp. 337 f.

36. *Ibid.*; (*G.W.*, vol. III, p. 8; *O.O.*, vol. III, p. 138):

> 'It is told of Georg Joachim Rheticus, in the time of our fathers, a well known pupil of Copernicus—who was the first to dare to yearn for a restoration of Astronomy, and sought to achieve it by most valuable observations and inventions—that, when he studied the motions of Mars, he was astonished; and being unable to resolve his difficulties, he appealed to the oracle of his familiar genius, either [with the intention of] testing its knowledge (if the gods were so disposed), or [because he was impelled] by an eager longing for truth; his harsh protector was exasperated thereby, and seized the unfortunate enquirer by the hair, bumped his head against the ceiling, then released him, and let him fall to the floor, saying: "There you have the motion of Mars." Public rumour is deplorable, and nothing is more harmful to a good reputation: for rumour is no less persistent in relating trumped-up and shameful stories as in proclaiming the truth. . . . Nevertheless, it is not beyond belief, that Rheticus might have become deranged in mind, seeing that his speculations came to naught, and in a fit of rage bumped his head against the walls.'

37. Kepler is referring to the disruption of the team formed by Tycho Brahe; to the rivalry and obstruction of his son-in-law, Tengnagel; to his

domestic troubles; and to the loss of time caused by his study of the *Stella Nova in Pede Serpentarii* (1606). The loss of time was merely relative, for the treatise in question is most interesting. In it, Kepler publicly stated for the first time that the Sun rotates on its axis, and hence, by analogy, that all the planetary bodies, except the Moon, do so too.

38. On 13 January 1606, Kepler announced to Herwart von Hohenburg that his work was finished (*O.O.*, vol. III, *Proemium*, p. 9); however, Book V was only finished during 1606; in June 1606, he wrote to say, that he proposed writing to the Emperor to ask permission to seek another patron, as the money for publication was not forthcoming (*ibid.*). At last, in December 1606, the Emperor granted 400 gulden, part of which sum Kepler used for his own needs. The year 1607 was spent in putting the manuscript into shape. Kepler had the wood blocks for the diagrams made at Prague in the summer of 1607, and sent them to Frankfurt in August of that year; a copy of the work was sent to the printer (F. Vögelin of Leipzig) in December 1607; printing started in 1608 (at Heidelberg, where Vögelin had a second printing office) after the Emperor had granted a further sum of 500 gulden; printing was completed in 1609. In consideration of his having granted a subvention, the Emperor reserved to himself the ownership of the work, and forbade Kepler to dispose of it without previous approval, stipulating (29 December 1606) that '*Er one Unser vorwissen und bewilligung nymanden Kain Exemplar davon gebe*'; (*O.O.*, vol. III, *Proemium*, p. 9). However, Kepler did not obey the Imperial injunction; the treasury of the Empire being once more empty and his salary (for the year 1609) remaining unpaid, he sold the work to his printer for 3 gulden per copy, as he told Magini (1 February 1610).

PART III. CHAPTER I.

THE *EPITOME ASTRONOMIAE COPERNICANAE*

1. This extension to all the planets as well as to the satellites of Jupiter occurred, *expressis verbis*, only in the *Epitome Astronomiae Copernicanae* (*G.W.*, vol. VII, p. 318; *O.O.*, vol. VI, p. 361; cf. *supra*, pp. 297 f.). However, it is stated very clearly by Kepler in the *Astronomia Nova*: cf. *supra*, p. 215, and the introduction to Lib. IV (*G.W.*, vol. III, p. 272; *O.O.*, vol. III, p. 326):

> '*Pars Qvarta: Investigatio verae mensvrae primae inaeqvalitatis ex cavsis physicis et propria sententia. Quae tertia parte demonstrata sunt, ad omnes planetas pertinet: unde non injuria clavis astronomiae penitioris dici possunt. Quam tanto magis gaudere debemus inventam, quanto certius estnulla alia ratione investigare potuisse, praeterquam per stellae Martis observationes.*'

2. The vague concept of 'moles' was replaced by the precise one of *mass* (quantity of matter): furthermore, as pointed out by Max Caspar (cf.

Astronomia Nova; G.W., vol. III, p. 481), Kepler states his dynamic law of planetary motion (the law of distances) more accurately by noting that it applies, strictly speaking, to the perpendicular component of the radius-vector of motion, and not to the orbital motion itself. This was an extremely important modification; but Caspar is in error in construing this as Kepler's discovery of the true law; in fact, the true law is based on the dynamics of inertia; (linear) *acceleration* is proportional to the force, whereas Kepler's law (or laws) is based on Aristotle's dynamics: (circular or linear) *velocity* is proportional to the motive force; cf. *supra*, p. 319.

3. Such as those of ellipticity of the path and the law of areas. Furthermore, Kepler, now completely freed from obsession with circularity, abandons the whole epicyclic mechanism and restricts himself to libration; on the other hand, he reduces to a minimum the discussions, which are so prominent in the *Astronomia Nova*, on the possibility, or the impossibility, of explaining the celestial motions by the action of planetary souls or intelligences (all the same, he does not deny the animation of celestial bodies, especially of the Earth), and he straightway puts forward a 'magnetic' explanation of them: he even gives a diagram which is reproduced in Fig. 23, p. 299. Then again, in the *Epitome*, he redevelops the theory of the five regular bodies (with appropriate corrections); cf. *Epitome*, Lib. IV, pars. I; (*G.W.*, vol. VII, pp. 273 f.; *O.O.*, vol. VI, pp. 319 f.).

4. I cannot avoid them entirely. The *Epitome* offers a system whose parts are linked together. Furthermore, the persistence in Kepler's mind of certain fixed themes is important, and cannot be ignored by the historian.

5. EPITOME ASTRONOMIAE COPERNICANAE, *Usitatâ formâ Quaestionum et Responsionum conscripta, inq; VII. Libros digesta, quorum* TRES *hi priores sunt de* DOCTRINA SPHAERICÂ ... Lentijs ad Danubium ... Anno MDCXVIII; EPITOMES ASTRONOMIAE COPERNICANAE ... *Liber quartus, Doctrinae Theoricae Primus: Quo Physica Coelestis, Hoc Est, Omnium in Coelo Magnitudinum, motuum, proportionumque, causae vel Naturales vel Archetypicae explicantur* ... Anno MCDXX; ... Libri V, VI, VII, *Quibus propriè* DOCTRINA THEORICA (*post principia libro IV. praemissa*) *comprehenditur.* Avthore Ioanne Keplero, Anno MDCXXI. In Frisch's edition the *Epitome* appears in vol. VI (1860); and in Caspar's edition it appears in vol. VII (1953).

6. The *Epitome Astronomiae Copernicanae* was developed over a number of years, and when published it included the *Harmonice Mundi*; discoveries are used also in the *Doctrina Theorica.* The *Epitome* is both a summary and a text-book. As a summary, Kepler reviews the new astronomy, which, in spite of the upsets which he had caused to the views of Copernicus, he regards as Copernican because, in his opinion, these upsets—the introduction of dynamics into astronomy, abandonment of the principle of circularity in celestial motions—are merely corrections to his tenets, and

are of no more than secondary importance when compared with the true revolution which Copernicus accomplished by asserting the immobility of the Sun and the motion of the Earth. As a text-book, it was intended to replace the text-commentaries of Sacrobosco and of Peurbach which perpetuated the tradition of geocentrism, and which were used in university teaching. Consequently, he adopted in his text-book the 'usual form' of question (objection) and answer, and he took the utmost care to give a systematic (pedagogical) order to his exposition, so that everything should be quite clear; and he avoided bothering his reader with autobiographical confidences and the story of his mistakes and misgivings; in short, his procedure in the *Epitome* is the reverse of that which he adopted in the *Astronomia Nova*.

7. In the foreword of Lib. IV of the *Epitome* (*G.W.*, vol. VII, p. 251; *O.O.*, vol. vi, p. 303), Kepler says:

> 'Ten years ago I published my *Commentaria de Motibus Stellae Martis*. Seeing that this book was printed in a small edition only, and that I had, so to speak, concealed the doctrine of the causes of celestial [motions] in a multitude of numbers (calculations) and the rest of the astronomical display, and as the harder to please [readers] were put off by the price, it seemed to my friends that it would be right and consistent with my duty to write the *Epitome* in which I should give a summary of the doctrine, both physical and astronomical, of the sky, and explain it in simple, clear language, sparing [my reader] the tediousness of proofs.'

Kepler undoubtedly exaggerated the effect of the price on the diffusion, or non-diffusion, of the *Astronomia Nova*. As for the edition being limited to a small number of copies, I regret to say that Kepler has given wrong information here: 1000 copies were, in fact, printed. On the other hand, the obvious disorder of the *Astronomia Nova*—though it was indeed an order dictated by Providence—and its technical difficulty were enough to repel the reader, as Kepler himself realized.

8. Kepler conceived the idea of the *Epitome* at a rather early date, in fact, immediately after publication of the *Astronomia Nova* (1609); but the practical realization of his plan was hindered, first of all by the publication of Galileo's great astronomical discoveries (the *Sidereus Nuncius* appeared in 1609) to which Kepler replied with his *Dissertatio cum Nuncio Sidereo* (1610) and his *Dioptrice* (1611); and secondly, by outside events, the war and the abdication of Rudolph II, as a result of which he lost his position and was obliged to leave Prague in order to settle at Linz (1612). It was not till 1612 that he was able to start work. In 1613, the first part (*Doctrina Sphaerica*) together with much of the *Doctrina Theorica* were ready, and the work was even announced in the catalogue of the Leipzig Fair in the autumn of 1614. In consequence of considerable difficulties, the printing of the first part, which contains the spherical astronomy besides '*physicam*

accuratam explicationem Motus Terrae diurni, ortusque ex eo circulorum Sphaere' in three books, was not completed till 1617. As for the theoretical part (celestial physics and the theory of lunar and planetary motions), it did not appear till 1620 (first part, Book IV) and 1621 (Books V, VI and VII). Kepler took advantage of this delay to include therein his 'third law', which he had discovered in 1618.

9. The 'archetypical' and 'harmonic' number 720, which is the number of divisions of a monochord needed to give the major and minor scales, determines the whole structure of the solar system; cf. *infra*, p. 347.

10. *Epitome*, Lib. IV, pars. I, 1. *De partibus mundi praecipuis*; (*G.W.*, vol. VII, p. 258; *O.O.*, vol. VI, p. 310).

11. *Ibid.*, Lib. I, pars. II; (*G.W.*, vol. VII, p. 51; *O.O.*, vol. VI, p. 143):

> '*Prove that the sphere is found in the image of the Trinity whom we should worship.* In the sphere, there are three [things]: the centre, the surface and the uniformity of the distance [of the radii] which are [so linked together that] if one of them be omitted, then the others will collapse, and yet they are so distinct among themselves that one is not the other. The centre is, as it were, the origin of the spherical figure, for the surface is not conceived as emanating from the centre, but as being derived from the innumerable straight lines drawn in all directions in the intervening gap [in such a way] that no trace of it remains, the [central] point communicating itself in the amplitude of the sphere to the extent of the uniformity of all the distances; this method of generation is quite different from that used by geometers to make themselves understood. You should note that the creation of a single straight line [considered above] is the finite image of this infinite generation of the spherical surface starting from the centre. The centre, by its very nature, is invisible and unknowable, but through the mediation of the uniformity of the intervening gap, it is apparent everywhere in the supremely uniform curvature of the surface. Therefore, then, the surface is the expression and the image of the centre, and in a way its emanation and the path leading to it; he who views the surface, thereby sees the centre, and not otherwise.'

Geometers employ a method of generation which goes from the line to the circle, and from the circle to the sphere; cf. *supra*, p. 144.

12. *Ibid.;* (*G.W.*, vol. VII, p. 259; *O.O.*, vol. VI, pp. 310 f.).

13. Cf. *supra*, p. 154 and n. 58 where Kepler says that without the sphere of the fixed stars there would be no motion, for it is that sphere which defines the space in which these bodies move.

14. As we shall see later (cf. p. 329), the 'phenomena' of planetary motions constitute a harmony with respect to the Sun in so far as they form the object of a 'vision', and therefore are considered only with

respect to the angles. Kepler concludes that this is capable of being seen by an attentive intelligence present 'where the Sun is placed'.

15. Notwithstanding the rôle which he subsequently attributed to the archetypical numbers 720 and 360, Kepler held fast to his view of number and did not ascribe any intrinsic reality to it; cf. *supra*, p. 139 and p. 346.

16. In German the word for Sun—*die Sonne*—is feminine.

17. *Epitome;* (*G.W.*, vol. VII, p. 313; *O.O.*, vol. VI, p. 261). It is interesting to note that Kepler is fully conscious of the reversal of the relationships brought about by Copernicus.

18. *Ibid.;* (*G.W.*, vol. VII, p. 298; *O.O.*, vol. VI, p. 343):

> 'Telescopic observations reveal that the body of the Sun is covered by spots which cross the Sun's disc in 12, 13 or 14 days, moving more slowly at the beginning and end than in the middle of the passage; this clearly proves that the spots are attached to the Sun's surface.'

Kepler is making an implicit criticism here of C. Scheiner, who identified sunspots with clouds of small bodies rotating round the Sun. Scheiner's view was certainly wrong; but Kepler's criticism is valueless.

19. *Ibid.;* (*G.W.*, vol. VII, p. 294; *O.O.*, vol. VI, p. 341).

20. *Ibid.;* (*G.W.*, vol. VII, p. 294; *O.O.*, vol. VI, p. 340):

> 'If there were no solid spheres, it appears that it would be even more necessary to have intelligences to control the motions in the heavens, provided that they are not gods, but are angels or some other rational creature.'

21. *Ibid.;* (*G.W.*, vol. VII, p. 295; *O.O.*, vol. VI, p. 340); cf. *supra*, p. 222.

22. *Ibid.;* (*G.W.*, vol. VII, p. 295; *O.O.*, vol. VI, p. 341). This is what would happen, if God had created the Universe only taking into account pure geometrical relationships.

23. The circle is defined with respect to its centre, the ellipse with respect to its foci.

24. The law of the balance—or of the lever—was the fundamental law of action of material forces in Kepler's view (cf. *supra*, p. 190 and p. 308), and, as such, asserted itself in God's creative act, being a necessity which explains the deviations between empirical reality and the archetypical model. Consequently, in the case of planetary orbits, the law of the balance permitted *concentric* circular orbits only, or elliptical orbits, but excluded eccentric circular orbits. If, therefore, the Creator decided against concentric spheres, for reasons of harmony, he could not avoid ellipses. Here is another example: for archetypical reasons the Earth should complete its circuit round the Sun in 360 days (its rotation on its own axis should be accomplished in one three hundred and sixtieth part of its orbital motion), and the Moon should complete its circuit round the Earth in thirty days. Now, as a result of excitation by light and heat from

the Sun—an inevitable effect because of the very nature of this body—the Earth rotates quicker than it ought, with the result that we have the quite absurd number of 365 (plus a fraction) days in one year. Similarly, with regard to the Moon: the motive action of the Earth, which should be uniform, in fact is not so, seeing that the Sun periodically 'excites' the Earth and reinforces its action: as a result there is inequality in the Moon's motion.

The intervention of material necessity even makes itself felt in the realization of cosmic harmony, particularly in the ratios of the periodic times:

> '*It seems that the proportion between the periodic times is the work of a mind, and not material necessity.* The agreement, so exact and harmonious in itself, between the extreme motions, the slowest and the fastest, of each of the planets, is the work of the supreme and adorable mind or creative wisdom: but the duration of the periodic times, if they were the work of a mind, would have something of the beauty with which the rational proportions are endowed: double, triple and so on. Now, the proportions of the periodic times are ineffable (irrational), and therefore participate in the infinitude in which there is no mental beauty, because there is no ending. Furthermore, these times cannot be the work of a mind (I speak not of the Creator, but of the nature of the motive force) because the time of one period is composed of the unequal times in different parts of the circle. Now, these inequalities in time have their origin in material necessity, and, as it were, in accordance with the principle of the balance.'

25. Cf. *supra*, p. 203.

26. *Epitome;* (*G.W.*, vol. VII, pp. 295 f.; *O.O.*, vol. VI, pp. 341 f.).

27. Cf. *supra*, p. 136. Life is determined by the ability for self-motion; or, more exactly, by the faculty of a vital soul to move the members of its body.

28. It is interesting to note that rotation of the planets about their axes is now asserted as a fact, whereas in *De Stella Nova in pede Serpentarii* (cf. *supra*, p. 436, n. 37) it was put forward only as a possibility or probability. Galileo's astronomical discoveries allowed him to assume that all the planets have satellites—an impossibility if they did not rotate. It is equally of interest to note that, in Kepler's view, the Moon, which has no satellite, does not rotate.

29. *Epitome;* (*G.W.*, vol. VII, p. 297; *O.O.*, vol. VI, p. 342).

30. 'Without positive example' = without any corresponding terrestrial model.

31. Motion of *rotation*, contrary to that of *translation*, not only can be, but is in fact, produced by a soul.

32. *Epitome;* (*G.W.*, vol. VII, p. 297; *O.O.* vol. VI, p. 342).

33. Sesquialteral = in the ratio of 3 to 2. Here, it is the square of the time to the cube of the distance, T^2/R^3; cf. *supra*, p. 338, the history of the discovery of the 'third law'.

34. Double proportion = inversely proportional to the square of the distance; in fact, the motive force increases proportionally (inversely as the distance) as the planet approaches the Sun, and the length along the orbit decreases in the same proportion. It is not the planet's linear velocity, but its angular velocity, which increases in the double proportion.

35. The light and heat emanating from the Sun do not move the planets; a particular *species* would be needed for the purpose; but they reinforce the planets' own motive forces and souls which ensure rotation on their axes.

36. Cf. *supra*, p. 205 and p. 208.

37. *Epitome; (G.W.*, vol. VII, pp. 92 f.; *O.O.*, vol. VI, pp. 178 f.).

38. *Ibid.; (G.W.*, vol. VII, p. 89; *O.O.*, vol. VI, p. 175).

39. The dynamics of *impetus* provides the basis of the discussion and refutation of the objections made by Tycho Brahe and Fabricius against the Earth's motion according to which a cannon-ball could never hit its target, because during its flight the target would have been carried to another position. Cf. my *Études Galiléennes*, III, pp. 26 f.

40. *Epitome; (G.W.*, vol. VII, p. 89; *O.O.*, vol. VI, p. 176).

41. Example from Nicholas of Cusa; cf. Pierre Duhem, *Études sur Léonard de Vinci*, Paris, 1907, vol. II, pp. 187 f.

42. The material inertia of a body corresponds to its mass. According to the theory of *impetus*, as Kepler reminds us, the former is more readily held in a body when its mass and density are greater.

43. The structure of planetary bodies is (as we realize) rather complex (and will become still more so, as will be seen later); it comprises two kinds of fibres (to which a third is added subsequently); they are: (a) straight fibres, parallel to the planet's axis of rotation, whose direction remains constant, and with respect to which the globe is at rest; (b) circular fibres pointing in the direction of its rotatory motion, and which impart motion to the globe; cf. *supra*, p. 208.

44. *Epitome, loc. cit.*

45. *Ibid.; (G.W.*, vol. VII, pp. 90 f.; *O.O.*, vol. VI, pp. 177 f.).

46. *Ibid.; (G.W.*, vol. VII, p. 298; *O.O.*, vol. VI, p. 343).

47. *Ibid.; (G.W.*, vol. VII, p. 298; *O.O.*, vol. VI, p. 343). Cf. *Astronomia Pars Optica*, cap. VI, 1; (*G.W.*, vol. II, p. 199; *O.O.*, vol. II, p. 270):

> '*Quia Solis officium im mundo . . . hoc est, quod cordis in animali . . . vt vitam scilicet huic aspectabili mundo dispenset; animam quoque tanti muneris administram, seu malis facultatem vitalem, in corpore Solis inesse necesse est. Ex huius igitur inhabitatione in corpore densissimo et purissimo, eiusque potentissima vivificatione seu informatione, victoriâ nempe animae*

et subiugatione contumacissimae materiae, lucem resultare consentaneum est, incertum quâ ratione, certum tamen exemplis multarum rerum sublunarium.'

48. The harmony of the planetary motions is *perceptible* only to an intelligence looking at them from the Sun.

49. *Epitome;* (*G.W.*, vol. VII, pp. 299 f.; *O.O.*, vol. VI, pp. 344 f.).

50. *Ibid., loc. cit.* '*On the rotation of the Sun's body about its axis, and the effect on planetary motions.*'

51. *Ibid;* (*G.W.*, vol. VII, p. 300; *O.O.*, vol. VI, p. 345).

52. *Ibid., loc. cit.*

53. It should be noted, however, that Kepler never considered forces as acting as a couple: attraction-repulsion as acting at the same time on the two 'sides' of the planet. In his view, attraction and repulsion combine algebraically, as it were, like + and −, and it is only the resultant which comes into play, *i.e.*, either attraction, or repulsion, but never the two together.

54. *Ibid.*

55. Cf. *supra*, p. 216.

56. *Epitome;* (*G.W.*, vol. VII, p. 301; *O.O.*, vol. VI, p. 345).

57. *Ibid.;* (*G.W.*, vol. VII, p. 302; *O.O.*, vol. VI, pp. 346 f.):

'*Now, if everything be accomplished through natural powers, which oppose and resist the inertia of the material substance, how can the planets maintain their periodic times so that they are always exactly the same with respect to each other?* More easily than if directed by a mind: because the ratio of the motive force to the material substance of the globe which is moved is invariable, it follows that the periodic times are perpetually the same.'

58. *Ibid., loc. cit.*

59. Cf. *supra*, pp. 355 f. and p. 373 on the mistake in Kepler's reasoning.

60. *Epitome;* (*G.W.*, vol. vii, p. 333; *O.O.*, vol. vi, p. 374):

'Quae causa est, cur Sol non aequè fortiter prenset planetam eminus atque comminus?—*Attenuatio ipsa speciei corporis Solaris, major in effluxu longiori quam in breviori: quae attenuatio quamvis sit in proportione intervallorum duplicatâ, hoc est tam in longum quam in latum: operatur tamen solùm in proportione simpla, hoc est, secundùm solam longitudinem: caussae supra sunt dictae.*'

61. *Epitome;* (*G.W.*, vol. VII, p. 302; *O.O.*, vol. VI, p. 346).

62. *Ibid.;* (*G.W.*, vol. VII, pp. 333 f.; *O.O.*, vol. VI, pp. 375 f.).

63. *Ibid.;* (*G.W.*, vol. VII, pp. 335 f.; *O.O.*, vol. VI, pp. 376 f.). These magnetic fibres (indicated on the diagram on p. 306) constitute the third

kind of fibre and must not be confused with the other two kinds already mentioned; cf. n. 43.

64. *Ibid.*

65. *Ibid.;* (*G.W.*, vol. VII, p. 336; *O.O.*, vol. VI, p. 377).

66. This is common sense; but, curiously enough, Descartes expressed the same opinion: a small body on collision with a larger body cannot impart motion to the latter.

67. *Epitome;* (*G.W.*, vol. VII, pp. 337 f.; *O.O.*, vol. VI, pp. 377 f.).

68. The magnetic fibres of the planet have their own proper inertia, a ἀδυναμίαν distinct from ἀδυναμία, peculiar to matter as such, and in virtue of which they tend to preserve the [absolute] direction which God imparted to them at the creation. Kepler leaves philosophy to decide if it be a question here of a true ἀδυναμία, or, on the other hand, a δύναμις.

69. At the stage to which we have now come, Kepler considers that the magnetic fibres rigidly preserve their direction. That side of the planet which is 'friendly towards the Sun' finds itself 'turned towards the Sun' solely on account of being displaced, and not because the planet has been 'converted'.

70. We repeat; in Kepler's dynamics, velocity is proportional to force.

71. Kepler had already assumed this in the *Astronomia Nova*; cf. *supra*, p. 252.

72. Consequently, the 'converting' action as well as the attracting action are exerted during a proportionally shorter time.

73. In the ellipse, by the distance between the two foci. Cf. *Epitome;* (*G.W.*, vol. VII, p. 341; *O.O.*, vol. VI, p. 380): 'The angles of conversion completed by the planets, as well as the forces acting on them are measured by the *sines.* For this reason the ratio of the mean distance *PB* (or *NA* in the ellipse) to one-half of the libration *BA*, performed in one single (the first) quadrant, which is equal to the eccentricity, is the same as that of the semi-diameter of the planetary globe *NQ* taken as the *sinus totus*, to the *sine* of the greatest angle of inclination *MNQ*, which is attained in the same time that the planet moves from its furthest positions *P* to *N*.'

74. *Epitome;* (*G.W.*, vol. VII, p. 366; *O.O.*, vol. VI, p. 403).

75. Kepler's expression may seem strange: is not the planet always affected by the same beam of lines of force (motive rays) which falls on it from the Sun? Does not the inclination of the fibres (or its *sine*) measure, on the whole, the changes in the effect of these forces? No doubt; but in Kepler's view, the effect (motion) is strictly proportional to the cause (force); furthermore, if the beam has fixed dimensions, then one fibre perpendicular to this beam will receive it completely, but if inclined to it will receive it only partially.

76. Kepler is using here the argument of the hypothesis on p. 304, namely, assuming absolute constancy in the direction of the fibres, and ignoring the fact that this is not the case.

77. Kepler ought not to have used these terms: *Sun-seeking* and *Sun-fugitive* which imply a tendency, or real activity, on the part of the fibres, that was definitely eliminated in the *Epitome* and replaced by unilateral action on the part of the Sun. He ought rather to have spoken of fibres, or of *Sun-attracting* and *Sun-repelling* ends.

78. In the quadrants at *I* and *E*.

79. The *sinus totus* is equal to the radius.

80. And not only Keplerian. As I have already had occasion to point out (cf. *supra*, p. 190) the law of the lever, or of the balance, fundamental both to statics and dynamics, the first of the laws known to mathematical physics, was accepted as a model by all who attempted the mathematization of knowledge in the sixteenth and seventeenth centuries. Very often, the results were disastrous. . . . Even Kepler himself refers to the *sine* not only when he really means the *sine*, but also when he means the *sinus complementi* (*cosine*) of the angles in question.

81. Kepler keeps to half the length of the fibres because, as we have seen already, he does not assume the simultaneous action of the Sun's attracting and repelling forces on the planet, or on its fibres; therefore, he considers one half at a time.

82. *Epitome*; (*G.W.*, vol. VII, p. 369; *O.O.*, vol. VI, p. 406). The force of libration having been found to be equal to the sine of the true anomaly (*PAI*), its magnitude will be *AB* (eccentricity) sin *PAI*; the resulting orbit will not be an ellipse; for an elliptical orbit, the force of libration must be proportional to the sine of the eccentric anomaly (*PBI*), and not to the sine of the true anomaly (*PAI*).

83. Angle *MNQ* is the complement of angle *PAN*.

84. Cf. *supra*, p. 306.

85. Caspar, in a note in his edition of the *Epitome*; (*G.W.*, vol. VII, p. 595), expresses Kepler's argument in the following way. In Fig. 24 on p. 306, let the radius of the circle be unity, and designate angle *PAI* by v_1, and angle *PAN* by v_2, then in triangles *BIA* and *BNA*

$$\frac{IB}{AB} = \frac{\sin v_1}{\sin AIB} \quad \text{and} \quad \frac{NB}{BA} = \frac{\sin v_2}{\sin ANB}$$

Because $IB = NB = 1$, therefore,

$$\frac{\sin v_1}{\sin v_2} = \frac{\sin AIB}{\sin ANB}$$

From previous determinations

$$\frac{\sin HIS}{\sin ANB} = \frac{\sin v_1}{\sin v_2}$$

whence angle $HIS <$ angle AIB. Therefore the angle between the radius vector *IA* and the fibres *IH* is $AIH < BIS - \beta$ (eccentric anomaly). As it has been previously assumed that the libration is proportional to the sine of the complement (cosine) of the angle between the radius vector and the direction of the fibres, it now follows that this libration is no longer proportional to the sine of the true anomaly, but to the sine of the eccentric anomaly; therefore, it will be equal to AB (eccentricity) $\sin \beta$, which corresponds to an elliptical orbit.

86. *Epitome;* (*G.W.*, vol. VII, pp. 370 f.; *O.O.*, vol. VI, pp. 406 f.).

87. Always assumed to be infinitely small.

88. Corresponding to the arcs *KX*, *GF*, etc., of the circular orbit.

89. *Epitome;* (*G.W.*, vol. VII, p. 373; *O.O.*, vol. VI, p. 407). With regard to this passage, Caspar comments as follows (*G.W.*, vol. VII, p. 596). In modern notation, Kepler wishes to find the value of

$$\lim_{n \to \infty} \sum_{i=1}^{i=n} \sin \frac{i\beta}{\pi}$$

he relates this to *versed sine* β or $(1 - \cos \beta)$; this last expression is the solution to a definite integral,

$$\int_0^\beta \sin \beta \, d\beta = (1 - \cos \beta)$$

90. Cf. *supra*, p. 229; *Epitome;* (*G.W.*, vol. VII, p. 371; *O.O.*, vol. VI, p. 407):

'*Demonstrationem eiusdem theorematis per numeros et anatomiam circuli, vide tentatam in Comment. Martis, capite. LVII. Ibi loci videbatur haec proportio nonnihil deficere, quia Pappvm nondum legeram. Sed causa fuit, quia primam sagittam sumpsi arcus non satis parui; quod perinde est, ac si in Pappo diuideres superficiem sphaericam in partes non minutiores quàm vnius gradus latitudine. Tunc enim minimae zonae latitudo necessariò prodiret dupla eius, quod verum esset.*'

In any case, the arcs of a circle are equal, but those of an ellipse are not. Does not this destroy the strictness of the proof? Kepler thinks not, and even if there were an error (*turbela*), it would be negligible.

91. *Epitome;* (*G.W.*, vol. VII, p. 371; *O.O.*, vol. VI, p. 407).

92. *Ibid.;* (*G.W.*, vol. VII, p. 372; *O.O.*, vol. VI, p. 408).

93. Kepler introduced this term in connection with the ellipse.

94. Kepler's proof, carried out in the purest style of Apollonius, is rather long; in modern notation it reduces to the following. It is required to find the relationship between *HF* and *GF*. Let *PB* (the semi-axis major) equal unity; designate the eccentricity *BA* by e, the eccentric anomaly *PBG* by β, and the distance of the planet from the Sun by r. Then

$$HF^2 = HA^2 - AF^2 = r^2 - (e + \cos \beta)^2.$$

At aphelion $r = 1 + e$, and the total libration at that point in the path corresponding to the eccentric anomaly β has been proved to be equal to $e(1 - \cos\beta)$; therefore, r (at this point) is equal to

$$1 + e - e(1 - \cos\beta) = 1 + e\cos\beta.$$

Therefore,

$$HF^2 = (1 + e\cos\beta)^2 - (e + \cos\beta)^2 = (1 - e)\sin^2\beta.$$

$GF = \sin\beta$, therefore,

$$\frac{HF}{GF} = \sqrt{1 - e^2},$$

that is to say, the ratio of HF to GF is the same as that of the minor axis to the major axis. Cf. Caspar's note in the *Epitome;* (*G.W.*, vol. VII, p. 597).

95. This approximation, already mentioned by Kepler in the *Astronomia Nova*, was adequate for the astronomical needs of the period.

96. I am conscious of the fact that this expression is modern and not Keplerian.

97. *Epitome*, Lib. V, pars. I, c. 4; (*G.W.*, vol. VII, pp. 375 f.; *O.O.*, vol. VI, pp. 410 f.): '*De mensura temporis seu morae planetae in quodlibet arcu orbitae.*'

98. *Ibid.*

99. In the *Astronomia Nova* (cf. *supra*, pp. 270 f.), Kepler had already made use of the equivalence between an ellipse divided into unequal parts and a circle divided into equal parts. In that instance, he used the method to obtain a measure of the motive forces; in the present instance, he proceeds directly.

100. *Epitome;* (*G.W.*, vol. VII, p. 376; *O.O.*, vol. VI, p. 411). In other words, their lengths will be in the inverse ratio of the minor axis to the major axis.

101. *Ibid., loc. cit.:* '*What is the consequence of this division of the elliptical orbit into unequal arcs?*'

102. *Ibid.*

103. It is curious to note that this incompatibility seems not to have been perceived by the supporters of astronomy of the ellipse; consequently, they seem not to have noticed Kepler's correction. As a result, not only Boulliau and Seth Ward, but also Robert Hooke, and even Leibniz, adhered to the uncorrected formula: velocity inversely proportional to the distance. The same remark applies to the historians of Kepler, with the sole exception of Max Caspar.

104. Unfortunately, Kepler does not tell us *when* he noticed the need to correct his law of velocities. Probably, it happened only when he was writing the *Epitome*. In fact, the *Harmonice Mundi* still makes use of the relationship—orbital velocity proportional to distance: cf. *supra*, p. 333.

105. *Epitome; (G.W.*, vol. VII, p. 377; *O.O.*, vol. VI, pp. 411 f.):

> '*Dictum quidem est in superioribus, divisâ orbita in particulas minutissimas aequales: accrescere iis moras planetae per eas, in proportione interuallorum inter eas et Solem. Id verò intelligendum est non de omnimoda portium aequalitate, sed de iis potissimum, quae rectâ obiiciuntur Soli, vt de* PC, RG *vbi recti sunt anguli* APC, ARG*; in caeteris verò obliquè obiectis intelligendum est hoc de eo solùm, quod de qualibet illarum portionum competit motui circa Solem. Nam quia orbita planetae est eccentrica, miscentur igitur ad eam efformandum duo motûs elementa, vt hactenus fuit demonstratum, alterum est circumlationis versus Solem virtute Solis vnâ, reliquum librationis versus Solem virtute Solis aliâ distinctâ a priori.*'

106. *Ibid.*

107. On the other hand, it would apply to a circular (eccentric) orbit described in accordance with the mechanism of the equant.

108. *Ibid.*; (*G.W.*, vol. VII, p. 377; *O.O.*, vol. VI, p. 412 (diagram on p. 411)).

109. *Ibid.*; (*G.W.*, vol. VII, p. 378; *O.O.*, vol. VI, p. 142).

110. *Ibid., loc. cit.*

111. *AC* is smaller than *AP*; *AG* is greater than *AR*; but their sum is equal to *AP* + *AR*.

112. *G.W.*, vol. VII, p. 598.

113. In the *Epitome*, Kepler no longer gives any formula for the direct determination of orbital velocities.

114. Naturally, this does not imply that anyone can understand every mathematical proof.

115. In Kepler's case, it is the arc and the chord which merge together.

116. *Epitome*; (*G.W.*, vol. VII, p. 378; *O.O.*, vol. VI, p. 413).

117. 'Equal effect' with respect to the motion of circumvolution round the Sun.

118. Cf. *supra*, p. 269. The 'obscurity' resulted, as we now know, from the mistake concerning the relationship between velocity and distance, which mistake Kepler has just corrected.

119. *Epitome*; (*G.W.*, vol. VII, pp. 379 f.; *O.O.*, vol. VI, pp. 413 f.). Cf. *supra*, p. 266.

120. *Ibid.*; (*G.W.*, vol. VII, p. 380; *O.O.*, vol. VI, p. 414):

> '*Why is it necessary to reject the equant?*—Because the equant never gives the truth in a completely [correct manner] unless we ascribe an unequal libratory motion to its centre. For this reason we should depart from the simplicity of Hypotheses and should frame a much more complicated and difficult astronomy for practice than that given in

> Books IV and V [which proceed by] explaining causes; (2) because the part played by this Ptolemaic equant is different in [the theory of] the superior planets from what it is in [that] of the Sun. However, the area of the eccentric circle is applicable everywhere in the same manner; (3) because the circle is far removed from the true causes of the motion, which is closely represented by the area of the eccentric circle, seeing that it [the eccentric circle] is of the same nature as an Ellipse.'

Similarly, according to Kepler, with regard to the other inventions 'that the admirable power of human intelligence' can invent—or has invented—for the representation of phenomena, such as the theory of Fabricius, or the double epicycle of Copernicus, even though they were mathematically equivalent to the new astronomy: they ignore physical causality, and consequently deviate from reality.

121. Attraction is not a general cosmic force; the Sun does not attract the planets as the Earth attracts the Moon; cf. *supra*, p. 413. The only instance where Kepler seems to assume a force of attraction 'similar to a magnetic force' between the Sun and the Earth concerns tides: the waters of the Earth are attracted both by the Moon and the Sun; cf. notes to *Somnium seu de Astronomia Lunaris*; (*O.O.*, vol. VIII, p. 61, n. 202).

122. It does not even seem that a force of attraction keeps the Moon close to the Earth and compels it to follow the latter in its motion round the Sun.

123. Cf. C. Frisch, *Proemium* to the *Epitome*; (*O.O.*, vol. VI, p. 8): '*Hanc vim, non plane magneticam, sed ei simillimam, propter attractionem et repulsionem, inesse putat etiam corporibus planetarum, et Solem trahendo et repellendo retinere illorum corpora retinendoque circumducere, dum se ipsum convertat.*'

124. Kepler's insistence on the Sun's immobility at the *centre of the Universe* reveals the wide gap which separates him from Newton, who, laying down reciprocity of action and reaction as a universal principle, calmly accepted the Sun's mobility. The Newtonian Universe is quite different from Kepler's Cosmos; cf. my *From the Closed World to the Infinite Universe*, Baltimore, 1957.

125. In view of persistent, and recent, misinterpretations (*e.g.*, Lynn Thorndyke, *A History of Magic and Experimental Science*, New York, vol. VII, 1958, p. 30), it is desirable to emphasize this point. It is easy to understand *why* Kepler refused to transform gravific attraction into a cosmic force; it, too, involves a *reciprocal* relationship: (cf. *Somnium*; *O.O.*, vol. VIII, p. 47) '*Gravitatem ego definio virtute magneticae simili, attractionis mutuae.*' which would imply planetary action on the Sun, and, therefore, instability of the latter. Perhaps, that is the reason why the *Epitome* does not refer to the mutual attraction between the Earth and the Moon: the stability of the Earth in its orbit would not be affected thereby.

126. The system described in the *Epitome* could be considered as a

prefiguration of the Universe of Descartes with its planetary vortices steeped in the vast whirlpool of the Sun.

127. Orbital velocity in the *Astronomia Nova*; lateral velocity, impressed on the planet by the *species motrix* (radius vector) in the *Epitome*; cf. *supra*, p. 319.

128. As previously pointed out (p. 272), this expression is non-Keplerian.

CHAPTER II. THE *HARMONICE MUNDI*

1. Iohannis Keppleri, *Harmonices Mundi Libri V. Qvovrvm Primus Geometricvs, De Figurarum Regularium, quae Proportiones Harmonicas constituunt, ortu & demonstrationibus. Secundus Architectonicvs, seu ex Geometria Figvrata, De Figurarum Regularium Congruentia in plano vel solido: Tertius propriè Harmonicvs, De Proportionum Harmonicarum ortu ex Figuris; deque Naturâ & Differentiis rerum ad cantum pertinentium, contra Veteres: Quartus Metaphysicvs, Psychologicvs & Astrologicvs, De Harmoniarum mentali Essentiâ earumque generibus in Mundo; praesertim de Harmonia radiorum, ex corporibus coelestibus in Terram descendentibus, eiusque effectu in Natura seu Anima sublunari Humana: Quintus Astronomicvs Metaphysicvs, De Harmoniis absolutissimis motuum coelestium, ortuque Eccentricitatum ex proportionibus Harmonicis* . . . Lincii Austriae, MDCXIX. The *Harmonices Mundi Libri* V were published by C. Frisch in vol. v of *O.O.*, and by M. Caspar in vol. VI of *G.W.*

As we have already seen, Kepler was always interested in 'harmonic' problems and their application to cosmology (cf. *supra*, p. 396).

In 1600 (14 July) he informed Herwart von Hohenburg that he would have already finished his book on the Harmony of the Universe if he had not been too busy with his work with Tycho Brahe. In 1605, in a letter to Heydonus at London, in which Kepler informed his correspondent of his decision to dedicate his work to James I of England, he wrote: 'May God deliver me from Astronomy, so that I can devote all my attention to my work on the Harmonies'; (*G.W.*, vol. XIV, p. 233). However, he had to wait till 1618 (February) before he could seriously resume his project. Success was not long delayed.

2. Cf. *Epitome*; (*G.W.*, vol. III, p. 275; *O.O.*, vol. III, p. 323):

'*Mundi mobilis archetypus constat non tantum ex quinque figuris regularibus, quibus curricula planetarum et cursorum numerus defineretur, sed etiam ex proportionis harmonicis, quibus cursus ipsi ad quandam veluti musicae caelestis ideam seu consensus harmonici sex vocum attemperandi fuerent.*'

3. Kepler had already spoken about it in 1600 in a letter to Herwart von Hohenburg. Cf. *Harmonice Mundi*, Lib. V, cap. I; (*G.W.*, vol. VI, p. 293; *O.O.*, vol. v, p. 272) and *ibid.*, Lib. II; (*G.W.*, vol. VI, pp. 78 f.; *O.O.*, vol. v, pp. 126 f.). Cf. *supra*, p. 332.

4. According to Kepler, music is based on geometry, and not on arithmetic; but is not to be identified with it, for geometry embraces all relationships between magnitudes (expressible and non-expressible, and infinite in number), whereas music is based on a small number, namely, seven very simple, fundamental ratios between the lengths of strings, or the divisions of a monochord. Seeing that these numerical ratios do not form the basis of harmony—according to Kepler, nothing is based on pure numbers which represent only the expressible ('effable'), and therefore rational and cognizable, part of geometrical magnitudes and ratios, in contrast with 'ineffable', and therefore irrational and incognizable ratios and magnitudes—Kepler sought the justification for harmony in geometrical structures, particularly, in the division of the circle (which represents the monochord) by inscribed polygons. Certain of these polygons are 'constructible' (by means of ruler and compass), others, not. On the one hand, this constructibility is identified by Kepler with rationality—we 'know' only that which is constructible, he said in 1605; as for that which is not constructible, it is impossible, not only for us, but even for God, to know (*scire*) it. Consequently, we cannot 'know' the side of the heptagon. On the other hand, the constructibility of the figures corresponds to the 'consonance' of sounds. By restricting himself to 'directly constructible' polygons, *i.e.*, those which do not imply the previous construction of another figure, Kepler was finally able to co-ordinate the notes and harmonic ratios with the sides of the constructible polygons. Then, he declared that only these ratios are to be taken into account for the construction of the Universe, and that God *could not* make use of any others.

Kepler's theory of harmony and his musicology are most interesting, but an account of them would take us too far from the proper subject of our study.

5. Thus, in particular, he was obliged to allot them elliptical orbits, instead of circular ones.

6. At this time (in 1599), Kepler had not yet realized that the harmony of the spheres should be related to the Sun, and was not, first and foremost, only a solar *phenomenon*, that is to say, that the planetary motions constituted a harmony only in so far as they were *viewed* from the Sun, and consequently, that it was angular velocities that should be taken into account. Furthermore, he had made the mistake of taking the ratio between the *orbital* velocities of the planets, allotting to them the following numbers: 3 (Saturn), 4 (Jupiter), 8 (Mars), 10 (Earth), 12 (Venus), 16 (Mercury), and interposing the following musical intervals: an octave between Jupiter and Mars, an octave plus a major sixth between Saturn and Earth, a minor sixth between Earth and Mercury, etc. From the velocities, the periodic times, and the harmonic ratios he calculated the ratios of the distances of the planets from the Sun; and, in fact, obtained slightly better results than those in the *Mysterium Cosmographicum* (letter to Maestlin

dated 29 August 1599). He was so enchanted thereby, that when Herwart von Hohenburg objected that it was based merely on suppositions and *Ahnungen* (hunches), he proudly replied that suppositions and hunches are not necessarily sources of error, for 'man is the image of God, and it is quite possible that in regard to certain things that make the ornament of the Universe, he has the same opinions as God. For the Universe partakes of quantity, and the mind of man—something which transcends the Universe—apprehends nothing so well as quantities, for a knowledge of which he was obviously created.' (Letter to Herwart von Hohenburg dated 14 September 1599; (*G.W.*, vol. XIV, p. 62); cf. *supra*, p. 379, n. 15).

7. *Harmonice Mundi*, Lib. V, cap. III; (*G.W.*, vol. VI, pp. 296 f.; *O.O.*, vol. V, pp. 274 f.).

8. A difficult undertaking! As Kepler reminds us in the preface to Book IV of the *Epitome*; (*G.W.*, vol. VII, p. 253; *O.O.*, vol. VI, p. 305): 'Academies are established to govern the studies of students, and are opposed to frequent changes in the curricula; consequently, it often happens that choice falls not on things that are nearer the truth, but on those that are simpler.' Geocentric astronomy was much simpler than that of Copernicus, at least for a sixteenth century mind; cf. *supra*, p. 104, n. 3.

9. Undoubtedly, this refers to faith in the power of reason.

10. A curious statement, which is not historically true.

11. Diagram from *G.W.*, vol. VI, p. 298; *O.O.*, vol. V, p. 276. Cf. the scheme from the *Mysterium Cosmographicum*, p. 146 and p. 148.

12. Another curious statement, which is just as inaccurate as the former one: the planetary eccentricities, except those of Mercury and Mars, are extremely small, and the changes in distance are insignificant.

13. Cf. *supra*, p. 143 and *Epitome*; (*G.W.*, vol. VII, pp. 275 f.; *O.O.*, vol. VI, pp. 323 f.).

14. Cf. *Epitome*; (*G.W.*, vol. VII, pp. 273 f.; *O.O.*, vol. VI, pp. 321 f.).

15. Divergent and convergent intervals; cf. *supra*, Table A, p. 335. A consideration of the maximum and minimum distances is equivalent to that of the eccentricity, and had already occurred in the *Mysterium Cosmographicum* and the attempts of 1599–1600.

16. Cf. *Epitome*, Lib. IV, pars. I; (*G.W.*, vol. VII, p. 275):

> '*If the intervals approach so closely to the ratio of the figures, why is there still a certain difference?*—Because the archetype of the mobile Universe is formed not only from the five regular solids from which the planetary paths and the number of motions would be fixed; but is also formed by the harmonic proportions of the six parts in agreement with which the motions themselves have been tuned to the idea of a certain celestial music, or of a six-part harmonic chord. Seeing that this musical ornament required some differentiation between the slowest and the quickest motions of each of the planets, which differentiation was achieved by a

change in the distance between the planet and the Sun; and as the amount or proportion of this change must be different for the different planets, it was necessary for the intervals [distances] fixed by the figures [regular solids] without any change in uniformity to be modified somewhat by the licence of the composer in order to realize the harmony of the motions. This was all the more necessary as the distances fixed by the regular solids were "ineffable".'

17. Kepler remains faithful to his dynamical principle. Cf. *supra*, p. 318. Cf. Max Caspar, *Nachbericht*, to his edition of the *Harmonice Mundi*; (*G.W.*, vol. VI, p. 545):

'*Es mag auffallen, dass Kepler hier, wo er seine Vorstellungen über den Mechanismus der Planetenbewegungen entwicklet, um die für seine späteren Ausführungen notwendigen Unterlagen zu bekommen, statt seines* Flächensatzes *den Satz anführt: Die Geschwindigkeit des Planeten in seiner Bahn ist umgekehrt proportional seinem Abstand von der Sonne. Die beiden Sätze—der letztere mag kurz als* Radiensatz *bezeichnet werden—decken sich natürlich nicht. Die dem Radiensatz zugrunde liegende Vorstellung ist gewesen und hat ihn zu seinen glänzenden Entdeckungen geführt; sie begleitet ihn sein Leben lang und war, wie die vorliegende Stelle beweist, in ihm stets lebendig. Den Flächensatz führte er in der* Astronomia Nova *nur als bequemen Ersatz für den zu practischer Rechnung höchst ungeeigneten Radiensatz ein, wobei er sich aber des Unterschiedes beiden Sätze sehr wohl bewusst war. Den theoretischen, nicht nur practischen Ubergang zum Flächensatz vollzog er erst im Buch V seiner 1621 erschienenen* Epitome Astronomiae Copernicanae. *für die Bewegung der Planeten in den Apsiden, die bei Keplers vorligenden Untersuchungen die Hauptrolle spielen, fallen übrigens die beiden Sätze zusammen.*'

Cf. *G.W.*, vol. III, pp. 442 and 481; cf. fig. 28, *supra*, p. 314.

18. This is equivalent to the law of areas.

19. The distances correspond to the lengths of a vibrating string; the velocities correspond to the number of vibrations. In fact, there is no harmony between the distances, nor between the orbital velocities of the planets, but only between their angular velocities with respect to the Sun, that is to say, between the angular velocities of their motions as *seen* from the Sun. Seeing that it is inconceivable that God should have created a harmony perceptible to an intellect without having at the same time created an intelligence capable of discerning the harmony, its presence in the Sun is a necessary consequence; and it must be either the Sun itself, or intelligent beings that inhabit the Sun as we do the Earth.

20. My expression, not Kepler's. In Kepler's view, there are no real 'sounds', *i.e.*, audible ones, in the heavens; cf. *Harmonice Mundi*; (*G.W.*, vol. VI, p. 311; *O.O.*, vol. V, p. 286): '*Soni in coelo nulli existunt.*'

21. Cf. *Harmonice Mundi*; (*G.W.*, vol. VI, p. 312; *O.O.*, vol. V, p. 287); cf. also J. L. E. Dreyer, *History of the Planetary Systems*, Cambridge, 1906, p. 406.

22. *Ibid*; (*G.W.*, vol. VI, p. 318; *O.O.*, vol. V, p. 291). The table given in the text has been quoted from J. L. E. Dreyer, *op. cit.*, p. 408. In the original it appears as follows:

Motus	*Prim.*	*Sec.*
Perihelii ☿ septimum subdupla, seu 128va	*3.*	*0*
Aphelii ☿ sextum subdupla, seu 64ta	*2.*	*34–*
Perihelii ♀ quintum subdeupla, seu 32da	*3.*	*3. +*
Aphelii Veneris quintum subdupla, seu 32da	*2.*	*58–*
Perihelii Terrae quintum subdupla, seu 32da	*1.*	*55–*
Aphelii Terrae quintum subdupla, seu 32da	*1.*	*47–*
Perihelii Martis quartum subdupla, seu 16ma	*2.*	*23–*
Aphelii Martis tertium subdupla, seu 8va	*3.*	*17–*
Perihelius Jovis subdupla	*2.*	*45*
Aphelius Jovis subdupla	*2.*	*15*
Perihelius Saturni	*2.*	*15*
Aphelius Saturni	*1.*	*46*

23. Cf. J. L. E. Dreyer, *op. cit.*, p. 408.

[From further consideration of the celestial music, Kepler concluded that the various authentic and plagal modes were distributed amongst the planets. After some 'adjustments', he suggested that each planet expressed itself best in the following modes:

Saturn in the seventh or eighth; Jupiter in the first or second; Mars in the fifth or sixth; Earth in the third or fourth; Mercury in any of the modes, because of the wide range of its tune; but Venus, with its extremely limited melodic range of less than a minor semitone, cannot be allotted to any particular mode, though Kepler favoured the third or fourth modes, because, compared with the other planets, Venus is pitched on the note *E*. The corresponding tunes appear in the original (*G.W.*, vol. VI, p. 422; *O.O.*, vol. V, p. 294) as follows:

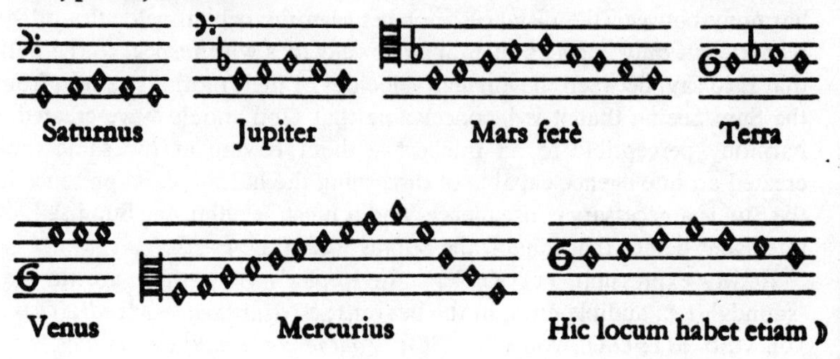

It is to be noted that the tunes given in modern notation on p. 337 are not equivalent in every case to those given above: the two sets of tunes are related to different conditions. The one relates to the individual planetary tunes as deduced from Table B (p. 337); the other relates to the same tunes after fitting them to the modes. (Translator's note.)]

In a marginal note, Kepler remarks that the tune given by Earth, and which is limited to the notes *MI* and *FA* is most apt: for is not *MI-seria* and *FA-mes* (Misery and Famine) our lot on Earth?

24. This table, together with others given by Kepler, contains much information on celestial music which is not given here. Though extremely interesting, it has no direct bearing on astronomy, and will not be considered further.

25. *Harmonice Mundi;* (*G.W.*, vol. VI, p. 302; *O.O.*, vol. V, p. 279):

'*Hic aliqua pars mei* Mysterij Cosmographici, *suspensa ante 22 annos, quia nondum liquebat, absolvenda, et huc inferenda est. Inventis enim veris Orbium intervallis, per observationes* Brahei, *plurimi temporis labore continuo; tandem, genuina proportio Temporum periodicorum ad proportionem Orbium*

sera quidem respexit inertem.
Respexit tamen et longo pòst tempore venit;

eaque si temporis articulos petis, 8 Mart. hujus anni millesimi sexcentesimi decimi octavi animo concepta, sed infoeliciter ad calculos vocata, eòque pro falsâ rejecta, denique 15. Maji reversa, novo capto impetu, expugnavit Mentis meae tenebras; tantâ comprobatione et laboris mei septendecennalis in Observationibus Braheanis, et meditationis hujus, in unum conspirantium; ut somniare me, et praesumere quaesitum inter principia, primò crederem. Sed res est certissima exactissimaque, quòd proportio quae est inter binorum quorumcunque Planetarum tempora periodica, sit praecisè sesquialtera proportionis mediarum distantiarum, *id est Orbium ipsorum.*'

26. In modern notation $T^2/R^3 = C$.

27. Cf. J. B. Delambre, *Histoire de l'Astronomie Moderne*, vol. I, p. 356:

'*Cette loi qui a couté tant de temps et de calculs à Kepler se trouverait aujourd'hui avec une extrême facilité. Il se proposait de trouver le rapport des mouvements avec les distances ou*

$$\frac{T}{t} : \frac{R}{r}; \quad soit \quad \left(\frac{T}{t}\right) = \left(\frac{R}{r}\right)^x.$$

Kepler essaya x = 1, 2, 3, etc., il essaya des nombres fractionnaires . . .;'

R. Small, *An Account of the Astronomical Discoveries of Kepler*, p. 299: 'His trials for this purpose were various and repeated: he first employed himself in comparing the ratios of the simple distances, or times, with those of the regular solids in geometry, and with the division of musical chords; it next occurred to him, on the 8th of March 1618, that, instead of comparing together the simple distances and times, he should compare the numbers expressing their similar powers, such as their squares or their cubes, etc.; and lastly he made the very comparison on which his discovery was founded, between the squares of the times and the cubes of the distances.'

28. The motive force, and therefore the velocity, decreases proportionally with the distance, and the distance to be traversed increases proportionally with the latter.

29. It was more natural for Kepler to do so, than it would be for us, seeing that exponential notation was not known to him.

30. All the more so, seeing that Kepler, in the *Astronomia Nova*, was explaining the differences between planetary velocities by the differences between their *moles*; cf. *supra*, p. 203.

31. Cf. *supra*, pp. 354 f.

32. Cf. *Harmonice Mundi*; (*G.W.*, vol. VI, p. 308; *O.O.*, vol. v, p. 384):

'. . . *conjectura probabilissima (quippe nixa demonstrationibus geometricis et doctrina de causis motuum planetariorum, tradita in Commentaris Martis) planetarum moles corporum esse in proportione periodicorum temporum, sic ut Saturni globus sit trigecuplo circiter major globo Telluris. Jupiter duodecuplo, Mars minus duplo, Terra globi Veneris sesquialtero major, Mercurialis globi quadruplo major.*'

Kepler adds that there is nothing harmonic about these ratios.

33. In the *Harmonice Mundi* as well as in the *Epitome*, *moles* is used with the precise meaning: *volume* = *spatium.*

34. Cf. *supra*, pp. 353 f.

35. *Harmonice Mundi*; (*G.W.*, vol. VI, pp. 357 f.; *O.O.*, vol. v, p. 318). Cf. J. L. E. Dreyer, *History of the Planetary Systems*, Cambridge, 1906, p. 409; E. F. Apelt, *Johann Keplers astronomische Weltansicht*, pp. 93 f.

36. *Harmonice Mundi*, Lib. V, cap. IX; (*G.W.*, vol. VI, p. 361; *O.O.*, vol. v, p. 322).

37. *Ibid.*; (*G.W.*, vol. VI, pp. 362 f.; *O.O.*, vol. v, p. 323):

'*Gratias ago tibi Creator Domine, quia delectasti me in facturâ tuâ, et in operibus manuum tuarum exultavi. En nunc opus consummavi professionis meae, tantis usus ingenij viribus, quantas mihi dedisti; manifestavi gloriam operum tuorum hominibus, istas demonstrationes lecturis, quantum de illius infinitate capere potuerunt angustiae Mentis meae; promptus mihi fuit animus ad emendatissimè philosophandum: si quid indignum tuis consiliis prolatum à me, vermiculo, in volutabro peccatorum nato et innutrito, quod scire velis homines: id quoque inspires, ut emendem: si tuorum operum admirabili pulchritudine in temeritatem prolectus sum, aut si gloriam propriam apud homines amavi, dum progredior in opere tuae gloriae destinato; mitis et misericors condona; denique ut demonstrationes istae tuae gloriae et Animarum salati cedant, nec ei ullatenus obsint, propitius efficere digneris.'*

The book closes as follows (*G.W.*, vol. VI, p. 368; *O.O.*, vol. V, p. 327):

> '*Magnus Dominus noster, et magna virtus ejus et Sapientiae ejus non est numerus; laudate eum coeli, laudate eum Sol, Luna et planetae, quocunque sensu ad percipiendum, quacunque linguâ ad eloquendum Creatorem vestrum utamini; Laudate eum Harmoniae coelestes, laudate eum vos Harmoniarum detectarum arbitri: (Tuque ante omnes Moestline foelici senecta, namque tu solebas has dictis animare speque curas): lauda et tu anima mea dominum Creatorem tuum, quamdiu fuero: namque ex ipso et per ipsum et in ipso sunt omnia χαὶ τὰ αἰσθητὰ χαὶ τὰ νοεορὰ; tam ea quae ignoramus penitus, quàm ea quae scimus, minima illorum pars; quia adhuc plus ultrà est. Ipsi laus, honor et gloria in saecula saeculorum. Amen. Absolutum est hoc opus' die 17.27. Maji, anno MDCXVIII.*

38. *Ibid.*, preface; (*G.W.*, vol. VI, p. 289; *O.O.*, vol. V, pp. 268 f.):

> '*Quod ante duos et viginti annos, primum atque figuras quinque solidas inter orbes coelestas reperi, sum auguratus . . . totam Harmonices naturam, quantaquanta est, cum omnibus suis partibus, libro III. explicatis, inter Motus coelestes reperiri; non eo quidem modo, quem ego conceperam animo; pars haec est non postrema mei gaudij; sed diversissimo alio, simulque et praestantissimo et perfectissimo . . . Jam postquàm à mensibus octodecim prima lux, à tribus dies justa, à paucissimis verò diebus, Sol ipse merus illuxit contemplationis admirabilissimae; nihil me retinet, lubet indulgere sacro furori, lubet insultare mortalibus confessione ingenuâ, me vasa aurea Aegyptiorum furari, ut Deo meo Tabernaculum ex ijs construam, longissimè ab Aegypti finibus. Si ignoscitis, gaudebo; si succensetis, feram: jacio en aleam, librumque scribo, seu praesentibus, seu posteris legendum, nihil interest: expectet ille suum lectorem per annos centum; si Deus ipse per annorum sena millia contemplatorem praestolatus est.*'

CHAPTER III. HARMONY OF THE UNIVERSE IN THE *EPITOME*

1. Namely, his solving of the problem of the relationships between the distances, and the discovery of the *Harmony*.

2. *Epitome*, Lib. IV, Pars. prima, cap. IV: *De praecipuorum mundi corporum inter se proportionibus*; (*G.W.*, vol. VII, pp. 276 f.; *O.O.*, vol. VI, pp. 324 f.).

3. It must be remembered that the *Epitome* is a textbook, written in the classical form of question (by the *discipulus*) and answer (by the *magister*).

4. Cf. *supra*, p. 153. The whole of chap. x (Epilogue) of Book V of the *Harmonice Mundi* is a Hymn to the Sun, which has been borrowed almost entirely from Proclus. As regards the *Epitome*, we have seen that Kepler repeats there the theologico-mystical considerations (the Universe as an expression of the Holy Trinity) of his *Mysterium Cosmographicum*.

5. *Epitome*; (*G.W.*, vol. VII, p. 277; *O.O.*, vol. VI, p. 325).

6. *Ibid.*; (*G.W.*, vol. VII, p. 277; *O.O.*, vol. VI, p. 325).

7. *Ibid.*

8. This means that it is 'irrational', 'ineffable', unknowable. Cf. *Harmonice Mundi* (*G.W.*, vol. VI, p. 61; *O.O.*, vol. V, p. 111); cf. *supra*, p. 450, n. 4.

9. *Ibid.*; (*G.W.*, vol. VI, pp. 136 f.; *O.O.*, vol. V, pp. 159 f.).

10. *Epitome*; (*G.W.*, vol. VII, pp. 278 f.; *O.O.*, vol. V, p. 326).

11. If we follow Caspar (*G.W.*, vol. VI; *Nachbericht*, p. 587) and designate the semi-diameters of the Sun, the Earth and the Moon by C_s, C_t, C_l respectively, and those of the Earth and the Moon by R and r respectively, then

$$\frac{R}{C_s} = \frac{r}{C_l} = 229; \qquad \frac{C_s{}^3}{C_t{}^3} = \frac{R}{C_t}; \qquad \frac{C_t{}^3}{C_l{}^3} = \frac{r}{C_t}.$$

The value obtained by Kepler for the distance between the Earth and the Sun is better than that known to the Ancients. Nevertheless, it is still about one-seventh of the true value.

12. *Epitome*; (*G.W.*, vol. VII, p. 279; *O.O.*, vol. VI, p. 326).

13. Cf. the Introduction to J. L. E. Dreyer's edition of Tychonis Brahe, *Opera Omnia*, vol. i, p. xl.

14. It is strange that the 'superior' planets remain *superior* for the Copernican Kepler.

15. *Epitome*; (*G.W.*, vol. VII, p. 279; *O.O.*, vol. VI, p. 327). It should be noted that Kepler does not speak here of attraction between the Earth and the Moon.

16. 'The body of the Earth' and 'the body of the Moon' mean their volumes; cf. n. 11.

17. Cf. the diagram in *G.W.*, vol. VII, p. 280. In the *Mysterium Cosmographicum*, Kepler regarded Mercury as being the Moon of the Sun.

18. Kepler's *Hipparchos* remained unfinished, and was first published by Frisch (*O.O.*, vol. III).

19. *Epitome*; (*G.W.*, vol. VII, p. 281; *O.O.*, vol. VI, p. 328). Here, as everywhere, Kepler's union of the firmest apriorism with the strictest epiricism is to be noted.

20. *Ibid.*

21. The surfaces are proportional to the square, and the bodies [volumes] are proportional to the cube of the radii.

22. Kepler had assumed this in the *Harmonice Mundi*. Cf. *supra*, p. 326: the volumes (cube) would be proportional to the surfaces (square), and the radii proportional to the intervals.

23. *Epitome*; (*G.W.*, vol. VII, p. 282; *O.O.*, vol. VI, p. 329).

24. *Ibid.*

25. The non-similarity of celestial bodies, that is to say, their qualitative difference, explains the absence of mutual attraction between them.

26. The density and rarity of celestial bodies had already been taken into consideration by Kepler in the *Astronomia Nova* (cf. *supra*, pp. 194 f.); but had not been systematically distinguished from, and compared with, the size (*moles*) of these bodies.

27. Kepler's conception: mass = density × volume = quantity of matter, is exactly the same as that of Newton; in both instances it is proportional to the weight and inertia (or *vice versa*) of the body in question; but, whereas with Newton the mass of the body is inversely proportional to its acceleration (produced by a given force), with Kepler it is inversely proportional to the velocity. We could say that when dealing with infinitesimals, that is to say, at the start of the motion, the two concepts are equivalent. On the other hand, we should add that Keplerian inertia being resistance to *motion*, and Newtonian inertia being resistance to the change of *state* of motion, it follows that acceleration can only be positive according to Kepler, whereas according to Newton it can be negative as well, *i.e.*, it can be a retardation.

28. *Epitome*; (*G.W.*, vol. VII, p. 283; *O.O.*, vol. VI, p. 330).

29. Cf. *supra*, p. 342.

30. *Epitome*; (*G.W.*, vol. VII, p. 284; *O.O.*, vol. VI, p. 330); cf. *supra*, p. 456, n. 36.

31. Remus Quietanus, the Emperor's physician.

32. Kepler wrote *ponderosior*, which is not quite correct. In fact, it is not a question of the *weight* of the planet (strictly speaking, a planet has none, seeing that *weight* results from the attraction of one body by another; cf. *supra*, p. 194, and n. 33), but of a *quasi*-weight, *i.e.*, of *inertia*, depending on the mass = quantity of matter in Saturn.

33. Keplerian matter is not the same as Aristotelian matter, which is only potentiality; whereas the former is extended and impenetrable, although capable of having different densities.

34. The ratios between the volumes (v), masses (m), densities (d) and distances (r) (from the Sun) are therefore as follows:

$$\frac{v_1}{v_2} = \frac{r_1}{r_2}; \qquad \frac{m_1}{m_2} = \frac{\sqrt{r_1}}{\sqrt{r_2}}; \qquad \frac{d_1}{d_2} = \frac{\sqrt{r_2}}{\sqrt{r_1}}.$$

35. Kepler is referring to his *Auszug aus der Messekunst Archimedis*, published at Linz, 1616; (*O.O.*, vol. V, pp. 609 f.).

36. *Epitome*, Lib. IV, Pars. II; (*G.W.*, vol. VII, p. 291; *O.O.*, vol. VI, p. 337):

'*In what way is the measure of the periodic times which you have assigned to the moving bodies related to the orbits in which they move?*—The ratio between the times is not equal to the ratio between the orbits, but is greater for the primary planets, the former being in the sesquilateral [$\frac{3}{2}$] proportion of the latter. That is to say, if the cube root of 30 years for Saturn and 12 years for Jupiter be multiplied (raised) as the square, then the squares of these numbers will give the exact measure of the orbits of Saturn and Jupiter. Similarly, if those orbits which are not close to each other be compared, the result is the same. For example, Saturn revolves in 30 years and the Earth in one. The cube root of 30 is very nearly 3·11 and the cube root of 1 is 1; the squares of these roots are 9·672 and 1. Therefore, the orbit of Saturn is to the orbit of Earth as 9672 to 1000; if more accurate times be taken, then a more accurate result will be obtained.'

Cf. *supra*, p. 340.

37. *Ibid.*; (*G.W.*, vol. VII, pp. 306 f.; *O.O.*, vol. VI, pp. 350 f.).

38. Proportion divided into two parts = proportional to the square roots.

39. *Ibid.*; (*G.W.*, vol. VII, p. 307; *O.O.*, vol. VI, p. 351).

40. *Ibid.* It is strange that Kepler, as in the *Astronomia Nova* (cf. *supra*, p. 411, n. 25), does not take into account the linear velocity of the *species* or motive force. The latter increases in direct proportion to the distance from the Sun. It follows that the quantity of the force passing through a point or segment of the sphere is everywhere the same. If Kepler had taken this fact into account, he would have been able to deduce therefrom the attenuation of the motive force in *direct* proportion to the distance from the Sun without infringing the laws of optics according to which light is attenuated in the double proportion; on the other hand, by taking into account the volumes of the planets, it follows that they all experience the same action from the *species*.

41. *Rationes Copernici*—method of calculating planetary distances.

42. *Quasi*—stationary with respect to the Moon.

43. The sphere of the Earth.

44. This is the explanation of Keplerian *inertia*; it is, therefore, a function of the immobility of the celestial vault, which is the place of the Universe.

45. $\frac{\text{Radius of Sun}}{\text{Radius of solar system}} = \frac{\text{Radius of solar system}}{\text{Radius of the sphere of the fixed stars}}$

46. Use of the telescope, which has diminished, or even reduced the apparent size of the (fixed) stars to a point by depriving them of their halo, has made it possible to decrease their true size (calculated from the

apparent dimensions) also. Kepler refused to assimilate the stars to the Sun, as did Giordano Bruno, and strenuously maintained the uniqueness of the latter; cf. my *From the Closed World . . .*, pp. 80 f.

47. Cf. *Epitome*; (*G.W.*, vol. VII, p. 288; *O.O.*, vol. VI, p. 334).

48. The idea that there is as much matter in the Sun as in the whole of the mobile Universe was an old conviction of Kepler's: it had previously appeared in *Astronomiae Pars Optica*, cap. VI, 1; (*G.W.*, vol. II, p. 199; *O.O.*, vol. II, p. 270); but he had not, at that time, extended it to the sphere of fixed stars:

> '*Quòd autem certam materiae in corpore Solis quantitatem defino, aequiparans eam residuo materiae, qua secundum diuinum Mosen extensio, seu insufflatio facta est inter aquas et aquas: id videtur sic requirere proportionis concinnitas: vt cuius vis vniversum illud spatium permeare debuit, idem tantum receperit corporis, quantum in vniverso illo spatio inest.*'

49. Cf. *Epitome*; (*G.W.*, vol. VII; p. 288; *O.O.*, vol. VI, p. 334; cf. *G.W.*, vol. VII, p. 52; *O.O.*, vol. VI, p. 144).

> '*Every spatial form exists in matter; what, then, is the matter by which the Universe is encompassed exteriorly?*—This question cannot be decided by the act of seeing; it is right, therefore, that we follow authority, which teaches us that the stars are all [contained] within a thickness called *raquia* in Hebrew, whose upper [surface] is covered with waters, that is to say, that the stars are [contained] in an aqueous sphere above the extremely rarified aetheric aura; and if anyone should wish to affirm that this water is concrete and crystalline as a result of the freezing caused by its great distance from the Sun, then he is allowed to do so by Copernican astronomy, provided that he restricts himself to appearances and does not use this sphere [to explain planetary motions].'

CONCLUSION

1. *Mysterium Cosmographicum*; preface (letter of dedication) to the second edition, 20 June 1621 (*O.O.*, vol. I, p. 102):

> '*Etsi vero tunc oppido juvenis eram, primumque hoc astronomicae professionis tyrocinium edebam, successus tamen ipsi consecutorum temporum elata voce testantur, nullum admirabilius, nullum felicius, nullum scilicet in materia digniori positum esse unquam a quoquam tyrocinium. Non enim haberi debet illud nudum ingenii mei commentum (absit hujus rei jactantia a meis, admiratio a lectoris sensibus, dum sapientiae creatricis tangimus Psalterium heptachordum) quandoquidem non secus ac si dictatum mihi fuisset ad calamum oraculum coelitus delapsum, ita omnia vulgati libelli capita praecipua et verissima statim (quod solent opera Dei manifesta) fuerunt agnita ab intelligentibus: et per hoc 25 annos mihi telam pertexenti restaurationis astronomiae (coeptam a Tychone Brahe e nobilitate*

Danica celebratissimo astronomo) facem non unam praetulerunt: denique quidquid fere librorum astronomicorum ex illo tempore edidi, id ad unum aliquod praecipuorum capitum hoc libello propositorum referri potuit, cujus aut illustrationem aut integrationem contineret.'

2. Cf. *supra*, p. 262.

3. Letter from Maestlin to Kepler, 21 September 1616; (*G.W.*, vol. XVII, p. 187; *O.O.*, vol. VI, p. 16):

'*Existimo autem . . . à causis physicis abstinendum esse, et Astronomica astronomicè, per causas et hypotheses astronomicas, non physicas esse tractanda. Calculus enim fundamenta Astronomica ex Geometria et Arithmetica, suis videlicet alis, postulat, non coniectures physicas, quae lectorem magis perturbant, quam informant.*' To which Kepler replied (letter to Maestlin, 12/22 December 1616; *ibid.*, p. 202) setting forth his own concept of astronomy as a science of reality, as opposed to the purely computational concept, which '*nec vera est, nec ipse auctor veram credit. Ego verò si minus veram, saltem mea opinione veram trado causam. . . . Dico enim librarj planetam in linea rectâ versus Solem appetentiâ magneticâ, quae in diversis fibrarum inclinationibus sit diversae fortitudinis; metior hanc fortitudinem per sinus, effectum per Versos anomaliae Eccentrj; pro ratione vero distantiarum ad Solem facio tardum vel velocem reverâ. Hypothesis est physica, quia habet exempla physica magnetis, est physica hoc est naturalis, quia vera, et educta ex ipsa natura interna corporum planetae et Solis, est physica, quia modos omnes tenet motionum naturalium. Est tamen astronomica, quia compendiosissime potest computari aequatio et distantia Planetae.*'

4. Leibniz declared that the vortical cosmology of Descartes was inspired by Kepler.

It is an attractive hypothesis, which has led some historians of Kepler astray. It is not impossible that Descartes may have had some knowledge of Kepler's ideas, but it is a fact that his vortices, which are physical and carry the planets along, have little in common with the *immaterial* whirlpool of the *species motrix* which *pushes* the planets, but does not carry them along. It seems to me that the Cartesian vortices could be derived just as well from the spheres of ancient, or Copernican, astronomy, from which they differ only by their fluidity. Moreover, even if Descartes did borrow his vortices from Kepler, this should not make us forget that he borrowed nothing else, especially, neither his celestial physics, nor, what is even more important, any of his laws. Cf. J. Pelseneer, 'Gilbert, Bacon, Galilée, Descartes; leurs relations', *Isis* (1923).

5. This is known as the 'simple elliptical theory'; cf. *infra*, pp. 476 f., account of Boulliau's concept by Borelli.

6. Jeremiae Horroccii, Liverpoliensis Angli, ex Palatinatu Lancastriae,

Opera Posthuma. . . . Londini, MDCLXXIII: I. Astronomia Kepleriana defensa & promota. Prolegomena, p. 7 s.:

'*Quoniam autem non omnibus illud est otii, ut Astronomiam veram inter tam multa incerta, proprio sudore acquirant; dedi ego operam, ut in hoc Libro objectionibus omnibus, in quantum possim, occurerem; & veritatem Astronomiae* Keplerianae *adversus omnes, maxime* Lansbergium, *defenderem*'.

Philip Lansberg, a semi-Copernican, author of *Tabulae Coelestium Motuum Perpetuae*, Middleburg, MDCXXXII, had made a sharp attack on Kepler and the *Tabulae Rudolphinae*, instead of which he proposed his own.

7. Cf. J. Horroccii . . . , *Opera Posthuma*, pp. 183 f.

8. *Ibid.*, p. 11:

'*Credi poterit, excentricitates Harmonicas in rerum primordio exquisite constitutas jam tandem antiquitate violatas esse, & per causas Physicas accidentarias non nihil à pristinâ mensurâ immutatas.*'

It must be noted, however, that in his defense of Keplerian truth, Horrocks confines himself to generalities and to the calculation of tables; he does not explain the mathematical structure of the system, nor does he mention the second and third laws which govern planetary motion.

9. Seth Ward, *In Ismaelis Bullialdi Astronomia Philolaica Fundamenta Inquisitio Brevis*, Oxoniae, 1653; cf. *infra*, p. 518, n. 14.

10. Seth Ward, *Astronomia Geometrica* . . . , Londini, MDCLVI, p. 1:

AOTRONOMIA SOLARIS. LIBER PRIMUS. PARS I. CAPUT I. Astronomiae Ellipticae Principia quædam Generalia. *Astronomia Elliptico-Copernicana, supponit solem in Centro (seu Nodo Communi) omnium planetarum, fixum & immobilem. Planetae autem singuli circa solem, orbitam eandem perpetuis vicibus describentes moventur. Orbitae hujus perimeter est figuarae Ellipticae [Evincentibus hoc, Keplero in motibus Martis, multisque aliis argumentis, quae forsan aliquando proferentur]. Hujus Ellipsoes, cùm focus alter sit sol (Motuum planetariorum verum atque physicum Instrumentum) super alterum interim focum, ita temperatur planetae cujusque motus, ut temporibus aequalibus, aequales illic angulos absolvat. Quare cûm super focum* unum *Ellipseos sit motus* aequalis, *necesse est ut sit super* alterum, *atque etiam in ipsâ Ellipsi,* inaequalis*: neque in Motu Elliptico (quantum ad orbitam suam) alia quaerenda est inaequalitas praeter illam quae à motu medio (seu aequali) regulatur*'.

Starting from true Solar Astronomy, Seth Ward engages in the diversion of developing planetary astronomies, that is to say, explaining how various celestial phenomena appear to the inhabitants of other planets which they assume to be stationary. It is strange that he does not mention Kepler's third law. In passing, this fact shows that Newton derived his knowledge of Kepler's laws from Kepler himself, and not from Seth Ward.

11. J. A. Borelli, *Theoricae Medicearum Planetarum ex causis physisis dedactae*, Florence, 1666.

12. Cf., my 'An Unpublished Letter of R. Hooke to I. Newton', *Isis* (1952).

13. 'Harmonic motion.'

III

J. A. BORELLI AND CELESTIAL MECHANICS

Introduction

Giovanni Alfonso Borelli is remembered today chiefly as the author of the posthumously published work, *De motu animalium.*[1] This is understandable, because we have there a work of great importance in which the movement of living creatures is interpreted, fully and systematically, from a strictly mechanical point of view, and in which the locomotory apparatus of animals—at least, of vertebrates, for Borelli was concerned only with them—is regarded as an assembly of connecting-rods and levers.

On the other hand, his mathematical,[2] physical[3] and astronomical[4] works are much less known. They are, moreover, far less important. Though he was an honest and reliable scholar, who made some contribution to every branch of knowledge he touched, Borelli was certainly not exceptionally gifted. He was not a Copernicus, a Kepler, a Galileo or a Descartes. Historians of physics and astronomy constantly mention his works with approval, but they are very rarely read.[5] After all, that is not surprising: history does not stop to consider those who have themselves stopped half-way along the path of discovery. Furthermore, Borelli was an atrociously bad writer, whose interminable, heavy phrases are calculated to discourage the most sympathetic reader.[6]

Nevertheless, as I had occasion to say in a communication to the *VIe Congrès d'Histoire des Sciences*, where I outlined the history of the discovery of the law of universal gravitation,[7] I believe that a study of Borelli's cosmological work is of very great interest indeed to the historian of scientific thought.

I say this, not only because he almost certainly influenced Robert Hooke, and because Newton did him the signal honour of citing him amongst his predecessors[8]; or, again, because his work provides the clearest evidence of Kepler's influence on the Italian school (and of the reaction of the latter to some of Kepler's fundamental ideas)[9]; or, finally, because it offers an extremely interesting view of the relation between experiment (or observation) and theory in science—a view which is all the more significant seeing that it came from one of the most active and most influential members of the famous *Accademia del Cimento*, which, as is well known, was the first school, or study centre, for the experimental method in Europe. Undoubtedly, all this is quite true, but there is something more. In the work of Borelli we find a fulfilment—imperfect, but nonetheless *decisive*—of that identification of celestial physics with terrestrial physics which was the dream of modern science, and which Kepler and Descartes thought they had achieved, but only Newton realized. It appears in the work of Borelli by an admission that celestial motions (circular planetary motions) produce centrifugal forces, as on Earth; this is something that they do not do according to Copernicus, Kepler, or even Galileo. Last, but not least, Borelli's work reveals in a striking manner how a too great demand for intellectual clarity can lead to frustration, and the discreet renunciation of a theory can end in a dilemma.

Borelli did not write a *Systema Mundi* nor a *Physica Coelestis*. His cosmological views were put forward by him with a certain amount of bias on the occasion of a study on the 'Medicean planets' (Jupiter's satellites). These satellites had been observed by him, besides others, at Florence for a long time, consequent upon the acquisition by Ferdinand II, Grand Duke of Tuscany, of a telescope 'of enormous size and admirable perfection'.[10] Observations were first of all made on Saturn, a satellite of which had just been discovered by Huygens, and then attention was directed towards Jupiter. Borelli tells us, that a theory of the motion of these satellites gradually took shape, and that the Grand Duke (who had taken an active part in the observations) as well as friends advised him to publish it. This was the origin, according to Borelli, of his *Theoricae mediceorum planetarum*.

This is all very possible, and even rather plausible. The observations provided the *occasionem scribendi;* the desire to flatter the Grand Duke[11] fixed the title of the work, which was dedicated, of course, to the Grand Duke himself. There is no reason to doubt that. However, the structure of the work, as pointed out very pertinently by E. Goldbeck in his excellent, but little known, study of Borelli,[12] is not entirely explained on that score. There is something else to which we shall refer later.

In the preface to his book Borelli gives a brief account of the discovery of the Medicean planets by Galileo. Galileo discovered that four small planets, which he named the Medicean planets, revolve round Jupiter in exactly the same way that the Moon revolves round the Earth.[13] He noted the same phases that are observed with our Moon: Full Moon, crescent, New Moon, occultation. He determined the order of their succession, the size of their orbits, their period; but he was not able to observe the numerous anomalies that must exist in their motion, as in the case of all other planets, and whose existence he assumed. Subsequently, much attention was given to the Medicean planets (by Borelli, as well as others), but very little progress was made, in spite of improvements in the instruments available for observations.

For this reason Borelli decided to reverse the *modus procedendi*, and to attack the problem theoretically, seeing that the observations did not provide the desired conclusion; that is to say, he developed first of all *a priori*, a theory of periodic motion for planets, as well as for their satellites or moons, starting from certain data or physical requirements, and then made the appropriate deductions. These deductions were then compared with the empirical data from observations. By considering the observations *after*, instead of *before*, working out his theory, his task was greatly facilitated, for he knew what to look for; and knowing it, could easily find it.

Borelli's undertaking, namely, astronomy *a priori*, may seem absurd, or at least unduly ambitious; and just as inordinately ambitious as the Cartesian project to deduce, *a priori*, the position of the fixed stars in the sky. In fact, it is not so; and a change of terms will immediately enable us to understand the intent of Borelli's project. He wanted to develop a theoretical astronomy, or, if preferred, a rational celestial mechanics, to serve as a basis for observational astronomy in general, and for that of the Medicean planets in particular, seeing that their motion ought to be similar to that of the other planets. Not only is this perfectly reasonable, but it is also fully in agreement with the teaching and spirit of the great Galileo, whose conscious and intelligent pupil Borelli showed himself to be in this undertaking.[14]

In Galilean science, which is experimental and not experiential, theory precedes and guides experiment (*experimentum*), which confirms, or invalidates, the theory, and provides firm data relating to the matter under investigation. However, theory constitutes science;

and in the same way that Galileo, on the famous occasion concerning the cannon-ball falling from the top of a mast of a ship in full sail, was able to proclaim that he was such a good physicist that he could predict the behaviour of the cannon-ball, *a priori*, without making any experiment, so Borelli could have said that he was such a good astronomer that he could predict the general nature of planetary orbits, *a priori*, without observing planetary motions.

Borelli, in fact, did not say so; but that was how he acted, and the way in which he justified his undertaking is most interesting. Naturally, he starts by saying that all our scientific knowledge derives from our senses and must be based on experiment. Then, exactly like Galileo, he invokes the principle of uniformity in nature which proceeds everywhere by the simplest and easiest ways, and has no pleasure in following other ways that lead to the same end; it always makes use of the same causes to produce the same effects. Consequently, in spite of apparent diversity, there is perfect conformity in planetary motion, and a no less perfect similarity with the internal structure of animals, even of animals from extremely different regions and climates. Also, we are able to make assertions, *a priori*, about quite a number of things concerning animals that we have never seen.

For this reason we are quite justified in applying to a study of the motion of the Medicean planets (Jupiter's satellites), a theory that has been confirmed by observations relative to the Moon's motion.

Borelli, therefore, describes the anomalies of the Moon's motion, and simply transfers them to the Medicean planets, where (as we have said) it had not yet been possible to observe them. Then, proud of his triumph, he sets forth the theoretical basis of his astronomy.

We might very well ask what business the Medicean planets have here, and why Borelli did not confine himself to an exposition, on the basis of his celestial mechanics, of a theory of lunar motion, which is so much easier to check by observation. E. Goldbeck makes two suggestions. First of all, he believes[15] that Borelli was aware of the inadequacy of his theory and his mathematical ability, and so realized that he could not achieve his purpose; consequently, he preferred to speak of Jupiter's satellites instead of the Moon, for the very good reason that it would be much more difficult to make any check. Personally, I do not think Goldbeck's hypothesis is very probable: however weak Borelli's theory may be, he was, nevertheless, convinced of its truth. He dealt with the Medicean planets for the purpose of showing the fertility of his method (as well as of flattering the Grand Duke), and of opening a fresh path for future observers. Surely, it was the sterility of observation, pure and simple, that made him realize the need to start with theory. Goldbeck's second suggestion, on the other hand, seems to be quite convincing. So long as he limited his discussion to Jupiter's moons, and although he made Jupiter revolve round the Sun (as did Tycho Brahe and even Giambatista Ricciolo) together with all the other planets (as did Tycho Brahe, but not Riccioli), Borelli was able to avoid a charge of Copernicanism by the very simple means of avoiding mention of the Earth and placing it amongst the planets.[16] Furthermore, it was enough to avoid saying that the Sun is at the centre of the Universe, and to mention its motion only once or twice. In this way he conformed literally to the condemnation of Copernicus, and suggested to the reader (especially one who was not too attentive) that he adhered to Tycho Brahe's system, or to something similar, which at any rate had not been condemned by the Church.[17]

That would have been quite impossible had he wished to deal with the Moon. It would have been all the more so, seeing that in Borelli's celestial mechanics (as in that of Kepler's, except that Borelli accepted the principle of inertia) planetary motion is explained in the last analysis by the Sun's rotation on its axis. The motion of satellites (in which respect they are strictly distinguished from primary planets) is explained by the rotation of their central body; but—and this is of prime importance—whereas the primary planets are moved by the Sun, and by the Sun only, the satellites are subjected to the simultaneous influence of their central body and of the Sun. Whence we have further anomalies in their motions.[18] Consequently, Borelli could not deal with the motion of our Moon without putting himself in jeopardy[19]; therefore, he made no attempt to explain it.

We can now turn to a consideration of Borelli's cosmology in its broadest aspects. A detailed study would take us too far away from our main purpose, and would not be particularly profitable. Borelli's importance for the history of scientific thought lies in his attempt to elaborate a Galilean system of celestial mechanics, and not in his having been mistaken—or otherwise—in its specific applications.

I

The Problem of Planetary Motion

The great problem that dominated Borelli's mind, and which had been brought into the limelight by Kepler, may be summarized thus: *a quo moventur planetae ?*, and why do they remain in their places? I have already had occasion to explain that this problem, however important it became subsequently, did not arise for pre-Copernican astronomy, nor even for Copernicus; or perhaps, one ought to say that it was solved before it arose. The motion of the planets was regarded as being intimately linked with that of the solid celestial spheres in which they were set; and the motion of the latter was quite naturally explained by the action of the souls or intelligences associated with them. Copernicus, too, although he never mentioned intelligences or souls in the celestial spheres, still believed in the existence of these spheres, and even succeeded in imparting to them the uniform circular motion that Ptolemaic astronomy had been obliged to abandon.[1]

It is only when we come to Tycho Brahe—or, as Kepler put it, after the 'destruction' of solid spheres and circles by Tycho Brahe—that the problem of planetary motion, namely, its cause, acquired real interest; and it must be said that its current interest was much less than we should have expected it to be. For example, Tycho Brahe, hardly mentions it. Kepler does deal with it in his great work, *Astronomia Nova sive Physica Coelestis*,[2] but Galileo ignores it completely. So do Riccioli, Fabri, and, what is more surprising, Ismaël Boulliau[3] also. Borelli is the exception.

This state of affairs, so strange from our point of view, may be explained, as I see it, by the persistence of an extremely powerful tradition, or what is the same thing, by the effects of intellectual inertia. Astronomers, as I have said, never raised the problem of the physical causes of celestial motions[4]—the peculiar nature of celestial reality removed it from the action of causes which are effective in terrestrial reality; furthermore, they were accustomed to consider celestial motions only from the kinematic,[5] and not the dynamic, point of view. Consequently, we can understand how the problem arose for the first time with Kepler. He united celestial physics with terrestrial physics, and so ascribed exactly the same *inertia, i.e.*, resistance to motion as found in sublunary bodies, to the planets which move freely in space, being no longer set in the spheres. Furthermore, and this is a most important point, his planets no longer move in circles, but describe ellipses with a variable, non-uniform velocity.[6] A search for a physical explanation of such strange phenomena then became imperative.

The search was much less imperative, or even unnecessary, for those who did not admit the elliptical nature of planetary orbits; even if they rejected the antithesis of two worlds—sublunary and superlunary—by accepting the unification provided by physics. Persistence in the belief of the natural character of circular motion—a belief to which assent was given by Kepler[7] and even Galileo, who never entirely freed himself from it[8]—ended in a paradoxical and curious result, namely, that it was precisely Galileo and his followers who evinced no need to seek the cause of planetary circular motions. Transferring their *principle of inertia, i.e.*, the principle of the conservation of motion and velocity, to planetary motions, they regarded it as quite 'natural' for these motions to continue of their own accord, once they had started. For them, as for Kepler and Tycho Brahe, the Universe was always *finite*, and in a finite universe, circular motion is certainly invested with special advantages. For that reason, they, as well as Kepler, did not regard planetary motion to be a source of centrifugal forces.

It is greatly to Borelli's credit that he recognized the importance of Kepler's work and that he frankly accepted the Keplerian revolution—the ellipticity of planetary orbits—and, at the same time, definitely jettisoned the privileged notion of circularity. In his view, motion in a *straight line*, and linear velocity, persist in the skies exactly as they do on Earth. Being a better Galilean than Galileo himself, he could apply all the progress achieved by the Galilean revolution to a profitable study of the Keplerian problem; but he was to pay a price for his faithfulness to Galilean principles.

We shall now see how Borelli, taking his stand on 'facts', namely, the ellipticity of planetary orbits, develops his theoretical astronomy.[9]

'Having given a general description of the various and numerous inqualities [anomalies] in the motions of the Medicean planets, it becomes necessary to give an exact explanation of the shape of the circumvolution performed by the above mentioned planets, how it is disposed, and where it is placed. Taking our starting point in higher regions, I shall assume in the first place that the skies themselves are either completely void, or at least filled with an extremely fluid aetherial substance, which is far more tenuous than the air surrounding our terrestrial globe; this [is deduced] from the most exact experiments of modern [philosophers] which everyone accepts without hesitation or contradiction.

'Furthermore, it is certain that all the wandering stars (the planets) are set in different places of the aforesaid aetherial fluid region, where they are balanced, as it were, by nature, and revolve about one of the large bodies of the Universe, namely, about the Sun, the Earth, Jupiter, or Saturn. It is equally clear that the aforesaid planets make circuits about the said large bodies, and revolve about them in everlasting circles; that they never forsake their path once it has been adopted, and that they never come too near to, nor move too far from, [the central body], nor do they deviate from their adopted path.

'From both old and recent observations it is also clear that the body about which they revolve is not exactly at the centre of their orbits, but that these planets rise [in the sky] more on one side than on the opposite side by describing a certain eccentric path, which for some time past has been admitted not to be circular, but to resemble an ellipse.

'Finally, observations have shown that the line of apsides of the aforesaid eccentric is not fixed, *i.e.*, is not parallel to itself, for which reason it does not always point to the same fixed stars, but moves continuously through the signs [of the Zodiac] in sequence. Jo. Kepler was the first, by his boldness and in opposition to the ancient philosophers, to give the order which banished perfectly circular orbits from the sky; he proved most clearly from Tycho Brahe's observations that confirmation was provided in the case of the orbit of Mars; then he noticed also that it [ellipticity of the orbit] ought of necessity to occur in the case of Mercury; but in the case of the Sun [10] he was unable to demonstrate the elliptical shape [of its orbit], although he ascribed such to all the planets on the basis of extremely ingenious, but false,[11] physical arguments. This view was so agreeable to all *savants* that it was readily accepted, especially as the most learned and famous astronomer Bullialdus [12] had largely perfected it, even though he had deduced it from different principles, and had formed the said elliptical figure with different elements. For this reason, it will perhaps be useful to give a short account of what Bullialdus put forward, so that later on we can show in what manner this doctrine of ellipticity is confirmed (Fig. 1).

'Imagine a scalene cone with its apex at *A* and having a circular base of diameter *BC*. Let its axis be *AI*; and let the triangle through the axis perpendicular to the circular base be *ABC*, so that angle *AIC* is acute, and its complementary angle is obtuse. Draw the straight line *EK* to subtend an angle at the apex so that *EK* is divided into two equal parts in the point *X* by the straight line *VT*, which is equal to *EK*, is parallel to the base *BC*, and cuts the axis in the point *Z*. It follows that the triangle *MXZ* will be isosceles, having *MX* equal to *ZX*. Therefore the triangle *AEK* will not be sub-contrary to triangle *ABC*. Through the straight line

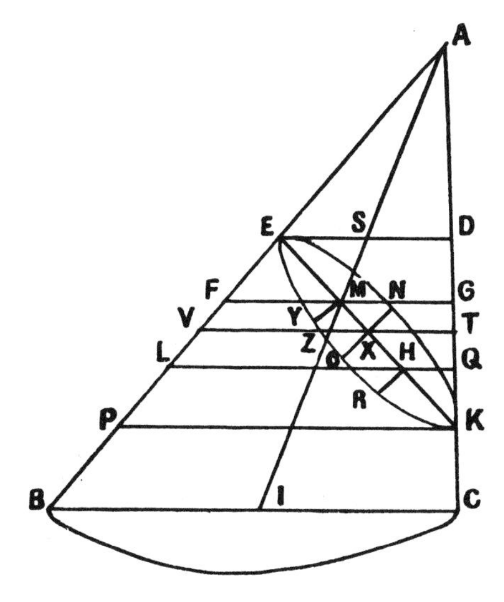

Figure 1.

EK raise a plane surface [perpendicularly] to the plane of triangle *ABC*, which surface will develop the ellipse *ERK* in the section of the cone. The transverse axis of this ellipse will be *EK*, its conjugate axis will be *ON*, its centre will be *X*, and one of the foci or poles will be [the point] *M* on the axis of the cone. [The segment] *XH* being equal to *XM*, the other focus of the ellipse will be *H*. This being presupposed, Bullialdus then assumes that the Sun is at the point *H*, and that the planet moves with uniform motion about the axis *AMI* of the cone in circles which are always equidistant from the circle of the base *BC* of the cone, which circles may be called equant circles. The point *M*, or rather the entire axis will be called the centre of uniform motion.[13] Seeing that it is the nature of uniform circular motion to sweep out equal angles at the centre in equal times; and seeing that these angles at the centre belong to similar circumferences, which are, nevertheless, proportional to their radii, it follows that when the planet passes through the point *E* belonging to the circle whose semi-diameter is *SE*, then its motion will be slowest, because this circle is the smallest of all those which are described by the celestial body in its proper period; but when the celestial body arrives at the point *Y* and describes the circumference of circle *FG* on the cone, (which circle passes through the focus *M*), its motion will be rapid, because this circle is greater. Later, when it traverses the peripheries of other circles, the greatest of which passes through *K*, its motion will be fastest because it will describe the periphery of the largest circle *PK*. In this position [*K*] the planet is nearest to the pole *H*. Consequently, from aphelion *E* to perihelion *K*, the celestial body will traverse the peripheries of innumerable circles which increase successively: for this reason the uniform motion which sweeps out equal angles about the axis *AI* in equal times is represented on the elliptical periphery *ERK* by increases which will have the same relationship to each other as the radii of the said circles above the minimum [$SE = HK$].

'Although, by deducing the physical equation as well as the optical [equation] from this hypothesis, Bullialdus, as pointed out by Seth Ward,[14] omitted certain things concerning the mean motion, nevertheless, it cannot be denied that this first invention of his is admirable, ingenious and praiseworthy. It is no doubt true that the Astrophilosophers raise two objections; in the first place, these cones for each planet are fictitious, consequently, it is not clear how it comes about that the planet moves on a certain conical surface which has no existence anywhere in the Universe; and in the second place, it seems quite contrary to fact that all these motions are performed about a [certain] point [*M*] and a [certain] line of uniform motion passing through *M*, for this point is indivisible and assumed by the imagination [as being] in the aether, and has absolutely no substance or [physical] faculty. Consequently, there is absolutely no reason why the planet should revolve about the said point and the said imaginary line in accordance with a perfectly constant law, and, on the contrary, move in a quite irregular manner with respect to the very large globe of the Sun itself, placed at the point *H*, as if the main object of the star were not to turn round the Sun itself, but [to turn round] the said imaginary, fantastic point, which is without perfection or faculty. Indeed, this is such an adequate objection, that it is very difficult to answer it.

'With regard to the first [point], I believe that I can not only deal with it satisfactorily, but, perhaps, also give enlightenment on some things concerning the secrets of nature. In the first place, we shall imagine the planet to move under two motions, the one, circular, the other, on the contrary, linear, and we shall show that from these two motions [taken] as elements, an elliptical motion can result. Let us assume that the Sun is at *H*, and that the planet to begin with is at aphelion *E*, but having two motions, the one, orbicular round the Sun, and the other, linear in the direction from *A* towards *P*; [let us assume also] that the said motions are commensurable with each other in such a manner that when the planet describes a semi-circle starting from E, it must go from *E* to *P* in the same time with a linear motion; but that during the following semi-circle the planet returns from *P* to *E*. We must assume also that the plane *ED* of the circular motion is always inclined with respect to the line *EP* of linear motion, whence it follows that the star, by its motion in a straight line, will traverse the peripheries of innumerable circles which are always equidistant from each other, and if, during this time, the circular motion were uniform, especially if equal angles with respect to the centre were swept out in equal times, then the star would describe an elliptical orbit, as we have said above. We see, therefore, even though no real cone be supposed [to exist] in the Universe, how it is nevertheless possible for elliptical motion to take place in exactly the same way as it would if we assumed the existence of a solid cone of that kind.

'It can be shown that the hypothesis of the aforesaid two motions is possible, in the first place, by the example of all those planets that have orbits similar to circles, but never follow the exact periphery of circles; furthermore, the curves described by them incline more uniformly to the plane of the planet's orbit than is postulated by the inclination on which its latitude depends.

'Seeing that the plane of the solar whirlpool is definitely inclined to the plane of the ecliptic, then, if it were true, as Kepler thought, that the planets are in some way caught up and carried round the Sun by the solar rays as they revolve, they [the planets] should necessarily be carried round along the circle *LQ* and others parallel to it. In fact, if we imagine the Sun [to remain] at the point *H* [and] to describe circles parallel to *LQ* by its own whirlpool in such a way that the axis of this whirlpool is raised [perpendicularly] to the plane *LQ* of the whirlpool, the solar rays would move in the plane of the said circle *LQ* and others parallel to it; also, the planets, being carried around by these rays, would advance on this plane, and, as in the meantime the planet would execute its proper motion from *E* to *P* and from *P* to *E*, it would be obliged to travel first in the circle *PK*, then in [the circle] *LQ* and so on, without remaining for any definite time in any one of them (because the motion on *EP* is supposed to be continuous). Consequently, they [the planets] must be carried round along the peripheries of innumerable circles of different sizes, which provide a measure of the irregular velocity with which the planet moves round the Sun; and we see that such a motion, far from being impossible, is on the whole rational and probable, provided that it does not run into other objections.

'Furthermore, it must be noted here that, although the solar body always revolves in the same place, nevertheless, the equidistant circles described by its rays whilst [the Sun] itself revolves on its own axis are always in the same plane with respect to the situation and extent of the

Universe. Therefore, in order that the elliptical paths described by the different planets may be saved, we need only [a] place the aphelia at various distances from the Sun in various places in the starry sky and Zodiac in such a way that the line of aphelion drawn from the Sun to the planet is more or less inclined with respect to the plane of the Sun's whirlpool; and [b] ascribe to each of them a motion along the straight line inclined to the plane of the solar whirlpool and having the exact value necessary to form an ellipse suitable to the planet in question with its periods and all other circumstances which have been observed in its motion.

'The other difficulty remains, namely, if it be possible for the planet to move round the focus of the ellipse, or the point of equality, whilst the Sun is placed at the other focus or pole. Most certainly, it seems difficult and incomprehensible that the planet, whether it be carried around by its own innate power, or by some exterior power, should be moved around this point of equality which possesses no power or entity. This view is indeed refuted by the reality of non-uniform velocity of planetary motion, which velocity, according to this hypothesis, should increase in exactly the same proportion as the distances drawn from the axis of the cone through the focus increase; or else to the same extent as the semi-diameters of the equant circles increase. This definitely cannot occur either [in the case] where the motion of the planet on the circumferences of the equant circles is effected by a proper force existing in the planet itself; or [in the case where it results from] an external faculty which propels it [the planet]; for in the former case the velocity of the planet should always be uniform and constant, whereas in the latter case the velocity of the planet should decrease in proportion to the increase in the semi-diameters and peripheries of the equant circles, as will be proved later. For these reasons we have been obliged to abandon the above-mentioned hypothesis and, if possible, [to try] and find another more probable one, or instead to show that the same elliptical path of planetary motion can be retained, but only on the basis of firmer hypotheses more conformable to physical arguments.'

Before doing so, Borelli thought it necessary to formulate several lemmas (eight in all), which it is unnecessary to consider in detail. The purpose of these lemmas is to demonstrate the equivalence of Kepler's view with that of Bullialdus (Lemma VI *et seq.*) by showing that two motions about two centres (the foci of the ellipse), one of which becomes slower in ratio to the increase in distance (increase in length of the radius-vector) and the other in the inverse ratio to the decrease in distance (decrease in length of the radius-vector), will describe similar (elliptical) orbits.

Having done so, Borelli goes on to expound his celestial mechanics:

'In the first place, we ask[15] by what necessity the planets never break away from the circles they have once embarked upon; neither by retreating from the globe about which they revolve in order to traverse other parts of universal space, nor by approaching [the central globe] until they unite [with it]. We are informed by natural intelligence that this question can be resolved in three ways:

'*Primo*[16]: if the planets were attached to the said central body by some corporeal link, such as a cord or a solid wheel, these, being fixed to the Sun itself, would never allow the planet to move further away than the length of the cord or the radius of the solid wheel would permit.[17]

'*Secundo*[18]: [the question may also be resolved] if we were to suppose, as some do, that the Sun is surrounded by a certain aetheric aura, like a *confinium*, having a fixed consistency and density similar to sea-water, so that the body of the planet swims in the exterior layer of this aether in the same way that a ship floats in water. Although it may seem difficult [to accept] that a solid and dense body such as a planet can be supported in the highest and most attenuated aetheric region, nevertheless, this result can be regarded as probable in several ways: *primo*, by supposing that the quasi-magnetic force by whose influence the planet tends to approach the solar body,[19] whilst it [the planet] is immersed in the said layer, does not increase according to the laws of gravity, [laws] according to which the *conatus* of a body to move towards the Sun increases in the same degree as its dense and material part increases. We can base this argument on the fact that if we take a hollow ball of iron containing air, we find that it approaches a magnet with much *impetus* and energy, but that this does not take place with a ball of marble or of gold, even though very heavy. *Secundo*; it is not impossible for the quasi-magnetic force by which the planet is compelled to approach the Sun to be weaker and of less power than the force of the aether in question, if we imagine this compulsion to originate from the larger or smaller magnetic faculty,[20] instead of from the larger or smaller quantity of matter in the planet itself. Hence, it would not be impossible for the body of the planet to swim in the highest region of the aether as though in some ocean; and seeing that this aether can have a variable density, therefore the various planets could swim at different depths and at different distances from the Sun; and similarly, the four Medicean planets would be at different distances from Jupiter. This notion can be confirmed by many examples, and especially by the experiment in which a vessel is filled partly with oil of stones and partly with spirit of tartar, or with similar substances, and then small glass bubbles containing air are immersed in the liquid; of the assorted glass bubbles one particular bubble, but not the others, will float at each level of the said mixed liquids.[21]

'It must be assumed in addition that the ultimate surface of each of these aetheric oceans is not exactly spherical in shape, but is enlarged more in one part than the other[22]; the motion of the floating planet will then be eccentric. However, it is obvious that the former of these two [explanations] is already denounced as false by observations on the part of our senses, and the latter is similarly denounced by the numerous difficulties that it implies.[23] Consequently, we are obliged to abandon both of them, and to see if the appearance can be saved in an easier and more certain manner without postulating the absurd solidity of the celestial substance and the above mentioned aetheric oceans.[24] For that reason we shall proceed by assuming something that it would seem impossible to deny, namely, that the planets have a certain natural desire to unite with the globe about which they revolve in the Universe, and which they tend to approach with all their power: the planets to the Sun, the Medicean stars to Jupiter.[25] Furthermore, it is certain, that the circular motion confers on the moving body an *impetus* to move away from the centre of revolution, as we know from experience by spinning a wheel, or whirling a sling, whereby a stone acquires the *impetus* to move away from the centre of revolution.[26] We shall assume, therefore, that the planet tends to approach the Sun, whilst at the same time it acquires the *impetus* to move away from the solar centre through the *impetus* of circular motion: then, so long as the opposing forces remain equal (the one is in fact compensated by the other), [the planet] cannot come closer to, nor move further away from, the Sun, and must remain within a certain, fixed space; consequently, the planet will appear to be in equilibrium and floating.'

The scheme is quite clear: assimilation of celestial mechanics to terrestrial mechanics; the introduction, undoubtedly after the manner of Galileo,[27] but much more consistently, of the concept of the gravitation (natural tendency) of celestial bodies towards their central body (planets towards the Sun, satellites towards the primary planet), as well as that of centrifugal force produced by their motion about the primary, enabled Borelli to solve the problem of the stability of the solar system in an extremely elegant way, and, in principle, quite correctly. In order that the celestial bodies should remain at the same distance from the Sun (and satellites from their central body), it is necessary and sufficient for the centrifugal force to be equal to the centripetal force.

A terrestrial experiment will prove that it is possible. Borelli then continued as follows[28]:

'The above mentioned assertion can be confirmed by the following experiment.[29] Take a wooden circle *ABC* fitted with the ruler *AB* across its diameter, and having a small pin or rod *DE* rising from the plane of the circle *ABC* at its centre *D* where a small magnet *F* is placed with its south pole pointing towards *A*. This whole assembly is then floated on water in the vessel *RS*. At *G* place a piece of cork on which a small ball of iron *I* is resting, so that the whole assembly can float freely on the water. Then push the cork towards the magnet until it comes within the sphere of influence of the said magnet,[30] that is to say, until it gets into a position where the small ball of iron in question starts to move slowly towards the said magnet. The tip *E* of the rod is then turned horizontally with the hand so that the cork *G* together with the small ball of iron is moved by the semi-diameter *AD*. Let us assume that the ball of iron *I*

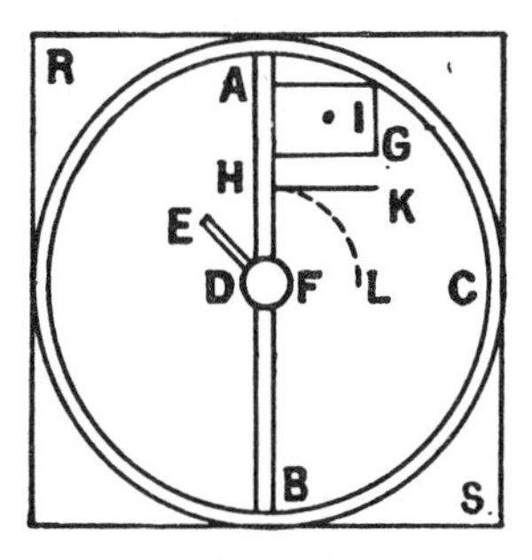

Figure 2.

floating [in the vessel] can move from *G* to *H* during one second, getting closer to the magnet *F*; and let us assume also that in the same time the said ball of iron is carried over the arc *HL* by a circular motion which can be accomplished sufficiently quickly for the floating cork *I* to be pushed [during] one second along the tangent *HK*, as required by the nature of circular motion, and is repelled through the space *KL*, thus producing a movement equal to *GH*. Consequently, in the same time of one second, the movement *GH* towards the magnet will be equal to the displacement *LK* produced by the circular motion; as a result the distance [of the ball from the centre *D*] or the radius *DK* will be equal to the semi-diameter. However, seeing that the ball of iron *I* during the following periods of time, as well as during each of these instants of time, perseveres in the same *conatus*, that is to say, tends to approach the magnet *F* in the same manner, and that the circular motion perseveres similarly [in its action] repelling it [the ball], it follows that the floating ball will keep at the same distance from the magnet *F* and from the centre *D* during its entire circuit, and so will follow the periphery of the circle without any deviation; and the result will be the same even if the floating ball be not attached to any solid body, nor held in position by any external faculty, but swims in a liquid that offers no resistance to the magnet's approach to the centre *D*.'

The experiment is conclusive. Nevertheless, the use of a magnet to illustrate astronomical facts is dangerous, for the very good reason that it could suggest the existence of magnetic attraction in the solar system, and so bring us back to a concept (the second hypothesis) which we have already rejected. Borelli, therefore, provided another *exemplum*; this time, a purely mechanical one:

'However, the experiment can be carried out even more easily without the magnet, provided only that the wooden rod *AB* be bent at *D*, instead of being straight, so as to make an obtuse angle between the two radii *AD* and *DB* at the lowest point *D*, and provided that the small rod *ED* form the acute angles *EDA* and *EDB* [with the two radii *AD* and *DB*]. Then make a straight and smooth channel in the radius *AD* so that a heavy ball *G* can descend from *A* to *D* along this inclined channel. If, by turning the small rod *ED*, a rotation in the horizontal plane be imparted to the device *AB* whilst the ball *G* descends towards *D* along the small channel, and if the velocity of this whirling motion be so fixed that it repels the ball *G* towards *A* with the same force that it tends by its natural weight to descend towards the centre *D* along the channel, then, as a result of equality between the opposing forces, the ball will appear equally distant from the centre *D* of revolution during one complete revolution along the arc *HL*.

'In the same way, if we imagine the circular space to contain the planet *J* (which has a natural inclination to approach the Sun *D*),[31] and if at the same time [we imagine the planet] to move in a circle about the solar centre with a velocity that is sufficient to make the planet move back by a distance exactly equal to that by which it approaches the Sun at every instant, then there is no doubt but that, [being actuated by] these two opposing motions [which] are mutually compensating, the star *J* will not approach the Sun, nor move away from it, by a distance greater than the semi-diameter *DG*. For this reason, the planet will appear balanced and floating, or retained by some firm link, even though it be placed in the most limpid aether, and be supported on, or by, nothing. Now, all this can be produced without angelic or intellectual faculties, but simply by the forces of nature alone. Thus, the planets would be able to persist in their motion round the Sun or round Jupiter without any difficulty, and would be unable to deviate from their path, even though the aether were extremely fluid.'[32]

II

The Solar Whirlpool

Having thus laid the basis of his celestial mechanics, Borelli then turned to the problem of planetary motion—*a quo moventur planetae?* Once more, after rejecting 'animastic' solutions, he came to a purely mechanical solution, and one that was manifestly inspired by Kepler: it was even a solution that had been considered and rejected by Kepler. However, we must not anticipate, but listen to Borelli[1]:

'In the second place we must enquire by what virtue the planets are moved round the Sun, or round Jupiter, that is to say, [we must enquire] if this force come from a natural, internal principle, or from a violent, external [principle], or from both together; and if this principle be internal, we must ask if it be 'animastic',[2] like the principle of motion in animals, or natural, like the tendency of heavy bodies to fall, or the desire by which a magnet approaches iron[3]; but if, on the other hand, the above mentioned virtue be external, [we must] enquire if it be dependent on intelligences or on angelic spirits, or if it be similar to the motion of projectiles.

'Many [people] resort to a soul or intelligences as though to a sheet-anchor in order to account for planetary motions, which are not simple like the fall of a stone, but are accomplished with supreme artifice. Indeed, they cannot understand how a planet can move through the fluid aether according to a constant law, *i.e.*, in an eccentric circle without any deviation and yet with all the ingenious anomalies that are observed in its motion. On the other hand, we easily obtain this result by assuming that a guide, not only living, but even intelligent, is associated with the planet, and leads it on its path without any transgression.[4] Nevertheless, although this explanation is accepted by many [people], there is absolutely no-one who doubts that if it were possible to ascribe purely natural causes to planetary motion, then there would be no need to resort to a soul or to intelligences[5]; just as no-one (I am sure) will be convinced that the motion by which heavy bodies tend towards the Earth along the shortest path results from a soul, or even an intelligence, seated in the stones, and pushing them downwards. In fact, nature is able to accomplish that purpose with less work and expense by means of a simple natural faculty called gravity.[6]

'The chief reason for resorting to intelligence in respect of planetary motion is because the orbits are completed with such great skill and ingenuity that it does not seem probable that any simple, blind, natural faculty could keep the planets suspended steadily in the fluid aether, let alone enable them to make an eccentric circumvolution (and, what is still more surprising) even by an ellipse or by a line approaching an ellipse; could ensure the progression of their apogees and the regression of their nodes in succession [in accordance] with a fixed and stable law[7]; and could ensure no deviation from the path in spite of its diverse features. If, therefore, we could prove that all those things that we have described can be produced by a simple, natural power, whether it be internal or external, we should have no need of recourse to other agencies.'[8]

Borelli's celestial mechanics is based on the principle of conservation of motion and velocity, and in this respect is vastly superior to that of Kepler. However, nothing is more common than a lack of understanding of the implications of this principle, even by those who accept it: not to mention those who believe it to be valid for circular motion as much as for linear motion—the confusion between linear velocity and angular velocity is frequently encountered.[9] Now, it is extremely important to distinguish between them in order to make a correct application of the principle of inertia to the problem of planetary motion; and it was not without reason that Borelli applied himself to a refutation of the errors frequently made on this subject. In the first place there is the mistake of those who believe that when circular motion is produced by a given (constant) motive force, the motion is slower as the radius vector, or the circumference, increase. This belief seems to be strictly in accordance with experience, both celestial[10] and terrestrial.[11]

'For, in the first place, our very senses seem to show that circular motions (whether they be caused by a natural, internal power, or by a violent, external one) made in large circles (the motive force remaining constant) are always slower than those made in smaller circles. Thus, every heavy body performs its own oscillations about its point of suspension, and they are, without any doubt, produced by a natural, internal principle, namely, by the gravity of the said pendulum, which, spontaneously and of its own accord, without being pushed by any exterior force, performs its proper vibration. However, we note that if the cord by which the pendulum is suspended be increased in length

> whilst the pendulum is moving to and fro, then its motion immediately becomes slower; and if the cord be shortened, then the motion immediately becomes quicker. Therefore, in the case of circuits produced by a natural force, when motion has started on a larger circle, it is necessary for the moving body to take a longer time to cover the circuit and conversely. Similarly, if we were to make use of some external force to push the said pendulum over the periphery of the circle, we should see that if the cord were lengthened and the circuit increased, after applying such an impulse to the pendulum, then the motion would become slower. On the other hand, if the cord were shortened and the circuit decreased, then [the motion] would increase.
>
> 'The same thing happens in the case of the small toothed balance in a clock,[12] on which it is customary to place two weights equally spaced from the axis. Here, too, we see that the same thrusting force by the same weight turns the small balance, as well as all the other wheels, with a speed which is greater when the said weights are nearer to the centre of their proper revolution, and conversely.[13] Hence, circular motions produced by an inherent power as well as those produced by an external [power] complete the path [of the circuit] in a time which is longer in the same proportion as the semi-diameter of the circuit is greater, and conversely.'[14]

The theory being set forth, but not yet refuted, Borelli seems to have the advantage of being able to explain variations in the speed of planets in their orbits, that is to say, to explain why they move more rapidly when close to the Sun, and more slowly when farther away:

> 'Consequently, if we assume that the body of the planet turns round the Sun, or round Jupiter, through an internal force, or rather if it turns round [the Sun] through the impulsion of solar rays, whilst these latter turn together with the solar whirlpool, and similarly in the case of Jupiter,[15] then there is no difficulty [in explaining] the decrease in the planet's speed, for it will describe a larger circle the farther it is away from the Sun, and will slow down on that account. On the other hand when it is nearer to the Sun, it will describe a smaller circle, and will move faster on that account.'[16]

Let us be quite clear. In the concept which is criticized by Borelli, the planet does not move more slowly when far away from the Sun and more quickly when it is near, because the solar rays (Kepler's *species motrix*) act more feebly in the first instance, and more strongly in the second: the planet moves more slowly simply because it is farther away, and more quickly because it is nearer. In fact, Borelli took care to forewarn us that the motive force is regarded as constant in the concept under consideration. So, he continued by saying[17]:

'Although this proposition may seem adequately proved by the above mentioned experiments, nevertheless they do not suffice, nor are they free from errors. For this reason it is necessary for us to examine the matter more closely by a detailed analysis. In the first place, I say, that it is not true that the same moving body, which is always urged on by the same intrinsic motive force, and traverses at one time a path larger than a circle, and at another time one that is smaller, moves with a faster motion over the smaller circle than over the larger one. It proceeds, in fact, with the same velocity over the two unequal circles, that is to say, it covers equal distances in the same time, in such a manner that if the same moving body *B* conveyed by its internal principle, always equal to itself,[18] were moving now on the periphery of the circle *BC* whose radius is *AB*, and now on the circumference *BE* whose semi-diameter is *BD*, it would be wrong to say that the same moving body traverses a greater distance on the circumference *BC* in the same time. This is proved by a consideration based on natural reason as well as by experiment. In the first place, seeing that we have assumed that the motive faculty of the moving body *B* is constant, and that it never increases or decreases, it is certainly necessary for this faculty to traverse equal distances in equal times; in fact, a different direction, or inclination, or incurvation of the said motions bring no change to the afore-mentioned motive faculty, nor to its uniform operation, *i.e.*, to its velocity.[19] Consequently, the same moving body *B* is urged on with the same velocity in equal times, and so will traverse equal distances *BC* and *BE*. It is certainly true that the arc *BE* will be greater than if it were similar to the arc *BC*, that is to say that the angle *BDE* at the centre would be in the same reciprocal proportion to the angle *BAC* as the radii of the said circles are to each other, namely, as *AB* is to *BD*; because, if the angle *BDF* be made equal to angle *BAC*, the arcs *BC* and *BF* will be similar to each other; and as the arcs *BC* and *BE* are equal, they will bear the same proportion to the common arc *BF*. Arc *BC* is to the arc *BF* (to which it is similar) as the radius *AB* is to the radius *BD*. Therefore, as the arc *EB* is to the arc *FB*, so is the angle *EDB* to angle *FDB*, or to angle *BAC*; hence the angle *EDB* will bear the same ratio to angle *BAC* as the radius *AB* does to the radius *DB* (Fig. 3, p. 487).

'The most suitable experiment to confirm this truth is the following.[20] Take a pendulum *B*, suspended from a nail *A*, and let the distance *BD* be made less than the length of the cord *AB*; put another nail at *D*, so as to make the angle *CAB* equal to angle *BDF*. Then move the pendulum to the position *AC*, and let it fall freely towards the position *AB*, perpendicular to the horizon. There is absolutely no doubt but that the pendulum during its fall over the arc *BC* will have acquired a certain degree of velocity which will carry it to the point *G* on the circumference described by the shorter pendulum *BD*. It is acknowledged by experiment that the ratio of the angle *GDB* (subtended by the arc *BG*) to the angle *BAC* is equal to the reciprocal sub-duplicate proportion of the lengths of the pendulums[21]; that is to say, if *IB* be a mean proportional between *AB* and *BD*, then experiment proves that the angle *GDB* will bear the same ratio to angle *BAC* or angle *BDF* as *IB* does to *DB*. This experiment being accepted, we must prove that the velocity of the pendulum

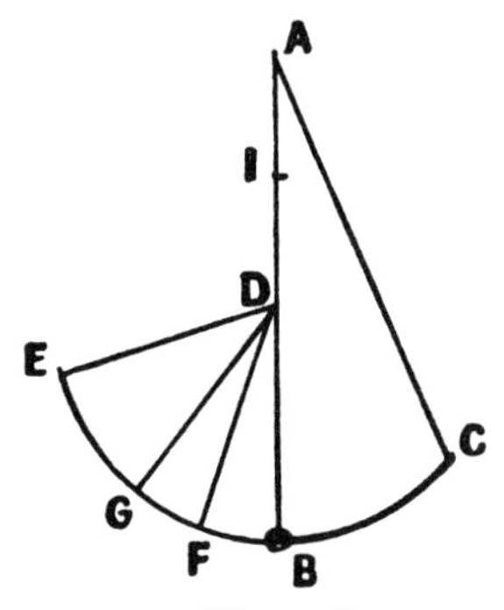

Figure 3.

AB is equal to the velocity of pendulum *DB*. Now, the peripheries of the circles, or similar arcs, are in the same ratio as the semi-diameters *AB* and *DB*. Therefore, the arc *BC* is to arc *BF* as *AB* is to *DB*; but the arc *GB* is to arc *FB* as *IB* is to *BD*; hence the arc *CB* is to arc *GB* as *AB* is to *IB*; but as the times of oscillation over the arcs *CB* and *BG* are in the sub-duplicate ratio[22] to the lengths of the pendulum (as I have shown elsewhere, and as is proved by experiment), it follows that the distance *CB* will bear the same ratio to *GB* as the times in which the afore-said paths are covered by the pendulum; but as the times are proportional to the distances that have been covered, the velocities are equal amongst themselves; consequently, the velocity of pendulum *AB* is the same as that of pendulum *BD*.'

Borelli was conscious of the fact that his laborious and abstract proof was too difficult for most of his readers, so he undertook to put it in a more concrete form[23]:

'For the benefit of those who do not perceive the force of this proof, the following experiment will possibly be convincing. Take a pendulum *AB* four times the length of pendulum *DB* and let it swing about the two centres *A* and *D*, that is to say, from *C* through *B* to *G*, and count four oscillations which we assume to be completed in exactly two seconds of time. Then, having removed the nail *D*, allow the said pendulum to swing freely over the circumference of the radius *AB*; it will be found that the said pendulum will complete three oscillations only in two seconds of time. Finally, having replaced the nail and attached the cord to it so that the pendulum will swing on this side over the circumference of circle *BG*, it will be found that the said pendulum will complete six oscillations in two seconds, whereas the pendulum *AB* completes three about the centre *A* in the same time. Consequently, in the time taken by the longer pendulum to complete one oscillation, the shorter pendulum completes two; and whilst the longer pendulum completes one-half oscillation over the arc *BC*, the shorter pendulum will complete one-half of its oscillation over the arc *BG* in one-half of the afore-said time. Consequently, whatever be the velocity with which it swings, the nature of the pendulum requires that the time of oscillation of pendulum *AB* shall be double the time of oscillation of pendulum *DB*, whenever the length of the latter is one-quarter the length of the former. In the case of the compound oscillations we have four semi-oscillations *CB* and four-oscillations *BG*, that is to say, we have two oscillations of the longer pendulum *AB* and two oscillations of the pendulum *BD*; now, two oscillations of the longer pendulum are completed in one second and 20 thirds, namely, in a time of two-thirds of a second, for this is the time in which it completes three oscillations; and two oscillations of the shorter pendulum *DB* are completed in one-third of the same time, for the latter oscillates six times. Therefore, the time taken by the shorter pendulum to traverse the arc *BG* is exactly one-half of the time taken in traversing the arc of the larger circle *BC*; and, consequently, the distance *CB* will have the same ratio to the distance *GB* as the time of passage over *CB* has to the time of passage over *GB*. Therefore, because the times are proportional to the distances traversed, the velocities are the same; and, so, they traverse equal distances in equal times.

'Whence it follows that its velocity is of necessity perpetually uniform,[25] seeing that the motive force is inherent to each moving body[24]; and as a result it covers equal distances in equal times, whatever be the shape of the path along which the moving body is carried. Consequently, whether the semi-diameter of the pendulum or of the circle be lengthened or shortened, it is, nevertheless, impossible for its velocity to undergo any alteration: in equal times it will always traverse equal distances. Seeing that our senses reveal that there is a real, physical inequality of motion in planetary motion,[26] namely, that the planets do not traverse equal distances in equal times on the path along which they are carried, we are obliged to seek another reason for the said inequality. Therefore, we must either concede that the planet's motive force does not remain always the same, but sometimes increases and sometimes deceases[27];

or we must have recourse to some external cause, in virtue of which the uniform course of the planet, to which it is naturally adapted, is accelerated or retarded.[28]

'With regard to the first [hypothesis], we see that the motion of a given planet never accelerates unless the planet approaches the Sun, and that the acceleration is greater the nearer the planet is to the Sun; conversely, when it is farther away, its velocity changes in the opposite way (in fact, its proper motion is physically retarded by reason of the increase in the afore-said distances). This being assumed, decided and admitted, we could say, seeing that the Sun is as it were the heart or vital source of the motive force of the planets, that the closer the planets approach this source the greater is the energy and strength they receive from it, and [consequently] they develop a greater [proper] motive force. For this reason, they are disposed to traverse their proper orbits in the vicinity of the Sun with a greater *impetus* and a [greater] velocity than they do in those regions remote from the Sun.[29]

'Before passing to [a discussion of] the second hypothesis, certain matters must be considered. In the first place, we shall show that any corporeal mass, no matter how big, floating in a fluid, suspended and balanced in such a way as to be completely unaffected by lateral motion,[30] can be moved transversely by any impulse, however small it be. Let *M* be a large sphere floating in the midst of the fluid *RS*, and so well balanced that it is quite unaffected by lateral motion towards *R* or *S*; then, if we suppress all external impediments, especially the density of the fluid in which the sphere is floating, and if some motive power, such as the atom *P* set in motion from *R* to *S* so as to collide with the sphere on the side *P*, should supervene, I say that the large sphere *M* will of necessity be pushed towards *S*. The sphere *M* whilst at rest is assumed to have no inclination for, nor aversion from, transverse motion towards *S*; consequently its inclination for motion towards *S* is just as much as its resistance [to this motion].[31] Therefore, if an impulse *P*, which is not indivisible[32] but is a given quantity (*quanta*), be added to it, it will necessarily overcome the resistance, which is no kind of force; and as a result the very large body *M* will be deprived of its primitive state of rest and will be pushed towards *S* (Fig. 4).'

Borelli's view, which is the same as Galileo's, was strenuously opposed by Descartes, who, strangely enough, though he formulated the principle of inertia much more clearly and concisely than Galileo, was quite unwilling to admit that a motionless body could be set in motion by any force, no matter how small, and who even improved the Aristotelian concept of the relationship between force and resistance by inventing his admirable but unfortunate theory of 'quantity of rest'. However, without naming him, Borelli proceeded to a refutation of this opinion 'of certain moderns', and to a demonstration that any force, provided that it be *quanta* and not 'indivisible', does confer on a moving body, however large, an impulse which is itself *quanta*.[33]

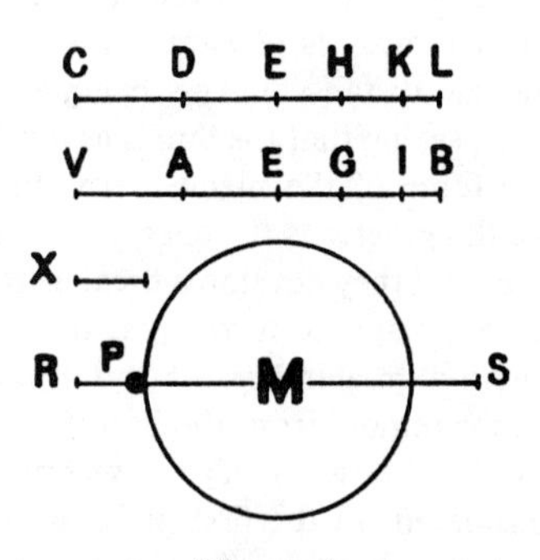

Figure 4.

'Seeing that certain moderns, without any reason but solely by their own authority, assert that a large body *M* cannot be deprived of its rest by a very small corpuscle *P* (being deceived, as I suspect, by the common experience of the property of the equal arms of a balance, where a small weight does not move a large one which is set against it, from which they deduce a general rule without noticing the difference between this operation and the other), the falseness of this assertion may be proved as follows. Because the corpuscle *P* is supposed not to be inert and motionless but in motion, and because motion cannot be conceived unless effected with a certain velocity, it follows that a certain velocity must be ascribed to the corpuscle *P*. Let the velocity be *VB*, and let us assume that the inertia of the body *M* on acquisition of motion[34] has the same ratio to the motive force of the corpuscle *P* as the corporeal bulk [mass] of *M* to the bulk [mass of] *P*, and that the velocity *VB* is to a velocity *X* as the bulk [mass] *M* to *P*, namely, as the inertia and resistance *M* to the impulse *P*. Then, if we assume that the bulk [mass] *M* be urged towards *R* with the velocity *X*, there would be two forces pushing mutually with contrary motions, and as [according to the rules of] mechanics the moment of the impulse *P* is composed not only of the degree of its motive force,[35] but also of the velocity *VB* with which it urges and pushes the moving body *M*, and similarly that the moment of the body *M* is composed of the degree of its resistance and of the velocity *X* with which it is urged against *P*. Now, the ratios of the forces and the velocities are the same, and reciprocally therefore the moment of the force *P* with the velocity *VB* will be exactly equal to the moment of the resistance *M* with the velocity *X*; and so neither will overcome the other, but the forces being equal, they will both be at rest.

'Whence it follows [that] if the velocity with which the moving body *M* is carried towards *M* were less than *X*, then most certainly the force *P* with the velocity *VB* would necessarily overcome the resistance of *M* [moving] with a velocity less than *X*, and so the sphere *M* would necessarily be transferred towards *S*. However, that follows in an even more obvious manner if the velocity of the resistance *M* be assumed to be not only diminished, but completely negligible and nil, as it is when the resistance is assumed to be at rest without any velocity; the moment of the force of the corpuscle *P* with the velocity *VB* will then overcome the resistance of the sphere *M* even more so, as it is not affected with any opposing velocity. Therefore, the great mass *M* will necessarily be deprived of rest by any corpuscle *P* whatsoever, and will be pushed towards *S*.'

Borelli's argument is most ingenious; and faultless, although in his *exemplum* it is not a question of the resistance of a body at rest to the 'acquisition of motion', but, on the contrary, of the resistance of a body in motion to the 'acquisition of rest'. In fact, if the 'moment' of a body be measured by the product of its mass and velocity, *i.e.*, if it be equal to mv, it is obvious that when v is zero, then the product $m.o$ is zero also, and that the motionless body, whatever its size, will offer no resistance to the 'impulse' it receives. This impulse is not an *impetus* or an 'internal motive force', but nothing more than motion itself, a motion which, once it has been impressed on the moving body, remains there indefinitely. Borelli says[36]:

'We have now to prove that any body in motion must, by its nature, move with a constant velocity. That is proved by experiment in the case of projectiles, where, after withdrawal of the propulsive force, the moving body being free, the motion persists in the same direction; and it is proved even more clearly by experiment with pendulums and floating bodies to which motion is imparted by any kind of slight, feeble impulse, [which indeed] persists in the moving body, even though the body giving the impulse withdraws and returns to rest.

'Nevertheless, it is now convenient to examine the nature of this *impulsus* and the manner in which it is impressed and retained. To this question, it is my practice to reply as follows: Suppose a piece of wood to be floating and at rest; suppose another body which is in motion to be present also: now, if the latter, whilst in motion, should collide with the floating, stationary piece of wood, which is unaffected by lateral motion, surely no-one will be surprised that the wood, when gripped and drawn by some instrument, or pushed by the moving body, is displaced and moved with the same velocity as that with which the moving body moves[37]; but here we see nothing more impressed on the wood, or the small boat, other than the motion itself which has been superadded [to it]. In truth, motion, by its nature, is nothing more than the migration [of the object which is moved] from one place to [another] place, and as such it is liable to transport the wood through such a distance in such a time. Consequently, although later on the body that has moved it in the first place does not remain in the vicinity and does not retain any link with it, nevertheless the little ship, according to the nature of the motion that it has already actually acquired, will be able to move through such a distance in such a time. Indeed, it is a law of nature that motion is nothing but motion, that is to say, a migration which would not be such if, by itself, it consumed itself without any opposing resistance. The metaphor used by Gassendi (very fine, no doubt, but hardly relevant) is of no value.[38] Gassendi says, in effect, that the small boat is accustomed by a certain apprenticeship to carry out the motion whilst it is being pushed by the moving body; as if the wood were capable of acquiring knowledge and education. It is preferable to say that motion by its very nature is a state which is quite easily acquired, foreasmuch as that which is in the act of moving, when it strikes a body unaffected by motion, cannot but help displacing it [the body] together with itself; and seeing that it is the nature of this displacement to continue perpetually until some hindrance or destructive cause arises, it follows that it is the nature of any motion whatsoever, deriving from any principle whatsoever, to be able to continue and persist, that is to say, to be able to traverse equal distances as in the preceding equal periods of time.'

A body, when it has been set in motion, retains its motion, *i.e.*, its velocity. Consequently, if another motion be imparted to the body in question, it is superadded to the former and increases the velocity; therefore, through a succession of impulses, no matter how weak (for example, collision with an atom, or thrust from a ray of light) any moving body, no matter how large (a satellite, or even a planet) can be brought to move with as great a velocity as we wish, provided that the number of impulses is sufficiently high—naturally, the final velocity cannot exceed the velocity of the impulses (shocks) themselves. Borelli then continued[39]:

'I say that the motion of the mass M, although it be slow and imperceptible at the start, can nevertheless increase until the mass moves with the same velocity as that of the force P which gives the impulse, provided always that the velocity of the latter is always the same.

'Let *VB* be the velocity of the thrust from *P*, and let us assume that during the first small fraction of time the mass *M* is moved with a small and imperceptible velocity *CD*; draw *VA* equal to *CD*. Seeing that during the second small fraction of time the moving body *M* does not resist, nor is averse from, the velocity of *P*, assuming that it acts in the same direction, the force *P* will press on the mass *M* with the excess of its velocity over the velocity *CD*, and as a result the energy compounded of the motive force *P* and the velocity *AB* will be less than the force by which it thrust the mass *M* during the first small fraction of time. Therefore, the velocity *DF* will be superadded [to the velocity *CD*] and will be slightly less than *CD*. Seeing that the preceding velocity *CD* by its nature persists in the moving body *M*, it follows that during the second small fraction of time the mass *M* will move with a total velocity *CF*. Draw *AE* equal to *DF*. During the third period of time the moving body *M* will be pushed by another degree of superadded velocity *FH*, less than *DF*; thus, the total velocity will increase, though by unequal increments. In this way the velocity *CL* will be [progressively] accumulated [in the moving body *M*], and will finally become exactly equal to the velocity *VB* possessed by the thrusting force *P*.

'I say, then, that the velocity *M* will remain perpetually in this state, and will not increase subsequently. Because the velocity *VB* of the force *P* is equal to the velocity *CL*, therefore the mass *M* flees away with the same speed as that with which the impulsive force *P* pursues it. Therefore, the *impulsus* completely disappears; in fact, if someone with a sword drawn from its sheath tries to touch the collar-bone of another who is running away, and if both are running with the same speed, then it is obvious that the attacker will not inflict any wound on the fugitive. Hence, the ultimate velocity, which is finally impressed on the mass *M*, will be the greatest and will not increase subsequently, although the force *P* continues to follow the moving body *M* and even touches its rear; the force will no longer produce a thrust; and the motion of *M*, if it were circular, would be able to continue perpetually in repeated revolutions.'

However, a difficulty remains. We have assumed, in the course of our argument, that the power of the motive force *P* remains constant and does not give out, nor even diminish, whilst moving the moving body. Obviously, this is impossible: a body that pushes another body loses as much motion as it imparts to the other. Therefore, a correction is needed:

'In the course of our argument, we have assumed that the body *P* pushes the sphere *M* and produces its motion perpetually with the same velocity, which does not seem possible, for, although the body *P*, before it strikes *M*, moves with the maximum velocity *VB*, nevertheless, seeing that it is subsequently obliged to move together with the slow machine *M*, this maximum velocity with which it was previously endowed will not persist. This difficulty must be resolved by showing quite clearly in what manner our hypothesis could be verified. Imagine the corpuscle *P* to be one of those innumerable [particles] of which fluids, such as water, air or fire, are composed; and imagine that all these particles are acting together with the same velocity *VB* in the direction *R* to *S*, as is the case with running water, or with the wind. Certainly, the first particles striking the surface of the globe *M* and producing the first *impulsus* rebound to the side, but they are followed by other small drops which push the mass *M* afresh with the velocity *VB*, and so on, as occurs in the wheels of watermills, and other similar machines.

'Consequently, in any instant of time, particles of water, or air, strike and push the moving body *M* with the same velocity. Seeing that all these particles have the same force with which, one after another, they push the machine in the same direction, it is permissible to imagine and to use [in the argument] a single force *P*, pushing the machine *M* successively with the same force and the same velocity *VB*. . . .

'This being admitted,[40] it is clear that the Sun is the centre of the planetary system and that it turns on its axis, as is proved by the rotation of the spots. Now the [Sun's] rays, which are most efficient, are certainly able to seize and push the planetary bodies in a solar whirlpool; for if light be a corporeal substance diffused by the solar body in the manner of a perpetual wind, this radiating substance should turn in a circle in the same way that the solar body turns; in which case [it would be] not only possible but necessary for planetary bodies, balanced and floating in the celestial aetherial aura, to be pushed by these corporeal rays in motion of translation.[41]

'Although this is not accepted by some who assert that light is an incorporeal substance,[42] nevertheless we should not doubt but that it possesses some motive force and energy, for we see terrestrial bodies moved and agitated by these rays of light: they move vegetable particles by separating them, and raising them together with other vapours and exhalations. We see that the flowers of plants, too, are set in local motion by these same solar rays, as may be observed in the flowers of the fields. It follows that the motive faculty of the solar rays, however weak and small we assume it to be, will be capable, in accordance with the aforesaid example, to push and move the planetary bodies, because the planetary globes are perfectly suspended and balanced in the extremely fluid aether, and have neither inclination to transverse motion, nor aversion from it; consequently, they will be able to be moved by any motive force that pushes them. For this reason the rays of the Sun, however weak we imagine them to be, will be able to push the planetary bodies; and although this motive force at first seems able to impress only a small and insignificant motion, nevertheless the motion will be able to increase progressively to an appreciable velocity. No doubt, at a given first instant of time, the solar rays in turning push the planets but feebly and imperceptibly, but this minimum velocity does not disappear, it remains impressed [in the body of the planet] in the manner demanded by the nature of motion. The first thrust is succeeded by a very weak second *impulsus* from the same solar rays, which doubles the *impetus* of the planet; the third *impulsus* does the same; and so do the fourth and all the succeeding ones. The final result is the maximum velocity that the Sun's rays which are moved in circuit together with the Sun are able to impress.'[43]

The theoretical possibility of the astro-optical mechanism having been established, Borelli, according to his usual practice, sought to confirm it by experimental data, or more precisely, by an analogous example found, or failing that, imagined, in terrestrial reality.[44]

'If I may say so, numerous examples of a similar mechanism offer themselves [to us]; to be brief, we shall choose one of them only, namely, that of a ship, which we shall imagine to be very large and floating on a very calm sea. There is no doubt but that if it were pulled by the most slender hair, or driven by a breath of wind, it would be able to be moved from one place to another. Although, at the start, this motion would be so weak and slow as not to be observable, and consequently the ship would appear to be at rest, nevertheless it is clear that each of these minimum *impulsus* by which it is affected is impressed on it; and having been impressed there, the *impulsus* remains there together with the whole sequence of successive *impulsus*, and so finally produces a force which becomes apparent and manifest, and makes the motion of the said ship observable.

'If, then, we see this happening to a very large body, which, in order that it can move, must divide large amounts of water and drive them away on all sides by displacing them from one place to another, and even overcome the repulsion and friction of the surrounding water, how much more likely would this occur if the said ship were not floating in water, (which is a rather firm and tenacious body), but in a supremely liquid and fluid sea having no tenacity, such as the aether. Anyone can see that the aforesaid ship would be able to be moved from one place to another by an incomparably weaker force than that provided by a woman's hair or the least puff of wind.'

The analogy is obvious; the conclusion *a fortiori* is therefore permissible. Borelli triumphantly concludes as follows:

'The proofs given [above] together with the example of a ship driven by a feeble puff [of wind] are sufficient for my present purpose. By such an example, by the enlightenment of nature itself, we are persuaded that the Sun's rays, however small be the force they possess, have nevertheless been able to take hold of the bodies of the planets balanced in the supremely fluid aether and carry them along. Similarly, the motive rays emitted by Jupiter, however weak be the power ascribed to them, and which are turned around its proper axis like a whirlpool, will nevertheless carry the four Medicean stars around with them in a circle in the extremely fluid aether.'[45]

III

Celestial Mechanics. Conclusion

The mechanism devised by Borelli was most ingenious. Unfortunately, its operation was to lead to effects not exactly in agreement with the astronomical data from which it originated: in fact, by the successive accumulation of innumerable impulses from the luminous rays (or motive force), the planets and satellites should end by moving with a (linear) velocity which is *absolutely constant* for each of the celestial bodies, the velocity being strictly equal to that of the rays themselves at that particular point; furthermore, all the planets (and all the satellites of a central body) should move similarly with an angular velocity equal to that of the rotation of the Sun (or of the central body).

Was Borelli aware of the consequences of his theory? It is difficult to say. Perhaps they escaped his notice. It is even possible, that in order to avoid them, he developed his explanation of the 'real and physical' variation in the velocity of motion of celestial bodies in orbit:—another explanation that derived from Kepler. Borelli says:

> 'It now remains for us to show in what manner and for what reason the motive power which resides in the Sun, or in Jupiter, although it be perpetually of the same degree and uniform in itself, can nevertheless impart a greater velocity to one planet, and a less one to another, according as the planet is close to, or remote from, the Sun, or Jupiter. Now, this can be done very easily starting from certain principles of mechanics which we shall briefly enumerate.'[1]

The mechanical principles to which Borelli appeals are those of the lever and the balance.[2] He imagined the action of the motive rays, as did Kepler, to be similar to that of a lever with its centre of rotation at the centre of the Sun, and the point of application of the force on the surface of the Sun (or of the primary planet): it is obvious that the action of this ray-lever is weaker when the ray is longer; indeed, it is inversely proportional to its length. Here then is the required explanation: the motive rays act more strongly on the planets when they are near to the Sun, and less strongly when they are far away. Their distance constantly varies. Consequently, the velocity varies also, and inversely as the distance.

Now, this explanation, though perfectly reasonable in the Aristotelian dynamics of Kepler, according to which bodies (even celestial ones) are endowed with *inertia* or *resistance to motion*, and a tendency to come to rest, and according to which velocity (*that does not maintain itself in the same state*) is proportional to the motive force, is not at all reasonable in the dynamics of Borelli. In the latter case, indeed, bodies offer no resistance to motion and do not tend to come to rest; their velocity, therefore, is not a function of the motive force which is exerted on them along their orbits; velocity is maintained, and as a result celestial bodies should move with uniform motion with the velocity they have acquired, in spite of momentary, or temporary, weakening of the motive force.

Finally, even without taking into consideration the conservation of accumulated velocity, the Keplerian argument is unacceptable in Borelli's mechanics. In fact, Borelli knew perfectly well, and strange to say he expressly appealed to this principle, that 'impulse' is a function both of the 'power' of the agent and its velocity. . . . Now, the velocity of the agent in question, that is to say, the linear velocity of the point of the motive ray which encounters the planet is greater when its 'power' is weaker. However, the planets go more or less quickly. The reason is that they offer resistance to the *impulsus* of the motive solar rays which is greater the further away they are from the centre of their motion. To explain the existence of these 'resistances' Borelli wrote as follows:[3]

'In the first place, therefore, it must be assumed that for the body of a planet to be deprived of rest by any force whatsoever and to acquire a certain degree of velocity, even though it is indifferent to transverse motion and is placed in the supremely fluid aether it is necessary for this force to consist of *quanta* and not to be indivisible. This is because every body at rest resists in a certain way the force by which it must be moved.[4] For [in order that it shall be moved] this resistance must be equal to the motive force, or even slightly weaker, as is true in a general way, if we disregard the places occupied by the motive force and the moving body, as well as the distances [separating them] from the centre of revolution. For this reason, if we imagine a balance, or a bar, *ABC*, moving round the centre or the support *S*, and assume that the force at the point *A* remains the same, but that one and the same resistance is placed first at *B* and then at *C* so that the distance *BS* is less than *CS*, then there is no question but that the force at *A* required to balance and move the resistance at *B* will be less than that necessary to balance the resistance when placed at the greater distance at *C*, because the resistance of the same moving body at *B* and *C* is proportional to the distances *BS* and *CS* (Fig. 5).[5]

'From the science [of] mechanics I borrow also [the proposition] that when we consider the action and motion of some force on some resistance, it is necessary to take into account not only the degree of the force, but also the velocities with which the force and the resistance are moving. Let *D* be the velocity of force *A*, and *E* the velocity of the

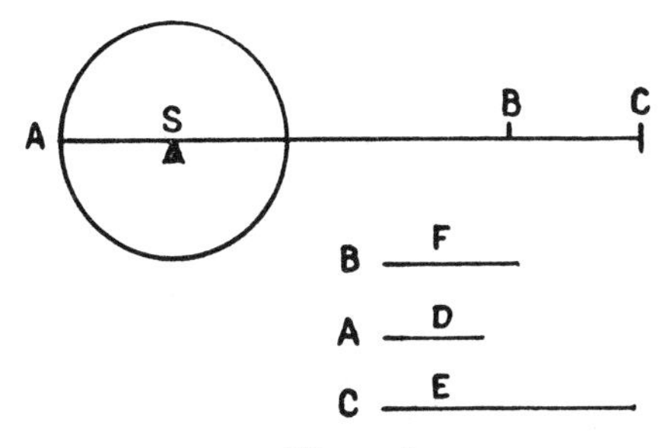

Figure 5.

resistance *B*, and *F* the velocity of the resistance *C*. There is no question but that the moment or energy with which the force and resistance operate will be composed of these two factors in such a way that the degree of the force *A*, together with its celerity, must equal and even exceed somewhat the degree of resistance of the moving body *B* similarly taken with its velocity. Their moments will then certainly be equal reciprocally, because the ratio of the force *A* to the resistance *B* will be reciprocally the same as the ratio of the velocity *E* of the moving body *B* to the velocity of the other one *A*. Similarly, in order that the moment of *A* shall be equal to the moment *C*, it is necessary for the force *A* to be in the same ratio to the resistance *C* as the velocity *F* is to the velocity *D*. Seeing that the moment of the moving body *B* is equal to the moment *A*, and similarly that the moment of the same *A* [is] equal to the moment of *C*, the resistance *B* must be to the force *A* as the velocity *D* is to the velocity *E*, and also the force *A* must be to the resistance *C* as the velocity *F* is to the velocity *D*.'

We realize that the foregoing is nothing more than an exposition of the conditions of equilibrium for the Roman steelyard; or, in other words, an exposition of the theory of the lever. A clumsy exposition, no doubt, and even rather confused; but correct for all that.

On the other hand, the following is much less satisfactory. Up to now, Borelli was considering the case where different weights (resistances) placed in different positions kept the same force (weight) in equilibrium. He now goes on to a study of the case where the same weight is placed at different points on the steelyard where the equilibrium is upset. He says:

> 'The equality being disturbed, seeing that the [ratio of the] resistance of the same moving body at *B* to its resistance at *C* will be the same as that of the velocity *F* of *C* to the velocity *E* of *B*; but the [ratio of the] resistance of the same moving body at *B* to its resistance at *C* is the same as that of the distance *BC* to the distance *CS*; therefore the velocity *F* with which this same moving body must move at *C*, in order to equalize the moment of *A* which is constant, to the velocity *E* which it must have at *B* will be as *BS* to *CS*; and, consequently, the measure of the velocity that it will have at *C* will be *BS*, and the measure [of the velocity] of the same moving body at *B* will be *CS*.'

In order to restore equilibrium, that is to say (according to Borelli), to re-establish equality between the 'moments' of the force and the resistance, in spite of moving the body *B* from *B* to *C*, it is necessary and sufficient for its velocity at *C* to be less that its velocity at *B*, and it must be exactly inversely proportional to its distance from *S*.

Borelli's reasoning which replaces the 'force' (acting downwards) of a weight placed on the long arm of a steelyard (that is to say, which replaces its moment) by its resistance to motion upwards enables him (at least, so he thought) to reverse the lines of Kepler's reasoning: it is not the motive force that decreases as it gets further away from the centre of revolution, but the resistance of the body which is moved that increases. He has only one more step to take: namely, to compare the motions of the planets (which move freely in the aether) with those of weights sliding on the arm of a steelyard, then to assign to them a resistance to motion, and to impose on them maintenance of this strange equilibrium, the conditions for which have been deduced above. Triumphantly he says:[6]

> 'This being assumed, imagine [that] the solar, or Jovian, body *AS* turns about its own proper centre, and [that] the globe of a given planet is close to the Sun at one time, and then further away at *C* at another time. Now, seeing that the force with which the Sun acts is measured by the

force of its rays (which are constant and have always the same energy) and by the velocity of its own whirlpool, which similarly remains unchanged[7]; and seeing that its moment is composed of these two factors, [and] that this moment must equal the two resistances of the same planet at B and C, it is necessary for the moment to act with greater efficiency against the smaller resistance of the planet at B (and, therefore, turn the planet with greater velocity) than that with which it acts against the greater resistance of the same planet situated at a greater distance at C; consequently, it is moved with a motion that is slower in the reciprocal ratio of the resistances, or of the distances, as we have said.'

Thus, the resistance resulting from increased distance plays the same rôle for Borelli, as does attenuation of the *species motrix* for Kepler; and it is this resistance that explains not only the decrease in effect from the thrust of the solar rays (or motive forces) which move the planets, but also the fact that the planets do not maintain their velocity once it has been acquired, and that they slow down in the course of their progression in proportion as they get further away from the Sun (or from Jupiter).

Borelli's theory is absurd; but we must not be too severe on the Italian *savant*: the task that he had set himself, namely, to transform Kepler's celestial dynamics in such a way as to make it acceptable for a Galilean, was really very difficult; indeed, too difficult. Strictly speaking, it was impossible. However, the credit and exemplary value of Borelli's work lie in this attempt—and in its failure. We must now pass to a consideration of his deduction of planetary orbits, for we are not yet at the end of our labours. Indeed:[8]

'We have explained why the motion of a planet, which passes round the terrestrial globe at different distances, can be effected at different velocities reciprocally proportional to the said distances.[9] It still remains to be shown how, and by what necessity, the planets approach and recede from the globe about which they turn, namely the Sun or Jupiter. Now, this will be the second element of which the elliptic motion of planets is composed. In order to take account of these approaches and recessions Kepler imagined that one face of the planet was friendly towards the Sun, and the other, hostile, in the same way that a magnet has one part that attracts iron, and another that repels it. This concept, however ingenious it may be, seems quite unable to be adapted to the appearances of the planets, and as a result we are obliged to abandon it, especially as Nature can produce these effects by other means.'

Borelli does not say in what respect Kepler's theory does not agree with the 'appearances'. Personally, I very much doubt if that were the reason why he found himself obliged to abandon it. The fact is that he had never accepted it: a concept based on the existence and action of forces of attraction and repulsion (even magnetic ones) could never find favour in the eyes of Borelli, the Galilean. His own theory has precisely the advantage of avoiding these magic forces,[10] and also of demonstrating how unnecessary it was to postulate their existence.

Kepler's theory, in spite of all its weaknesses—real weaknesses which I have analysed in detail in their proper place—nevertheless, contained one element of very great value: it subjected the planets to the action of forces of attraction and repulsion depending on their distance from the Sun. Gravity or the natural tendency of planets to come closer to the Sun (or of satellites to come closer to their central body) was, on the contrary, *a constant force* for Borelli: why, indeed, should this tendency change according as the distance from the central body becomes more or less? Therefore, starting from a set of constant forces producing constant effects, Borelli undertook to deduce the wide range of sidereal motions.

Borelli's solution is extremely simple and elegant, and his principle may be formulated as follows: constant, equal forces acting in opposite directions produce, generally speaking, a state of equilibrium. However, when the equilibrium is disturbed in favour of one of the forces, periodic variations result, for the interaction of these forces ends in establishing a corresponding contrary state of disequilibrium, from which the series of operations starts again. For example, a weight suspended by a cord does not move. Displace it from its vertical position, it descends, but does not stop in its former position; on the contrary, it rises to the same height from which it started to fall. It then descends once again, and this series of motions would be reproduced for ever if the retarding resistances were completely eliminated. Plunge a wooden cylinder into a vessel full of water, in a certain position it will be in equilibrium and will remain at rest being partly above, and partly below, the level of the water. Raise it above its equilibrium position, it will start to descend, but having arrived at the equilibrium position, it will not stop there. On the contrary, it will continue to descend until it attains a state of disequilibrium similar and corresponding (but in the opposite direction) to that from which it started.[11]

The explanation of these phenomena is obvious: every moving body in falling to the equilibrium position acquires a degree of velocity which is capable of carrying it to the height from which it started; this means that it is able to overcome an equivalent resistance. Thus, a pendulum overcomes the resistance offered by gravity to its motion upwards; and the wooden cylinder, when placed in a vessel full of water, overcomes the pressure of the water which pushes it upwards during its movement downwards. In both cases, the motion ceases only when the acquired velocity is completely absorbed by the resistance, that is to say, when the new position of disequilibrium (equivalent to that at the starting point) is reached.

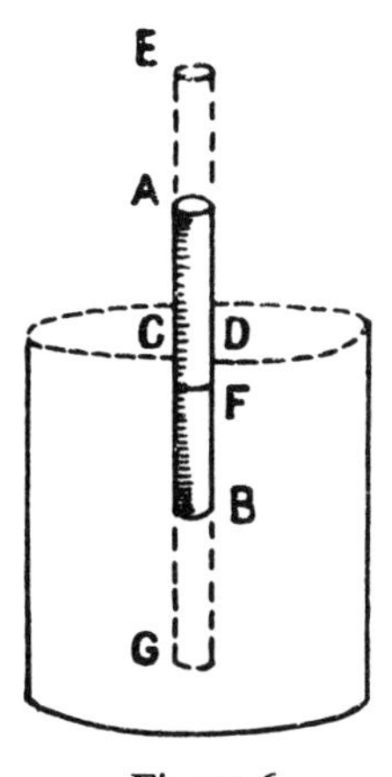

Figure 6.

Gravity then acts freely again on the pendulum, which falls (and ascends to the initial position); the pressure of the water then pushes the cylinder upwards again, and the cylinder in virtue of the velocity acquired in this ascent does not stop at the equilibrium position, but continues its upward movement until it reaches the initial position (Fig. 6).

Borelli starts by recalling and ruling out 'animastic' explanations of the periodic variations in the distances between planets and their motive centres:[12]

> 'We see that all animals are endowed with perpetual pulsation in the heart, that is to say, a certain systole and diastole which are also observed in the arteries: similarly, all parts of an animal are endowed with a certain peristaltic motion by which they dilate and contract. We might assume that planets have a similar perturbation by means of which they approach and move away from their proper vital source about which

they move in orbit, and so perform a pulsation somewhat similar to that of the heart.

'In truth, other natural operations in which we see similar effects produced by blind necessity present themselves to those to whom these views are not acceptable on account of their *animastic* character. Such, for example, is the behaviour of pendulums through which they spend long hours performing their proper oscillations which would continue indefinitely if the retarding influences were completely removed; but it is preferable to consider another natural operation, more like the operation of the planets. Let the wooden cylinder *AB* be arranged in such a way that it floats in the vessel *RS* perpendicularly to the water-level *CD*; and so that portion *CB* is immersed and the portion *AD* emerges above the water-level, when the cylinder is in equilibrium with the water, in which position the cylinder stays motionless. Then, let the cylinder be raised and transferred to the position *EF* so that the emergent portion is *ED*; it is obvious that the said cylinder raised to such a position will not stay there, but descending into the water will become submerged. Seeing that the fall of any heavy body supplies it with more and more *impetus* and degrees of velocity beyond that which is appropriate to its weight in every instant of time during which it falls, therefore, when the said cylinder arrives at the position *AB*, it will not remain stationary there, but will be pushed downwards by the degrees of velocity acquired during the preceding descent, and will be carried, let us say, to the point *G*; and seeing that the weight of the water resists such a fall in an ever increasing proportion, and pushes in the oppostite direction, that is to say, upwards, it follows that the degrees of velocity acquired by the moving body during the preceding descent are continuously exhausted, until, when they are entirely exhausted, the force thrusting on the surrounding water starts to act, that is to say, pushes the cylinder upwards, *i.e.*, from *G* towards the water-level *CD*; then, because the said cylinder continuously acquires more degrees of velocity during the whole time in which it is rising and is constrained to rise still higher, therefore, when it reaches the position *BA*, it does not stop there, but, in virtue of the degrees of velocity already acquired, it will rise [still] higher, though not to the position *EF*. Thereafter, for the reasons given above, it will fall and rise until, after all these emersions and immersions, it remains stationary in the position *AB* with which it is naturally connected.

'However, it must be noted here that if we could completely suppress the influences and accidental causes that continuously diminish and weaken the above-mentioned up-and-down oscillations, and which are composed mainly of the tenacity and corpulence of those parts of the water surrounding the cylinder, then there is no doubt but that the oscillations would not only persist for a very long time but would even be everlasting. In fact, we know that a plumb-line continues to swing to-and-fro in air for a very long time by virtue of the same natural causes, whereas, on the other hand, its motion in water is spent after very few oscillations. Consequently, if the aforesaid cylinder were floating in the supremely fluid aether, the cause of the decrease in the rise and fall would be completely suppressed[13]; and therefore this motion ought to be perpetual so long as no other causes arise to retard it.'

It is obvious, moreover, that whilst the pendulum, or the wooden cylinder, carry out their oscillatory motion, their velocities do not remain constant, but change continuously from one instant to the next, and in such a way that there is first of all an increase from zero up to a certain *maximum* which is then followed by a decrease to zero.

Now, this is exactly what happens in the case of planetary motion. There, too, we are dealing with opposing forces—gravitational force and centrifugal force—whose initial disequilibrium is perpetuated and reproduced eternally by virtue of simple mechanical principles: the same principles that come into play in any kind of motion. The elliptical orbit of planets is merely a strictly unavoidable consequence.

Borelli was very proud of his discovery, as we can easily understand, especially if we consider it *a parte ante* and not *a parte post*, and if we compare it not only with the solutions put forward by Hooke and by Newton, but also with those of a Boulliau and a Roberval. Furthermore, Borelli, like Newton and Leibniz after him, did not regard the reduction of celestial physics to terrestrial physics at all dangerous from the religious point of view: on the contrary, by exposing the perfection of the work, the supreme perfection and supreme wisdom of the Divine Architect are revealed to us. So he writes:[14]

'It seems to me that philosophers have excellent reasons to praise that passage in Aristotle where he enquires into the ingenuity and wisdom with which animal functions are accomplished, for example, the digestion of food; the formation of chyle, blood and [animal] spirits; their distribution in their proper places; the propagation of the species; all of which are performed for definite purposes by supreme and mysterious art and intelligence. The philosopher says, in fact, that it is not necessary for the animastic faculty to be intelligent and architectonic, and that it accomplishes its work in every part of the animal, but that it suffices for it to produce its operations in the same manner as is done in a well-ordered Republic, in which it is not necessary for the prince and supreme legislator, as well as the magistrate and judges themselves to accomplish and give effect everywhere to everything that they have prescribed; but it suffices for them to have available the means and agents disposed orderly and wisely everywhere so that they operate by themselves and conform to the established order. We shall have something similar to say about the admirable order by which the [celestial] machine is moved and directed. There is no doubt but that a supreme and admirable art is revealed by the motion of the planets, because this is the chief feature of the Republic of the Universe disposed and established in wonderful order by the infinite wisdom of the Divine Architect; and it does not appear necessary for spirits, intelligences or souls to produce everywhere the motions which He Himself has prescribed, and that they should, as it were, take the starry globes by the hand; for, on the contrary, the Divine Architect has been able to order and dispose all things with such wonderful skill, that having done so, everything conforms without hesitation or deviation to the divine commands with His general assistance only: this seems to me more worthy of the infinite wisdom. Indeed, more intelligence and greater skill are needed to construct a [self-]moving machine than an inert one. Thus, if there were two architects, one of them having at his disposal a machine with various wheels driven by the force of weights with such skill that it shows the time, the course of the Sun and the Moon, and with the help of musical instruments performed various symphonies besides similar things; whilst the machine of the other architect could do exactly the same, though not automatically but by the agency and work of servants who show the time with their own hands, trace out the course of the Sun and the Moon, and make music with the help of sound and voice, surely no-one will deny that the work of the first architect is superior by its dexterity and ingenuity; and if we knew of our own certain knowledge that the architect [of the second machine] was the cleverest and shrewdest, we should certainly charge him with negligence, seeing that his machine is so inert that it requires the help and constant stimulus of servants, and is unable to function by itself. In the same way, therefore, seeing that we have admitted that this excellent work, the Universe, has been made by the best, greatest and cleverest Artist; and furthermore, seeing that it is obvious that the planetary motions have been arranged with such diligence and skill that they work by themselves like a clock,[15] it seems quite incredible and absurd that the Divine Architect should have wished to work with less skill, that is to say, by making the planets completely inert so that they would need guides and would have to be thrust around their proper orbits by the hands of servants. Verily, if such an argument be without appeal to the most learned, then I am not the one who would wish to defend his opinion at all costs,[16] but it will be enough for me to have indicated (whilst keeping within the limits of natural reasoning) an easy and possible way of arranging the planetary whirlpools with all the contrivances that are observed in them.

'Imagine the solar globe turning on its proper axis from West to East, at *S*; and the body of a planet at *A*, which by a natural instinct tends to approach the Sun by a direct motion, in the same way that we know that all heavy bodies have a natural instinct to approach our Earth, namely, impelled by the force of gravity which is naturally united with them, and [in the same way that we know that] iron moves in a straight line towards a magnet.[17] For this reason, it would not be impossible for the body of the planet to have a certain faculty, similar to the magnetic faculty, by means of which it moves towards the solar globe; and, in fact, as we know [that] the planet never leaves the Sun, nor departs beyond a certain fixed space of its confines, this seems to show more than adequately that it is retained by a kind of magnetic force, which, moreover, cannot be received [by the planet] unless we imagine that the planet also possesses a certain natural instinct to unite with the Sun and to move towards it.[18] Now, therefore, this will be the first element of which the eccentric revolution of the planets is composed.[19]

'In the second place, we shall assume that the aforesaid planet is carried in orbit round the Sun by the whirlpool of solar rays moving from West to East on the periphery of the circles; and because, as we have said, circular motion naturally impresses a certain *impetus* on the moving body by which it moves away from the centre and is repelled from it, as may be observed in the case of a sling or a wheel, it follows that, as the aforesaid planet moves circularly, it will move away from the centre of the Sun *S*.[20] Furthermore, such a repulsion will take place with a greater or less *impetus* according as the circumduction of the planet is more or less rapid; and, in fact, as we have hinted above, this circumduction is faster when the planet is nearer to the Sun. Therefore, so long as the planet remains in the highest and most distant position *A*, this force or *impetus* to move away from the Sun is extremely weak; but when the planet approaches the Sun [and arrives] at the point *B*, its circular motion becomes faster, and, as a result, the planet is repelled from the Sun towards *A* by a greater *impetus*. For the same reason, when it has arrived at the point *C*, it will be repelled by the Sun by a greater force; [and this force of repulsion will be] still greater at the point *D*, and will reach its maximum at the point *E*. There will now be two linear motions opposing each other: the one, perpetual and uniform, by which the planet *A*, pushed by its own inherent magnetic force, progressively approaches the solar body[21]; the other, on the contrary, non-uniform and continually decreasing, by which the planet is repelled by the Sun through the force of circular motion; [this repulsion will be] particularly strong at *E*, of medium strength at *C*, and negligible at *A*. Thus, as we have proved above, by compounding the said motions we shall have a certain force and a compound *impetus* on which the period (variation) in velocity acquired by the planet depends, which, from the furthermost limit *A* up to the nearest at *E*, will increase in the same ratio as the distances decrease.[22] Let us assume [that] the force of approach is equal to the force with which the planet is moved away from the Sun, as is the case at the intermediate [point] *C*. If, then, we imagine that the Author of Nature had placed the planet at the point *C* in the beginning, the moving (thrusting) force by virtue of which at that point the planet approaches the Sun would be exactly equal to the force of repulsion by which it is moved away from the Sun, which [force] comes from the circular whirlpool *CGL*.[23] Consequently, the planet [placed] in such a position would remain there perpetually, and would be carried round on the circumference of the circle *CGLN* about the solar centre *S*; and this motion would be uniform, namely, in equal periods of time it would traverse equal arcs on the circle. This would be necessary, because the thrusting force of the planet would be unable to overcome the opposing

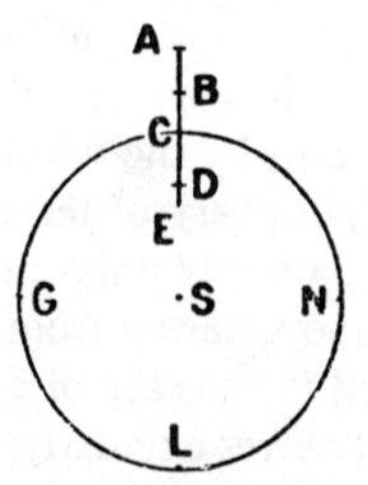

Figure 7.

repelling force, and *vice versa*. Now, if these forces were equal (as we have assumed) the result would be that the planet *C*, although placed in the most fluid aether offering no resistance, would never be able to waver, nor depart from the periphery of the circle *CGLN*, but would persist in this motion with perfect constancy as though attached to a solid sphere, or floating in a spherical ocean; and because all the motive forces are assumed to act perpetually in the same manner, there would be no reason for this circular motion to slow down, or to depart from its initial course; and, consequently, it would of necessity complete its natural circuits round the Sun perpetually (Fig. 7).

'Let us now assume that the Divine Wisdom for very lofty and inscrutable purposes had decreed that the motion of the planets round the Sun should be eccentric and not circular in shape, but elliptical. [This end could be realized] with supreme economy, for nothing more would be necessary than to create and place the planet in the beginning, not at the point *C*, but at the more distant point *A*.[24] Eccentric and elliptical revolution of this planet round the Sun would then result automatically. In fact, as we have said, the motive force of the planet at

A is compounded of circular impulsion, [corresponding] to the uniform thrusting force, and the weaker one [corresponding to] the repelling force[25]; therefore, as a result of the excess of the former over the latter,[26] the planet must approach the Sun; for which reason the said planet [in moving] in its circular whirlpool will get closer and closer to the Sun in all subsequent instants of time. It is consequently impossible for the body of the planet during an indefinite time, such as *AB*, to move over the same circular periphery *AG*; it must pass from a larger circle to a smaller one until it comes to *F*. For the same reason, [in the case of motion] in the circle *BF* during the following period of time *BC*, the thrusting force, being still greater (but to a less extent) than the repelling force, continually pushes the planet towards the Sun[27]; for this reason, on leaving *F*, [it] will not be able to continue its motion along the circle *FX*, but must pass to other smaller and smaller circles until it comes to the limit *G* on circle *CG*, which is mid-way from the Sun. We have already pointed out that during the descent *AC* the planet acquires additional degrees of velocity, which are indelibly impressed on it[28]; and so, even if the thrusting force were to disappear completely and were reduced to nothing and consequently no longer be able to act, the planet would not stop at *G*, but would approach still closer to the Sun, continuing its progression thereto, because the degrees of velocity of motion acquired during the fall exceed the repelling force.[29] Therefore, the planet will not move on the circumference *GT*, but on other smaller ones, until it arrives at *H* at the end of time *CD*, whence, seeing that a certain portion of the impressed force acquired during the fall is still fully active, it will advance still further towards the Sun, though with a slower motion[30]; for which reason it will not travel over the periphery *HV*, but over others, smaller and still smaller, until all the *impetus* being completely exhausted, and all approach coming to an end as a result, the planet arrives at perihelion[31]; but here, as a result of the very great velocity of motion[32] of the planet on a very small circle, a very large repelling force comes into play, and so, during the time *ED* the planet progresses backwards in ever larger circles in such a manner that it arrives again at *N*, then at *O*, and finally at the supreme limit *A*. It is clear from what has been proved that not only the angles *HSP* and *PSM* must be equal, but also the distances (lines representing the movement of the rays) *SH* and *SM* [must] be equal also, and similarly at *G*, *N*, *P* and *O* (Fig. 8).'

Thus, the constant force of gravity is opposed by the variable force of centrifugal repulsion, and as a result the initial disequilibrium produces a motion by which it is recreated. Will the orbit so described be an ellipse? Borelli thought so, and we can easily understand why. In the first place, he was unable to calculate the orbit that would really result from the hypotheses he assumed;[33] secondly, he had so carefully drafted his theory on Kepler's that, rightly or wrongly, he was firmly convinced of their complete (mathematical) identity. In fact, in both theories the (linear) velocities of the planets are inversely proportional to their distances from the Sun; for Borelli, if not for Kepler, this was a necessary and adequate reason for the elliptical shape of their orbits. As for the distances themselves, whether they vary as a function of the combined action of the forces of magnetic attraction and of repulsion, or vary as a function of the combined centripetal tendency of planets to move towards their central (fixed)

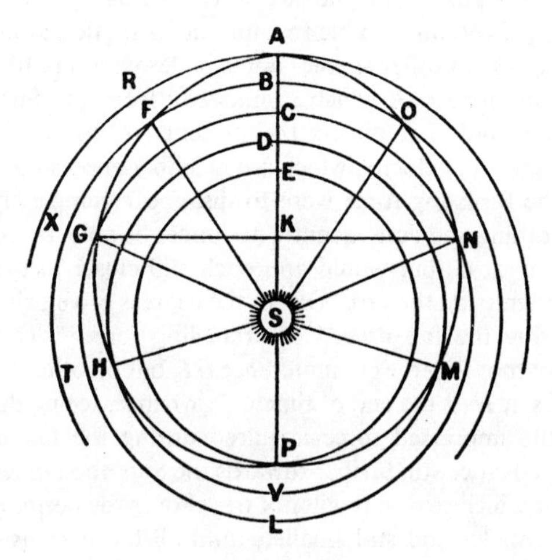

Figure 8.

body and of the centrifugal force (variable also, like the velocity, inversely as the distance), that does not affect the working of the system. If Kepler be right, then Borelli is right, too. Kepler, as a mathematician, was certainly right, seeing that Boulliau himself, who criticised Kepler's celestial mechanics so shrewdly and pertinently, took Kepler's analyses as a basis for his theory.[34] Mathematically, the principles of Boulliau and of Borelli are equivalent; the former being merely kinetic, the latter being truly dynamic,[35] as Borelli triumphantly declared.

'We must now show that the orbit *AFGH* of the planet is not a circle, but an ellipse. Let us go back to the elliptical orbit *EOK*, ascribed to the planet by Bullialdus, in [one] focus of which is placed the Sun, the motion of equality taking place about the remaining [empty] focus in such a way that all the equant circles on which the planet is carried are described about the axis of the cone drawn through the focus *M*. Then, the semi-diameters of the said unequal circles are exactly equal to the polar radii *ME*, *MY*, *MO*, *MR*. Because the velocity of the planet increases by increments that equal those of the circumferences of the equant circles, and because the latter are proportional to the said polar radii, the motion of the planet accelerates from its least speed at *E* (aphelion) successively by the same increments as the polar radii increase in length.[36]

'On the other hand, we have shown that the same planet moves round the Sun *S* in its orbit *AGP*, and not round an unreal, imaginative point of equality *M*, but round the body of the Sun *S*; and we have shown that the motion accelerates progressively from the least speed of the planet at aphelion *A*, and all the more so as the polar radii *SF*, *SG*, etc., become shorter, in accordance with the laws of motion depending on physical causes, namely, the closeness of the planet to the Sun, and its repulsion by the whirlpool of solar rays which thrust it in such a way that the velocity of the planet is inversely proportional to the periphery of the various circles which it traverses, or to the radii of the [same] circles which are the successively decreasing polar distances from the Sun. Furthermore, at each of these positions the increments in the polar radii *ME*, *MY*, *MO* in the theory of Bullialdus, or the decrements in the radii *SA*, *SF*, *SG* in our theory, are proportional to the versed sines of the angles that the motion of the planet subtends in its orbit. Furthermore, in both theories respecting the same planet, the eccentricity *NX* and *SV* is exactly the same, and hence the least polar distance *ME* in the theory of Bullialdus will be equal to the distance *NK* of the planet at perihelion, namely, equal to our perihelion *SP*. Because the greatest polar distance *MK* will be equal to the distance *NE* from aphelion, therefore *MK* will be equal to *SA* seeing that the planet is at the same distance and perihelion.

'Furthermore, the slowest speed of the planet at *E* (aphelion) will be exactly the same according to both theories, and similarly the greatest speed at perihelion will be the same. Finally, at one and the same fixed instant of time, the same planet will be at *Y* according to the theory of Bullialdus, but at *F* according to ours; consequently, the distances *NY* and *SF* from the Sun will be equal, and so, too, will be the distances of the radii from the points *Y* and *F* from aphelion *E* or from *A*. It will be the same, when the planet is at *O* and *G* at some later instant of time; and so on, because the polar radii *ME*, *MY*, *MO* in the ellipse *EOK* increase in the same ratio as the polar radii *NE*, *NI*, *NO* decrease (Fig. 9). The latter are equal to the solar radii *SA*, *SF*, *SG*; consequently, the polar radii of Buallialdus increase in the same time in the same ratio as the solar radii *SA*, *SF*, *SG* decrease. As a result, two motions take place about the centres *M* and *S*; which motions start with the same degree of slowness at the aphelia *E* and *A*, and each of them acquires equal degrees of velocity in equal times; for example, at *Y* and *F*, or at *O* and *G*. During one motion, the velocities increase in the same ratio as do the polar radii; and during the other, in the ratio by which the solar radii decrease. Seeing that the greatest [distances] *MK* and *SA* are equal, and similarly that the least [distances] *ME* and *SP* [are] equal, it follows that the paths *EOK* and *AGP* are exactly equal and similar; and as the former, *EOK*, was proved by Bullialdus to be elliptical, the path *AGP* of this planet will be an ellipse also, equal and similar to *EOK*. Therefore, etc., . . . Hence the path described by the planet will be excentric [with

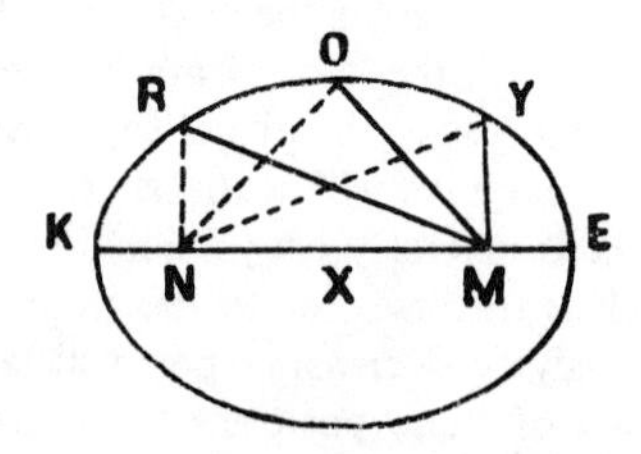

Figure 9.

respect] to the Sun, and not circular, but elliptical. The time required to cover the straight path from *A* to *E* must, therefore, be the same as that required to complete one-half revolution *AGP*: but for that to take place, nothing more is needed except to assume that the said rectilinear motion depends on the magnetic force thrusting from *A* towards *E*, and that the circular motion of the planet round the Sun depends on the thrust from the solar rays; and because, after one completed revolution, the same causes, being stable and perpetual, come into play again, the period . . . necessarily reproduces itself perpetually. . . .'

Poor Borelli! he was really on the track of a great discovery. . . . If he did not succeed, but failed to reach the position attained by Hooke in this very same year of 1665, in which he (Borelli) wrote his *Theoricae Mediceorum Planetarum*, it was because he was blinded by the brilliance of Galileo, and was unable to recognize the fruitful germs of a gravitational theory in the work of Kepler. He repudiated the concept of attraction which inspired Hooke and Newton in spite of its seeming—and real—obscurity, and, like Galileo, he held fast to the *fact* of gravity. Thereby, he renounced any theory going beyond bare fact derived from experience, and as a result barred his own path to progress. Hooke and Newton were more courageous. Newton's intellectual daring, no less than his genius, enabled him to overcome the obstacles that hindered Borelli, and to discover in celestial phenomena a precedent for events that happen on Earth.

Notes

INTRODUCTION

1. J. A. Borellus, *De motu animalium*, 2 vols., Rome, 1680, 1681; second edition, Leyden, 1685.

2. *Euclides restitutus*, Pisa, 1658; and *Elementa conica Apollonii Pergaei et Archimedis opera nova breviori methodo demonstrata*, Rome, 1679.

3. *De vi percussionis*, Bologna, 1667; *De motionibus naturalibus a gravitate pendentibus*, Reggio di Calabria, 1670; second edition, Leyden, 1686.

4. *Theoricae mediceorum planetarum ex causis physicis deducta*, Florence, 1666. Cf. Kurd Lasswitz, *Geschichte des Atomismus*, Hamburg and Leipzig, 1890, vol. II; see also my '*De motu gravium naturaliter candentium in hypothesi terrae motae*', *American Philosophical Society, Transactions*, Philadelphia, 1955, vol. 45, p. 4.

5. As a result, opinions never held by him and which he even opposed have not infrequently been attributed to him, for example, that of gravific attraction. The article, '"Borell's Hypothesis" and the Rise of Celestial Mechanics', by Angus Armitage in *Annals of Science*, 1950, vol. 6, p. 268, is an exception in this respect.

6. The reader will be able to judge from the examples which will be given later. However, it must be pointed out that bad writing was not peculiar to Borelli. Most people in the seventeenth century wrote atrociously badly. Writers of the standard of Galileo and Torricelli were quite rare exceptions, and that is why they alone are still read . . . and that provides the reason why I have thought it necessary to quote Borelli as fully as possible.

7. Cf. 'La gravitation universelle, de Kepler à Newton', *Archives d'histoire des sciences*, 1951, p. 138; and my article 'La mécanique céleste de Borelli', *Revue d'histoire des sciences*, 1952, vol. 5, p. 101.

8. Cf. W. W. Rouse Ball, *An Essay on Newton's Principia*, London, 1893, p. 159. Letter from Newton to Halley dated 20 June 1686: 'Borelli did something in it and wrote modestly.' Cf. my 'An unpublished letter of Robert Hooke to Isaac Newton', *Isis*, 1955.

9. Borelli, in fact, took into account celestial reality as found in Kepler's

discoveries, which were ignored by Galileo and Descartes. With regard to Galileo, see Erwin Panofsky, *Galileo as a Critic of the Arts*, La Haye, 1954, and *Isis*, 1955, as well as my review 'Attitude esthétique et pensée scientifique', *Critique*, 1955; for Descartes, see Jean Pelseneer, 'Gilbert, Bacon, Galilée, Kepler, Harvey et Descartes: leurs relations.', *Isis*, January, 1932.

10. J. A. Borelli, *Theoricae mediceorum planetarum* . . . , *Ad Lectorem* (pp. vi f.):

'*Scito igitur aestate huius anni telescopium ingens, ac mirae perfectionis industrij, ac solertissimi Iosephi Cãpani Serenissimo Magno Duci Haetruriae missum fuisse, hoc admirabili instrumento primo Saturnum, posteà Iovem observare cœpimus, tunc iussu suae Celsitudinis ex tabulis Galilei aephemerides Mediceorum calculis deduxi, ut quotidiè vespertinis horis praedicto telescopio situs eorumdem precogniti observarentur, interim quamplurima de motibus, positionibusque Mediceorum disserebantur, vnde accidit, ut me non aduertentem, & reluctantem, in eiusmodi speculationibus implicuerim, ac paulatim, vt fit, vna speculatione reliquam sibi connexam trahente, factum est, vt hoc opusculum è manibus exciderit, quod cum ostendissem Serenissimo, Sapientissimoque Principi Leopoldo, eiusque acerrimo iudicio submisissem, censuit ipse, pariterque alij amici, vt quam primum ederetur, indeque post meum à Florentia discessum amici excudendum curarunt.*'

11. Cf. *ibid.*, p. iii:

'*Non dissimile templum magnus ille Galilaeus preclara sua, & admirabili doctrina edidit, & regali, ac glorioso nomini Mediceo, non in hisce infimis, ac terrenis regionibus, sed in supremo ipso Caelo, ac Ioviali solio è fundamentis erexit, atque tuae gloriosissimae prosapiae dicauit, gloriamque, & nomen amplissimum in illa serena, ac beata regione perpetuò duraturum facnauit* . . .'

12. 'Ernst Goldbeck, Die Gravitationshypothese bei Galilei und Borelli,' *Wissenschaftliche Beilage zum Jahresbericht des Luisenstädtischen Gymnasiums zu Berlin, Ostern*, 1897; Berlin, 1897.

13. J. A. Borelli, *Theoricae mediceorum planetarum* . . . , *Proemium*, pp. 1 f.:

'*Primus omnium in suo nuncio sydereo doctissimus Galileus diuulgauit, quatuor planetas circa Iovem in orbe agi, quos ipse Medicea sydera nuncupauit, qui non secùs, ac Luna Terram, Iovis corpus circumeunt, cuius respectu phases omnes, quas nobis Luna exhibet, representant, plenas scilicet, silentes, falcatas dico tmas, gibbas, unà cum suis omnibus mirandis ecclipsium phaenomenis, quae Lunae ipsi contingunt respectu nostri: Ordinem quoque, ac magnitudinem orbium ipsorum planetarum eorumque reuolutiones, ac reuolutionum periodes adinuenit; verùm, licet sagacissimum illud ingenium agnouisset in praedictis paruis planetis, varias illas atque multiplices anomalias esse debere quae in alijs etiam errantibus syderibus observantur, breui tamen suae vitae tempore eas inuenire non potuit; At postquam ille fato concessit, quamuis multi in hoc insudarint, nunquam tamen ne minimam quidem notitiam acquirere potuerunt praeter illam quam nobis ipsemet tradidit Galileus. Cumque mihi tam sanè ardua, ac perdifficilis, quàm longa ac laboriosa praedictarum eccentricitatum, ac periodorum videatur inquisitio, contemplatricem partem sum aggressus, animaduertens, non paucos extitisse, qui cum multum circa eiusmodi materiam studij ac operae contulissent, nedum aliquid noui non indagarunt, verum quid ipsi etiam exquirere debuissent ignorauerunt, quod à ianua profecto aberrare, est, vti aegregiè Aristoteles asserebat; si ego itaque praedicta mediceorum planetarum admiranda phaenomena non reperi, non inutile tamen, ac periucundum futurum existimaui, si iter alijs ostenderem rectè circa haec phaenomena meditandi, patefaciendo scilicet quos habituri sunt motus, & habitudines satellites Iouis, dummodo circa ipsos sit eodem modo philosophandum, quo natura in alijs Stellis erraticis operatur; tales autem motus relationes, ac habitudines physicis necessitatibus arguendas duxi, à quibus tales effectus veresimiliter pendent; tandemque ostendi modos varios, ac artificia quibus varietates praedicta reperiantur*'.

14. J. A. Borelli, *op. cit.*, cap. I, p. 3: *Mediceorum syderum motus similes esse debere motibus caeterorum planetarum:*

'*Ad Mediceorum syderum theorias ritè inquirendas, & explicandas necessariò profectò esset exacta quaedam, ac omnimoda praecognitio eorum motuum, & multiplicis anomaliae, quorum equidem vestigatio, cum nobis tam breui tempore ab eorum prima cognitione tradita à Galileo minimè permissa sit, circa ipsos motus vlterior contemplatio non dari videtur, omnis enim nostra cognitio scientifica, ac discursiua, à sensibus, atque experimentis oriri debet, sed licet id verum sit, ostendam nihilominus nobis dictorum mediceorum syderum dispostiones motionesque perquirere datum esse, paucis tantum illis obseruationibus praeeuntibus, quae nobis vsque adhuc innotuerunt, vice enim earum, quae nobis desunt permutltis satisque notis caeterorum planetarum obseruationibus vti possumus, cum quibus iouialia sydera in omnibus motuum generibus debent vniuersim conuenire; idque ex hoc primo principio, ac axiomate deduci potest, naturam scilicet ad omnia sua munia obeunda simplicissima semper facillimaque media adhibere, tandemque eam varietate non delectari, diuersisque rationibus operandi, verùm constanti semper perseueranti ijsdem organis vti, a instrumentis eademque methodo, cum effectus inter se similes operatur. Huius rei innumera propè dixerim exempla suppeterent, verùm animalium plantarumque structuram actionesque perpendere satis erit, in quibus easdem naturales, vitales, ac animasticas operationes producit, eadem prorsus organa, ac motus adhibendo*'.

15. Cf. E. Goldbeck, n. 12, *supra*, *op. cit.*, pp. 18 f.

16. *Ibid.*, p. 19.

17. This does not seem to have been accepted without opposition on the part of the ecclesiastical authorities. The dedication to the Grand Duke, Ferdinand II, is dated 10 October 1665, but the *imprimatur* is dated 26 February 1666. Furthermore, as pointed out by Goldbeck, the 'I' in 'MDCLXVI' on the title-page appears to have been added subsequently. Borelli's Copernicanism was rather too flagrant for the authorities to ignore it. . . .

18. Planetary motion was explained by the pressure from the whirlpool of solar rays; that of the satellites by a pressure from the motive-rays coming from the primary planets. The satellites were subjected to the pressure of both whirlpools, which pushed them first in the same direction, and then in opposite directions. This theory, which was obviously inspired by Kepler, is quite interesting. In fact, it foreshadows the explanation given by Newton (admittedly on the basis of attraction and not of physically applied force) of certain anomalies in the Moon's motion. A detailed study of this is beyond the scope of the present work.

19. He ought, in fact, to have linked the Moon's motion in its orbit (round the Earth) to the Earth's rotation on its axis, and even to its orbital motion.

CHAPTER I. THE PROBLEM OF PLANETARY MOTION

1. Cf. my introduction to N. Copernic *Des Révolutions des Orbes Célestes*, Livre I, Paris, 1934; and *supra*, p. 57 f.

2. *Astronomia Nova ΑΙΤΙΟΛΟΓΗΤΟΣ sive Physica Coelestis, tradita commentariis de motibus stellae Martis*, [Heidelberg], 1609. In the title of Borelli's work, the words *ex causis physicis deducta* correspond to *ΑΙΤΙΟΛΟΓΗΤΟΣ* of Kepler.

3. Ismael Bullialdus, *Astronomia Philolaica*, Paris, 1645.

4. Celestial matter offering no resistance to motion, a consideration of motive forces was pointless.

5. Astronomers traditionally regarded their science as a computing and pragmatic discipline whose aim was *salvare apparentias* and to permit forecasts without postulating the reality of the 'circles' or 'spheres' they employed. Cf. P. Duhem, *La Théorie Physique, son Objet, sa Structure*, Paris, 1908.

6. In fact, the persistence of rest or of uniform circular motion needs no physical explanation; but the constant change in velocity does.

7. If planetary motion, even circular, could be explained in Kepler's view only by the action of a physical cause, nevertheless, it did not produce centrifugal forces. In order to prevent the Moon from falling on to the Earth an 'animal' force was necessary.

8. Cf. my *Études Galiléennes*, Paris, 1940; and the introduction to G. de Santillana, *Dialogue on the Great World Systems* of Galileo, Chicago, 1953.

9. J. A. Borelli. *Theoricae Mediceorum Planetarum . . .*, pp. 31 f.

10. Borelli is pushing prudence rather far!

11. The mistake in Kepler's theories consists in the adoption of a theory of attraction. In other respects, Borelli follows Kepler very closely, as will be seen later.

12. Ismael Bullialdus (Boulliaud). Boulliaud's astronomical theories are set forth in his *Astronomia Philolaica*, Paris, 1645. The criticism of Kepler will be found on pp. 210 f. I have reproduced it above in Appendix III to the section dealing with Kepler; see p. 371.

13. Nothing is more curious or significant than this obstinate desire to safeguard the principle of uniform circular motion as a basis of celestial mechanics, even to the extent of such an artificial construction as that of the imaginary cone *ABC*.

14. Seth Ward; first Savilian Professor of Astronomy in the University of Oxford; afterwards Bishop of Salisbury and F.R.S. Cf. *In Ismaelis Bullialdi Astronomia Philolaica Fundamenta, Inquisitio Brevis*, Authore Setho Wardo . . . Oxford, 1654; (this is the book to which Borelli referred).

Seth Ward wrote: '*Keplerus, vir Ingenio, industria, & (quod omnium instar) felicitate inventorum, Admirabilis, ex Planetarum Theoriis circulos amovit; Excentricorum atq; Epicyclorum strues penitus sustulit; vnà simplici linea eaq; Ellipticâ motus Planetarum proprios absolvi magno labore evincit.*' Somewhat later in his *Astronomia Geometrica . . . Opus Astronomis hactenus desideratum*, London, 1656, dedicated with admirable eclecticism to *Nobilissimis Astronomis Paulo Nelio, Equiti Aurato Anglo, Johanni Hevelio, Dantiscano, Petro Gassendo, Matheseos Professori Regio apud Gallos, Ismaeli Bullialdo, Astronomiae Philolaicae Autori, Jo. Bap. Ricciolo, Astr. Professori, Bononiae*, Seth Ward resolutely declared himself to be a Keplerian, and thoroughly upholds and adopts Kepler's law of planetary motion, the ellipticity of planetary orbits, without ascribing any physical cause to it.

15. Borelli, *Theoricae Mediceorum Planetarum . . .*, cap. XI, pp. 45 f.: '*Propositions ou principes philosophiques nécessarires pour l'intelligence des causes des excentricités des orbes des astres médicéens, et des figures elliptiques décrites par ceux-ci.*'

16. Nobody ever maintained this hypothesis in the form given by Borelli. It is equivalent, in his mind, to the theory of homocentric spheres.

17. It is strange to note that, whereas Kepler was preoccupied with the question: why the planets do not unite with the Sun (and the Moon with the Earth), it is the contrary question which bothers Borelli: why do they stay in the vicinity, instead of moving off? The reason is that Kepler in his celestial mechanics ignored centrifugal forces, whereas Borelli took them into account.

18. The second hypothesis is a mixture of Keplerian concepts with those set forth by Roberval in his *Aristarchus Samius*, Paris, 1647.

19. This quasi-magnetic force which appears here is not gravity; neither is it a force of attraction, but only a tendency.

20. According to Kepler, the mutual attraction of similar material bodies (Earth and Moon), which explains the phenomenon of gravity, is proportional to the *moles* (mass) of these bodies; but the magnetic attraction of the planets by the Sun is not a function of their mass.

21. Experiments of this kind were carried out by the *Accademia del Cimento* which was greatly interested in the determination of specific weights.

22. Like the vortices of Descartes.

23. Borelli does not say what these difficulties are.

24. Borelli is exaggerating! The 'absurd solidity' of the spheres had been admitted for centuries.

25. Borelli is faithful to the Galilean concept of heaviness: gravity (whose nature is unknown) is a natural tendency of bodies: stones tend towards the Earth, the planets tend towards the Sun, but they are not attracted. The difference between these two concepts has frequently been misunderstood by modern historians; for us, however, gravitation and attraction are identical. It was not so for thinkers of the seventeenth century; for them, they were two quite contrary things. Hence, Borelli (following Galileo) is able to assert that the planets *tend* or *gravitate* towards the Sun, and even that this gravitation is the effect of a quasi-magnetic force (cf. *supra*, p. 507), whilst at the same time denying that they are attracted by it; in general, he even denies the existence of attraction in the Universe; cf., Joh. Al. Borellus, *De Motionibus naturalibus a gravitate pendentibus*, Leyden, 1686, cap. VI, pp. 166 f.: '*Nullam Attractionem, nec vim Tractivam in Natura dari.*'

26. It has sometimes been asked (for example by E. Goldbeck, *Die Gravitations-hypothese bei Galilei und Borelli*, Berlin, 1897, p. 22) if Borelli really realized that centrifugal force results from the tangential inertia of bodies in circular motion. The diagram (Fig. 2) used by Borelli to illustrate his theory—it was missing from Goldbeck's copy—and the accompanying text (cf. p. 481) leave no doubt on this score. This is not to his credit seeing that Galileo (to say nothing of Descartes and Huygens) already had quite correct ideas on the matter.

27. Galileo, as we know, assumed a gravitation of the planets towards the Sun, at least at the moment of creation: the planets fell towards the Sun in order to acquire the velocity with which they move (cf. *Dialogo* in *Opere* (Ed. Naz., vol. VII, p. 44)); but the action of gravitation no longer functions since their motion of fall was altered by God into circular motion. For Borelli, this gravitation is a constant factor in the Universe. It should be noted, however, that this gravitation operates only between the planets

and the Sun, satellites and the central body; it is nowise universal: satellites do not tend towards the Sun, and the planets do not tend towards each other. The generalization of gravitation was the achievement of Hooke and Newton.

28. J. A. Borelli, *Theoricae Mediceorum Planetarum . . .*, pp. 47 f.

29. Borelli's experiments are most curious, and deserve to be seen in their historical reality. For that reason, I have given a full description of them (quoting Borelli himself) instead of restricting myself to a modernized outline, as has been done by those historians who have dealt with the matter. Cf. Borelli's diagram, *supra*, p. 481.

30. Like everyone else, and particularly Kepler, Borelli assumes that a magnet acts only within a fixed 'sphere of action'.

31. 'Natural inclination' or 'natural instinct' is contrasted here—as everywhere else—with magnetic or quasi-magnetic *attraction*, although it is itself a quasi-magnetic force.

32. In company with Bruno and Kepler, Borelli does not admit the existence of an interstellar vacuum: space is filled with aether.

CHAPTER II. THE SOLAR WHIRLPOOL

1. Borelli, *Theoricae Mediceorum Planetarum ex causis physicis deducta*, p. 51.

2. I retain 'animastic', used by Borelli, instead of using the general term 'animal'.

3. Borelli's hostility to the notion of attraction was so great that he preferred to say: *the magnet approaches iron*, and not: *the magnet attracts iron*.

4. The explanation of planetary motions by the action of 'animastic' forces was still current in the seventeenth century in anti-Copernican quarters. Cf. J. B. Riccioli, *Almagestum Novum*, Bologna, 1651, vol. II, Lib. IX.

5. Kepler had already said the same thing: it is pointless to have recourse to 'animal' forces when physical forces are adequate. Kepler had added that the existence of guiding intelligences, obliged to pass their lives in calculating the continuous changes in direction and speed of motion of the celestial bodies, would be extremely distressing, *valde misera*; cf. *supra*, p. 225.

6. There never was a God of gravity—except, perhaps, the God of Newton.

7. Borelli speaks the language of pre-Copernican astronomy—knowingly no doubt.

8. Borelli's problem was exactly the same as that of Kepler: to a sixteenth (and even seventeenth) century mind, a circular motion (or combined

circular motions) could, if need be, be explained by the action of non-intelligent causes; but it was a quite different matter when the motion was not circular or linear.

9. Even for a mind of the stature of Fermat's, the distinction was far from clear. Cf. Fermat, *Œuvres*, vol. IV, Supplément, pp. 33 f.

10. Borelli could have invoked the authority of Copernicus (cf. *supra*, p. 52) for whom a retardation in motion of the planetary spheres as they increased in distance from the Sun was the fundamental law of the system of the Universe. Perhaps it was too dangerous to do so.

11. Borelli, *Theoricae Mediceorum Planetarum* . . ., pp. 52 f.

12. The foliot.

13. It is strange that Borelli, who so clearly saw the error in the argument he was criticizing, made the same mistake himself; cf. *supra*, p. 500.

14. This is quite correct. The mistake that Borelli was contesting consists in not seeing that it is the distance travelled (and hence the time of travel), and not the velocity of the moving body, that varies.

15. As I have already previously noted, Borelli's theory was developed under the influence of Kepler's. It differs from the latter by the fact that, in the case of the Sun, Borelli ascribed the motive function to the rays of light; a solution which Kepler had been obliged to reject because, in his semi-Aristotelian dynamics in which the persistence of motion implies continuous action of the driving force, the occultation of one planet by another would bring about an arrest of motion of the former. Furthermore, the action of solar rays, which are attentuated as a function of the square of the distance of the planet from the Sun, would be much too weak and would be unable to maintain, or produce, a motion whose velocity is inversely proportional to the distance. Because Borelli adopted the law of conservation of velocity, he was able to ignore these difficulties; but he was obliged to introduce into his theory a duality that did not exist in Kepler's: if the Sun move the planets by its rays of light, then the latter move their satellites by motive rays *sui generis*. Kepler's *species motrix* took charge of everything. Cf. *supra*, p. 199.

16. The idea underlying this argument consists in an erroneous application of the simile of a pendulum to planetary motion: the planets are regarded as executing round the Sun a motion similar to that of a circular pendulum whose radius-vector is shortened and subsequently lengthened, and the mistake is precisely that of confusing angular velocity with linear velocity. Furthermore, and this is a view shared by Borelli with Ismaël Boulliau, the curvilinear orbit of the planet is regarded as being composed of an infinite number of infinitesimal circular elements (cf. *supra*, Fig. 8): the planet, even when describing an ellipse, moves in fact in an infinite number of circles, constantly passing from one to the other.

17. Borelli, *Theoricae Mediceorum Planetarum* . . ., p. 53.

18. Borelli—and Galileo too, moreover, though he denied it—makes a

strict distinction between internal principle of motion, as for example, gravity, and external principle, namely, thrust or traction.

19. Borelli's exposition is dynamically conceived, or, at least, is formulated in pre- or semi-Galilean terminology: velocity is said to be an operation of the motive power, and uniform motive power is associated with uniform velocity. In fact, Borelli was well aware that a constant motive force produces accelerated motion, and that once a velocity has been enforced on a moving body it is conserved by the body. Cf. *supra*, p. 491. The motive power in question here is therefore nothing more than motion itself, or, better still, quantity of motion.

20. Borelli's use of Galileo's famous experiment is most ingenious: Galileo had designed it to demonstrate that a body falling from a certain height along a path of any length acquires sufficient velocity to enable it to regain the same height (by any path whatsoever). Borelli draws the conclusion therefrom that the linear velocities of the bodies in question are the same, seeing that the times of travel are proportional to the lengths of the paths.

21. It is inversely proportional to the square root of the lengths.

22. Proportional to the square root.

23. Borelli, *Theoricae Mediceorum Planetarum* . . ., p. 54.

24. *Ibid.*, p. 54.

25. Borelli's argument, taken literally, is quite wrong: the motion of a pendulum is nothing less than uniform motion precisely because the 'intrinsic motive force' of the moving body, namely, gravity, is constant. In fact, we have here an archaistic translation of the Galilean thesis: the velocity acquired by a moving body during fall depends only on the height from which it falls.

26. The planets, or, in pre-Copernican astronomy, the centres of their epicycles, do not move with uniform (linear) velocity; this results in the necessity of introducing 'equant' circles or points in the mechanism of their motions. Copernicus, as we have seen, suppressed the 'equants'; Kepler, on the other hand, recognized afresh their necessity, at least, so long as planetary motions continued to be represented by means of circular motions (cf. *supra*, p. 227), for the variations in planetary velocities are real and not apparent. In astronomy of the 'ellipse', the equant point becomes the second focus of the ellipse.

27. As always, Borelli says, 'the motive force of the planet', and not simply its motion, or its velocity; cf. *supra*, p. 485.

28. The variation in velocity of planetary motion was not a physical problem in pre-Keplerian astronomy, because it resulted from a combination of uniform (in principle, if not in fact), circular motions, and especially because pre-Keplerian astronomy did without dynamics. It became the main problem of Keplerian and post-Keplerian astronomy.

29. Borelli, *Theoricae Mediceorum Planetarum* . . ., pp. 56 f. Borelli is

expressing himself in pre-Galilean language, by seeming to accept the axiom of Aristotelian dynamics: a constant force produces a constant velocity. In fact as I have already point out (cf. *supra*, p. 479) and as we shall see later (p. 491), the motive force is something quite different in his view from velocity itself.

30. The restriction to lateral 'motion' expresses the condition of hydrostatical experience; furthermore, for Borelli as for Galileo, the upward and downward directions were always qualitatively distinct.

31. The Aristotelian language should not conceal from us Borelli's faithfulness to Galilean concepts: inclination as well as resistance to motion are equal to zero, or at least, smaller than any finite value (*quanta*).

32. Infinitely small.

33. Borelli, *Theorica Mediceorum Planetarum* . . ., p. 57.

34. Note the subtlety of Borelli's expression: inertia, *i.e.*, opposition or resistance not to *motion*, but to the *acquisition* of motion.

35. The 'motive force' of the force P and the 'resistance of the body M, correspond to their masses, and consequently their 'moments' are equal to $P.VB$ and $M.X$ respectively. Borelli's terminology is explained by the fact that his mind remains dominated by the model of the balance.

36. Borelli, *Theoricae Mediceorum Planetarum* . . ., pp. 58 f.

37. Borelli has in mind the case where the driving force is linked with the moving body.

38. Cf. P. Gassendi, *De motu impresso a motore translato*, Paris, 1642.

39. Borelli, *Theoricae Mediceorum Planetarum* . . ., p. 59; cf. Fig. 4, p. 490.

40. *Ibid.*, pp. 61 f.

41. Borelli's views on light—they are similar to those of Kepler and undoubtedly derive from him—are rather curious; and it is desirable to elucidate them, because they differ both from modern views and from the views of the schoolmen. In fact, for Borelli as well as for Kepler, luminous rays, which are propagated in space with infinite velocity, are not the result of a succession of emissions, but are stable and permanent corporeal (although not material), entities, which persist and remain attached to the source from which they are emitted. Consequently, these rays, rectilinear and rigid—physical manifestations of the linear rays of geometrical optics—participate in all the motions of the source, and turn with it when it is endowed with a rotatory motion (the beam of light from a lantern sweeps out a path in space). Consequently, when an object turns before our eyes and presents different faces in succession, fresh rays of light do not continuously strike our eyes; they are the same rays, which have already been emitted, and which move *laterally* with the object itself. As a result, the rays that move in this manner exert a lateral pressure on objects and carry them along. It is this whirlpool of rays which carries along with it the 'extremely subtle aether' which fills interstellar space, and it is that also

which progressively confers on planets their own particular velocity. The action of the motive rays from non-luminous celestial bodies (the primary planets) is conceived on exactly the same model; cf. *supra*, p. 477 and n. 15.

42. The incorporeal nature of light was generally admitted in medieval optics on account of its immaterial nature. It was precisely against this identification of corporeality with materiality that Kepler took his stand; in his view, light was both immaterial and corporeal.

43. Cf. *supra*, p. 491. The rays of the Sun continuously impress additional velocity on the planets, though it is always extremely small, and these additions of velocity are conserved until the motion of the planets has accelerated to a velocity equal to that of the solar rays themselves.

44. The 'experiments' quoted by Borelli were frequently imaginary experiments; see my article mentioned above, p. 574, n. 4.

45. The concept of rigid rays (cf. *supra*, p. 494 and n. 41) underlies this curious (Keplerian) notion of motive rays.

CHAPTER III. CELESTIAL MECHANICS. CONCLUSION

1. Borelli, *Theoricae Mediceorum Planetarum*, p. 63.

2. The misinterpretation of the principles of the lever and the balance, as well as their application to conditions with which they had no connection, was quite common in the seventeenth century. Kepler, the geostatists and Fermat vie in incomprehension on this matter. So, we must not be too hard on Borelli.

3. *Op. cit.*, pp. 63 f.; cf., *supra*, p. 499.

4. This assertion seems to contradict Borelli's fundamental thesis according to which the resistance to motion is zero. No doubt, he had confused in his mind the resistance of bodies to the acquisition of motion (acceleration) and the resistance of weights on a balance (for a weight to be raised, the force must be greater than the resistance).

5. From this correct statement Borelli drew the wrong conclusion, namely, that it is the resistance of the moving body itself which increases with distance. Borelli's diagram is wrong: *E* should be put instead of *F*, and *vice versa*.

6. *Op. cit.*, p. 65.

7. Borelli here makes the same mistake as that with which he charges those who believe that the circular motion of a given body varies simply with its distance from the centre of revolution, that is to say, he confuses angular velocity with linear velocity. In the absence of this mistake, and assuming that the 'power' of the solar rays is the same everywhere, we must conclude that their action does not remain the same, but, on the contrary, increases with the distance . . . In fact, the *linear* velocity of these rays is proportional to distance, and therefore their 'moment' is greater as the latter increases. Borelli would have arrived at a better result

if he had followed Kepler more closely and had accepted a falling off in the action of the solar (or motive) rays as the distance increases. By applying Kepler's formula for the attenuation of *light* with distance (it varies inversely as the square of the latter) and combining it with the acceleration of the lateral motion of the rays, he would have obtained a falling off in the 'moment' exactly proportional to the elongation of the ray.

8. Borelli, *Theoricae Mediceorum Planetarum*, pp. 64 f.

9. Borelli adopted Kepler's incorrect calculation according to which the planetary velocities are inversely proportional to the distance, *i.e.*, to the radius vector.

10. Naturally, Borelli does not deny the existence of magnetic forces; nor even their relationship with gravity—it must not be forgotten that Galileo, and Gilbert before him, had declared the Earth to be a large magnet. Borelli can have no hesitation, therefore, in speaking of the magnetic or quasi-magnetic force of gravitation, but he regards it always as a *tendency*, besides being *constant*.

11. If the Earth were pierced from one side to the other by a shaft passing through its centre, and if a heavy body were thrown into this shaft, then this body, according to medieval dynamics, would not stop at the centre of the Earth, but would rise to the surface at the other end of the shaft; it would then fall again to the centre and rise to its starting point, and so on. Nevertheless, this to-and-fro motion (like the swings of a pendulum) would, according to medieval dynamics, decrease in amplitude in virtue of the internal resistance of the body to (violent) motion, and therefore would of necessity end at the centre of the Earth. In the Galilean dynamics of Borelli these oscillations could persist indefinitely.

12. Borelli, *op. cit.*, pp. 66 f.

13. Borelli does not ask himself if the 'supremely fluid aether' would offer enough resistance to the fall of the cylinder in question, and would have enough force to push it upwards. . . .

14. Borelli, *op. cit.*, pp. 74 f.; cap. III: *On the necessity of an elliptical shape for planetary orbits.*

15. The comparison of the *machina mundi* with a clock was borrowed from Kepler; cf. *supra*, p. 378, n. 8, but by his insistence on the perfection of the machine Borelli was a forerunner of Leibniz.

16. Borelli is being cautious. It must not be forgotten that Riccioli in 1651 had declared, and in 1665 had upheld, that belief in motive intelligences was consistent with Holy Writ.

17. It is characteristic of Borelli—and reveals the depth of his aversion from the idea of attraction—that he does not say: iron is *attracted* by a magnet, but: iron *moves towards* a magnet. Cf. *supra*, p. 520, n. 3.

18. The natural instinct of planets to unite with the Sun and to move towards it was assumed by Galileo; (cf. *supra*, p. 519, n. 27) but not by Kepler, who (like Copernicus) believed that there was a great qualitative

difference between the nature of the Sun and the planets, whereas there was no difference, or practically none, in the view of Galileo and his followers.

19. It is interesting to note that if Borelli speaks here of gravity as being a natural instinct of bodies to approach the Earth (as did Copernicus also), and not to the centre of the Universe, he completely avoids ascribing to the planets any tendency to unite with their central body.

20. The introduction of centrifugal forces into the mechanism of planetary motion was Borelli's great innovation. It implies the concept of the infinite extent of the Universe, about which he naturally has nothing to say

21. Borelli's *argument* opposes the action of a constant force (gravitation) by a variable one (centrifugal force); but the language in which it is couched seems once again to assume proportionality between the motive force and the velocity, and to imply a constant velocity of approach (or fall) of the planet to the Sun. However, he has only expressed himself badly; in fact, he had already explained that a constant force produces accelerated motion (*Theoricae Mediceorum Planetarum*, pp. 70 f.).

22. Borelli's idea of the nature of centrifugal forces is very imperfect. In his view, centrifugal force is inversely proportional to the distance (radius vector), not in the case of constant linear velocity (which would be right), but in the case where the velocity itself varies in that manner, because centrifugal force seems to him to be proportional to velocity, and moreover thereby seems to ensure perfect equilibrium in the system.

23. Borelli is mistaken. For this effect to take place, God must not only place the planet at the point *A*, but in addition give it an appropriate lateral velocity.

24. We have already seen that a state of disequilibrium causes an oscillatory motion which recreates the initial state of affairs, and so on indefinitely.

25. Repelling force—centrifugal force.

26. Excess of the thrusting force over the repelling force—excess of the centripetal force over the centrifugal force.

27. The centrifugal force increases, but never manages to counterbalance the natural tendency of the planet to move towards the Sun.

28. As in the case of the pendulum, the velocity, or more correctly, the *velocities* acquired during fall are conserved.

29. In the same way that a pendulum does not stop in the vertical position, or the wooden cylinder stays in its equilibrium position.

30. Impressed force, *impetus* are nothing more than the motion and the velocity themselves; cf. *supra*, pp. 491 f.

31. During this approach, the linear velocity of the planet round the Sun is continually increasing, and as a result, the centrifugal force increases also.

32. At perihelion the initial situation is reversed: the centrifugal force is greater than the centripetal force of gravity. A disequilibrium, equivalent and in the opposite sense to the initial state, is thereby produced, and the

oscillatory motion recommences and goes through the same phases in the inverse direction.

33. We cannot blame him for that. The calculation is extremely difficult and was not accomplished even by Newton.

34. Borelli had a great admiration for Boulliau, who certainly possessed superior mathematical ability.

35. Borelli, *Theoricae Mediceorum Planetarum*, pp. 79 f.; cf. the diagram on p. 510.

36. Borelli puts the Sun at the focus N in Fig. 9. The focus M is then the point of equality.

Name Index

A CATALOG OF SELECTED

DOVER BOOKS

IN ALL FIELDS OF INTEREST

A CATALOG OF SELECTED DOVER BOOKS IN ALL FIELDS OF INTEREST

DRAWINGS OF REMBRANDT, edited by Seymour Slive. Updated Lippmann, Hofstede de Groot edition, with definitive scholarly apparatus. All portraits, biblical sketches, landscapes, nudes. Oriental figures, classical studies, together with selection of work by followers. 550 illustrations. Total of 630pp. 9⅛ × 12¼.
21485-0, 21486-9 Pa., Two-vol. set $29.90

GHOST AND HORROR STORIES OF AMBROSE BIERCE, Ambrose Bierce. 24 tales vividly imagined, strangely prophetic, and decades ahead of their time in technical skill: "The Damned Thing," "An Inhabitant of Carcosa," "The Eyes of the Panther," "Moxon's Master," and 20 more. 199pp. 5⅜ × 8½. 20767-6 Pa. $4.95

ETHICAL WRITINGS OF MAIMONIDES, Maimonides. Most significant ethical works of great medieval sage, newly translated for utmost precision, readability. Laws Concerning Character Traits, Eight Chapters, more. 192pp. 5⅜ × 8½.
24522-5 Pa. $4.50

THE EXPLORATION OF THE COLORADO RIVER AND ITS CANYONS, J. W. Powell. Full text of Powell's 1,000-mile expedition down the fabled Colorado in 1869. Superb account of terrain, geology, vegetation, Indians, famine, mutiny, treacherous rapids, mighty canyons, during exploration of last unknown part of continental U.S. 400pp. 5⅜ × 8½. 20094-9 Pa. $7.95

HISTORY OF PHILOSOPHY, Julián Marías. Clearest one-volume history on the market. Every major philosopher and dozens of others, to Existentialism and later. 505pp. 5⅜ × 8½. 21739-6 Pa. $9.95

ALL ABOUT LIGHTNING, Martin A. Uman. Highly readable non-technical survey of nature and causes of lightning, thunderstorms, ball lightning, St. Elmo's Fire, much more. Illustrated. 192pp. 5⅜ × 8½. 25237-X Pa. $5.95

SAILING ALONE AROUND THE WORLD, Captain Joshua Slocum. First man to sail around the world, alone, in small boat. One of great feats of seamanship told in delightful manner. 67 illustrations. 294pp. 5⅜ × 8½. 20326-3 Pa. $4.95

LETTERS AND NOTES ON THE MANNERS, CUSTOMS AND CONDITIONS OF THE NORTH AMERICAN INDIANS, George Catlin. Classic account of life among Plains Indians: ceremonies, hunt, warfare, etc. 312 plates. 572pp. of text. 6⅛ × 9¼. 22118-0, 22119-9, Pa. Two-vol. set $17.90

ALASKA: The Harriman Expedition, 1899, John Burroughs, John Muir, et al. Informative, engrossing accounts of two-month, 9,000-mile expedition. Native peoples, wildlife, forests, geography, salmon industry, glaciers, more. Profusely illustrated. 240 black-and-white line drawings. 124 black-and-white photographs. 3 maps. Index. 576pp. 5⅜ × 8½. 25109-8 Pa. $11.95

THE BOOK OF BEASTS: Being a Translation from a Latin Bestiary of the Twelfth Century, T. H. White. Wonderful catalog real and fanciful beasts: manticore, griffin, phoenix, amphivius, jaculus, many more. White's witty erudite commentary on scientific, historical aspects. Fascinating glimpse of medieval mind. Illustrated. 296pp. 5⅜ × 8¼. (Available in U.S. only) 24609-4 Pa. $6.95

FRANK LLOYD WRIGHT: ARCHITECTURE AND NATURE With 160 Illustrations, Donald Hoffmann. Profusely illustrated study of influence of nature—especially prairie—on Wright's designs for Fallingwater, Robie House, Guggenheim Museum, other masterpieces. 96pp. 9¼ × 10¾. 25098-9 Pa. $8.95

FRANK LLOYD WRIGHT'S FALLINGWATER, Donald Hoffmann. Wright's famous waterfall house: planning and construction of organic idea. History of site, owners, Wright's personal involvement. Photographs of various stages of building. Preface by Edgar Kaufmann, Jr. 100 illustrations. 112pp. 9¼ × 10.
23671-4 Pa. $8.95

YEARS WITH FRANK LLOYD WRIGHT: Apprentice to Genius, Edgar Tafel. Insightful memoir by a former apprentice presents a revealing portrait of Wright the man, the inspired teacher, the greatest American architect. 372 black-and-white illustrations. Preface. Index. vi + 228pp. 8¼ × 11. 24801-1 Pa. $10.95

THE STORY OF KING ARTHUR AND HIS KNIGHTS, Howard Pyle. Enchanting version of King Arthur fable has delighted generations with imaginative narratives of exciting adventures and unforgettable illustrations by the author. 41 illustrations. xviii + 313pp. 6⅛ × 9¼. 21445-1 Pa. $6.95

THE GODS OF THE EGYPTIANS, E. A. Wallis Budge. Thorough coverage of numerous gods of ancient Egypt by foremost Egyptologist. Information on evolution of cults, rites and gods; the cult of Osiris; the Book of the Dead and its rites; the sacred animals and birds; Heaven and Hell; and more. 956pp. 6⅛ × 9¼.
22055-9, 22056-7 Pa., Two-vol. set $21.90

A THEOLOGICO-POLITICAL TREATISE, Benedict Spinoza. Also contains unfinished *Political Treatise*. Great classic on religious liberty, theory of government on common consent. R. Elwes translation. Total of 421pp. 5⅜ × 8½.
20249-6 Pa. $7.95

INCIDENTS OF TRAVEL IN CENTRAL AMERICA, CHIAPAS, AND YUCATAN, John L. Stephens. Almost single-handed discovery of Maya culture; exploration of ruined cities, monuments, temples; customs of Indians. 115 drawings. 892pp. 5⅜ × 8½. 22404-X, 22405-8 Pa., Two-vol. set $15.90

LOS CAPRICHOS, Francisco Goya. 80 plates of wild, grotesque monsters and caricatures. Prado manuscript included. 183pp. 6⅜ × 9⅝. 22384-1 Pa. $5.95

AUTOBIOGRAPHY: The Story of My Experiments with Truth, Mohandas K. Gandhi. Not hagiography, but Gandhi in his own words. Boyhood, legal studies, purification, the growth of the Satyagraha (nonviolent protest) movement. Critical, inspiring work of the man who freed India. 480pp. 5⅜ × 8½. (Available in U.S. only)
24593-4 Pa. $6.95

ILLUSTRATED DICTIONARY OF HISTORIC ARCHITECTURE, edited by Cyril M. Harris. Extraordinary compendium of clear, concise definitions for over 5,000 important architectural terms complemented by over 2,000 line drawings. Covers full spectrum of architecture from ancient ruins to 20th-century Modernism. Preface. 592pp. 7½ × 9⅜. 24444-X Pa. $15.95

THE NIGHT BEFORE CHRISTMAS, Clement Moore. Full text, and woodcuts from original 1848 book. Also critical, historical material. 19 illustrations. 40pp. 4⅝ × 6. 22797-9 Pa. $2.50

THE LESSON OF JAPANESE ARCHITECTURE: 165 Photographs, Jiro Harada. Memorable gallery of 165 photographs taken in the 1930's of exquisite Japanese homes of the well-to-do and historic buildings. 13 line diagrams. 192pp. 8⅜ × 11¼. 24778-3 Pa. $10.95

THE AUTOBIOGRAPHY OF CHARLES DARWIN AND SELECTED LETTERS, edited by Francis Darwin. The fascinating life of eccentric genius composed of an intimate memoir by Darwin (intended for his children); commentary by his son, Francis; hundreds of fragments from notebooks, journals, papers; and letters to and from Lyell, Hooker, Huxley, Wallace and Henslow. xi + 365pp. 5⅜ × 8. 20479-0 Pa. $6.95

WONDERS OF THE SKY: Observing Rainbows, Comets, Eclipses, the Stars and Other Phenomena, Fred Schaaf. Charming, easy-to-read poetic guide to all manner of celestial events visible to the naked eye. Mock suns, glories, Belt of Venus, more. Illustrated. 299pp. 5¼ × 8¼. 24402-4 Pa. $7.95

BURNHAM'S CELESTIAL HANDBOOK, Robert Burnham, Jr. Thorough guide to the stars beyond our solar system. Exhaustive treatment. Alphabetical by constellation: Andromeda to Cetus in Vol. 1; Chamaeleon to Orion in Vol. 2; and Pavo to Vulpecula in Vol. 3. Hundreds of illustrations. Index in Vol. 3. 2,000pp. 6⅛ × 9¼. 23567-X, 23568-8, 23673-0 Pa., Three-vol. set $41.85

STAR NAMES: Their Lore and Meaning, Richard Hinckley Allen. Fascinating history of names various cultures have given to constellations and literary and folkloristic uses that have been made of stars. Indexes to subjects. Arabic and Greek names. Biblical references. Bibliography. 563pp. 5⅜ × 8½. 21079-0 Pa. $8.95

THIRTY YEARS THAT SHOOK PHYSICS: The Story of Quantum Theory, George Gamow. Lucid, accessible introduction to influential theory of energy and matter. Careful explanations of Dirac's anti-particles, Bohr's model of the atom, much more. 12 plates. Numerous drawings. 240pp. 5⅜ × 8½. 24895-X Pa. $5.95

CHINESE DOMESTIC FURNITURE IN PHOTOGRAPHS AND MEASURED DRAWINGS, Gustav Ecke. A rare volume, now affordably priced for antique collectors, furniture buffs and art historians. Detailed review of styles ranging from early Shang to late Ming. Unabridged republication. 161 black-and-white drawings, photos. Total of 224pp. 8⅜ × 11¼. (Available in U.S. only) 25171-3 Pa. $13.95

VINCENT VAN GOGH: A Biography, Julius Meier-Graefe. Dynamic, penetrating study of artist's life, relationship with brother, Theo, painting techniques, travels, more. Readable, engrossing. 160pp. 5⅜ × 8½. (Available in U.S. only) 25253-1 Pa. $4.95

HOW TO WRITE, Gertrude Stein. Gertrude Stein claimed anyone could understand her unconventional writing—here are clues to help. Fascinating improvisations, language experiments, explanations illuminate Stein's craft and the art of writing. Total of 414pp. 4⅜ × 6⅜. 23144-5 Pa. $6.95

ADVENTURES AT SEA IN THE GREAT AGE OF SAIL: Five Firsthand Narratives, edited by Elliot Snow. Rare true accounts of exploration, whaling, shipwreck, fierce natives, trade, shipboard life, more. 33 illustrations. Introduction. 353pp. 5⅜ × 8½. 25177-2 Pa. $8.95

THE HERBAL OR GENERAL HISTORY OF PLANTS, John Gerard. Classic descriptions of about 2,850 plants—with over 2,700 illustrations—includes Latin and English names, physical descriptions, varieties, time and place of growth, more. 2,706 illustrations. xlv + 1,678pp. 8½ × 12¼. 23147-X Cloth. $75.00

DOROTHY AND THE WIZARD IN OZ, L. Frank Baum. Dorothy and the Wizard visit the center of the Earth, where people are vegetables, glass houses grow and Oz characters reappear. Classic sequel to *Wizard of Oz*. 256pp. 5⅜ × 8. 24714-7 Pa. $5.95

SONGS OF EXPERIENCE: Facsimile Reproduction with 26 Plates in Full Color, William Blake. This facsimile of Blake's original "Illuminated Book" reproduces 26 full-color plates from a rare 1826 edition. Includes "The Tyger," "London," "Holy Thursday," and other immortal poems. 26 color plates. Printed text of poems. 48pp. 5¼ × 7. 24636-1 Pa. $3.95

SONGS OF INNOCENCE, William Blake. The first and most popular of Blake's famous "Illuminated Books," in a facsimile edition reproducing all 31 brightly colored plates. Additional printed text of each poem. 64pp. 5¼ × 7. 22764-2 Pa. $3.95

PRECIOUS STONES, Max Bauer. Classic, thorough study of diamonds, rubies, emeralds, garnets, etc.: physical character, occurrence, properties, use, similar topics. 20 plates, 8 in color. 94 figures. 659pp. 6⅛ × 9¼. 21910-0, 21911-9 Pa., Two-vol. set $15.90

ENCYCLOPEDIA OF VICTORIAN NEEDLEWORK, S. F. A. Caulfeild and Blanche Saward. Full, precise descriptions of stitches, techniques for dozens of needlecrafts—most exhaustive reference of its kind. Over 800 figures. Total of 679pp. 8⅛ × 11. Two volumes. Vol. 1 22800-2 Pa. $11.95
Vol. 2 22801-0 Pa. $11.95

THE MARVELOUS LAND OF OZ, L. Frank Baum. Second Oz book, the Scarecrow and Tin Woodman are back with hero named Tip, Oz magic. 136 illustrations. 287pp. 5⅜ × 8½. 20692-0 Pa. $5.95

WILD FOWL DECOYS, Joel Barber. Basic book on the subject, by foremost authority and collector. Reveals history of decoy making and rigging, place in American culture, different kinds of decoys, how to make them, and how to use them. 140 plates. 156pp. 7⅞ × 10¾. 20011-6 Pa. $8.95

HISTORY OF LACE, Mrs. Bury Palliser. Definitive, profusely illustrated chronicle of lace from earliest times to late 19th century. Laces of Italy, Greece, England, France, Belgium, etc. Landmark of needlework scholarship. 266 illustrations. 672pp. 6⅛ × 9¼. 24742-2 Pa. $14.95

ILLUSTRATED GUIDE TO SHAKER FURNITURE, Robert Meader. All furniture and appurtenances, with much on unknown local styles. 235 photos. 146pp. 9 × 12. 22819-3 Pa. $8.95

WHALE SHIPS AND WHALING: A Pictorial Survey, George Francis Dow. Over 200 vintage engravings, drawings, photographs of barks, brigs, cutters, other vessels. Also harpoons, lances, whaling guns, many other artifacts. Comprehensive text by foremost authority. 207 black-and-white illustrations. 288pp. 6 × 9. 24808-9 Pa. $9.95

THE BERTRAMS, Anthony Trollope. Powerful portrayal of blind self-will and thwarted ambition includes one of Trollope's most heartrending love stories. 497pp. 5⅜ × 8½. 25119-5 Pa. $9.95

ADVENTURES WITH A HAND LENS, Richard Headstrom. Clearly written guide to observing and studying flowers and grasses, fish scales, moth and insect wings, egg cases, buds, feathers, seeds, leaf scars, moss, molds, ferns, common crystals, etc.—all with an ordinary, inexpensive magnifying glass. 209 exact line drawings aid in your discoveries. 220pp. 5⅜ × 8½. 23330-8 Pa. $4.95

RODIN ON ART AND ARTISTS, Auguste Rodin. Great sculptor's candid, wide-ranging comments on meaning of art; great artists; relation of sculpture to poetry, painting, music; philosophy of life, more. 76 superb black-and-white illustrations of Rodin's sculpture, drawings and prints. 119pp. 8⅜ × 11¼. 24487-3 Pa. $7.95

FIFTY CLASSIC FRENCH FILMS, 1912–1982: A Pictorial Record, Anthony Slide. Memorable stills from Grand Illusion, Beauty and the Beast, Hiroshima, Mon Amour, many more. Credits, plot synopses, reviews, etc. 160pp. 8¼ × 11. 25256-6 Pa. $11.95

THE PRINCIPLES OF PSYCHOLOGY, William James. Famous long course complete, unabridged. Stream of thought, time perception, memory, experimental methods; great work decades ahead of its time. 94 figures. 1,391pp. 5⅜ × 8½. 20381-6, 20382-4 Pa., Two-vol. set $23.90

BODIES IN A BOOKSHOP, R. T. Campbell. Challenging mystery of blackmail and murder with ingenious plot and superbly drawn characters. In the best tradition of British suspense fiction. 192pp. 5⅜ × 8½. 24720-1 Pa. $4.95

CALLAS: PORTRAIT OF A PRIMA DONNA, George Jellinek. Renowned commentator on the musical scene chronicles incredible career and life of the most controversial, fascinating, influential operatic personality of our time. 64 black-and-white photographs. 416pp. 5⅜ × 8¼. 25047-4 Pa. $8.95

GEOMETRY, RELATIVITY AND THE FOURTH DIMENSION, Rudolph Rucker. Exposition of fourth dimension, concepts of relativity as Flatland characters continue adventures. Popular, easily followed yet accurate, profound. 141 illustrations. 133pp. 5⅜ × 8½. 23400-2 Pa. $4.95

HOUSEHOLD STORIES BY THE BROTHERS GRIMM, with pictures by Walter Crane. 53 classic stories—Rumpelstiltskin, Rapunzel, Hansel and Gretel, the Fisherman and his Wife, Snow White, Tom Thumb, Sleeping Beauty, Cinderella, and so much more—lavishly illustrated with original 19th century drawings. 114 illustrations. x + 269pp. 5⅜ × 8½. 21080-4 Pa. $4.95

SUNDIALS, Albert Waugh. Far and away the best, most thorough coverage of ideas, mathematics concerned, types, construction, adjusting anywhere. Over 100 illustrations. 230pp. 5⅜ × 8½. 22947-5 Pa. $5.95

PICTURE HISTORY OF THE NORMANDIE: With 190 Illustrations, Frank O. Braynard. Full story of legendary French ocean liner: Art Deco interiors, design innovations, furnishings, celebrities, maiden voyage, tragic fire, much more. Extensive text. 144pp. 8⅞ × 11¾. 25257-4 Pa. $10.95

THE FIRST AMERICAN COOKBOOK: A Facsimile of "American Cookery," 1796, Amelia Simmons. Facsimile of the first American-written cookbook published in the United States contains authentic recipes for colonial favorites—pumpkin pudding, winter squash pudding, spruce beer, Indian slapjacks, and more. Introductory Essay and Glossary of colonial cooking terms. 80pp. 5⅜ × 8½. 24710-4 Pa. $3.50

101 PUZZLES IN THOUGHT AND LOGIC, C. R. Wylie, Jr. Solve murders and robberies, find out which fishermen are liars, how a blind man could possibly identify a color—purely by your own reasoning! 107pp. 5⅜ × 8½. 20367-0 Pa. $2.50

ANCIENT EGYPTIAN MYTHS AND LEGENDS, Lewis Spence. Examines animism, totemism, fetishism, creation myths, deities, alchemy, art and magic, other topics. Over 50 illustrations. 432pp. 5⅜ × 8½. 26525-0 Pa. $8.95

ANTHROPOLOGY AND MODERN LIFE, Franz Boas. Great anthropologist's classic treatise on race and culture. Introduction by Ruth Bunzel. Only inexpensive paperback edition. 255pp. 5⅜ × 8½. 25245-0 Pa. $6.95

THE TALE OF PETER RABBIT, Beatrix Potter. The inimitable Peter's terrifying adventure in Mr. McGregor's garden, with all 27 wonderful, full-color Potter illustrations. 55pp. 4¼ × 5½. (Available in U.S. only) 22827-4 Pa. $1.75

THREE PROPHETIC SCIENCE FICTION NOVELS, H. G. Wells. *When the Sleeper Wakes, A Story of the Days to Come* and *The Time Machine* (full version). 335pp. 5⅜ × 8½. (Available in U.S. only) 20605-X Pa. $6.95

APICIUS COOKERY AND DINING IN IMPERIAL ROME, edited and translated by Joseph Dommers Vehling. Oldest known cookbook in existence offers readers a clear picture of what foods Romans ate, how they prepared them, etc. 49 illustrations. 301pp. 6⅛ × 9¼. 23563-7 Pa. $7.95

SHAKESPEARE LEXICON AND QUOTATION DICTIONARY, Alexander Schmidt. Full definitions, locations, shades of meaning of every word in plays and poems. More than 50,000 exact quotations. 1,485pp. 6½ × 9¼. 22726-X, 22727-8 Pa., Two-vol. set $31.90

THE WORLD'S GREAT SPEECHES, edited by Lewis Copeland and Lawrence W. Lamm. Vast collection of 278 speeches from Greeks to 1970. Powerful and effective models; unique look at history. 842pp. 5⅜ × 8½. 20468-5 Pa. $12.95

THE BLUE FAIRY BOOK, Andrew Lang. The first, most famous collection, with many familiar tales: Little Red Riding Hood, Aladdin and the Wonderful Lamp, Puss in Boots, Sleeping Beauty, Hansel and Gretel, Rumpelstiltskin; 37 in all. 138 illustrations. 390pp. 5⅜ × 8½. 21437-0 Pa. $6.95

THE STORY OF THE CHAMPIONS OF THE ROUND TABLE, Howard Pyle. Sir Launcelot, Sir Tristram and Sir Percival in spirited adventures of love and triumph retold in Pyle's inimitable style. 50 drawings, 31 full-page. xviii + 329pp. 6½ × 9¼. 21883-X Pa. $7.95

THE MYTHS OF THE NORTH AMERICAN INDIANS, Lewis Spence. Myths and legends of the Algonquins, Iroquois, Pawnees and Sioux with comprehensive historical and ethnological commentary. 36 illustrations. 5⅜ × 8½. 25967-6 Pa. $8.95

GREAT DINOSAUR HUNTERS AND THEIR DISCOVERIES, Edwin H. Colbert. Fascinating, lavishly illustrated chronicle of dinosaur research, 1820's to 1960. Achievements of Cope, Marsh, Brown, Buckland, Mantell, Huxley, many others. 384pp. 5¼ × 8¼. 24701-5 Pa. $7.95

THE TASTEMAKERS, Russell Lynes. Informal, illustrated social history of American taste 1850's–1950's. First popularized categories Highbrow, Lowbrow, Middlebrow. 129 illustrations. New (1979) afterword. 384pp. 6 × 9. 23993-4 Pa. $8.95

DOUBLE CROSS PURPOSES, Ronald A. Knox. A treasure hunt in the Scottish Highlands, an old map, unidentified corpse, surprise discoveries keep reader guessing in this cleverly intricate tale of financial skullduggery. 2 black-and-white maps. 320pp. 5⅜ × 8½. (Available in U.S. only) 25032-6 Pa. $6.95

AUTHENTIC VICTORIAN DECORATION AND ORNAMENTATION IN FULL COLOR: 46 Plates from "Studies in Design," Christopher Dresser. Superb full-color lithographs reproduced from rare original portfolio of a major Victorian designer. 48pp. 9¼ × 12¼. 25083-0 Pa. $7.95

PRIMITIVE ART, Franz Boas. Remains the best text ever prepared on subject, thoroughly discussing Indian, African, Asian, Australian, and, especially, Northern American primitive art. Over 950 illustrations show ceramics, masks, totem poles, weapons, textiles, paintings, much more. 376pp. 5⅜ × 8. 20025-6 Pa. $7.95

SIDELIGHTS ON RELATIVITY, Albert Einstein. Unabridged republication of two lectures delivered by the great physicist in 1920–21. *Ether and Relativity* and *Geometry and Experience*. Elegant ideas in non-mathematical form, accessible to intelligent layman. vi + 56pp. 5⅜ × 8½. 24511-X Pa. $2.95

THE WIT AND HUMOR OF OSCAR WILDE, edited by Alvin Redman. More than 1,000 ripostes, paradoxes, wisecracks: Work is the curse of the drinking classes, I can resist everything except temptation, etc. 258pp. 5⅜ × 8½. 20602-5 Pa. $4.95

ADVENTURES WITH A MICROSCOPE, Richard Headstrom. 59 adventures with clothing fibers, protozoa, ferns and lichens, roots and leaves, much more. 142 illustrations. 232pp. 5⅜ × 8½. 23471-1 Pa. $3.95

PLANTS OF THE BIBLE, Harold N. Moldenke and Alma L. Moldenke. Standard reference to all 230 plants mentioned in Scriptures. Latin name, biblical reference, uses, modern identity, much more. Unsurpassed encyclopedic resource for scholars, botanists, nature lovers, students of Bible. Bibliography. Indexes. 123 black-and-white illustrations. 384pp. 6 × 9. 25069-5 Pa. $8.95

FAMOUS AMERICAN WOMEN: A Biographical Dictionary from Colonial Times to the Present, Robert McHenry, ed. From Pocahontas to Rosa Parks, 1,035 distinguished American women documented in separate biographical entries. Accurate, up-to-date data, numerous categories, spans 400 years. Indices. 493pp. 6½ × 9¼. 24523-3 Pa. $10.95

THE FABULOUS INTERIORS OF THE GREAT OCEAN LINERS IN HISTORIC PHOTOGRAPHS, William H. Miller, Jr. Some 200 superb photographs capture exquisite interiors of world's great "floating palaces"—1890's to 1980's: *Titanic, Ile de France, Queen Elizabeth, United States, Europa,* more. Approx. 200 black-and-white photographs. Captions. Text. Introduction. 160pp. 8⅜ × 11¾. 24756-2 Pa. $9.95

THE GREAT LUXURY LINERS, 1927–1954: A Photographic Record, William H. Miller, Jr. Nostalgic tribute to heyday of ocean liners. 186 photos of Ile de France, Normandie, Leviathan, Queen Elizabeth, United States, many others. Interior and exterior views. Introduction. Captions. 160pp. 9 × 12. 24056-8 Pa. $10.95

A NATURAL HISTORY OF THE DUCKS, John Charles Phillips. Great landmark of ornithology offers complete detailed coverage of nearly 200 species and subspecies of ducks: gadwall, sheldrake, merganser, pintail, many more. 74 full-color plates, 102 black-and-white. Bibliography. Total of 1,920pp. 8⅜ × 11¼. 25141-1, 25142-X Cloth. Two-vol. set $100.00

THE SEAWEED HANDBOOK: An Illustrated Guide to Seaweeds from North Carolina to Canada, Thomas F. Lee. Concise reference covers 78 species. Scientific and common names, habitat, distribution, more. Finding keys for easy identification. 224pp. 5⅜ × 8½. 25215-9 Pa. $6.95

THE TEN BOOKS OF ARCHITECTURE: The 1755 Leoni Edition, Leon Battista Alberti. Rare classic helped introduce the glories of ancient architecture to the Renaissance. 68 black-and-white plates. 336pp. 8⅜ × 11¼. 25239-6 Pa. $14.95

MISS MACKENZIE, Anthony Trollope. Minor masterpieces by Victorian master unmasks many truths about life in 19th-century England. First inexpensive edition in years. 392pp. 5⅜ × 8½. 25201-9 Pa. $8.95

THE RIME OF THE ANCIENT MARINER, Gustave Doré, Samuel Taylor Coleridge. Dramatic engravings considered by many to be his greatest work. The terrifying space of the open sea, the storms and whirlpools of an unknown ocean, the ice of Antarctica, more—all rendered in a powerful, chilling manner. Full text. 38 plates. 77pp. 9¼ × 12. 22305-1 Pa. $4.95

THE EXPEDITIONS OF ZEBULON MONTGOMERY PIKE, Zebulon Montgomery Pike. Fascinating first-hand accounts (1805–6) of exploration of Mississippi River, Indian wars, capture by Spanish dragoons, much more. 1,088pp. 5⅜ × 8½. 25254-X, 25255-8 Pa. Two-vol. set $25.90

SIR HARRY HOTSPUR OF HUMBLETHWAITE, Anthony Trollope. Incisive, unconventional psychological study of a conflict between a wealthy baronet, his idealistic daughter, and their scapegrace cousin. The 1870 novel in its first inexpensive edition in years. 250pp. 5⅜ × 8½. 24953-0 Pa. $6.95

LASERS AND HOLOGRAPHY, Winston E. Kock. Sound introduction to burgeoning field, expanded (1981) for second edition. Wave patterns, coherence, lasers, diffraction, zone plates, properties of holograms, recent advances. 84 illustrations. 160pp. 5⅜ × 8¼. (Except in United Kingdom) 24041-X Pa. $3.95

INTRODUCTION TO ARTIFICIAL INTELLIGENCE: SECOND, ENLARGED EDITION, Philip C. Jackson, Jr. Comprehensive survey of artificial intelligence—the study of how machines (computers) can be made to act intelligently. Includes introductory and advanced material. Extensive notes updating the main text. 132 black-and-white illustrations. 512pp. 5⅜ × 8½. 24864-X Pa. $8.95

HISTORY OF INDIAN AND INDONESIAN ART, Ananda K. Coomaraswamy. Over 400 illustrations illuminate classic study of Indian art from earliest Harappa finds to early 20th century. Provides philosophical, religious and social insights. 304pp. 6⅛ × 9¼. 25005-9 Pa. $11.95

THE GOLEM, Gustav Meyrink. Most famous supernatural novel in modern European literature, set in Ghetto of Old Prague around 1890. Compelling story of mystical experiences, strange transformations, profound terror. 13 black-and-white illustrations. 224pp. 5⅜ × 8½. (Available in U.S. only) 25025-3 Pa. $6.95

PICTORIAL ENCYCLOPEDIA OF HISTORIC ARCHITECTURAL PLANS, DETAILS AND ELEMENTS: With 1,880 Line Drawings of Arches, Domes, Doorways, Facades, Gables, Windows, etc., John Theodore Haneman. Sourcebook of inspiration for architects, designers, others. Bibliography. Captions. 141pp. 9 × 12. 24605-1 Pa. $7.95

BENCHLEY LOST AND FOUND, Robert Benchley. Finest humor from early 30's, about pet peeves, child psychologists, post office and others. Mostly unavailable elsewhere. 73 illustrations by Peter Arno and others. 183pp. 5⅜ × 8½. 22410-4 Pa. $4.95

ERTÉ GRAPHICS, Erté. Collection of striking color graphics: *Seasons, Alphabet, Numerals, Aces* and *Precious Stones*. 50 plates, including 4 on covers. 48pp. 9⅜ × 12¼. 23580-7 Pa. $7.95

THE JOURNAL OF HENRY D. THOREAU, edited by Bradford Torrey, F. H. Allen. Complete reprinting of 14 volumes, 1837–61, over two million words; the sourcebooks for *Walden*, etc. Definitive. All original sketches, plus 75 photographs. 1,804pp. 8½ × 12¼. 20312-3, 20313-1 Cloth., Two-vol. set $125.00

CASTLES: THEIR CONSTRUCTION AND HISTORY, Sidney Toy. Traces castle development from ancient roots. Nearly 200 photographs and drawings illustrate moats, keeps, baileys, many other features. Caernarvon, Dover Castles, Hadrian's Wall, Tower of London, dozens more. 256pp. 5⅜ × 8¼. 24898-4 Pa. $6.95

AMERICAN CLIPPER SHIPS: 1833–1858, Octavius T. Howe & Frederick C. Matthews. Fully-illustrated, encyclopedic review of 352 clipper ships from the period of America's greatest maritime supremacy. Introduction. 109 halftones. 5 black-and-white line illustrations. Index. Total of 928pp. 5⅜ × 8½.
25115-2, 25116-0 Pa., Two-vol. set $17.90

TOWARDS A NEW ARCHITECTURE, Le Corbusier. Pioneering manifesto by great architect, near legendary founder of "International School." Technical and aesthetic theories, views on industry, economics, relation of form to function, "mass-production spirit," much more. Profusely illustrated. Unabridged translation of 13th French edition. Introduction by Frederick Etchells. 320pp. 6⅛ × 9¼. (Available in U.S. only) 25023-7 Pa. $8.95

THE BOOK OF KELLS, edited by Blanche Cirker. Inexpensive collection of 32 full-color, full-page plates from the greatest illuminated manuscript of the Middle Ages, painstakingly reproduced from rare facsimile edition. Publisher's Note. Captions. 32pp. 9⅜ × 12¼. 24345-1 Pa. $4.95

BEST SCIENCE FICTION STORIES OF H. G. WELLS, H. G. Wells. Full novel *The Invisible Man,* plus 17 short stories: "The Crystal Egg," "Aepyornis Island," "The Strange Orchid," etc. 303pp. 5⅜ × 8½. (Available in U.S. only)
21531-8 Pa. $6.95

AMERICAN SAILING SHIPS: Their Plans and History, Charles G. Davis. Photos, construction details of schooners, frigates, clippers, other sailcraft of 18th to early 20th centuries—plus entertaining discourse on design, rigging, nautical lore, much more. 137 black-and-white illustrations. 240pp. 6⅛ × 9¼.
24658-2 Pa. $6.95

ENTERTAINING MATHEMATICAL PUZZLES, Martin Gardner. Selection of author's favorite conundrums involving arithmetic, money, speed, etc., with lively commentary. Complete solutions. 112pp. 5⅜ × 8½. 25211-6 Pa. $2.95

THE WILL TO BELIEVE, HUMAN IMMORTALITY, William James. Two books bound together. Effect of irrational on logical, and arguments for human immortality. 402pp. 5⅜ × 8½. 20291-7 Pa. $7.95

THE HAUNTED MONASTERY and THE CHINESE MAZE MURDERS, Robert Van Gulik. 2 full novels by Van Gulik continue adventures of Judge Dee and his companions. An evil Taoist monastery, seemingly supernatural events; overgrown topiary maze that hides strange crimes. Set in 7th-century China. 27 illustrations. 328pp. 5⅜ × 8½. 23502-5 Pa. $6.95

CELEBRATED CASES OF JUDGE DEE (DEE GOONG AN), translated by Robert Van Gulik. Authentic 18th-century Chinese detective novel; Dee and associates solve three interlocked cases. Led to Van Gulik's own stories with same characters. Extensive introduction. 9 illustrations. 237pp. 5⅜ × 8½.
23337-5 Pa. $5.95

Prices subject to change without notice.

Available at your book dealer or write for free catalog to Dept. GI, Dover Publications, Inc., 31 East 2nd St., Mineola, N.Y. 11501. Dover publishes more than 175 books each year on science, elementary and advanced mathematics, biology, music, art, literary history, social sciences and other areas.